American Institute of Chemical Engineers

Topical Conference on Fuel Cell Technology 2003

Held at the 2003 AIChE Annual National Meeting

November 16-21, 2003
San Francisco, California, USA

Printed from e-media with permission by:

Curran Associates, Inc.
57 Morehouse Lane
Red Hook, NY 12571
www.proceedings.com

ISBN: 978-1-60423-483-1

Some format issues inherent in the e-media version may also appear in this print version.

American Institute of Chemical Engineers

Topical Conference on Fuel Cell Technology
2003

TABLE OF CONTENTS

* **These sessions were unavailable at the time of publication.**

Modification of Nanofiltration Membranes by Ion Implantation

Joshua O. Abitoye[1], **Parna Murkherjee**[2], **Kimberly L. Jones**[3]
[1]*Department of Chemical Engineering, Howard University, Washington DC 20059*
[2]*Department of Environmental Engineering, California State Polytechnic University, San Luis Obispo CA 93407*
[3]*Department of Civil Engineering, Howard University, Washington DC 20059*

Prepared for presentation at the AIChE 2003 annual meeting, San Francisco, CA. November 16-18, 2003
Poster Session: Advances in Membranes and Fuel Cell Technology
Unpublished

Abstract

A surface modification technique, ion implantation, was applied to commercially available nanofiltration membranes in order to increase the effective charge. The surface charge of a thin film composite (TFC) membrane; NF90 was modified by implanting fluoride ions (F^-) with ion doses of $1x10^{10}$ and $5x10^{10}$ atoms/cm^2. Permeation experiments of a mixture of 5mM KCl + 5mM $NaNO_3$ were carried out at a pressure and temperature of 100psi and 25°C respectively, in a dead end high-pressure filtration cell using the modified and unmodified membranes. Experimental results showed a decrease in permeate flux and an increase in rejection with increasing ion dose from the unmodified to modified membranes.

Keywords: Membranes; Nanofiltration; Ion Implantation; Charge; Rejection

1. Introduction

Nanofiltration (NF) is the most recently developed pressure-driven membrane separation process whose application has gained popularity during the last decade. Its properties lie between that of reverse osmosis (RO) and ultra filtration (UF). A great advantage of NF is that the operating pressures can be much lower than those required for RO, and the corresponding flux is also higher than that of RO. NF membranes are usually negatively charged, with a pore size of ~2nm which enables rejection of small organic molecules having molecular weights as low as 200 to 300 Daltons. NF membranes are usually composed of polymer layers with a selective layer of thickness ~1μm that controls rejection and flow properties.

Commercial NF membranes posses a fixed charge developed by dissociation of surface groups such as sulphonated or carboxy acids (Rautenbach and Groschl, 1990). These charges can repel co-ions therefore; the separation mechanism of charged membranes can be based upon, not only the "sieve effect" but also the "charge effect" (Tsuru *et al.,* 1991). As a result, NF membranes can reject ions much smaller than the membrane pore radii (Xu and Lebrun, 1999). Nanofiltration could be used as an effective technique for removal of environmentally undesirable and hazardous contaminants from aqueous systems such as the removal of hardness (e.g. divalent cations) from water, treatment of wastewater and demineralization of water. The pores in NF membranes are small enough so uncharged solutes may be highly rejected while the surface electrostatic properties allow monovalent ions to be reasonably well transmitted with multivalent ions mostly retained (Bowen and Welfoot, 2002).

Ion implantation is a technology that employs a high-energy ion transmitter to impart permanent charges into a polymer material. It has been shown that ion beam irradiation of polymers changes the properties and surface structures of polymer membranes bringing about effective surface charge (Toulemode et al, 1992 and Musumeci et al, 1995). Dworecki et al (2001) also investigated the electrical properties of polyethylene terephthalate (PET) membranes implanted with a variety of ions (C^{4+}, O^{6+}, S^{7+}) with the results showing substantial increase of the electrical conductivity of the membranes and exhibits of strong ability of charge accumulation.

The aim of this study is to manipulate the surface charge on the Nanofiltration membranes through ion implantation to allow increased rejection of monovalent and divalent ions. An expected outcome of this research is that NF membranes can reject monovalent ions and perform as well as RO membranes but with lower operating pressures, higher flux and consequently, lower cost.

2. Experimental

2.1. Membrane

An NF90 nanofiltration membrane (Film Tec. Inc.) was used in the permeation experiments. Before starting the rejection experiments, the membrane was soaked by floating it skin down in de-ionized water for 1-2h to remove the preservative on it. Membrane sheets were sent to Implant Sciences Corporation, Wakefield, Massachusetts, for ion implantation with fluoride ions. Specifications for implantation are summarized in Table 1.

Table 1: Membrane/Ion implantation Specification

NF90 (Film Tec)	**Specification**
Material	Polyamide
Ion Species	Fluoride Ion
Active Layer Thickness (micron)	0.2
pH range at 25°C	2-11
Temperature Range (°C)	0 – 135
Membrane Thickness (micron)	177.8
Implantation Depth (micron)	.0281
Ion Dose ($atoms/cm^2$)	$1 \times 10^{10} / 5 \times 10^{10}$
Ion Energy (keV)	10

2.2. Model Electrolyte and Solution Chemistry

A mixture of 5mM KCl + 5mM $NaNO_3$ salt solution was used in the experiments. The salt solution was prepared using ultra-pure water of pH 5.8, and the filtration experiments were carried out at room temperature (25°C).

2.3. *Permeation Experiment*

A 300 ml dead-end high-pressured stirred cell from Sterlitech™ (model HP-4750) was used for the experiments. The filtration apparatus is as shown in Figure 1. A clean water flux test was performed before and after each rejection experiment by mounting the membrane in the stirred cell, its glossy side toward solution, filtering ultra pure water at 100 psi for 2hrs., and determining the flux at specific time intervals. Next, the ultra pure water in the cell was replaced with the salt solution and filtered at a pressure of 100 psi. Inert nitrogen gas was used to pressurize the cell.

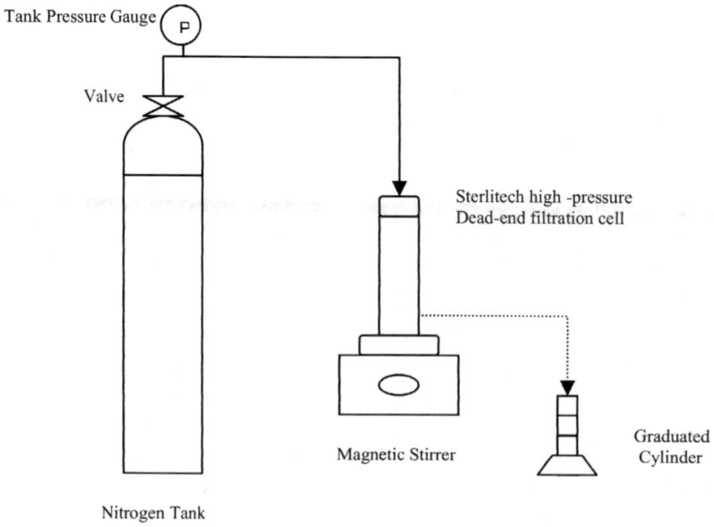

Fig. 1: Schematic of stirred cell NF System. Flux time was measured using a digital stop watch and graduated cylinder

The duration for each filtration run was 5-6h with samples collected at specific time intervals. Samples were analyzed to determine solute flux and concentration. After each run, the membrane was removed rinsed with distilled water and stored face down in a beaker of de-ionized water at a temperature of 3-4°C.

The product water flux (J) was calculated as:

$$J = \frac{V}{A \cdot t} \quad (1)$$

where V is the volume collected at a particular time (t) and A is the surface area of the membrane ($1.19 \times 10^{-3} m^2$).

The mass solute flux (J_s) was calculated as:

$$J_s = J \cdot C_p \quad (2)$$

where C_p is the concentration of permeate

2.4. Analytical Methods

The ion concentrations in the permeate stream were determined by ion chromatography (Dionex DX 120). The zeta potentials of the modified and unmodified membrane charges were determined using streaming potential measurements taken with an electro-kinetic analyzer (EKA) manufactured by Brookhaven Instruments Corp., New York. Several zeta potential measurements were taken at different pH values, using fresh membrane samples. The pH was adjusted by from pH 3-9 by adding drops of 0.1M KCl and 0.1KOH. The electrolyte used in the EKA was 1mM KCl solution.

3. Results and Discussion

3.1 Streaming Potential Measurements

The streaming potentials of the modified and unmodified membranes decreased with increasing pH. Figure 2 shows the zeta potentials for the modified and unmodified membranes, as calculated from the streaming potentials. All the membranes showed close iso-electric points (i.e.p) which was in the acidic range. The ieps for the unmodified, 1×10^{10} atoms/cm^2 modified and 5×10^{10} atoms/cm^2 modifed were at pH 4.2, 4.0 and 3.9 respectively. At pH greater than 6.3, the zeta potentials increased in order of ion dose. There was an inversion of zeta potential values between the two modified membranes at pH range of 4.2 – 6.3. These results show that the modified membranes were more electronegative than the unmodified at all pH.

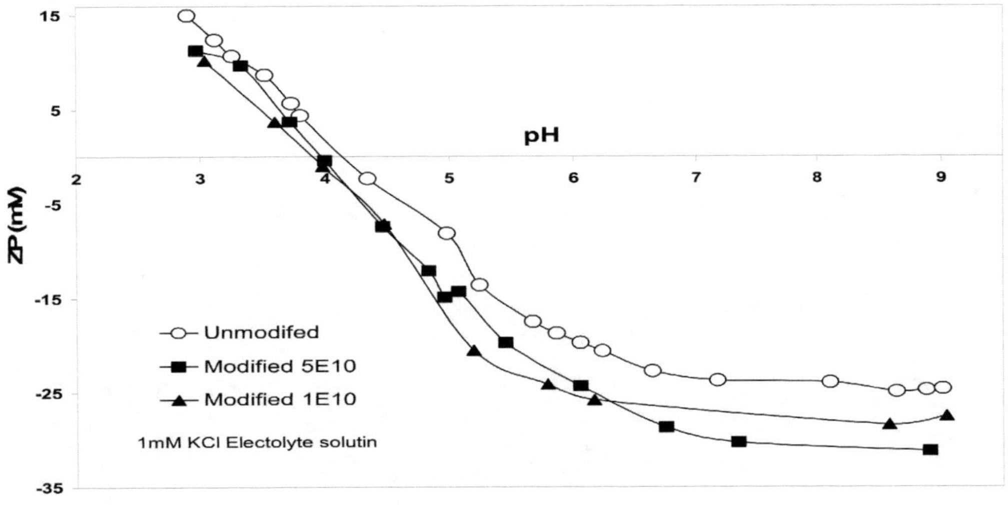

Fig. 2: Effect of pH on NF membrane charge

3.2. Stability of NF Membranes

The stability of the membranes, in order to ensure their integrity, was tested using the clean-water flux test. The test should show after stabilization, a constant flux throughout the length of the experiment. Figure 3 shows that membranes maintained constant clean water flux after a certain stabilization time. The unmodified membrane showed a decrease in clean water flux up to the 40th minute after which it the flux stabilized. The modified membranes had a constant clean water flux throughout the period of the experiment. The NF membranes could be considered stable during the experimental period.

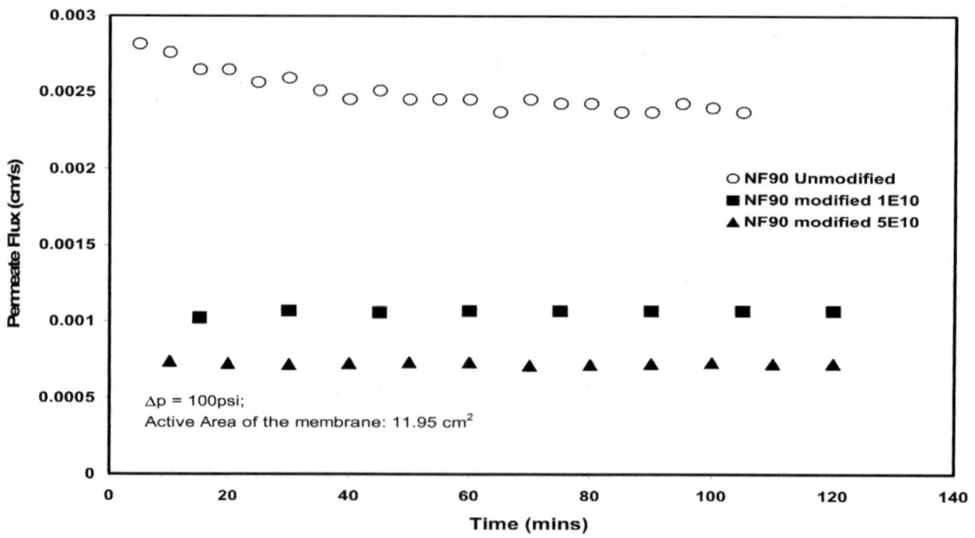

Fig. 3: Clean water flux test

3.3. Permeate Flux

Figures 4 and 5 show the permeate flux (with both salt solution and clean water) with the unmodified and modified membranes. The product water flux decreased from the unmodified membrane to the highly modified membrane. The unmodified membrane had highest water flux and the highly modified membrane (5 x10^{10} atoms/cm^2) had the lowest flux with the flux of the lowly modified membrane (1x 10^{10} atoms/cm^2) lying in between.

Generally, RO membrane produces less water flux (yield) than a NF membrane due to smaller pore sizes. The water flux through a similar commercially available RO membrane (BW30, Filmtec Inc.) was measured with same experimental conditions. It was observed that the permeate volume, hence permeate flux, through the BW30 membrane was much lower than even the highly modified membrane (5E10).

Ion implantation increases the electrostatic interactions at the membrane surface, which increases resistance to flow of a polar solvent (H_2O). This behavior causes NF membrane behavior to approach that of RO. This effect was most pronounced at the the high ion dose (5x10^{10} atoms/cm^2).

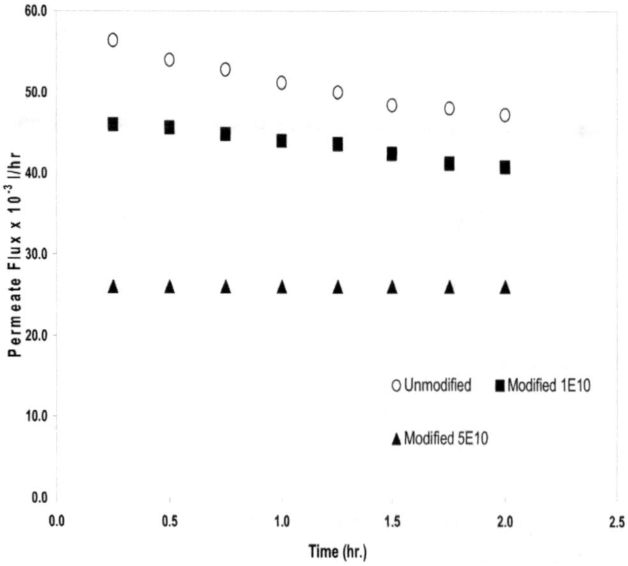

Fig. 4: Permeate Flux as a function of time

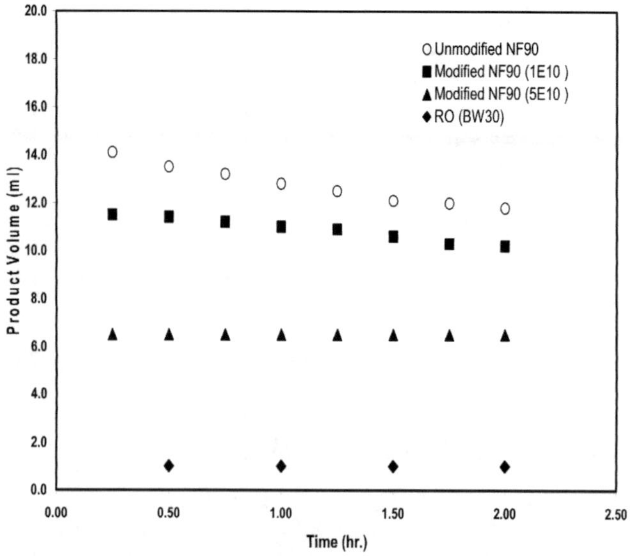

Fig. 5: Product volume as function of time

3.4. Salt Permeation

The salt permeation of all the individual ions (Cl^-, NO_3^-, Na^+ and K^+) is plotted as J_s (Solute mass flux = J x C_s) vs. Time (Fig. 6). The solute mass (salt) flux for the unmodified membrane and the low dose membrane ($1x10^{10}$ atoms/cm^2) are almost identical. However, there is significant decrease (about 50%) in the salt flux for all ions in the case of highly dose ($5x10^{10}$ atoms/cm^2) membrane.

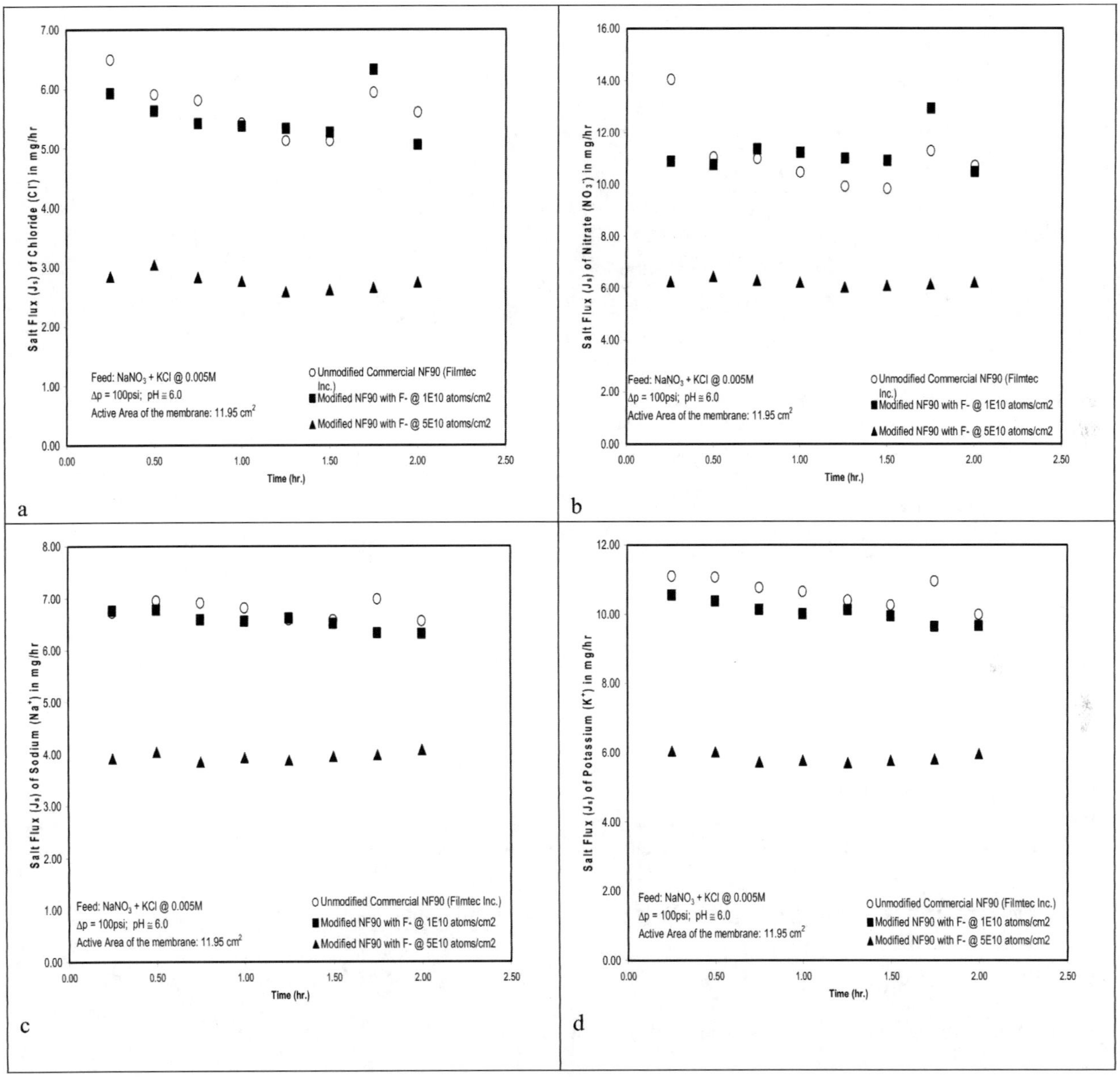

Fig 6: Salt flux as a function of time for a) Cl^-, b) NO_3^-, c) Na^+, d) K^+

Low modification ($1x10^{10}$atoms/cm) is not enough to produce significant electrostatic effect in NF90 membrane. The highly modified membrane ($5x10^{10}$atoms/cm) increases the electrostatic barrier of NF90 membranes significantly to reduce the ion flux through the membrane.

For all three membranes, the permeate samples had pH of 5.8 indicating that the electrostatic modification of the membrane did not change product water pH. This is as illustrated in Figure 7. The permeate samples were also analyzed for fluoride ions and there was no leaching of the implanted ion into the permeate.

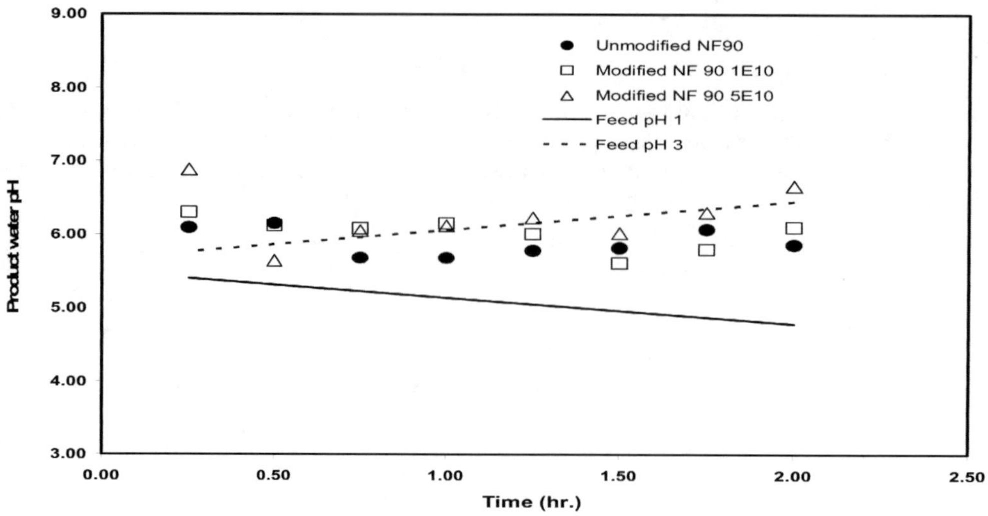

Fig 7: pH as a function of time

4. Conclusion

Solute flux decreased with increasing ion dose of the membranes. Interactions between implanted ions on membranes and ions in bulk solutions can affect solute flux and consequently rejection. Increased membrane charge density may increase rejection of certain ions, regardless of the pore size of the membrane.

Acknowledgements

The National Science Foundation (NSF) and Keck Foundation provided funding for this research.

References

Dworecki K., T. Hasegawa, K. Sudlitz, A. Slezak, and S. Wasik (2001): "Modification of electrical properties of polymer membranes by ion implantation (II)"; *Nuclear Intstruments and Methods in Physics Research B 185, 61-65*

Bowen W.R. and H. Mukhtar (1995): "Characterization and prediction of separation performance of nanofiltration membranes"; *Journ. Of Memb. Sc. 112, 263-274.*

Bowen W.R. and J.S. Welfoot (2002): "Modeling the performance of membrane nanofiltration – critical assessment and model development"; *Chemical Eng. Sc. 57*, 1121-1137.

Rautenbach R. and A. Groschl (1990): "Separation potential of nanofiltration membranes"; *Desalination, 77,* 73-84

Musumeci P., L. Calcagno, R. Percolla and G. Fotti (1995): *Journ. of Applied Physics 77 3766*

Toulemode M., C. Dunfour, and E. Paumier (1992): Phys. Rev. B 46 14362

Tsuru T., M. Urairi, S.I. Nakao, S. Kimura (1991): "Reverse osmosis of single and mixed electrolytes with charged membranes"; *Journ. of Chem. Eng. Jpn 24(4)* 518-524.

Xu Y. and R.E. Lebrun (1999): "Investigation of the solute separation by charged nanofiltration membrane: Effect of pH, ionic strength and solute type"; *Journ. of Memb. Sc. 158*, 83-104.

Synthesis of Nanoscale Metal Particles in Polyelectrolyte Membrane Matrix for Reductive Transformation of Halogenated Organic Compounds

Jian Xu and D. Bhattacharyya*
Dept. of Chemical & Materials Engineering
University of Kentucky, Lexington, KY 40506

* Corresponding author email: DB@ENGR.UKY.EDU

Prepared for presentation at AIChE 2003 Annual Meeting – [74] - Poster Session: Advances in Membranes and Fuel Cell Technology, November 18, 2003

Abstract

Nanoscale bimetallic particles (Fe^0/Ni^0) were synthesized in the polyacrylic acid (PAA)/polyether sulfone (PES) composite membrane matrix to degrade halogenated organic compounds (HOCs). In the presence of PAA, narrow size distribution of nanoparticles (10~30nm) was obtained. Rapid and complete destruction of trichloroethylene (TCE) was achieved by using nanoscale bimetallic Fe^0/Ni^0 particles. Typically 95% of a 10mg/L TCE feed was reduced within 1 h. Ethane was found in the headspace as the product, but no other intermediate byproducts were identified in both aqueous and headspace phase.

Keywords: Polyacrylic acid, Membrane, Nanoscale bimetallic particles, Halogenated organic compounds, Trichloroethylene, Dechlorination

Introduction

Nanoscale particles that are generally less than 50nm in diameter have aroused extensive interest in the past few years. The reason is that the nanoscale structures can have unique physical and chemical properties other than bulk particles. Various studies on nanoparticles have focused on the quantum effect, electronic, magnetism, catalytic reaction, and treatment of chlorinated organic contaminants of these nanoparticles. In the reductive dechlorination of halogenated organic compounds (HOCs) by zero-valent metals (Zn, Fe, Sn), nanoscale particles exhibit much better performance on the degradation of these HOCs compared to the bulk metals. In published studies, however, synthesis of zero-valent nanoscale metal particles for dehalogenation of HOCs was mainly achieved by reducing metal ions directly with sodium borohydride ($NaBH_4$) in the solution phase[1-6]. Without the presence of polymers or surfactants which can provide steric stabilization, the newly formed nanoparticles in the solution phase were unstable and easily aggregated into large particles, and finally resulted in big size and broad size distribution of particles. Most of the nanoparticles reported in the literature have the diameter of 1~100nm[1-2]. In this paper, nanosized (10~30nm) zero-valent metals were created in the membrane matrix by the use of polyelectrolytes or polyaminos containing polyfunctional metal binding ligands such as carboxylic group. Metal ions were captured in these membranes and then immobilized inside the membrane domain with the reduction by $NaBH_4$.

Experimental Section

Chemicals

Sodium borohydride (99.995%) was obtained from Aldrich. Ferrous chloride, nickel chloride, ethylene glycol, trichloroethylene, halogenated ethane standard were obtained from Fisher Scientific. Polyacrylic acid (MW 50,000) was obtained from Polysciences. Commercial polyether sulfone (PES, 0.22μm pore size, 47mm in diameter, 150~200μm thickness) porous membrane was purchased from Millipore.

Crosslinked PAA/PES Composite Membrane Preparation

PES microfiltration membrane obtained from Millipore was used as the support membrane layer. The PAA coating layer containing Fe^{2+} was added to the surface of PES support membrane by a dip-coating process. The crosslinked PAA/PES composite membrane was finally obtained by thermally heating the membrane in the oven under vacuum. The ether bond formed between carboxylic group and ethylene glycol generate the crosslinking PAA network structure. It is important to only do a partial cross-linking to leave free carboxylic groups for metal entrapment.

Synthesis of Nanoscale Metal Particles

Nanoscale Fe^0 particles were prepared by soaking the PAA/PES composite membrane into the $NaBH_4$ (2 wt%) aqueous solution for 10 minutes. Fe^{2+} in the PAA layer was reduced and immobilized inside the polyelectrolyte membrane matrix which can inhibit agglomeration of newly formed Fe^0 nanoparticles. Excess $NaBH_4$ was used to prevent oxidation of iron particles. The membrane surface was rinsed by large amount of deoxygenated DIUF water for three times to remove the residual reactant and product.

Bimetallic Ni^0/Fe^0 nanostructured particles were then produced by the following procedure. PAA/PES composite membrane containing Fe^0 nanoparticles was soaked into $NiCl_2$ solution (0.1wt %) for 10 minutes. Subsequent immersion of the membrane into the same $NaBH_4$ aqueous solutions again for 5 minutes results in the formation of Fe^0/Ni^0 (2) bimetallic nanoparticles. The membrane was also rinsed by large amount of deoxygenated DIUF water and then dried at room temperature for further dechlorination experiments.

Physical Characterization of Metal Particles

Particle size and size distribution were measured by Hitachi S-900 Scanning Electron Microscope (SEM) under a magnification of 100k.

Batch Experiments

Batch experiments for dechlorination of TCE were conducted in 42-mL serum glass vials. In each batch vial, 45 ± 7 mg metal particles (Fe^0/Ni^0) at the ratio of 4:1 (Fe: Ni) immobilized on 5 pieces of membranes were loaded into the vial containing 40-mL 10 ± 1mg/L TCE and 2-mL headspace. The serum glass vials were sealed with Teflon-lined silicon septa and placed on a wrist-action shaker throughout the duration of the experiment. Parallel blank experiments without membrane and metal particles, and control experiments with PES/PAA membrane (no particles) were also performed.

Method of Analyses

Organic analysis was performed using a Hewlett Packard 5890 Series II gas chromatograph equipped with a Series 6150 mass spectrometer. For the bimetallic experiments, an OI Analytical Model 4560 Purge-and-Trap (PT) was coupled to the GC/MS and used for direct analysis of the aqueous phase (0.1mg/L methane chloride was used as internal standard). To determine whether a significant amount of TCE was absorbed to the membrane, the total concentration of TCE in the membrane phase and aqueous phase were also measured. The concentration of chloride ion in the aqueous solution was determined by ion chromatograph (Dionex). Hydrocarbon products in the headspace were qualitatively identified with the GC/MS.

Results and Discussion

Characterization of Fe^0/Ni^0 Particles

Figure 1 is the SEM surface image of Millipore PES porous support membrane. The surface pore of the support membrane has the size of less than 200 nm. **Figure 2** shows a SEM image of nanoscale Fe^0/Ni^0 particles immobilized in the PAA/PES composite membrane surface. A quite narrow size distribution of Fe^0/Ni^0 nanoparticles was obtained because the particles were formed on the membrane phase and steric stabilization by PAA prevented the agglomeration of particles. Most of these particles had diameters in the range of 10~30 nm.

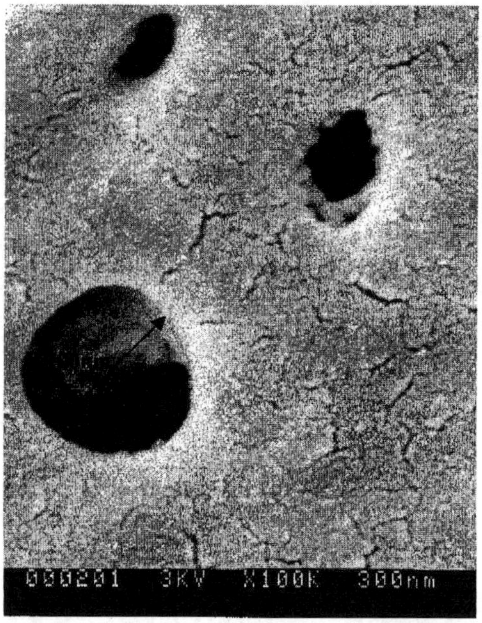

Figure 1. SEM image of PES support membrane surface (100,000×)

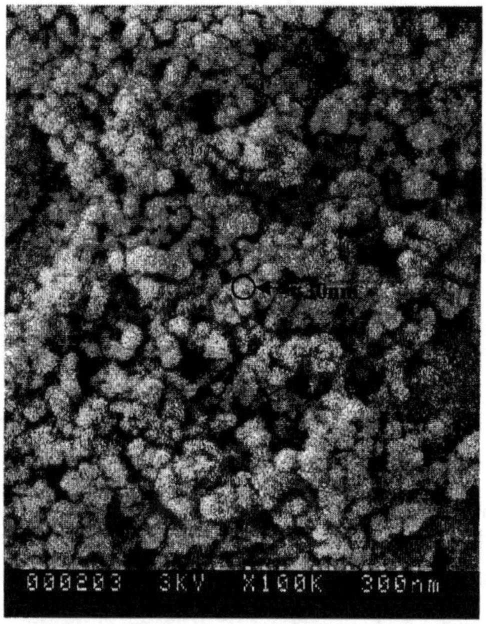

Figure 2. SEM image of nanoscale Fe^0/Ni^0 particles immobilized on the PAA/PES composite membrane surface (100,000×).

Dechlorination of 10mg/l TCE

Reduction results of 10mg/L TCE with nanoscale Fe^0/Ni^0 bimetallic particles is shown in **Figure 3**. 95% TCE was dechlorinated within 1 hour by Fe^0/Ni^0 bimetallic particles. Production of ethane was identified in the headspace. Chloride ion in the reacted solution was also analyzed. Assuming that for each degraded TCE molecular three chloride ions are released, 8.1mg/L chloride ions will be produced at the end of 10mg/L TCE dechlorination reaction. But only 3.0 mg/L chloride ions which equal to

37% Cl^- recovery were formed at the end of reaction. Lower Cl^- concentrations observed can be explained by the sorption of chloride ions on membranes and nanoparticles.

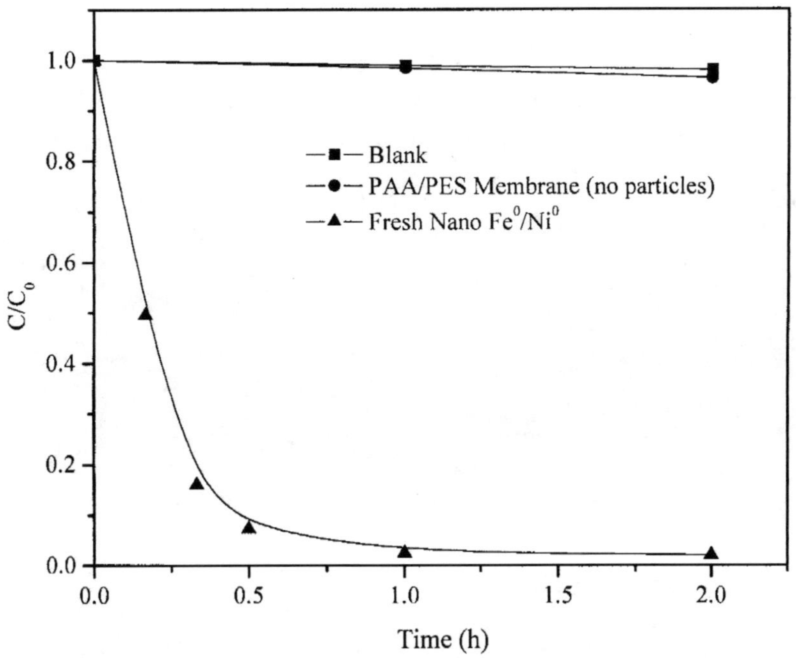

Figure 3. Degradation of 40-mL TCE by (▼) nanoscale Fe^0/Ni^0 (Fe^0 covered by Ni^0). Initial organic concentration was 10±1mg/L. Metal particle loading was 45±7 mg/40-mL for the Fe^0/Ni^0 at the ratio of 4:1 (Fe:Ni) in all experiments.

Conclusions

In summary, nanoscale Fe^0/Ni^0 bimetallic particles were successfully immobilized inside the PAA/PES composite membrane matrix for the dehalogenation of HOCs. These nanoparticles had a quite narrow size distribution due to the presence of polymer stabilization. Complete dechlorination of TCE was achieved within 1 hour by nanoscale Fe^0/Ni^0 bimetallic particles inside the PAA/PES composite membrane. Production of ethane (in the headspace) and Cl^- (solution) was identified.

Acknowledgements

The project is funded by the US EPA STAR Grant and their support is highly appreciated.

References

[1] S. Choe, S.H. Lee, Y.Y. Chang, K.Y. Hwang, J. Khim, Rapid reductive destruction of hazardous organic compounds by nanoscale Fe^0. Chemosphere. 2001, 42, 367-372.

[2] W.X. Zhang, C.B. Wang, H.L. Lien, Treatment of chlorinated organic contaminants with

nanoscale bimetallic particles. Catalysis Today. 1998, 40, 387-395.
[3] Y. Xu and W.X.. Zhang, Subcolloidal Fe/Ag Particles for Reductive Dehalogenation of Chlorinated Benzenes. Ind.Eng.Chem.Res. 2000, 39, 2238-2244.
[4] C.B. Wang and W.X.. Zhang, Synthesizing Nanoscale Iron Particles for Rapid and Complete Dechlorination of TCE and PCBs
[5] H.L. Lien, W.X. Zhang, Nanoscale iron particles for complete reduction of chlorinated ethenes. Colloids and Surfaces A: Physicochemical and Engineering Aspects 191 (2001) 97-105.
[6] B. Schrick, J.L. Blough, A.D. Jones, T.E. Mallouk, Hydrodechlorination of Trichloroethylene to Hydrocarbons Using Bimetallic Nickel-Iron Nanoparticles. Chem.Mater. 2002, 14, 5140-5147.

Carbon Monoxide (CO) Tolerance of High Temperature PEM Fuel Cells with Nafion® Composite Membrane Electrolyte

Ruichun Jiang, H. Russell Kunz, James M. Fenton

Department of Chemical Engineering
University of Connecticut

Prepared for Presentation at the 2003 AICHE Annual Meeting,
San Francisco, LA, Nov, 2003
Session [74]: Advances in Membranes and Fuel Cell Technology

November, 2003
Unpublished

Abstract

Carbon monoxide (CO) poisoning is an important obstacle for Proton Exchange Membrane (PEM) fuel cell applications. CO tolerance of PEM fuel cells was studied at relatively high temperature (T_{cell} > 100°C) operating conditions. ETEK 40% Pt-Ru/C was applied as the anode catalyst directly onto Nafion®-Teflon®-$Zr(HPO_4)_2$ membranes. These membranes have been previously shown to reduce IR losses due to membrane dehydration at high temperature.

Characterization methods included: CO striping voltammetry, anode polarization, AC impedance and cell performance testing. High temperature operation alleviates the CO poisoning effects. At high temperature operating conditions relative humidity also plays an important role in CO tolerance of the fuel cell. Lowering the anode catalyst loading was mildly lowered the performance on pure H_2, but strongly affects the H_2/CO over-potential.

Introduction

PEM fuel cell systems are emerging as promising electrical sources for transportation, stationary, and portable applications. Unfortunately, the PEM fuel cell system is only satisfactory when the fuel cell is fed with pure hydrogen fuel. For many practical applications, hydrogen from reformed fuels such as natural gas, gasoline, or alcohols is likely to be the fuel. In this case, the fuel will contain by-product carbon monoxide (CO) (1-2% CO if no CO cleanup following the reformer, 10–100 ppm if CO removal is used). For low temperature fuel cell operating conditions (T_{cell} < 100ºC), CO strongly adsorbs onto the anode platinum catalyst active surface, thus blocking the hydrogen oxidation reaction, and leading to serious CO poisoning [1]. Even using a CO tolerant catalyst, Pt-Ru/C, trace levels of CO in hydrogen can significantly deteriorate the fuel cell performance [2].

Since CO adsorption on the catalyst surface is an exothermic reaction, operation of the fuel cells at higher temperatures (T_{cell} > 100ºC) alleviates CO poisoning. However, at higher temperatures and ambient pressure, the popularly used Nafion® membrane becomes dry and leads to a high resistance due to the low relative humidity at atmospheric conditions. To reduce the IR losses at higher temperature/ambient pressure conditions, Nafion® based composite membranes were developed in our research group, i.e., Nafion®-Teflon®-PTA (phosphotungstic acid) membranes, Nafion®-Teflon®-$Zr(HPO_4)_2$ membranes and Nafion®-$Zr(HPO_4)_2$ membranes. These three membranes all show promising proton conductivity at higher temperature fuel cell operating conditions and make the study of the CO tolerance properties of higher temperature PEM fuel cells feasible.

In this study, CO tolerance of PEMFCs with Pt-Ru/C (ETEK, wt. 40%, Pt/Ru=1/1) anode catalysts applied onto Nafion®-Teflon®-$Zr(HPO_4)_2$ membranes were studied at various cell temperatures, CO concentrations, relative humidities and catalyst loadings.

Experimental

The Nafion®-Teflon®-$Zr(HPO_4)_2$ membranes were developed by impregnating a Teflon® matrix with a solution containing Nafion® and $Zr(HPO_4)_2$, which was generated in-situ according to the reaction: $ZrOCl_2$+ 2 H_3PO_4 $Zr(HPO4)_2$ + 2 HCl + H_2O. The membrane electrode assembles (MEAs) were prepared by spraying catalyst ink containing carbon-supported catalyst and Nafion® in methanol onto the membrane. MEAs with different anode catalyst loadings (0.05 mg Pt-Ru/cm^2 to 0.9 mg Pt-Ru/cm^2) were prepared for this study.

Using H_2 (H_2+CO)/Air (Oxygen) as the anode/cathode reactants, respectively, the MEAs were evaluated in a fuel cell environment by collecting performance data at different temperature and relative humidity conditions with different CO concentrations (from 10ppm CO in H_2 to 2% CO, 24% CO_2 in H_2) in the hydrogen fuel. The cell temperatures ranged from 80°C to 120°C, and the relative humidities ranged from 100% to 35%. By adjusting the system back pressure, the anode fuel or cathode reactant partial pressure was controlled to remain constant with different relative humidity conditions. Hence the effects of independently changing of the relative humidity and temperature on CO tolerance were studied.

Applying H_2 (H_2+CO)/H_2 as the anode/cathode gas flow enabled the electro-oxidation of H_2 or H_2 +CO mixtures under the above operating conditions to be characterized by anode polarization and electrochemical impedance spectroscopy (EIS). CO stripping voltammetry was also used to study the CO adsorption and oxidation properties on the anode catalyst surface.

Results

High temperature operation of PEM fuel cells dramatically improves the CO tolerance. From the performance data, the cell voltage losses ($V_{H2} - V_{H2+CO}$) at 105°C (1atm) and 120°C (1atm) conditions are much lower than that at 80°C conditions. The anode polarization results show that higher temperature operation can decrease the over-potential when oxidizing H_2+CO mixtures on the catalyst active surface. CO stripping voltammetry indicated that higher temperature leads to lower CO adsorption coverage on the anode catalyst active surface. This change was also certified in the EIS results by showing lower dissociative adsorption impedance of H_2 at higher temperature conditions. Higher relative humidity and higher anode catalyst loading are favorable for CO tolerance at high temperature conditions. Further results and discussion will be presented at the meeting.

Reference:

1. Shimshon Gottesfeld , Tom A. Zawodzinski, Polymer Electrolyte Fuel Cells, Advances in Electrochemical Science and Engineering, Vol. 5, 1997, WILEY-VCH.
2. Ruichun Jiang, Yongchao Si, Jung-Chou Lin, H. Russell Kunz, James M. Fenton, Abstracts of the 203rd Meeting of Electrochemical Society, Paris, May. 2003. (Abstract No. 176)

ANALYSIS OF MASS TRANSFER FOR MEMBRANE EXTRACTION IN FLAT-PLATE MODULES WITH THE CONSIDERATION OF CORRECTION FACTOR

Ho-Ming Yeh

Department of Chemical Engineering, Tamkang University

Tamsui 251, Taiwan, R.O.C.

Keywords□Membrane extraction, flat plate, correction factor

Preparated for presentation at 2003 Annual Meeting, San Franciso, CA, Nov. 16-21

Abstract

Mass transfer of solvent extraction in microporous-membrane mass exchangers can be analogous to heat transfer in heat exchangers. The correction factors for both cocurrent-flow and countercurrent-flow operations may be taken as unity when the logarithmic-mean concentration differences are properly selected. For cross-flow operation, the correction factors are function of flow rate, mass-transfer area, distribution coefficient and overall mass-transfer coefficient, and some values of them are given graphically.

Introduction

Recently, extensive studies on dispersion-free solvent extraction using microporous membranes have been carried out (Kiani et al., 1984;Yang and Cussler, 1986; Prasad and Sirkar, 1988; Ding and Cussler, 1991; Yeh et al., 1999; Yeh and Chen. 2000; 2001). However, the theories developed in previous works though are not difficult to be understood but rather complicated and cumbersome to be applied. Actually, parallel-plate membrane solvent extractors are like flat – plate heat exchangers except that a microporous membrane sheet is used in place of the heat-transfer medium. It is the purpose of present study to analogize the mass transfer in membrane extractors to the heat transfer in heat exchangers for making the engineering design and applications more simple to achieve.

Theory

Mass transfer rate

The extraction rate W in any one of the flat-plate membrane modules, as shown in Figs.1-3 with length L and width B, may be predicted from the mass-transfer point of view by

$$W = KBL \times (\text{averaged concentration difference}) \tag{1}$$

where K denotes the overall mass-transfer coefficient, while the averaged concentration differences between the raffinate (fluid a) and extract (fluid b) phases are varied with the flow patterns and membrane systems. For convenience, we may take its logarithemic means, $(\Delta C)_{lm,1}$ or $(\Delta C)_{lm,2}$, analogous to heat transfer in heat exchangers, as

$$W = KBLF_1(\Delta C)_{lm,1} \tag{2}$$

or

$$W = KBLF_2(\Delta C)_{lm,2} \tag{3}$$

where

$$(\Delta C)_{lm,1} = \frac{(H_{ac}C_{a,e} - H_{bc}C_{b,e}) - (H_{ac}C_{a,i} - H_{bc}C_{b,i})}{\ln[(H_{ac}C_{a,e} - H_{bc}C_{b,e})/(H_{ac}C_{a,i} - H_{bc}C_{b,i})]} \tag{4}$$

$$(\Delta C)_{lm,2} = \frac{(H_{ac}C_{a,i} - H_{bc}C_{b,e}) - (H_{ac}C_{a,e} - H_{bc}C_{b,i})}{\ln[(H_{ac}C_{a,i} - H_{bc}C_{b,e})/(H_{ac}C_{a,e} - H_{bc}C_{b,i})]} \tag{5}$$

and F_1 and F_2 are the so-called correction factors, to be determined, while H_{ac} and H_{bc} are the distribution coefficients of solute at the phase interfaces.

Unlike the known values of the inlet concentrations ($C_{a,i}$ and $C_{b,i}$), the outlet concentrations ($C_{a,e}$ and $C_{b,e}$) are not specified prior. Therefore, the calculation of the overall mass-transfer rate W should be also conducted with the use of the following overall mass-balance equations :

$$W = q_a(C_{a,i} - C_{a,e}) \tag{6}$$

$$= q_b(C_{b,e} - C_{b,i}) \tag{7}$$

and thus, the value of correction factors may be determined from Eqs. (2), (3) and (6) as

$$F_1 = \frac{q_a(C_{a,i} - C_{a,e})}{KBL(\Delta C)_{lm,1}} \tag{8}$$

$$F_2 = \frac{q_a(C_{a,i} - C_{a,e})}{KBL(\Delta C)_{lm,2}} \tag{9}$$

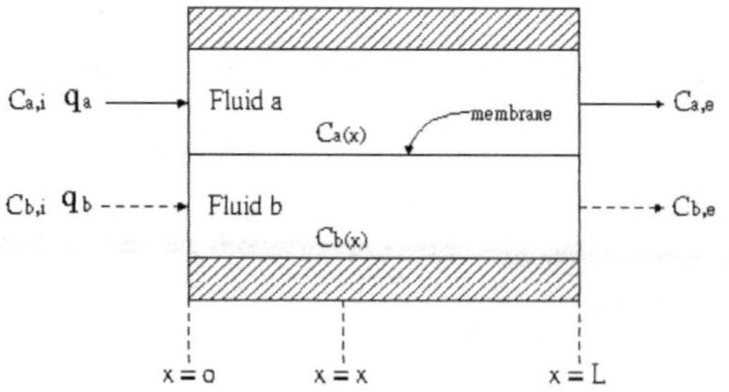

Fig.1 Schematic diagram of a concurrent-flow membrane extractor.

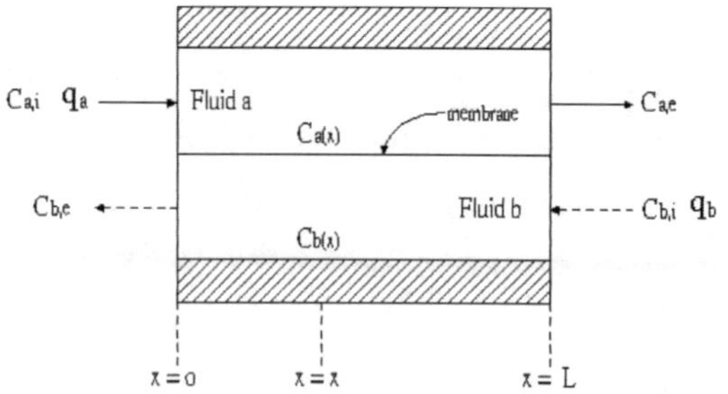

Fig.2 Schematic diagram of a countercurrent-flow membrane exactor.

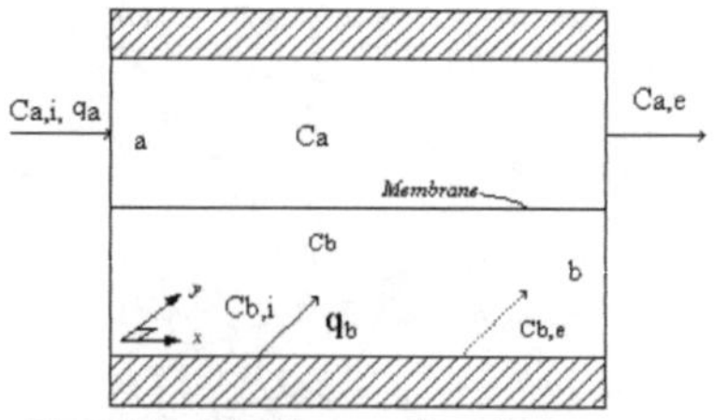

Fig.3 Schematic diagram of membrane solvent extraction in rectangular mass exchanger of cross flow.

Correction factor

The expressions for outlet concentrations, $C_{a,e}$ and $C_{b,e}$, were already derived in the previous work (Yeh and Hsu, 1999). Substitution of into these expressions Eqs.(8) and (9) gives the correction factors :

$F_1 = 1$ for cocurrent flow, (10)

$F_2 = 1$ for countercurrent flow, (37)

and F_1 and F_2 for cross flow are given graphically in Figs. (4) and (5).

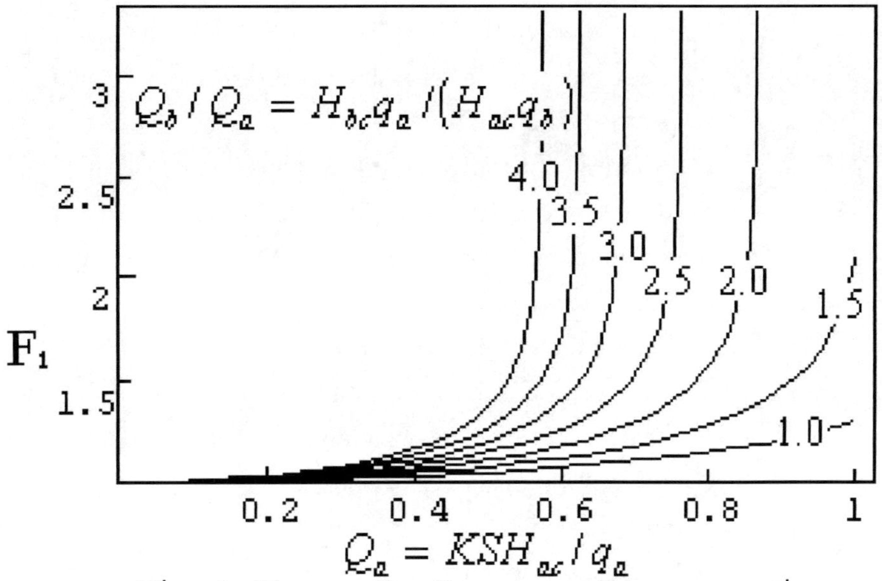

Fig. 4. F_1 vs. Qa for cross-flow operation.

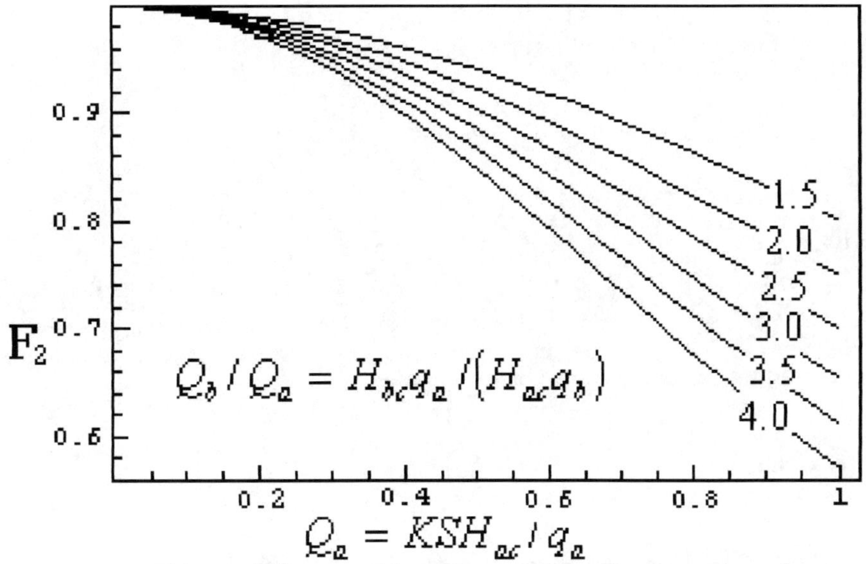

Fig. 5. F_2 vs. Qa for cross-flow operation

Conclusion

The expressions, Eqs.(2) and (3), for calculating the mass-transfer rates in flat-plate membrane extractors were derived with the consideration of correction factors. Eq.(2) is properly applicable to the cocurrent-flow operation with $F_1 = 1$ while for countercurrent-flow operation, Eq.(3) with $F_2 = 1$ is suitable to be used. When operating with cross-flow, either Eq.(2) or Eq.(3) is applicable with F_1 and F_2 given in Figs.4 and 5.

Nomenclature

B	membrane width, m
C_a, C_b	solute concentration distribution in phase a , in phase b, respectively, kg mol/m^3
C_{ai}, $C_{b\,i}$	C_a, C_b, respectively, at the inlet, kg mol/m^3
$C_{a,e}$, $C_{b,e}$	C_a, C_b, respectively, at the outlet, kg mol/m^3
$(\Delta C)_{lm,1}$, $(\Delta C)_{lm,2}$	logarithmic-mean concentration difference, defined by Eq.(4), by Eq.(5), kg mol/m^3
F_1, F_2	correction factor defined by Eq.(8), by Eq.(9)
H_{ac}, H_{bc}	distribution coefficient between phase a and phase c, between phase b and phase c, respectively
K	overall mass-transfer coefficient, m/s
L	length of the membrane sheet, m
q_a, q_b	volume flow rate in phase a, in phase b, m^3/s
S	overall mass-transfer area of a microporous membrane sheet, BL, m^2
W	total mass-transfer rate in a membrane extractor, kg mol/s

Acknowlegement

The author wishes to express his thanks to the National Science Council of the R. O. C. for financial aid under Grant No. NSC 91-2214-E-032-001.

Literature Cited

Ding, H. and E.L. Cussler, "Fractional Extraction with Hollow Fibers with Hydrogel-filled Walls, AIChE J., **37**, 855 (1991).

Kiani, A., R.R. Bhave and K.K. Sirkar, "Solvent Extraction with Immobilized Interfaces in Microporous Membrane", J. Membrane Sci., **20**, 125 (1984).

Prasad, R. and K.K. Sirkar, "Dispersion-Free Solvent Extraction with Microporous Hollow -Fiber Modules", AIChE J., **34**, 177 (1988).

Yang, M.C. and E.L. Cussler, "Designing Hollow-Fiber Contractors", AIChE J., **32**, 1910 (1986).

Yeh, H. M. and Y. S. Hsu, "Analysis of Membrane Extraction through Rectangular Mass Exchangers", Chem. Eng. Sci., **54**, 897 (1999).

Yeh, H. M., Y. Y. Peng and Y. K. Chen, "Solvent Extraction through a Double-Pass Parallel-Plate Membrane Channel with Recycle", J. Membrane Sci., **163**, 177 (1999).

Yeh, H. M. and Y. K. Chen, "Membrane Extraction through Cross-Flow Rectangular Modules", J, Membrane Sci, **170**, 235 (2000).

Yeh, H. M. and Y. K. Chen, "The Effect of Multipass Arrangement on the Performance in a Membrane Extraction of Fixed Configuration", Chem. Eng. Sci., **55**, 5873 (2001).

Paper number: 74d

Measurement and Prediction of Sorption and Transport Properties of EPDM Elastomer

Steven W. Rutherford and James E. Coons

Los Alamos National Laboratory, Engineering Sciences and Applications Division,

MS C930, Los Alamos, NM 87545.

Prepared for presentation at AIChE Meeting, 2003, November 18, session on Advances in Membranes and Fuel Cell Technology.

8/2003

Unpublished (LAUR03-2195)

Abstract

The equilibrium affinity and transport properties of gases and vapors for polymeric materials are important variables in the consideration of the membrane operations. At present, there are no available means to accurately, reliably and robustly predict (apriori) the affinity and transport (and hence permeability) properties of any given polymeric material and the dependence upon temperature and choice of probe gas. Hence these variables must be identified by the time consuming methods of experimental determination. In this investigation, a gas sorption apparatus was constructed and the solubility and diffusivity of methane, carbon dioxide, oxygen and argon were measured in EPDM elastomer. A previously proposed correlation is employed as a means for characterization of permeability for these gases and could possibly be employed for prediction of permeability for other gases of interest.

1. Introduction

The rate of chemical reactions that determine the degree of aging in polymeric materials can often be rate limited by the diffusion of reactant and product species into and out of the material- a situation known as diffusion limited aging (DLA). When DLA is observed, the changes in the mechanical properties and hence lifetime prediction can be determined only when transport properties of reactant and product species in the polymer of interest can be determined. Measurement of transport properties (diffusivity) and equilibrium properties (solubility) are important for the prediction of aging when DLA is observed. These variables are also important in the consideration of materials compatability and the interaction of chemical species in composite materials. At present, there are no available means to accurately predict, apriori, the affinity or transport properties of any given polymeric material and the dependence upon temperature and choice of probe gas. Hence these variables must be identified by the time consuming methods of experimental determination. However, measurement of these properties for a variety of gases in a variety of materials creates an experimental load which could be greatly reduced by establishment of a predictive means for estimation. In this investigation, a gas sorption apparatus was constructed and the solubility and diffusivity of methane, carbon dioxide, oxygen and argon were measured in EPDM elastomer. A correlation is proposed as a means for prediction of these properties for other gases of interest.

2. Experimental

In order to experimentally determine equilibrium and transport properties, a variable volume-pressure, batch sorption apparatus was constructed. The apparatus consists of a sample chamber connected to a series of vessels of various size up to 1 gallon by a series of valves. The operation of the volumetric adsorption apparatus is initiated by degassing the system to ultra high vacuum (less than 10^{-9} torr). Pure gas is injected into the isolated system volume that is subsequently expanded into the sample chamber. As sorption of the gas into the sample takes place, the pressure consequently drops. Pressure transducers monitor the dynamics of the pressure change which allow determination of the kinetics of adsorption and the final pressure yields equilibrium affinity data. The system volume is determined by helium expansion into a previously calibrated vessel. During the expansion, the system is thermostated by an oven in which the entire system is placed. All of the system components interface with Swagelock VCR face seal fittings. The system is housed inside a Despatch LAC 2-12 forced convection oven that maintains isothermality up to 260°C. The pressure transducers are MKS Baratron type 615A with a maximum pressure 20000 torr and 100 torr so that the apparatus is capable of measuring from low vacuum to high pressure. Vacuum is maintained by a varian turbo V70LP backed by a varian mechanical diaphragm pump-MDP 30. The dynamic pressure signal is logged to a PC using National Instruments hardware and software. The polymer studied in this investigation is an *EPDM (ethylene propylene diene monomer) elastomer* whose chemical structure is displayed in Figure 2. Thirty weight percent of a carbon black filler is contained in the polymer. A total of 42.37 grams of EPDM is cut into slabs of 0.0812 cm and placed in the sample chamber for analysis.

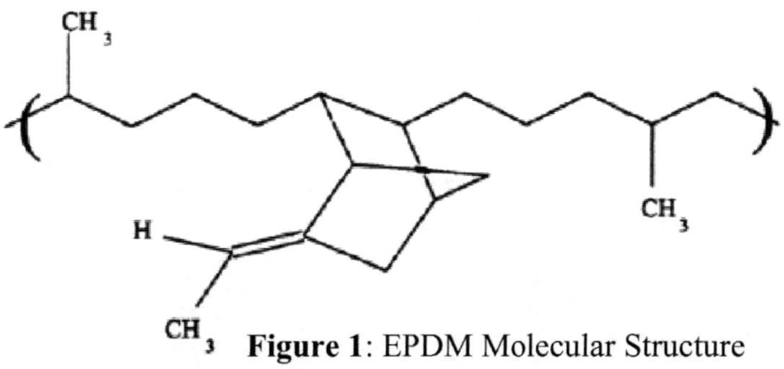

Figure 1: EPDM Molecular Structure

3. Results

Carbon dioxide, methane, argon and oxygen were chosen as probe gases capable of elucidating the transport and equilibrium properties of EPDM. Experiments were additionally undertaken at 293K in order to establish a correlation of transport and equilibrium properties against critical temperature and volume of the probe gas. Sorption experiments were performed by step changing the pressure in the sample chamber and allowing the system to return to equilibrium. Uptake (F) is defined as the normalized pressure change and under the conditions of constant boundary condition (imposed by matching the volume to allow the pressure change to be no greater than 10%) and Henry law equilibrium the uptake can be expressed as[1]:

$$F = 1 - \frac{8}{\pi^2} \sum_{n=0}^{\infty} \frac{1}{(2n+1)^2} e^{\frac{-D(2n+1)^2 \pi^2 t}{l^2}} \qquad (1)$$

where l is the half thickness and D is the diffusivity and is obtained via fitting of Equation (1) to the dynamic data.

3.1 Gas Uptake - Correlation with Gas Critical Properties

It has been shown by experimentation that, for any given polymeric material, the solubility of a sorbing species is related to critical temperature (T_c) according to:[2] $S = c_1 \, exp(c_2 T_c)$ (2). This correlation has had some success in characterizing the solubility for a series of gases in a variety of glassy and rubbery polymers. It has also been proposed that the diffusivity observed under Henry follows a relationship originally applied to describe gas diffusion in liquids. It is represented by the following power law[2]: $D = \frac{\tau}{V_c^{\eta}}$ (3) where V_c is the critical volume of the mobile species, τ is an adjustable parameter and η is a measure of the kinetic selectivity of the polymer. By measurement of solubility and diffusivity for a variety of gases the abovementioned correlations can be assessed for their predictive capability. To this end, the solubility and diffusivity of argon, methane carbon dioxide and oxygen have been measured at 293K in EPDM and the sorption isotherms are displayed in Figure 2. The isotherms of Figure 2 appear to be linear indicative of Henry law. The solubility, obtained from the slope of the plots, can be plotted against critical temperature of the probe gas for each species and this is presented in Figure 3. For the four probe gases chosen the correlation predicts the solubility reasonably well. Further study of a wider

---

[1] Crank J, The Mathematics of Diffusion, 1972.
[2] Merkel TC, Bondar VI, Nagai K, Freeman BD, Pinnau I, J. Poly. Sci., Part B- Poly. Phys., v. 38(#3) pp. 415-434, 2000

variety of gases is required in order to more robustly assess the correlation. The measured diffusivities for the four gases at 293K are plotted in Figure 4 against critical volume of the probe gas. The solid line represents the line of best fit for equation (3) and the value of the kinetic selectivity parameter obtained from the fit is given as 2.7. According to Table 1, EPDM is more selective than PDMS but less selective than PVC.

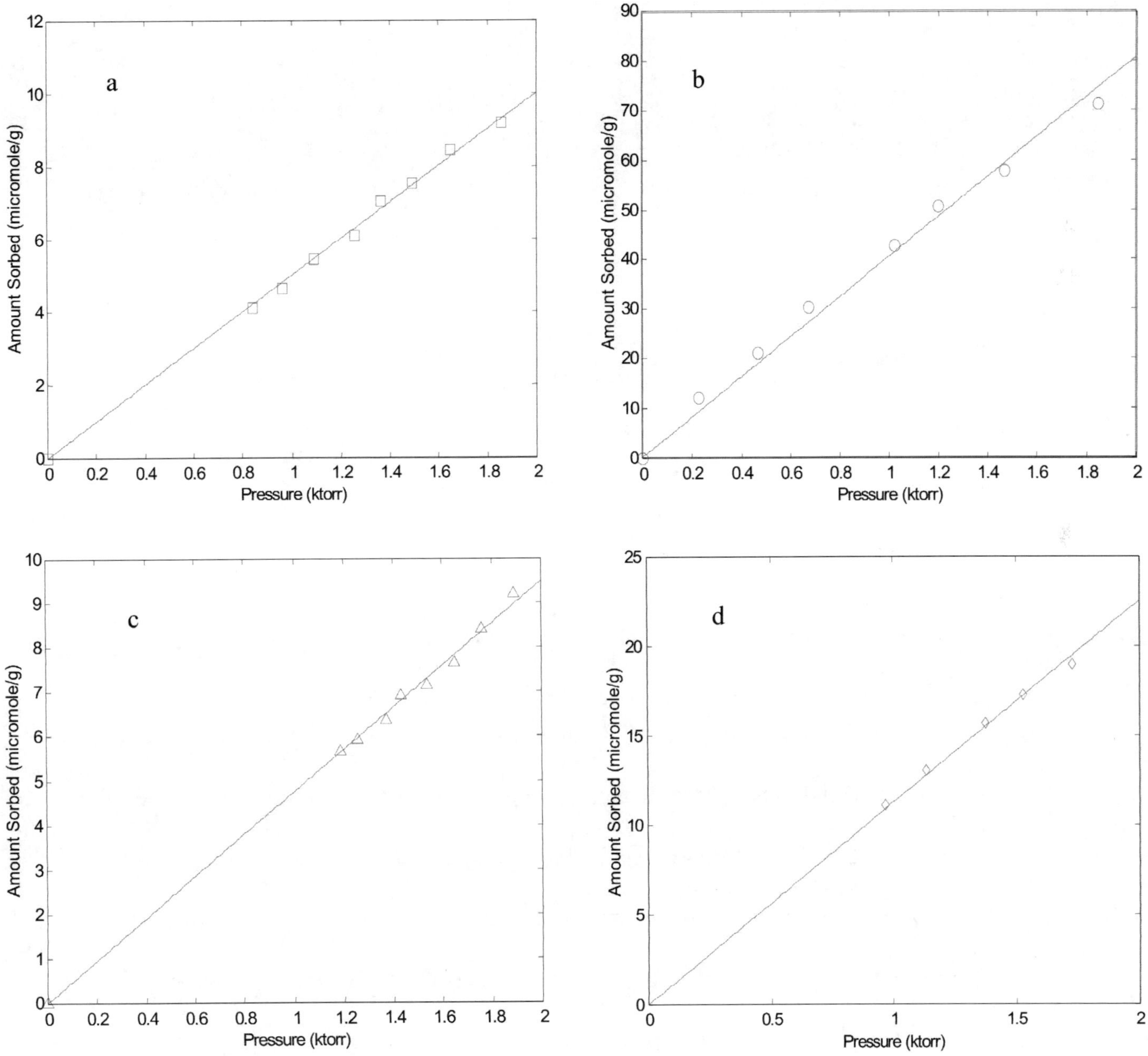

Figure 2: Sorption isotherms for oxygen (a), carbon dioxide (b), argon (c) and methane (d) in EPDM at 293K

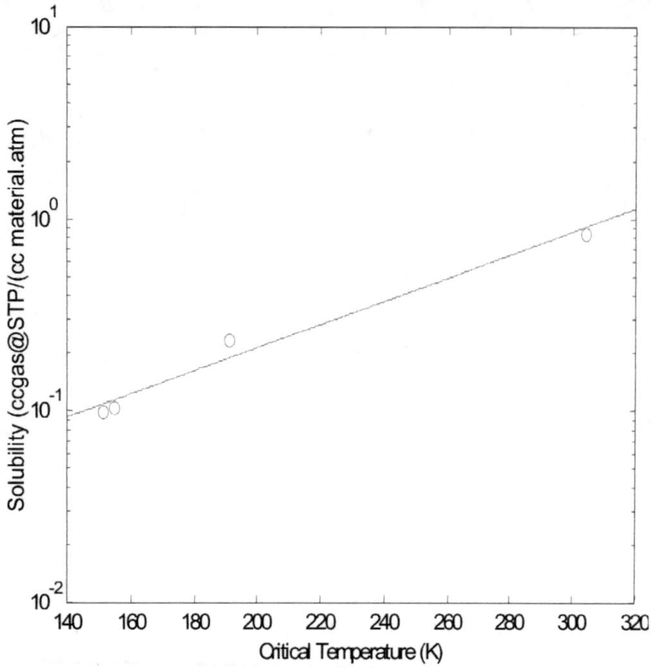

Figure 3: Correlation of solubility against critical temperature of the probe gas

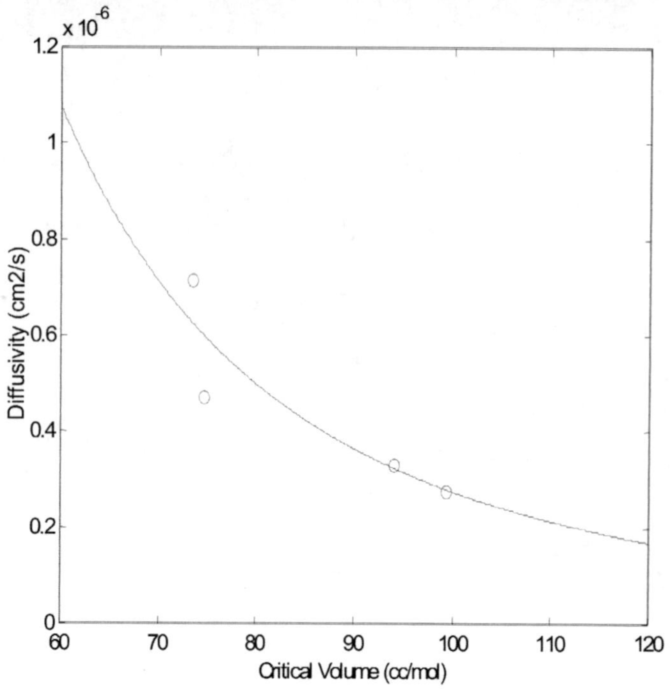

Figure 4: Correlation of diffusivity against critical volume of the probe gas

Polymer	η
PVC	10.5
PDMS	2.3
EPDM (this work)	2.7

Table 1: Values of the kinetic selectivity

5. Conclusions

Transport and equilibrium properties of a range of gases in an EPDM elastomer have been measured and the dependence of these properties upon temperature and choice of gas has been elucidated. A predictive model based on the critical properties of the probe gas has shown promise as a means to correlate the measured data and predict diffusivity and solubility for gases which have not or can not be measured. The results compare well with results obtained for other polymers. Further testing with a wider range of probe gases is recommended in order to more robustly establish the predictive model.

HIGH-PRESSURE, HIGH-TEMPERATURE PERMEANCE MEASUREMENTS OF PALLADIUM-COPPER ALLOYS

Bryan D. Morreale, Michael V. Ciocco

US DOE NETL Site Support Contractors, Parsons
PO Box 618, South Park, PA 15129
bryan.morreale@pp.netl.doe.gov

Bret H. Howard, Richard P. Killmeyer, Kurt S. Rothenberger

US DOE, National Energy Technology Laboratory
PO Box 10940, Pittsburgh, PA 15236

Robert M. Enick, Felipe Bustamante
NETL Research Associates, Department of Chemical and Petroleum Engineering
University of Pittsburgh, Room 1249 Benedum Hall, Pittsburgh, PA 15261

Key words: hydrogen, membrane, palladium-copper alloy, permeability

Prepared for Presentation at 2003 AIChE Annual Meeting, November 16th to 21st,
Topical Conference on Advanced Membrane-Based Separations -Advances in Membrane and Fuel Cell Technologies

Abstract

The hydrogen permeance of various palladium-copper alloy membranes (0, 40, 53, 60, 80, and 100wt % palladium) has been determined using feed gas mixtures both with and without H_2S present. The membrane samples were composed of thin metal foils (~100 microns) supported on high nickel alloy porous substrates, and studied at temperatures and pressures up to 1173 K and 2.6 kPa, respectively. Results of the study indicate that the crystalline phase (bcc vs fcc) of a particular alloy composition, induced by temperature has a significant impact on the resulting permeability, both with and without H_2S present.

All of the membranes studied exhibited infinite selectivity, with the pure palladium membrane exhibiting the highest permeability of the membranes studied in a hydrogen atmosphere. The permeability of the palladium membrane can be expressed in Arrhenius form yielding a pre-exponential constant and activation energy of permeation of 1.13×10^{-3} [mol m^{-1} s^{-1} $Pa^{-0.5}$] and 12.97 [J mol^{-1} K^{-1}], respectively, or fluxes as high as 62 cm^3 cm^{-2} min^{-1}.

The 60wt% palladium-copper alloy exhibited the highest permeance of the alloys tested in a hydrogen atmosphere at temperatures below 773 K, while the highest permeance achieved above 773 K was by the 80 wt% palladium-copper alloy. The difference in the maxima can be directly attributed to the temperature-induced bcc and fcc crystalline phases of the alloys.

The permeance behavior of the palladium-copper alloy membranes exposed to 1000 ppm H_2S was found to correlate to the alloy crystal structure. Sulfur poisoning resistance was exhibited for the alloys at temperatures corresponding to their fcc crystalline phase but there was a significant decrease in permeance when exposed to H_2S under bcc stability conditions.

Introduction

The production of hydrogen is expected to increase in the future as the world moves towards the widespread use of hydrogen as an energy carrier. Coal, which is considered to be the transition feedstock to a hydrogen economy through, produces large amounts of hydrogen mixed with carbon dioxide and trace amounts of other gases via gasification. Therefore, advances in separation technologies, such as membranes, have the potential to improve efficiency and recovery while decreasing the costs associated with hydrogen production. However, a significant technical barrier impeding hydrogen separation membrane development is the susceptibility of potential membrane materials to poisoning or deactivation by contaminating compounds present in gasifier effluent streams, most importantly sulfur compounds such as hydrogen sulfide. Past research in the area of hydrogen sulfide poisoning of membrane materials showed that even very dilute concentrations (ppm level) of sulfur-containing compounds can rapidly deactivate otherwise highly catalytic membrane candidates, such as palladium [1-4].

Thin palladium films deposited within porous alumina [5], platinum-coated palladium [2, 6-8], and palladium-copper alloys [7, 9, 10] have been identified as potential sulfur-resistant membranes candidates. Thus, the objective of this study is to characterize the permeance of several Pd-Cu alloys over a wide range of temperatures in the presence of pure-H_2 and H_2S/H_2 gas mixtures.

Experimental

Palladium-copper alloy foils, 99.9% purity, 0.1 mm thick, were fabricated by ACI Alloys. The alloy foils were produced by vacuum-arc melting palladium and copper powders together to form an ingot that was then rolled to a thickness of 0.1 mm (100μm). Membranes were prepared by hand-cutting roughly circular pieces from the as-received foil sheets. The foil disks were cleaned and mounted in specially designed membrane test units at the National Energy Technology Laboratory (NETL). The resulting mounted membranes had an active hydrogen permeation area of ~57 mm^2.

The "sulfur-free" permeance evaluation of the palladium-copper alloys of interest in this study was conducted in a high temperature, high pressure, steady-state hydrogen membrane test unit. The apparatus was designed to operate at temperatures and pressures up to 1173 K and 3 MPa, respectively. A detailed description of the unit can be found elsewhere [12].

Permeance analysis in the presence of sulfur-laden gases was conducted in a specially designed high temperature, low pressure transient screening unit at NETL. The retentate side of the membrane was charged with the test gas of interest, either pure hydrogen or a 1000 ppm H_2S/balance H_2 mixture. The assembly was placed within a Lindberg tube furnace used in conjunction with a dual element, type-K thermocouple. A concentric tube placed on the permeate side of the membrane allowed a constant argon sweep to evacuate the surface of the membrane of diffusing hydrogen. The assembly was then heated to the desired temperature and the retentate transient pressure data was used to calculate the rate of hydrogen transport through the membrane.

The permeance of the palladium-copper alloy in the absence (steady-state testing) and presence (transient testing) of sulfur-containing gases was determined using equations 1 and 2, respectively.

$$N_{H2} = -k'(P^{0.5}_{H2,Ret} - P^{0.5}_{H2,Per}) \qquad (1)$$

$$P_{H2,Ret}(t) = \left[P^{0.5}_{H2,Ret,Initial} - \frac{1}{2} A_M k' \left(\frac{RT}{V} \right) t \right]^2 \qquad (2)$$

Results and Discussion

The permeance results of the palladium-copper alloy membranes of this study, determined via steady-state testing in the absence of sulfur-containing gases, is presented in Figure 1 as a function of inverse absolute temperature.

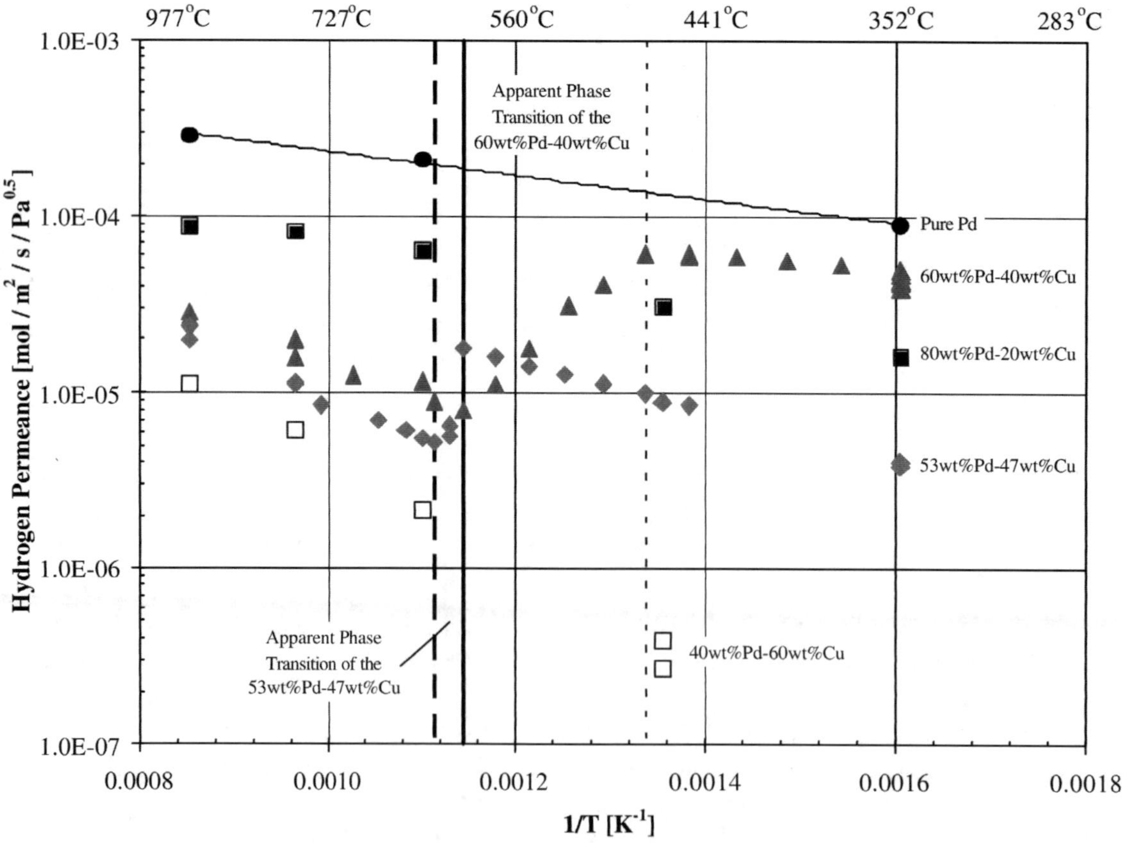

Figure 1. Permeance summary of the Pd/Cu alloy steady-state testing conducted in the absence of sulfur-containing gases at NETL as a function of temperature.

The steady-state membrane testing illustrated that the 80wt%Pd-20wt%Cu alloy resulted in the largest permeance values above 748K while the 60wt%Pd-40wt%Cu demonstrated the largest values below 748K. The 60wt%Pd-40wt%Cu and the 53wt%Pd-47wt%Cu exhibited significant decreases in permeance as the temperature increased from values corresponding to the body-centered-cubic (bcc) crystalline phase to values corresponding to the face-centered-cubic (fcc) phase.

Figure 2 illustrates the permeance results obtained from the low pressure, transient testing in the absence and presence of sulfur-laden gases. Transient testing of the membrane alloys demonstrated similar trends as the steady-state testing with respect to permeance as a function of temperature. Moreover, the temperature-induced crystalline phase of the palladium-copper alloys exhibited a direct correlation with resistance to sulfur poisoning. All of the alloys tested via the transient method exhibited significant resistance to sulfur poisoning at temperatures corresponding to their fcc crystalline phase while little sulfur resistance was observed for the bcc phase when exposed to H_2S.

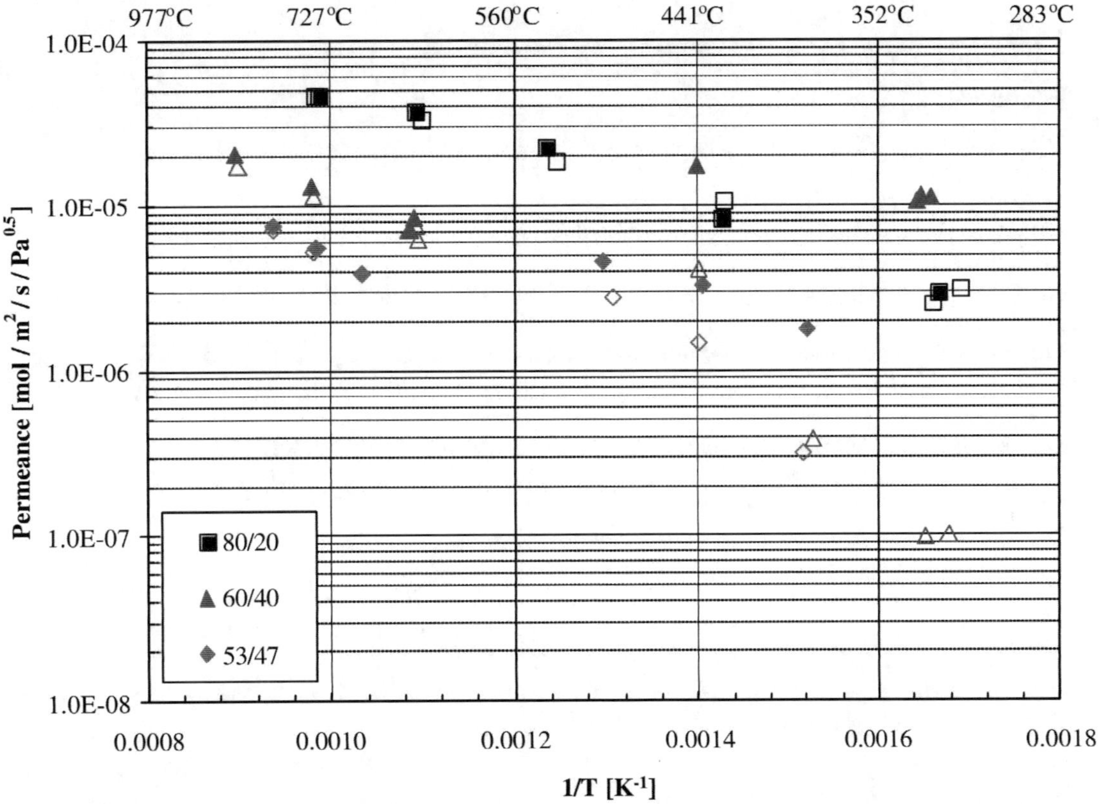

Figure 2. Permeance summary of the Pd/Cu alloy transient testing conducted in the absence (solid shapes) and presence (hollow shapes) of sulfur-containing gases at the NETL as a function of temperature.

Conclusions

- The palladium-copper alloys of this study exhibited hydrogen permeance values ranging from approximately one-half to one-tenth that of pure palladium over the 623 K to 1173 K and 0.1 MPa to 2.6 MPa hydrogen pressure range.
- A correlation between the temperature-induced crystalline phase of the palladium-copper alloy and its hydrogen permeance and sulfur tolerance was established:

- The 60wt%Pd-40wt%Cu (bcc phase) exhibited the highest permeance values at temperatures less than ~773 K.
- The 60wt%Pd-40wt%Cu and the 53wt%Pd-47wt%Cu showed significant decreases in permeance with increase in temperature, which can be traced to a phase change from the bcc phase to the fcc phase.
- The 80wt%Pd-20wt%Cu (fcc phase) exhibited the highest values at temperatures above ~773 K.
- The 80wt%Pd-20wt%Cu exhibited sulfur resistance over the entire temperature range of this study while the 60wt%Pd-40wt%Cu and the 53wt%Pd-47wt%Cu exhibited resistance at temperatures above 973K. The resistance can be attributed to the temperatures associated with the fcc crystalline phase of the alloys.

References

1. U. Feuerriegel, W. Klose, S. Sloboshanin, H. Goebel, J. Schaefer, *Langmuir* **10**, 3567 (1994)
2. D. Edlund, D. Friesen, B. Johnson, W. Pledger, *Gas Separation and Purification* **8**, 131 (1994)
3. P. Gravil, H. Toulhoat, *Surface Science* **430**, 176 (1999)
4. M. Kajiwara, S. Uemiya, T. Kojima, *International Journal of Hydrogen Energy* **24**, 839 (1999)
5. K. Kusakabe, S. Yokoyama, S. Morooka, J.I. Hayashi, H. Nagati, *Chemical Engineering Science* **51**, 3027 (1996)
6. D.J. Edlund, "A Catalytic Membrane Reactor for Facilitating the Water-Gas-Shift Reaction at High Temperatures, Phase I" *Final Report to the US DOE on Grant DE-FG03-91ER81229* Bend Research (1992)
7. D.J. Edlund, "A Catalytic Membrane Reactor for Facilitating the Water-Gas-Shift Reaction at High Temperatures, Phase II" *Final Report to the US DOE on Grant DE-FG03-91ER81229* Bend Research (1995)
8. D.J. Edlund, W.A. Pledger, *Journal of Membrane Science* **94**, 111 (1994)
9. D.L. McKinley, US Patent 3,350,845 (1967)
10. D.J. Edlund, Paper DOE/ER/81419-97/C0749; Contract DE-FG03-92ER81419
11. P.R. Subramanian, D.E. Laughlin, "Cu-Pd (Copper-Palladium)" in *Binary Alloy Phase Diagrams – Second Edition*, ed. T.B. Massalski (ASM International, 1990), pp. 1454-1456
12. B.D. Morreale, M.V. Ciocco, R.M. Enick, B.I. Morsi, B.H. Howard, A.V. Cugini, K.S. Rothenberger, *Journal of Membrane Science* **212** (2003) 87-97

Recovery of a Metal Working Fluid Using Ultrafiltration and Chelation

Brian E. Reed, Associate Professor
Department of Civil and Environmental Engineering
University of Maryland Baltimore County
1000 Hilltop Circle Baltimore, MD
reedb@umbc.edu

Prepared for Presentation at AIChE National Meeting, Nov 16-21, 2003
Session: Advances in Membranes and Fuel Cell Technology

ABSTRACT

Metal working (MW) fluids are used to control friction and temperature, improve workpiece surface quality, and reduce tool wear. The batch-life of the MW fluid investigated was controlled primarily by the concentration of metallic soaps (complexes between the oil emulsifier and Al/Mg). EDTA, a strong metal chelator, was added to the MW fluid to break the oil-Al/Mg complexes. The aqueous phase EDTA-Al/Mg complexes were then separated from the oil phase by ultrafiltration (UF). Al levels were reduced between 39 and 49% and Mg levels were reduced between 67 and 77%. Al transfer from the oil phase to the aqueous phase was slow and increased during UF concentration. Rinsing the concentrated MW fluid with deionized water decreased the ash content but had a lesser impact on Al/Mg removal. Al mass balances produced errors of 2.8% for run 3 and 6.2% for run 4. The UF system (0.11 μm ceramic membrane) effectively separated the oil and aqueous phases. Permeate turbidity was generally < 1 NTU and permeate flux ranged from 230 to 130 gal/ft^2-d depending on the degree of concentration.

INTRODUCTION

Metal working (MW) fluids are used to control friction and temperature, improve workpiece surface quality, and reduce tool wear. For the MW fluid investigated in this study, the batch-life is controlled primarily by the concentration of complexes of anionic emulsifiers (*e.g.*, oleic acid) and Al/Mg. These complexes are referred to as "metallic soaps". Al/Mg enter the MW fluid during normal milling operations through contact with the workpiece. Increasing the metallic soaps concentration increases the lubricating ability of the MW fluid. Initially, the increase in lubricating ability is desired but eventually the MW fluid becomes too "slippery" and the rolling process is adversely affecting because of unwanted slippage between the workpiece and the rolling mill.

If the concentration of metallic soaps can be reduced in the MW fluid, reuse of MW fluid is possible. The addition of a strong metal chelator (*e.g.*, EDTA) to a MW fluid containing metallic soaps should transfer the Al/Mg from the oil phase to the aqueous phase. The aqueous phase chelating agent-Al/Mg complexes and oil droplets can then be separated using a membrane filtration process. In Figure 1, a schematic of the chelation-ultrafiltration (CUF) process is presented. A batch of spent MW fluid enters a well-mixed feed tank where an appropriate amount of chelating agent, is added. When the aqueous phase Al and Mg concentrations approach a maximum, the emulsified oil droplets are separated using a UF membrane. UF permeate would contain high concentrations of chelating agent-Al/Mg complexes, very low concentrations of oil and is considered a waste. The concentrate would contain almost all the original oil/additives and would have a lower metallic soap concentration and can be reused. If necessary, the concentrate can be rinsed with deionized water to further remove aqueous phase chelating agent-Al/Mg complexes and to lower the conductivity. Deionized water would be added to the concentrated MW fluid after which the MW fluid is processed through the UF membrane. The rinsed/concentrated MW fluid can be brought back to its original oil content using deionized water and then reused. EDTA was chosen as the study chelating agent. A high shear rotary ultrafiltration (HSRUF) system, equipped with a 0.11 μm ceramic membrane, was used to separate emulsified oil from the aqueous phase. Goals of this study were to: 1) determine the ability of EDTA to sequester the Al/Mg associated with emulsified oil droplets and 2) ascertain the feasibility of using the HSRUF system to separate the aqueous phase EDTA-Al/Mg complexes from the oil droplets. Waste MW fluids were acquired from an actual rolling operation and several EDTA concentrations were used. Permeate Al and Mg concentrations, permeate flux and quality (turbidity and oil/grease) were measured. The effect of concentrating and rinsing the treated MW fluid was investigated. A mass balance on Al was also performed.

MATERIALS AND METHODS

Two batches of metal working (MW) fluid were received directly from a rolling mill operation in 55 gallon drums. The first batch was used for runs 1 and 2 and the second batch was used for runs 3 and 4. The MW fluid contained oil (≈5% v/v), oleic acid (an emulsifier), film strength additives and antifoaming agents. Specific contents of the mixture cannot be disclosed for proprietary reasons. The first batch of MW fluid contained 212 mg/L Al (7.86×10^{-3} M) and 14.8 mg/L Mg (6.1×10^{-4} M). The second batch contained 230 mg/L Al (8.52×10^{-3} M). Because the Mg concentrations were much lower than Al, only Al was monitored in runs 3 and 4. Al and Mg concentrations were determined by atomic absorption (AA) spectrophotometry. For initial MW fluid samples and concentrate from the UF process, samples were first digested using USEPA Method 3050 (USEPA, 1982) prior to AA analysis.

The high-shear rotary UF unit was equipped with a ceramic membrane (0.11 μm mean pore size; TiO_2-Al_2O_3; 492 cm^2) and operated under the following conditions: 1) transmembrane pressure = 40-42 psi, 2) membrane rotational speed = 1750 rpm, 3) recirculating flow rate = 4.5 L/min (1.2 gal/min), and 4) feed tank temperature = 43°C ± 1° C (110 ± 2 ° F). The system was either operated in recycle or batchdown mode. In recycle mode, permeate is returned to the feed tank and the concentrations of oil, Al/Mg, and EDTA remain constant although Al/Mg can be transferred from the oil phase to aqueous phase. During batchdown operation, permeate is removed from the system, increasing the oil content of the concentrate. The volumetrically-based oil concentration factor was calculated as:

$$CF = V_{initial}/(V_{initial} - V_{perm}) \qquad (1)$$

Where, CF = concentration factor, $V_{initial}$ = initial volume of MW fluid in the system, and V_{perm} = volume of permeate removed from system. The concentration factor is expressed as *n*X. An untreated MW fluid has a CF of 1X. If the system volume is reduced by half through permeate removal, the CF = 2X and so on. The volumetrically-based CF is equivalent to the mass-based CF if the amount of oil in the permeate is small relative to the amount of oil retained by the membrane.

The effectiveness of the CUF process was evaluated based on the following parameters: 1) Al chelation efficiency, 2) Al and Mg mass removal efficiencies, and 3) permeate flux and quality (turbidity, oil/grease (O/G), conductivity). The Al chelation efficiency was calculated as:

$$\text{Al Chelation Efficiency} = [Al]_{actual}/[Al]_{theory} \qquad (2)$$

where $[Al]_{actual}$ and $[Al]_{theory}$ are the actual and theoretical Al concentrations (mole/L) in the aqueous phase, respectively. $[Al]_{actual}$ is the measured Al concentration in the permeate while $[Al]_{theory}$ is calculated by

$$[Al]_{theory} = [EDTA]_{total} - [Mg]_{actual} \qquad (3)$$

Where $[EDTA]_{total}$ = total concentration of EDTA (mole/L) and $[Mg]_{actual}$ = permeate Mg concentration (mole/L) If $\{[EDTA]_{total} - [Mg]_{actual}\} > [Al]_{total}$ then Equation 3 does not apply and

$$[Al]_{theory} = [Al]_{total} \qquad (4)$$

where $[Al]_{total}$ is the total concentration of Al originally present in MW fluid. The chelation efficiency was based on Al because the Al concentration was much greater than Mg and the Al-oleic acid complex was found to be much stronger than the Mg-oleic acid complex. Al and Mg mass removal efficiencies were calculated as mass of Al/Mg in permeate that was removed from the system divided by the total mass of Al/Mg originally present in the MW fluid. Permeate flux was expressed as the volume of permeate/membrane area-time (gal/ft^2-d) while turbidity was used as a real-time measure of permeate quality. Permeate O/G (by freon extraction; Standard Methods, 1992) was measured less frequently because of the difficulty in acquiring freon.
very hour thereafter and analyzed for Al within 2 hours of sampling; when the permeate Al concentration stabilized for a given EDTA concentration, the next EDTA dose was added and 3) after final EDTA addition, the system was operated in batchdown mode until the CF reached 4X.

RESULTS AND DISCUSSION

Run 1

In Figure 2 permeate Al and Mg concentrations versus time are presented for run 1. Prior to the addition of EDTA, the permeate Mg and Al concentrations were 1.5 mg/L and < 0.1 mg/L, respectively. At an EDTA concentration of 10^{-4} M, the permeate Mg concentration doubled but there was little change in the Al concentration. Similar behavior was observed for 5×10^{-4} M and 10^{-3} M EDTA concentrations. Not until the EDTA concentration was increased to 5×10^{-3} M was there a significant amount of Al released from the oil phase. The initial Al and Mg concentrations in the MW fluid were 212 mg/L (7.86×10^{-3} M) and 14.8 mg/L (6.1×10^{-4}

M), respectively. The amount of Mg complexed at 5 x 10^{-3} M EDTA was approximately 10 mg/L (4.1 x 10^{-4} M) leaving 4.6 x 10^{-3} M of EDTA available for Al complexation. Assuming the EDTA-Al complex was 1:1 and the remaining EDTA is 100 percent efficient in sequestering Al, $[Al]_{theory}$ should equal 4.6 x 10^{-3} M (124 mg/L, see Equation 2). The highest Al concentration reached at 5 x 10^{-3} M EDTA was 91 mg/L (3.4 x 10^{-3} M) which corresponded to a chelating efficiency of about 74%. The percentage of the total Al and Mg residing in the aqueous phase at 5 x 10^{-3} M EDTA, were 43% and 67%, respectively.

At 10^{-2} M EDTA there was enough EDTA added to form 1:1 complexes with all Al and Mg present in the MW fluid. However, the permeate Al concentration increased by only 7 mg/L to 98 mg/L (3.6 x 10^{-3} M) while the Mg concentration decreased slightly. Under these conditions, the Al chelation efficiency decreased to about 46%. The percentage of the total Al and Mg residing in the aqueous phase at 10^{-2} M EDTA, were 46% and 61%, respectively. The incomplete sequesteration of Al and Mg when the EDTA/Al and EDTA/Mg ratios were greater than 1 gives some insight into the strength of the metallic soap complexes (especially for Al) but also may indicate that a portion of the Al and Mg is inaccessible to the aqueous phase EDTA. For example, oil may have coalesced around inert particles (*e.g.*, dust, workpiece fragments) shielding the oil droplet from contact with the aqueous phase.

Run 2

In Figure 3, permeate Al/Mg concentrations, permeate and concentrate pH, permeate turbidity, and flux and CF versus time are presented. As in run 1, prior to the addition of EDTA, the permeate Al concentration was negligible while the Mg concentration was approximately 2.5 mg/L. Based on manufacturer's information, the concentration of oleic acid was always larger than the combined Al/Mg content (*i.e.*, [oleic acid] > [Al + Mg]). Comparing permeate Mg (2 mg/L) and Al (< 0.1 mg/L) concentrations prior to EDTA addition, the formation constant for Mg-oleic acid is significantly less that for Al-oleic acid and the Mg-oleic acid complex should be easier to break. Following the first EDTA addition (10^{-3} M), the aqueous phase Mg increased immediately but then declined as the Al concentration increased from ≈0.1 mg/L to about 8 mg/L. This phenomenon was more apparent at 5 x 10^{-3} M EDTA. The Al-EDTA chelation efficiency for run 2 at CF = 1X was similar to that in run 1. The percentages of the total Al (Mg) residing in the aqueous phase at 5 x 10^{-3} M and 10^{-2} M EDTA were 39% (77%) and 46% (81%), respectively. Given the small increases in aqueous phase Al and Mg concentrations when the EDTA concentration was increased from 5 x 10^{-3} M to 10^{-2} M, an EDTA concentration of 5 x 10^{-3} M was chosen for use in runs 3 and 4.

Permeate Al and Mg concentrations increased with CF with the vast majority of total Al/Mg removal occurring as the CF was increased from 1X to 3X (see Table 1). Actual Al and Mg mass removal efficiencies, based on the mass of Al and Mg in the permeate collected as the CF was increased from 1X to 5X, were 49% for Al and 86% for Mg. These values were much higher than the removal efficiencies predicted using permeate Al and Mg concentrations observed during operation in recycle mode (37% and 64% for Al and Mg, respectively). Either additional Al/Mg were released as the oil content was increased or oil droplets containing Al/Mg passed through the membrane into the permeate. The later explanation is unlikely given that the permeate turbidity and O/G did not increase substantially during batchdown. It is more likely that during batchdown the oil droplets coalesced into larger droplets that required less emulsifier (oleic acid) to maintain emulsion stability. The formation of larger oil droplets caused the oil-bound Al and Mg to be released more easily.

Except for a few samples between 29 and 32 hr, permeate turbidity was generally less than 1 NTU and oil/grease < 200 mg/L, demonstrating the effectiveness of the ceramic membrane in rejecting oil droplets. The pH of the permeate and concentrate increased because the EDTA stock solution pH (made from Na_4EDTA) was 11.9. Increasing the pH should not affect the chelation of aqueous Al and Mg because Al and Mg are theoretically fully complexed with EDTA between pH = 7 to 9. However, pH could affect the relationship between oleic acid and Al/Mg as well as the chemistry of the oil droplet. The permeate flux increased from 195 gal/ft^2-d to greater than 250 gal/ft^2-d when the EDTA concentration was increased from 10^{-3} M to 5 x 10^{-3} M. Additional increases in EDTA concentration had little effect on the permeate flux. An increase in flux with conductivity (ionic strength) has been reported for the microfiltration of a domestic wastewater (Al-Malack and Anderson, 1996) and for fuel and crude oil contaminated waters (Tansel *et al.*, 1995). Al-Malack and Anderson (1996) reported an increase in flux with increasing salt concentration until an optimum dose was reached above which no improvement in flux was observed. During batchdown, permeate flux decreased with increasing CF (*i.e.*, increasing concentrate oil content). These results were consistent with earlier work on the same MW fluid (Reed *et al.*, 1997, 1998). As the oil concentration was increased, the solute boundary layer thickness increased due to 1) more oil being transported to the membrane surface via convection 2) decreased back diffusion of oil from the membrane surface because of a smaller oil concentration gradient between the membrane surface and bulk solution and 3) less turbulence at the membrane surface due to an increase in viscosity (Reynolds Number ~1/μ).

Runs 3 and 4

In Figure 4, permeate Al concentration, permeate and concentrate conductivity, and flux and CF versus time are presented for run 3. Results for run 4 were similar and will not be presented. Following the 5 x 10^{-3} M EDTA addition, the permeate Al concentration increased over a 5 hour period from near zero to 90 mg/L and then remained fairly steady for the remainder of the CF = 1X portion of the experiment. First and second order kinetic models ($r = k[Al]$ and $r = k[Al]^2$) were applied to the data in this 5 hour period but neither model adequately described the increase in Al with time. The following multi-step process may be occurring: (I) EDTA transport from the bulk aqueous solution to the stagnant liquid layer surrounding the oil droplet, (II) diffusion of the EDTA through the stagnant liquid layer, (III) reaction of the oleic acid-Al (-Mg) complexes with EDTA, (IV) diffusion of EDTA-Al and EDTA-Mg complexes through the stagnant liquid layer and (V) transport of EDTA-Al and EDTA-Mg complexes to the bulk solution. The bulk solution hydrodynamic conditions in the experimental system were turbulent as the MW fluid and EDTA were well mixed in a number of locations: 1) in the centrifugal pump that provided recirculation between the membrane chamber and feed tank, 2) within the high-shear membrane chamber, and 3) within the feed tank itself where a pneumatic mixer was located. The Reynolds number in the high-shear membrane chamber can be estimated by (Murkes and Carlsson, 1988):

$$R_e = \omega r^2/\mu \quad (7)$$

where ω = membrane rotational speed (rad/s), r = membrane disk radius (m) and, μ = absolute viscosity (kg/m-s). At 1750 rpm (183 rad/s) and 5% oil, R_e was > 1.8 x 10^6 (Viadero, 1997) which is much greater than 2.5 x 10^5, the cutoff between laminar and turbulent flow (Ketola and McGrew, 1968; Murkes and Carlsson, 1988). Thus, it is unlikely that steps I and V were rate limiting. Typically, reaction kinetics are not considered to be rate-limiting. However, little is known about the reaction between the metallic soaps and EDTA. As a side experiment, virgin MW fluid was contacted with Al to determine if a synthetically contaminated MW fluid could be developed. After several weeks of mixing, there was little change in the aqueous phase Al concentration. In the full-scale metal working process, it takes between 6 to 8 weeks before there is a buildup of metallic soaps. Thus, step III as well as steps II or IV may be rate-limiting. Complicating the transport of EDTA to the oil droplet and the transport of EDTA-Al (-Mg) complexes away from the oil droplet are the presence of negative charges on the oil droplet, and EDTA (-4), and EDTA-Al (-1) and EDTA-Mg(-2) complexes. The rapid increase in bulk aqueous phase Mg concentration compared with Al could be due to steps III and IV being faster for Mg – the more negative EDTA-Mg^{-2} complex may move away from the negatively charged oil droplet faster than the EDTA-Al^{-1} complex due to differences in electrostatic repulsive forces.

During the first batchdown (CF = 1X to 3X) the permeate Al increased from 90 mg/L to 127 mg/L and 30% of the Al was removed. Subsequent deionized water rinses removed little additional Al (4.7% for rinse 1, 4.3% for rinse 2). The total Al removal efficiency was 39% for run 3. The effectiveness of the rinse in removing ions from the aqueous phase is demonstrated by the decrease in permeate and concentrate conductivity. When the system was brought back to its original volume following the first batchdown, the concentrate conductivity was similar to the original value. This is important because the ash content of the MW fluid (defined by the MW industry as the residual remaining after heating at 770°C) would be unacceptably high, due to Na_4EDTA addition, unless the MW fluid was rinsed. During the first batchdown in run 4 (CF = 1X to 4.2X), 38% of the Al was removed from the system. In subsequent rinses, Al removal efficiencies for rinse 1 and 2 were 7.8% and 3.0 %, respectively. The total Al removal efficiency for run 4 was 49% and was higher than that observed in run 3 because a higher CF was reached (4.2X compared with 3X in run 3). Al mass balances, conducted at the end of the runs, produced errors of 2.8% for run 3 and 6.2% for run 4.

After the system was brought back to its original volume (CF = 1X), the flux took over 10 hours to reach its pre-batchdown level indicating that the solute boundary layer was somewhat stable. Similar behavior was observed during the second rinse. Stability of the solute boundary layer with changes in rotational speed were reported for an actual MW fluid (Reed *et al.*, 1997) and for a virgin MW fluid (Viadero, 1997).

SUMMARY

The aqueous phase EDTA-Al/Mg complexes and oil droplets were then separated using an ultrafiltration membrane process. The chelation-ultrafiltration (CUF) process was effective in removing oil-bound Al and Mg (metallic soaps). Removal efficiencies for Al (39 to 49%) and Mg (67 and 86%) were highest at 5 x 10^{-3} M EDTA. Because metallic soaps, up to a certain concentration, impart valuable lubrication properties, higher Al/Mg removal efficiencies may not be required. The UF system (0.11 μm ceramic membrane) was effective in separating the oil droplets from the aqueous phase. Permeate turbidity was generally less than 1 NTU and the O/G content was always less than 200 mg/L. The separation step does not appear to be the limiting factor in the effectiveness of the CUF process.

REFERENCES

Al-Malack, M.H. and G.K. Anderson (1996) "Coagulation-crossflow microfiltration of domestic wastewater," *J. of Membrane Science*, 121, 59-70.

Cheryan M., (1986). Ultrafiltration Handbook, Technomic Publishing Comp, Inc., Lancaster, PA.

Dean, J.A. (1979). Lange's Handbook of Chemistry, 12th Edition, McGraw-Hill, New York, NY.

Ketola, H. and J. McGrew (1968). "Pressure, Frictional Resistance and Flow Characteristics of the Partially Wetted Rotating Disk," *J. of Lubrication Tech.*, April, 395-404.

Laemmle, J.T. (1992). "Metal Working Lubricants." *Lubricants and Lubrication*, Institute of Petroleum, 139-148.

Murkes, J. and C. Carlsson (1988). Crossflow Filtration, John Wiley and Sons, Ltd., Bath, Great Britain.

Reed, B.E., P. Carriere, P., Lin, W., Roark G, and Viadero, R.C. (1998). "Treatment of an Aluminum Manufacturer's Oily Wastewater: Pilot-Scale Results and Preliminary Design," accepted for publication in *ASCE Practice Periodical of Hazardous, Toxic and Radioactive Waste Management.*

Reed, B.E., W. Lin, R.C. Viadero, and J. Young. (1997). "Treatment of Oily Wastes Using High-Shear Rotary Ultrafiltration," *ASCE J. of Environmental Engineering*, 123(12), 1234-1242.

Standard methods for the examination of water and wastewater. (1992). American Public Health Association (APHA), Washington, D.C.

Tansel, B., J. Regula, and R. Shalewits (1995) "Treatment of fuel oil and crude oil contaminated waters by ultrafiltration membranes," *Desalination*, 102, 301-311.

USEPA (1982), Testing Methods for Evaluating Solid Waste, US Environmental Protection Agency, Environmental Information Center, Cincinnati, OH.

Viadero, R.C. (1997). Effects of Membrane Rotation and Transmembrane Pressure on High-Shear Rotary Ultrafiltration of an Oily Wastewater," Ph.D. Dissertation in partial fulfillment of the Requirements for the degree of Doctor of Philosophy, Department of Civil and Environmental Engineering. West Virginia University.

Viadero, R. and B. Reed (1999). "Rotation and Concentration Effects in High-Shear Ultrafiltration," *ASCE J. of Environmental Engineering*, **125**, 7, 638-646.

Westall, J.C., Zachery, J.L., and Morel, F.M.M., (1976). "Mineql: A computer Program, for the Calculation of Chemical Equilibrium Composition of Aqueous Systems," Technical Note No. 18, Ralph M. Parsons Laboratory.

Table 1. Percentages of the Total Mass of Al and Mg Removed During Batchdown

CF Change	Percent of Total Al or Mg Present	
	Al	Mg
1X → 2X	61%	59%
2X → 3X	23%	23%
3X → 4X	12%	13%
4X → 5X	4%	5%

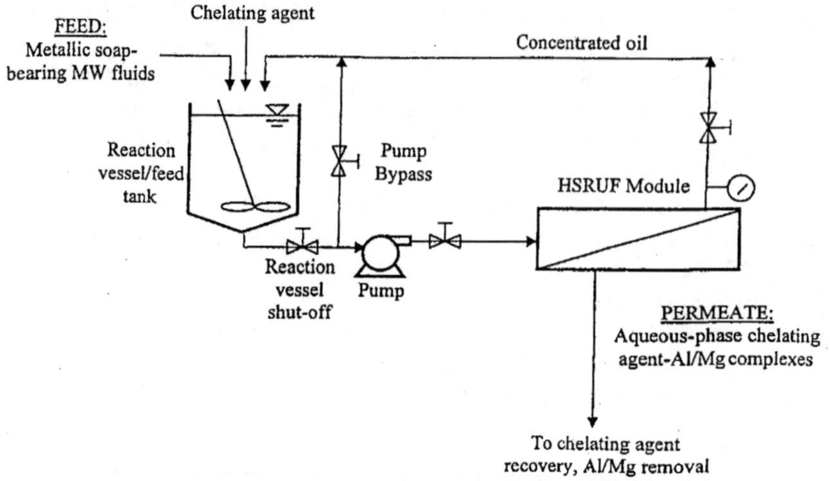

Figure 1. Schematic of the Chelation-Ultrafiltration (CUF) Process.

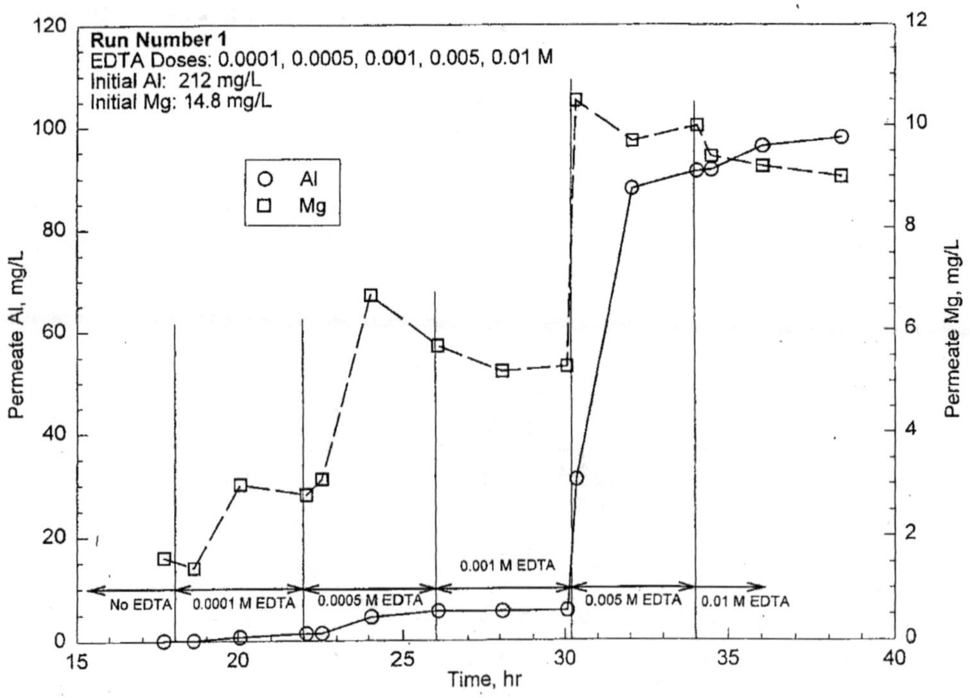

Figure 2. Permeate Al and Mg Concentrations Versus Time For Run 1.

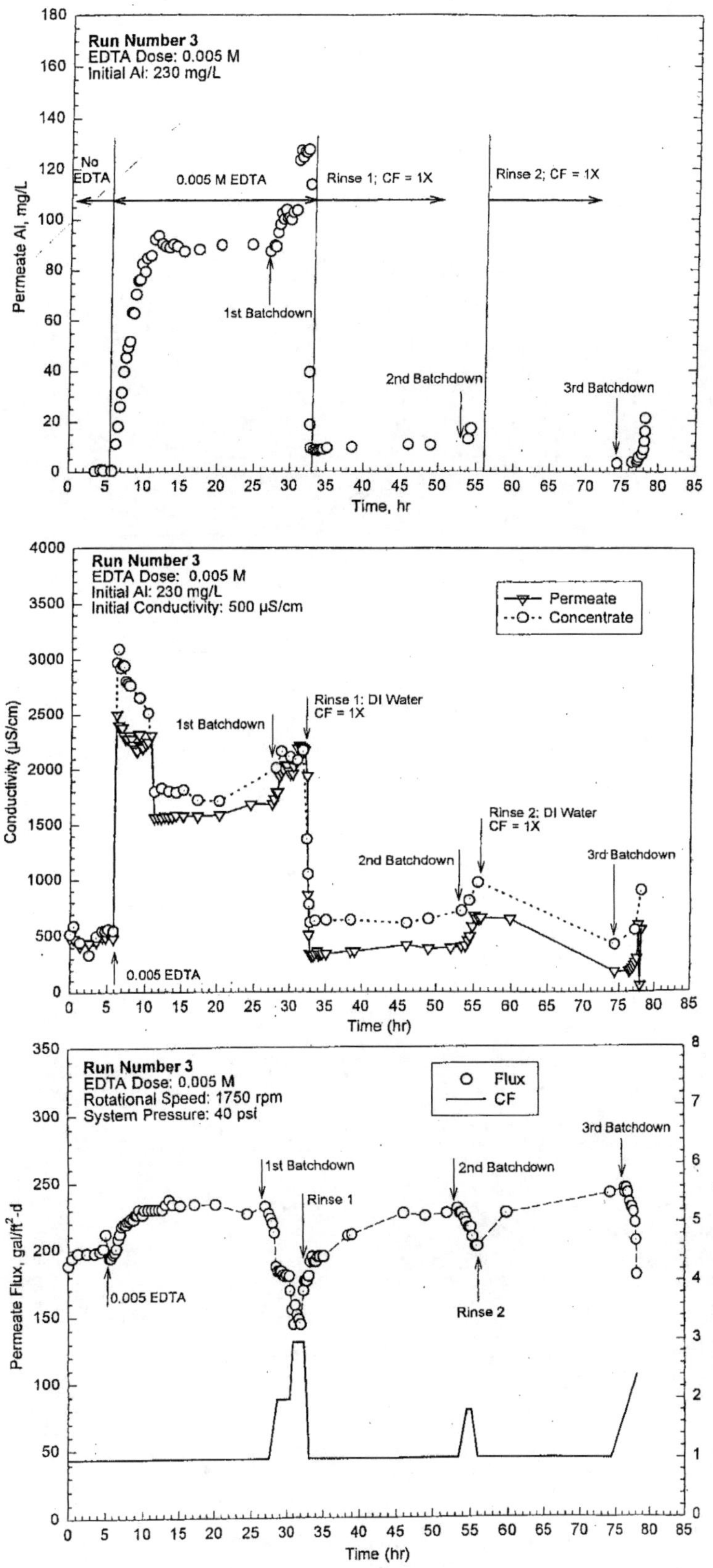

Figure 4. Run 4 Permeate Al Concentration (Top); Permeate and Concentrate Conductivity (Middle); and Permeate Flux and CF (Bottom).

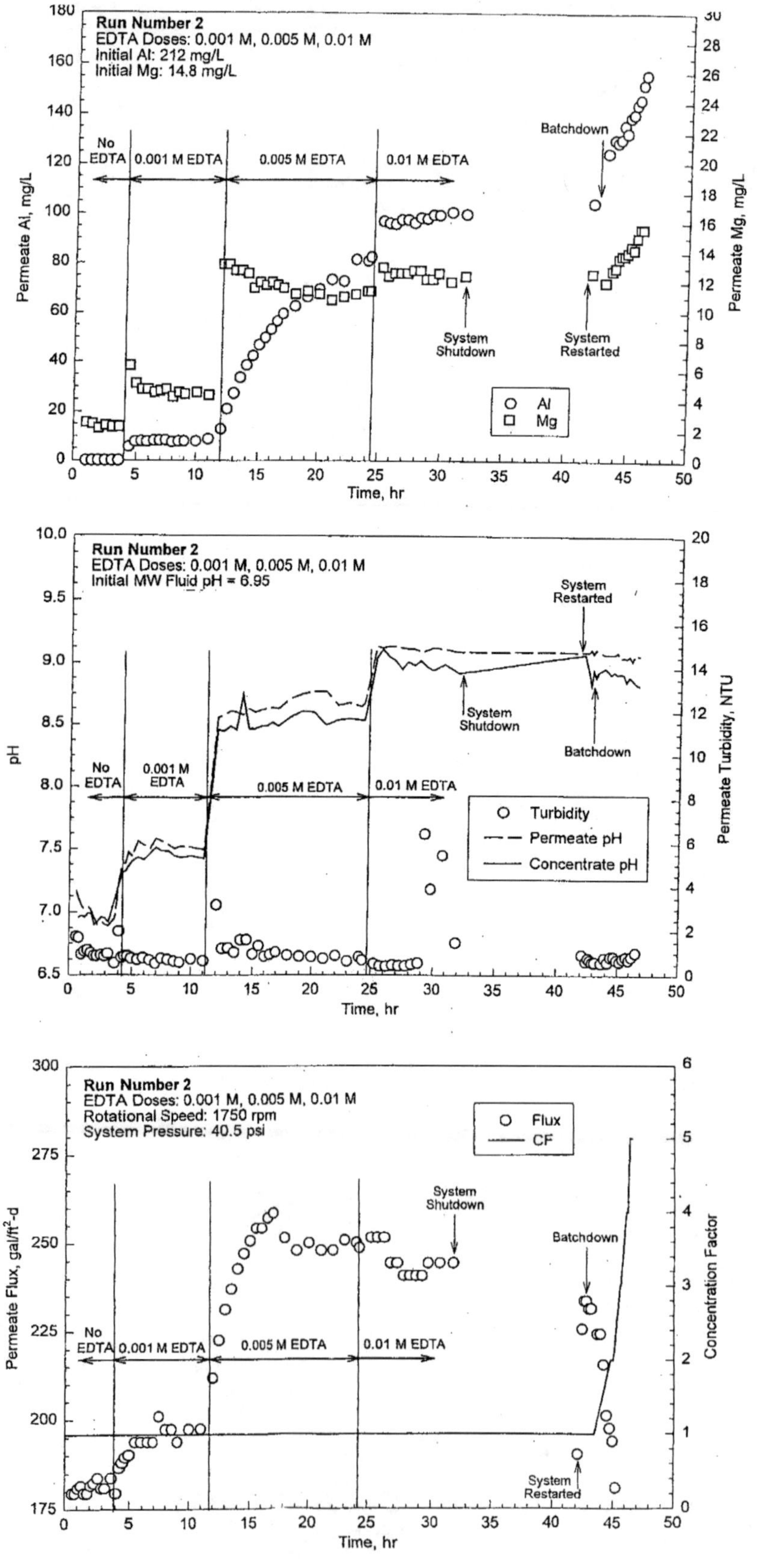

Figure 3. Run 2 Permeate Al/Mg concentrations (Top); Permeate and Concentrate pH and Turbidity (Middle); Permeate Flux and CF (Bottom)

Methane Steam Reforming over Metals Supported on Ce-ZrO_2 for Fuel Cell Application

Tomokazu Eda, Ken-Ichiro Sotowa and Katsuki Kusakabe

Department of Applied Chemistry
Kyushu University
Fukuoka 812-8581, Japan
kusaktcf@mbox.nc.kyushu-u.ac.jp

Keywords: Reforming, Methane, Hydrogen, Fuel Cell, Membrane Reactor, Rhodium

Prepared for presentation at the 2003 Annual meeting, San Francisco, CA, Nov. 16-21

Abstract

The steam reforming of methane to produce hydrogen as an energy source for use in a polymer electrolyte fuel cell over Ce-ZrO_2 supported Ni, Pt, Ru and Rh catalysts was studied. In order to select highly active and stable reforming catalysts for use in a membrane reactor, methane conversion and CO selectivity were determined in the low temperature ranges of 500-800°C. Among the structures of $Ce_{1-x}Zr_xO_2$ supports with a 10wt% Ni loading, the $Ce_{0.15}Zr_{0.85}O_2$ support showed the highest CH_4 conversion at 500-600°C and a high H_2/CO ratio due to the oxidation of CO via the water-gas shift reaction. The catalytic activity of $Ce_{1-x}Zr_xO_2$ supports can be further improved by loading with Pt, Ru and Ph. The highest activity was obtained for the Rh/$Ce_{0.15}Zr_{0.85}O_2$ catalyst for which the methane conversion was 28.1 % at 500°C.

1. Introduction

On-site hydrogen production has received considerable attention in recent years because hydrogen could be utilized as an energy source for a polymer electrolyte fuel cell (PEFC) for household use. The steam reforming of methane (SRM) is the conventional process used to convert natural gas into hydrogen or syngas (H_2 + CO).

$$CH_4 + H_2O \quad CO + 3H_2, \qquad \Delta H^0_{298} = 206 \text{kJ mol}^{-1} \qquad (1)$$

However, a large amount of energy is required because SRM is a highly endothermic reaction. The SRM process is typically carried out at a high temperature (>750 °C) and high pressure (> 2 MPa), and excess steam is also used to reduce carbon formation. The PEFC system with an reformer for household use would be desirable to reduce the size and the reaction conditions of SRM also should be moderated to some extent. In addition, a unit to remove CO from the reformate gas must also be incorporated into the system.

Membrane reactors, comprised of a packed bed of catalysts and hydrogen selective membranes, have been applied to SRM [1-3]. In SRM, the removal of hydrogen from the packed bed with the membranes shifts the thermodynamic equilibrium towards the product side. Lin et al. [4] examined SRM performance in a membrane reactor using a supported palladium membrane and industrial catalysts. The reaction temperature was lowered to 500°C to achieve a methane conversion of 45%, which is 15% higher than the thermodynamic equilibrium conversion. However, catalysts suitable for the SRM reaction in a membrane reactor, especially having activity at low temperature, have not been developed to any extent.

Ceria-zirconia mixed oxide catalysts [5-9] are of interest, particularly for their oxygen storage properties. Ceria also stabilizes metal dispersion and has a better catalytic activity at lower temperature. Recently, Ni loaded Ce-ZrO_2 were applied to the SRM reaction and the findings showed a high catalytic activity as well as stability compared with conventional catalysts [9-12].

In the present study, the performance of Ni/$Ce_{(1-x)}Zr_xO_2$ catalysts ($0.5<x<1$) in SRM was investigated at reaction temperatures in the range of 500-800°C. The catalytic activity of Pt, Ru or Rh catalysts on Ce-ZrO_2 substrate was also determined.

2. Experimental

Zirconia and CeO_2 modified zirconia supports were prepared by the urea hydrolysis method [5]. $ZrOCl_2 \cdot 8H_2O$ and $Ce(NO_3)_3 \cdot 6H_2O$ were dissolved in ion-exchanged water. The molar ratio of Ce:Zr in the solution was varied at 1:99, 15:85, 25:75 and 50:50. The total concentration of Ce and Zr was fixed at 0.1 mol$\cdot$dm^{-3}. A 0.4 mol$\cdot$dm^{-3} urea solution was added to the solution. The prepared solution was kept at 100°C for 50 h. The solution was filtered at room temperature and the filtrate was washed with ethanol to prevent particle agglomeration. The Ce-ZrO_2 support was then calcined at 500°C for 4 h. A γ-alumina support was used as a comparison. Platinum, rhodium and ruthenium and nickel were loaded by impregnating appropriate amounts of H_2PtCl_6, $RhCl_3$, $RuCl_3$ and $Ni(NO_3)_2$, respectively, on the supports for 1 h, followed by drying in a vacuum at 50°C for 2 h. After drying, the sample was calcined in air at 500°C for 5 h. The calcined samples were reduced with H_2 at 500°C for 2 h.

SRM experiments were carried out in a packed bed reactor. The pulverized catalysts (210-420 μm in particle size, 100 mg) were packed in the center of the reactor. The SRM reaction was performed at 500-800°C at a flow rate of 90 dm^3/min. The product gases were analyzed by an on-line TCD-GC and FID-GC with a methanizer. Considering only the SRM reaction indicated in equation (1), the CO selectivity is defined as shown in the following equation.

$$\text{CO selectivity} = (\text{moles of CO formed/moles of consumed } CH_4) \times 100 \qquad (2)$$

3. Results and discussion

SRM was carried out over the $Ni/Ce_{1-x}Zr_xO_2$ and Ni/Al_2O_3 catalysts with a Ni loading of 10wt% at 500-800°C. Fig. 1 shows the effect of reaction temperature on CH_4 conversion. Although there were no catalytic activities for the Ni/Al_2O_3 and $Ni/Ce_{0.5}Zr_{0.5}O_2$ catalysts in the temperature ranges of 500-600°C, they showed higher activities than Ni/ZrO_2, $Ni/Ce_{0.15}Zr_{0.85}O_2$ and $Ni/Ce_{0.25}Zr_{0.75}O_2$ at 800°C. The catalytic activity of the mixed oxide catalysts at high temperature is dependent on the their structural properties. It should be noted that the cubic-structure mixed oxide catalysts ($Ni/Ce_{0.5}Zr_{0.5}O_2$) is more active than the catalysts with tetragonal ($Ni/Ce_{0.15}Zr_{0.85}O_2$ and $Ni/Ce_{0.25}Zr_{0.75}O_2$) and monoclinic (Ni/ZrO_2) structures. Ce1-xZrxO$_2$ (X > 0.5) catalysts were recently reported to have high activities for CO oxidation [5], partial oxidation of methane [6] and CO2 reforming of methane [6]. The present studies were focused on the catalytic activity of SRM at low temperature and the findings show that $Ni/Ce_{0.15}Zr_{0.85}O_2$ has highest activity at 500-600°C.

Fig. 2 shows data on CO selectivity as a function of temperature. The CO selectivity over Ni/Al_2O_3 was in agreement with the equilibrium value calculated only from the SRM. CO selectivities over the $Ni/Ce_{1-x}Zr_xO_2$ catalysts were low because CeO_2 promotes CO oxidation through the water-gas shift reaction ($CO + H_2O \quad CO_2 + H_2$). An appropriate reaction mechanism for SRM over $Ni/Ce_{1-x}Zr_xO_2$ catalysts was proposed as follows. The reducibility of Ce-ZrO_2 makes the production of highly mobile oxygen species possible through a redox cycle of Ce^{4+}/Ce^{3+}. Partially reduced ceria sites produce active oxygen species from water molecules. Methane is reduced to hydrogen over the Ni sites, depositing carbon. The deposited carbon reacts with the oxygen spillover from the support onto the Ni sites.

Fig. 3 shows data on CH_4 conversion over $Ce_{0.15}$-$Zr_{0.85}O_2$ supported Pt, Ru and Rh catalysts. The supported Rh catalysts had the highest activity in the temperature range of 500-800°C. CH_4 conversion of 10wt%$Ni/Ce_{0.15}Zr_{0.85}O_2$ is also plotted in Fig.3. The catalytic activity of 10wt%$Ni/Ce_{0.15}Zr_{0.85}O_2$ is lower than those of noble metal catalysts in spite of its higher metal loading. The CO selectivities for SRM over the noble metal catalysts were higher than that over $Ni/Ce_{0.15}Zr_{0.85}O_2$, as shown in Fig. 4. Since the dehydrogenation of adsorbed CH_4 might be enhanced by using the noble metal catalysts, this reflects the higher CO selectivity. Based on CH_4 conversion measurements with time on stream over the supported noble metal catalysts at the conditions defined in Fig. 4, the catalysts had the outstanding stabilities.

When SRM is operated in a membrane reactor where the hydrogen produced in the catalytic packed bed permeates through a hydrogen selective membrane, the reaction temperature can be reduced through the influence of an increase in CH_4 conversion beyond equilibrium conditions. The dilution of hydrogen caused by selective hydrogen permeation from the catalytic packed bed favors coke formation. Therefore, we concluded that $Rh/Ce_{0.15}Zr_{0.85}O_2$, having catalytic properties that include high activities at low temperature and a high stability without coking can be employed as a catalyst for SRM in a membrane reactor.

4. Conclusions

From the present studies on methane steam reforming over the $Ce\text{-}ZrO_2$ supported metal catalysts at 500-800°C, the following conclusions have been drawn:

1. Among the structure of the $Ce_{1-x}Zr_xO_2$ supports with 10wt% Ni loading, the $Ce_{0.15}Zr_{0.85}O_2$ support showed the highest CH_4 conversion at 500-600°C and a high H_2/CO ratio due to the oxidation of CO via the water-gas shift reaction.
2. The catalytic activity of $Ce_{1-x}Zr_xO_2$ supports was further improved by the loading of Pt, Ru and Rh. The highest activity was obtained for the $Rh/Ce_{0.15}Zr_{0.85}O_2$ catalyst for which the methane conversion was 28.1 % at 500°C.

Lierature cited

[1] S. Uemiya, N. Sato, H. Ando, T. Matsuda, E. Kikuchi, Appl. Catal. 67 (1991) 223.

[2] J. Shu, B.P.A. Grandjean, S. Kaliaguine, Catal Today 25 (1995) 327.

[3] S.L. Jorgensen, P.E.H. Nielsen, P. Lehrmann, Catal Today 25 (1995) 303

[4] Y.-M. Lin, G.-L. Lee, M.-H. Rei, Catal Today, 44 (1998) 343.

[5] M. Thammachart, V. Meeyoo, T. Risksomboon, S. Osuwan, Catal. Today 68 (2001) 53.

[6] L.V. Mattos, E. Rodino, D.E. Resasco, F.B. Passos, F.B. Noronha, Fuel Process. Technol., 83 (2003) 147-161.

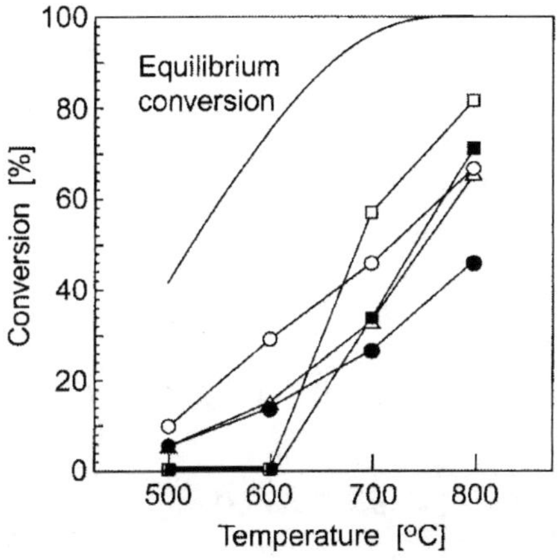

Fig. 1 Effect of temperature on CH_4 conversion over (○) Ni/$Ce_{0.15}Zr_{0.85}O_2$, (△) Ni/$Ce_{0.25}Zr_{0.75}O_2$, (□) Ni/$Ce_{0.5}Zr_{0.5}O_2$, (●) Ni/ZrO_2, and (■) Ni/γ-Al_2O_3. Reaction conditions: space velocity = 9000 mL-CH_4/g-cat·h, metal loading = 10wt%.

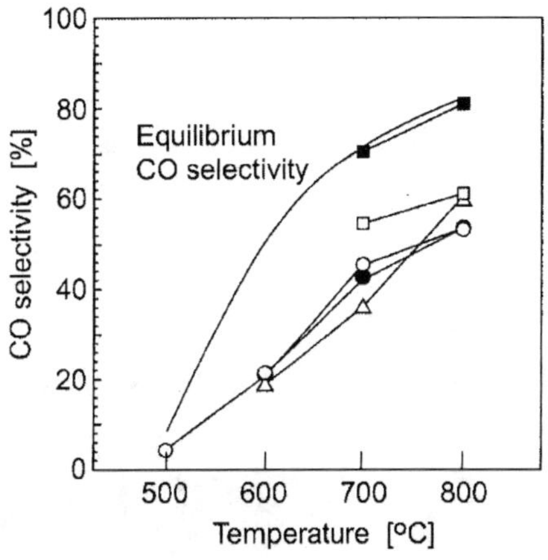

Fig. 2 CO selectivity over (○) Ni/$Ce_{0.15}Zr_{0.85}O_2$, (△) Ni/$Ce_{0.25}Zr_{0.75}O_2$, (□) Ni/$Ce_{0.5}Zr_{0.5}O_2$, (●) Ni/ZrO_2, and (■) Ni/γ-Al_2O_3. Reaction conditions: space velocity = 9000mL-CH_4/g-cat·h, metal loading = 10wt%.

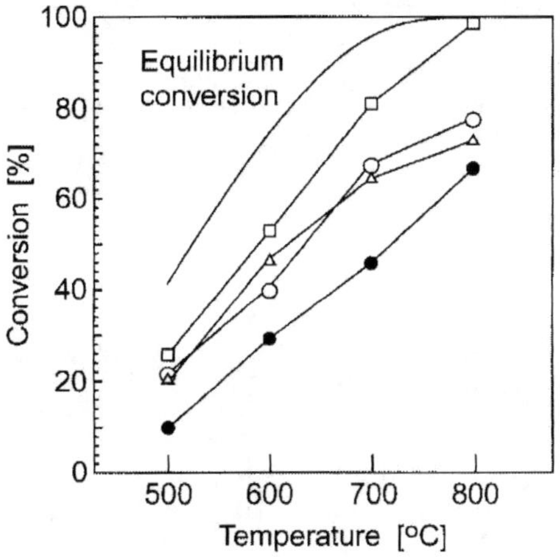

Fig. 3 Effect of temperature on CH_4 conversion over (○) 3wt%Ru/$Ce_{0.15}Zr_{0.85}O_2$, (△) 3wt%Pt/$Ce_{0.15}Zr_{0.85}O_2$, (□) 3wt%Rh/$Ce_{0.15}Zr_{0.85}O_2$, and (●) 10wt%Ni/$Ce_{0.15}Zr_{0.85}O_2$. Reaction conditions: space velocity =9000mL-CH_4/g-cat·h.

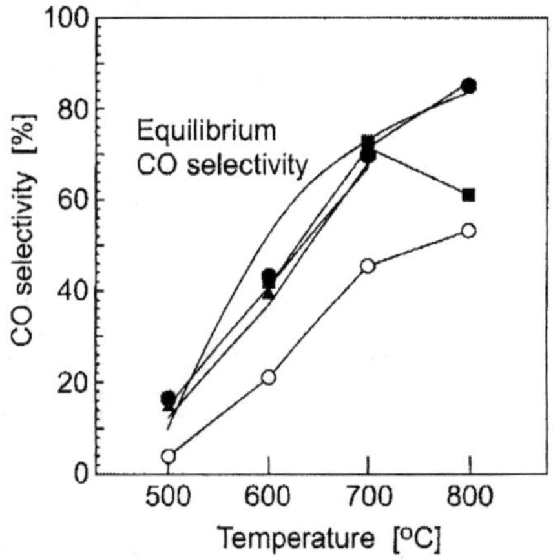

Fig. 4 CO selectivity over (○) 3wt%Ru/$Ce_{0.15}Zr_{0.85}O_2$, (△) 3wt%Pt/$Ce_{0.15}Zr_{0.85}O_2$, (□) 3wt%Rh/$Ce_{0.15}Zr_{0.85}O_2$, and (●) 10wt%Ni/$Ce_{0.15}Zr_{0.85}O_2$. Reaction conditions : space velocity = 9000mL-CH_4/g-cat·h.

Carbon Dioxide-Selective Membranes for Hydrogen Purification and Gas Separation

Jian Zou, Gaurav Shil*, and W.S. Winston Ho

Department of Chemical Engineering
The Ohio State University
140 W. 19th Ave., Columbus, OH 43210

*Department of Chemical and Materials Engineering
University of Kentucky, 40506

Key words: Carbon dioxide-selective membrane, interfacial polymerization, gas separation, hydrogen purification

Prepared for presentation at the 2003 Annual Meeting, San Francisco, CA, Nov. 16-21, 2003

Abstract

CO_2-selective membranes have numerous applications, including the removal of CO_2 from synthesis gas to obtain high purity hydrogen for fuel cells and the separation of CO_2 from flue gas for the greenhouse gas sequestration. In this study, two types of CO_2-selective composite membranes with facilitated transport were synthesized for these applications; one was interfacially polymerized membrane, and the other was cast polymer blend membrane.

For the interfacially polymerized membranes, microporous supports with a pore size range of 0.02-0.05 μm were made using polyetherimide and Matrimid, and a very thin film was then interfacially polymerized on the surface of the support using polyethylenimine and trimesoyl chloride. This membrane preparation gave defect-free membranes, and the average thickness of the interfacially polymerized layer was about 0.5 μm. The CO_2/N_2 selectivity data were obtained for temperatures ranging from 100°C to 170°C, and they showed a decrease from 80 at 100°C to 8 at 170°C. The permeability of CO_2 increased as temperature increased in that temperature range.

The cast polymer blend membranes consisted of crosslinked polyvinylalcohol with amino acid salt and polyamine as mobile carrier and fixed carrier, respectively. CO_2 transport was enhanced significantly due to its reactions with these mobile and fixed carriers. In the temperature range from 21 to 100°C, the permeability of CO_2 increased as temperature increased. However, the selectivity also increased as temperature increased. As polyamine concentration in the membrane increased, the selectivity increased while the permeability of CO_2 did not change significantly.

1. Introduction

Gas separation through synthetic membranes has gained considerable importance for more than two decades [1]. Membrane separation has the advantages of lower energy costs, simplicity of operation, and mechanical reliability over the conventional cryogenic distillation, absorption and adsorption.

CO_2-selective membranes show numerous unique application potentials, including the removal of CO_2 from synthesis gas to obtain high purity hydrogen for fuel cells and the removal of CO_2 from flue gas for the greenhouse gas sequestration. For fuel cells, the need to obtain high purity hydrogen from synthesis gas is critical. CO_2-selective membranes can remove CO_2 from synthesis gas containing CO_2, N_2 and H_2, thus producing high purity and high pressure H_2 at the retentate side of membrane. For environmental applications, the removal of CO_2 from flue gas containing 8-20% CO_2 has become increasingly important due to the more stringent environmental regulations. Membrane separation provides one of the potential methods to achieve this goal with reduced energy consumption.

Among membranes based on various types of mechanisms, facilitated transport membrane is currently attracting the interest of researchers because of its high selectivity and high flux. Facilitated transport is the molecular transport that combines diffusion and reaction. There are two main types of facilitated transport membranes, one with mobile carriers, and one with fixed carriers.

In this study, two types of CO_2-selective composite membranes with facilitated transport were synthesized for the applications mentioned above; one was interfacially polymerized membrane, and the other was cast polymer blend membrane.

2. Experimental

For the interfacially polymerized membranes, the microporous supports were made by means of the phase inversion process. Two different types of polymers, polyetherimide and Matrimid 5218, were used for preparation of the supports. Then interfacial polymerization was achieved on the microporous support. The interfacial polymerization consisted of two major steps: (1) immersion of the microporous support in aqueous polyethylenimine solution and (2) immersion in organic acid chloride solution. These steps were followed, in a similar way as in the synthesis of the interfacially polymerized reverse osmosis membranes [2 - 4], by post-treatment steps which included hydrocarbon vapor stripping, soaking in neutralization solution, rinsing with warm deionized water, immersing in glycerin solution, and finally drying at 90°C.

For the cast polymer blend membranes, an aqueous solution containing polyvinylalcohol (PVA), amino acid salt, and polyamine was cast onto a microporous Teflon support using a casting knife. The cast membranes were first dried at room temperature inside the hood overnight to remove most of the water and then heated at 120°C for 20 minutes.

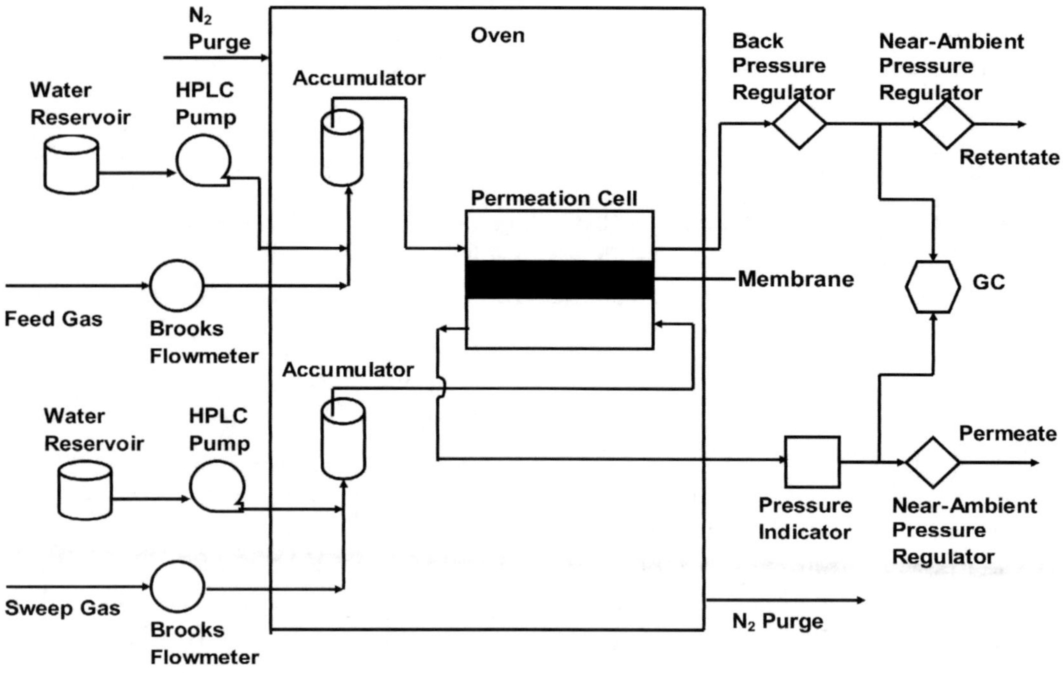

Figure 1. Schematic of gas permeation unit.

A gas permeation unit was tailor-made with data acquisition and analysis. A schematic of this unit is shown in Figure 1. An Agilent 6890N GC with a thermal conductivity detector was used to analyze the gas streams. The feed gas used was a certified gas mixture containing 20% carbon dioxide, 40% hydrogen, and 40% nitrogen. Helium or nitrogen was used as the sweep gas. A feed gas flow rate of 60 cc/min and a sweep gas flow rate of 30 cc/min were used. The feed gas and the sweep gas were passed counter-currently through permeation cell. Water was pumped into both sides with a flow rate of 0.028 g/min. Data for membrane performance were collected at different temperatures.

3. Results and Discussion

3.1 Interfacially Polymerized Membranes

By optimizing conditions for preparation, supports with proper pore size range (0.02-0.05 μm) were obtained. A very thin film was interfacially polymerized on the air-exposed surface of the support using polyethylenimine and trimesoyl chloride. From several scanning electron micrographs, the average thickness of the interfacially polymerized layer was measured to be about 0.5 μm. This is in the thickness range reported for interfacially polymerized membranes [2, 3]. A reaction scheme for polyethylenimine and trimesoyl chloride is presented in Figure 2. Figure 3 shows the scanning electron micrograph of the cross-section of interfacially polymerized composite membrane using 4% polyethylenimine.

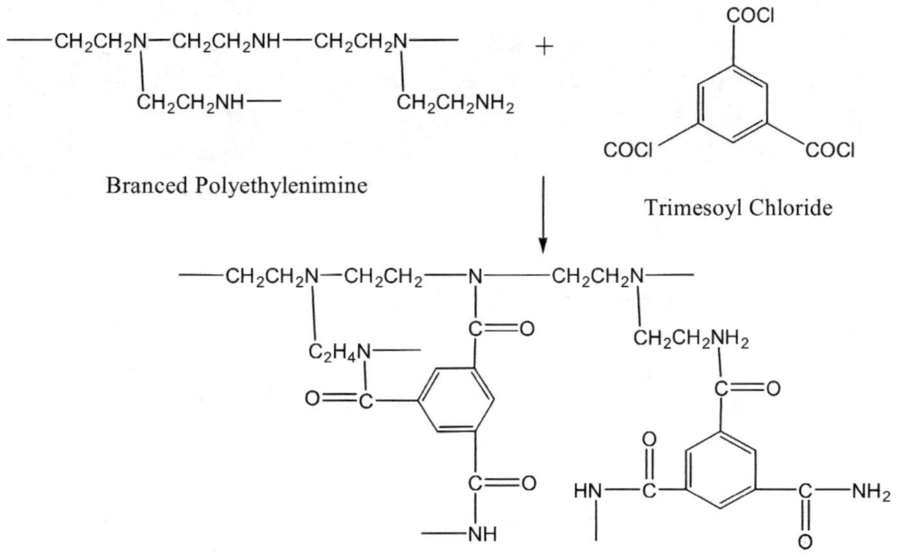

Figure 2. Reaction scheme for interfacial polymerization.

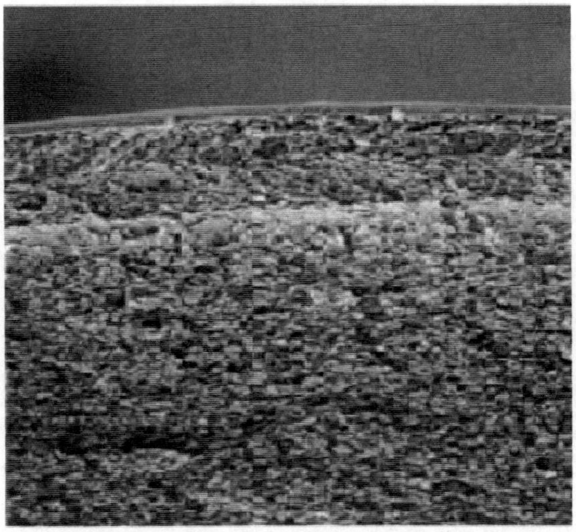

Figure 3. Scanning electron micrograph of the cross-section of interfacially polymerized composite membrane using 4% polyethylenimine.

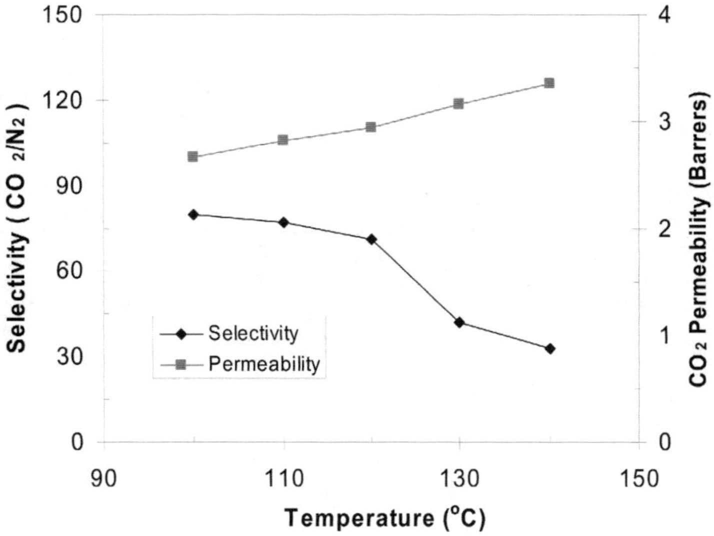

Figure 4. CO_2/N_2 selectivity vs. temperature.

The best results for CO_2/N_2 separation were obtained from 4% by weight aqueous polyethylenimine solution. The CO_2/N_2 selectivities obtained at 100, 110, 120, 130, 140, 150 and 170°C were 80, 77, 71, 42, 33, 15, and 8, respectively (Figure 4). Low selectivity was obtained at 170°C because this temperature approached the glass transition temperatures of the polyetherimide and Matrimid supports, which might lead to pore structure collapse. The permeability of CO_2 was also obtained in this temperature range. As expected, the permeability increased as the temperature increased.

3.2 Cast Polymer Blend Membranes

The cast polymer blend membranes consisted of crosslinked polyvinylalcohol with amino acid salt and polyamine as the mobile carrier and the fixed carrier, respectively [5]. For this study, glycine-lithium as the mobile carrier and polyethylenimine as the fixed carrier were examined. The active layer thickness of membranes obtained was about 30 to 40 μm. CO_2 transport was enhanced significantly due to its reactions with these mobile and fixed carriers. In the temperature range from 21 to 100°C, the permeability of CO_2 increased as temperature increased. The permeabilities obtained at 21, 50, and 80°C were 143, 464, and 1230, respectively. However, the selectivity also increased as temperature increased. The CO_2/H_2 selectivities obtained at 21, 50, and 80°C were 31, 50, and 78, respectively. The increase of both selectivity and permeability with temperature is unusual in view of typical membranes based on the solution-diffusion mechanism, for which an increase in permeability is always accompanied with a reduction in selectivity. The selectivity increase here can be explained by (1) higher CO_2 permeability due to the facilitated transport with a higher CO_2-amine reaction rate and (2) a lower H_2 permeability due to lower H_2 solubility in the membrane as temperature increases. From Figure 5, the activation energy for permeation of 7.71 kcal/mole was calculated through the Arrhenius equation (50 wt% glycine, 25 wt% polyethylenimine, 25 wt% PVA). The effects of polyamine in membranes at 21°C were also studied. As can be seen from Figure 6, when polyethylenimine concentration in the membrane increased, the selectivity increased while the permeability of CO_2 did not change significantly (50 wt% glycine-Li and polyethylenimine + PVA = 50 wt%).

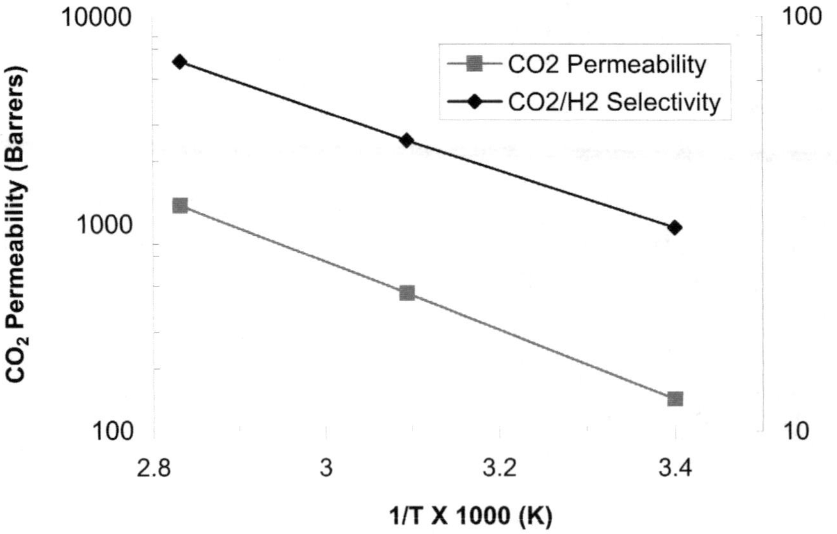

Figure 5. Effects of temperature on CO_2 permeability and CO_2/H_2 selectivity.

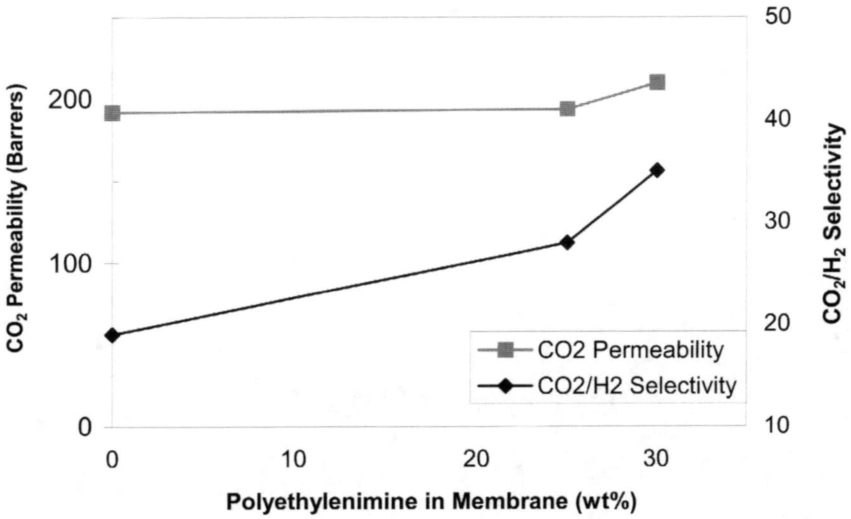

Figure 6. Effects of polyethylenimine concentration on CO_2 permeability and CO_2/H_2 selectivity at 21°C.

4. Conclusions

Two types of CO_2-selective composite membranes with facilitated transport were synthesized; one was interfacially polymerized membrane, and the other was cast polymer blend membrane. Membranes with interfacially polymerized layer of 0.5 μm were achieved through interfacial polymerization on microporous supports. The CO_2/N_2 selectivity and CO_2 permeability data were obtained for temperatures ranging from 100°C to 170°C. For the cast polymer blend membranes, CO_2 transport was enhanced significantly due to the facilitated transport. In the temperature range from 21 to 100°C, both the CO_2 permeability and CO_2/H_2 selectivity increased as temperature increased.

Acknowledgement

The authors acknowledge the U.S. Department of Energy and H2fuel, LLC for the financial support of this work.

Reference:

1. W. S. W. Ho and K. K. Sirkar, eds., Membrane Handbook, Chapman & Hall, New York (1992).
2. R. J. Petersen, "Composite reverse osmosis and nanofiltration membranes", J. Membr. Sci., 83, 81-150 (1993).
3. J. E. Cadotte, "Interfacially synthesized reverse osmosis membrane", U.S. Patent 4,277,344 (1981).
4. A. P. Rao, N. V. Desai, and R. Rangarajan, "Interfacially synthesized thin film composite RO membranes for seawater desalination", J. Membr. Sci., 124, 263-272 (1997).
5. W. S. W. Ho, "Membranes comprising aminoacid salts in polyamine polymers and blends", U.S. Patent 6,099,621 (2000).

THE USE OF ACTIVATED CARBON SUPPORTED CATALYSTS FOR SELECTIVE LOW TEMPERATURE CARBON MONOXIDE OXIDATION IN H_2-RICH GAS STREAMS

Şeyma Özkara, Ahmet Erhan Aksoylu

Department of Chemical Engineering, Boğaziçi University,

34342 Bebek, Istanbul, Turkey

Prepared for Presentation at Topical Conference: Advanced Membrane-Based Separations / 11.18.2003 / Poster Session: Advances in Membranes and Fuel Cell Technology

Abstract

Selective oxidation of CO in a H_2-rich gas stream was investigated over a series of Pt-Ce and Pt-Sn catalysts supported on activated carbon (AC). In order to study the effect of AC surface chemistry on the catalytic performance of AC supported catalysts, three types of activated carbon, namely (i) grinded and HCl washed activated carbon (AC1), (ii) air oxidized form of AC1 (AC2), and (iii) HNO_3 oxidized form of AC1 (AC3), were prepared and used as catalyst supports. The results have shown that the activities of AC supported catalysts increased with the increase in reduction temperature. Higher conversion levels were observed for 1% CeO_x-1% Pt /AC and 0.25% SnO_x – 1% Pt/AC catalysts supported on oxidized AC supports compared to those prepared on non-oxidized support.

Introduction

Fuel cells are receiving increased recognition as alternatives to existing power sources. The most promising fuel-cell technology for transport applications appears to be the Polymer Electrolyte Membrane Fuel Cell (PEMFC) fueled by hydrogen. In order to avoid storing high pressure H_2 on a vehicle, it has been proposed that hydrogen for automotive polymer electrolyte membrane fuel cells (PEMFC) can be produced in an onboard fuel processor [1].

Current technology impose that an economically feasible way of producing pure hydrogen has three catalytic processes in series:(i) *autothermal reforming* of a hydrocarbon (fuel + O_2 + H_2O $\Leftrightarrow$ CO_x + H_2) for producing hydrogen, (ii) the water gas shift reaction ($CO + H_2O \Leftrightarrow CO_2 + H_2$) maximizes H_2 concentration and lowers the CO concentration; and (iii) selective catalytic oxidation of CO in the H_2-rich reformate in order to reduce CO concentration less than 20 ppm [2].

In the present study [3], the Pt-Ce and Pt-Sn catalysts supported over oxidized (by air or by HNO_3 oxidation) and non-oxidized active carbons (AC) were tested in order to determine one or more catalysts that have sufficient low-temperature activity for the selective oxidation of carbon monoxide in hydrogen-rich streams. Three types of activated carbon, namely (i) grinded and HCl washed activated carbon (AC1), (ii) air oxidized AC1 (AC2), and (iii) HNO_3 oxidized AC1(AC3), were used as catalyst support to investigate the effect of surface chemistry of activated carbon on the catalytic performance. Several parameters including the active metal combinations, preparation methods, the physical and surface chemical properties of the catalysts and supports and optimum reaction conditions were tested and compared. The catalytic activity tests were conducted isothermally (393 K and 423 K) with a mixture of 1% CO, 1%O_2, 60% H_2 and He as balance at atmospheric pressure.

Results and Discussion

In order to develop the optimum pretreatment conditions for selective CO oxidation activities, coimpregnated and sequentially impregnated 1%CeO_x-1%Pt /AC3 samples were reduced at 623 K for 3 hours (reduction scheme R1) and at 673 K for 2 hours (reduction scheme R2) under 50 ml of H_2 flow prior to the reactions. For both preparation methods, it is observed that the activities of all samples increased with the increase in the reduction temperature applied prior to reaction. The reason why high temperature reduction procedure is beneficial in selective CO oxidation may be the increased metal-metal interaction during pretreatment of the Pt-Ce/AC catalysts. Therefore, all the catalysts were pretreated with reduction scheme R2, which was conducted at 673 K, prior to the activity tests performed for the rest of this study.

Besides, it has been proposed that the presence of Ce has several benefits such as improved noble metal dispersion, provision of lattice oxygen for reaction with adsorbed reactants and improved oxygen storage capacity [2, 4]. Between the samples prepared by coimpregnation and sequential impregnation, the catalyst prepared on HNO_3 oxidized support showed the highest activity, pointing out the beneficial effect of the acidic surface chemistry and the highest amount of surface oxygen bearing groups (Figure 1).

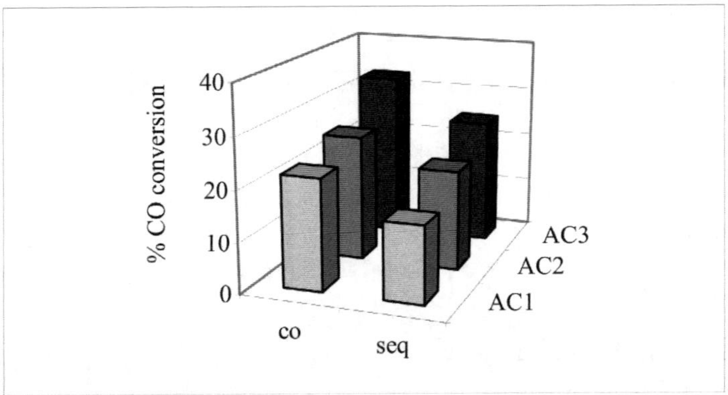

Figure 1. CO oxidation activities of coimpregnated and sequentially impregnated 1% CeO_x-1% Pt /AC samples reduced at 673 K (reduction profile R2)

The results of CO oxidation tests conducted on sequentially impregnated 0.25%SnO_x-1%Pt/AC samples reduced by reduction procedure R2 show that air oxidized support has the highest activity and selectivity level which points out to the beneficial effect of thermally stable surface oxygen bearing groups. However; the activity of AC3 (HNO_3 oxidized) supported sample is lower compared to the non-oxidized one although HNO_3 oxidized supports have the highest level of Pt dispersion and surface oxygen bearing groups, particularly carboxylic acid groups, which favors the formation of Pt_3-Sn alloys [5]. The unexpected low CO oxidation activity of 0.25%SnO_x-1%Pt/AC3 can be related to the high H_2 adsorption capacity of Pt_3-Sn alloys, whose formation is enhanced by the presence of carboxylic acid groups on HNO_3 oxidized support.

References

1. D. L. Trimm and Z. İ. Önsan, *Catalysis Reviews-Science and Engineering,* 43(2001)31.
2. I. H. Son and A. M. Lane, *Catalysis Letters,* 76(2001)151.
3. S. Özkara and A. E. Aksoylu, *Applied Catalysis A,* (2003), in press.
4. M. J. Tiernan and O. E. Finlayson, *Applied Catalysis B*, 19(1998) 23.
5. A. E. Aksoylu *et al.*, *Applied Catalysis A*, 192(2000) 29.

CATALYTIC COMBUSTION OF LIGHT HYDROCARBONS UNDER FUEL RICH CONDITIONS

Ahmet K. Avci[1], David L. Trimm[2], A. Erhan Aksoylu[1], Z. Ilsen Onsan[1]

[1] Department of Chemical Engineering, Bogazici University, 80815 Bebek, Istanbul,

Turkey.

[2] School of Chemical Engineering and Industrial Chemistry, University of New South Wales, Sydney 2052, Australia.

Prepared for Presentation at Topical Conference: Advanced Membrane-Based Separations / 11.18.2003 / Poster Session: Advances in Membranes and Fuel Cell Technology

Unpublished (some data were published)

Abstract

Oxidation of propane and n-butane was studied on Pt, Ni and Pt-Ni catalysts for fuel rich conditions. For both hydrocarbons, coke formation was not observed under the conditions applied. Light-off temperatures followed the order of Ni > Pt-Ni > Pt and of propane > n-butane.

Introduction

The integration of fuel processors, the catalytic units that can convert fossil fuels to hydrogen-rich streams, with hydrogen fuel cells is considered as a promising solution to the problems related to the availability, distribution and storage of hydrogen. Recent computer-based studies have demonstrated the possibility of efficient hydrogen production on-board the vehicles through catalytic indirect partial oxidation (total oxidation followed by steam reforming) of propane, a major constituent of LPG [1]. The model catalyst was bimetallic and included Pt (specific for exothermic total oxidation) and Ni (specific for endothermic hydrogen producing steam reforming) metals over the same support in order to drive micro level energy exchange between the reactions, leading to elevated amounts of hydrogen produced.

In this work [2], total oxidation of propane and n-butane (second major hydrocarbon in LPG) are investigated under fuel-rich conditions, i.e. at a set of fuel:oxygen ratios above the stoichiometric values, which can be found from the following reactions as 0.20 for propane and 0.15 for n-butane:

$C_3H_8 + 5O_2 = 3CO_2 + 4H_2O$
$n\text{-}C_4H_{10} + 6.5O_2 = 4CO_2 + 5H_2O$

Supported monometallic Pt/Al_2O_3 and Ni/Al_2O_3 prepared by incipient-to-wetness impregnation technique and bimetallic $Pt\text{-}Ni/Al_2O_3$ prepared by sequential impregnation technique are the catalysts of interest. The metal contents are 0.2 per cent (by weight) Pt and 15 per cent Ni. The catalytic activities are monitored through a series of reaction tests conducted in a continuous flow fixed-bed microreactor system equipped with mass flow and temperature controllers. Pretreatment of the catalysts at 500°C for 4h under 20 ml/min hydrogen flow is followed by reaction at temperatures increasing from 150°C to 385°C at a rate of 1°C/min. Inlet composition is set as $HC:O_2:N_2$ (ml/min)=1:x:139-x and total inlet flow is kept constant at 140 ml/min with x depending on the fuel:oxygen ratio. Product analysis is conducted through two gas chromatographs equipped with Porapak Q and Molecular Sieve 5A columns for hydrocarbon and fixed gas analysis, respectively. The catalyst samples are characterized by electron microscopy studies using a Philips XL30 ESEM-FEG system.

Results and Discussion

The effect of fuel:oxygen ratios on the total oxidation activities obtained over Pt, Ni and Pt-Ni catalysts are shown in Table 1. Catalytic activities are expressed in terms of light-off temperatures, which are defined as the temperatures at which 10 per cent conversion is obtained.

Table 1. Light-off temperatures of different fuel-catalyst combinations

Fuel:oxygen ratio		Light-off temperature (^oC)			
$C_3H_8:O_2$	n-$C_4H_{10}:O_2$	C_3H_8/Pt-Ni	n-C_4H_{10}/Pt-Ni	n-C_4H_{10}/Pt	n-C_4H_{10}/Ni
0.48	0.37	324	299	216	354
0.97	0.75	306	282	-	-
1.33	1.02	291	259	186	321
1.86	1.43	266	274	166	327
2.00	1.62	287	286	-	-
2.17	1.70	288	295	160	317
-	2.17	-	-	157	-

The results of n-butane oxidation indicate that the activities of the catalysts follow the order of Pt>Pt-Ni>Ni. Such an order is expected since Pt-based catalysts are known to be active in alkane oxidation [3], and poor activities of Ni are reported [4]. This trend is also confirmed by the fact that the light-off temperatures are lowered by ca. 50°C when Pt metal is added to Ni even at quantities as low as 0.2 per cent.

Although the amount of Pt metal is same in Pt/Al_2O_3 and Pt-Ni/Al_2O_3 catalysts, the former exhibits much higher activities. This difference can be explained by the degree of Pt dispersion. Electron microscopy studies showed that the particle size of Pt metal in Pt and Pt-Ni catalysts was in the range of 0.01-0.06 μm and of 0.07-0.30 μm, respectively. The studies also demonstrated big platinum clusters settled over Pt-Ni surface and the relatively small, well-dispersed Pt particles over Pt/Al_2O_3 surface. Light-off temperatures follow a unique pattern in propane and in n-butane oxidation over Pt-Ni/Al_2O_3 catalyst. As the amount of oxygen in the feed is reduced, i.e. fuel:oxygen ratio is increased, the temperature decreased down to a certain value (266°C-propane, 259°C-n-butane) and started to increase again. Formation of minimum values are due to possible synergistic interactions between Pt and Ni sites and Pt is likely to dominate in these interactions. The latter can be explained using the oxidation mechanism over Pt, which is based on the site competition between adsorbed oxygen and adsorbed hydrocarbon. It is reported that the sticking probability of hydrocarbons to Pt surface increases with the carbon number of the alkane [3]. Thus, the quantity of oxygen required to compete with n-butane is expected to be greater than that required in the case of propane. As a result, the fuel:oxygen ratio at which the light-off temperature is minimum is lower in n-butane oxidation ($[n\text{-}C_4H_{10}:O_2]_{min}=1.02$, $[C_3H_8:O_2]_{min}=1.86$).

References

1. A. K. Avci, D. L. Trimm, A. E. Aksoylu, Z. I. Onsan, *Catalysis Letters*, 88 (1-2) (2003) 17
2. A.K. Avci, Z.I. Onsan, D.L. Trimm, *Appl. Catal. A*, 216 (2001) 243.
3. G. Veser and L.D. Schmidt, *AIChE J.*, 42 (1996) 1077.
4. L. Ma, D.L. Trimm, C. Jiang, *Appl. Catal., A* 138 (1996) 275.

Removal of Cu(II) Using Recycled Membrane Phase from an Emulsion Liquid Membrane (ELM) in a Hybrid System

Richard Long, Jr. and Kyunghoan Lee
Dept ChE
New Mexico State University
Las Cruces, New Mexico 88003
rilong@nmsu.edu

Key Words: Emulsion liquid membrane, recycle, hybrid

Prepared for presentation at the 2003 Annual Meeting, San Francisco, CA, November 16-21

Abstract

Effective use of emulsion liquid membranes is a major technology for industrial applications. The most common shaft material of the agitator is iron. There are no reports of effect of the shaft for heavy metal separation with emulsion liquid membranes. But the iron shaft may affect the stability of the emulsion liquid membrane during the extraction process because the emulsion liquid membranes containing the heavy metals are observed to agglomerate along the shaft of the agitator. Emulsion droplets made with n-heptane are sensitive to the electric field but droplets made with hexane are not. Our research will focus on the effect of the electric field from the agitator shaft and design of a new system for copper extraction with recycled ELMs. The value of our research are: reducing the amount of ELMs for heavy metal separation with ELMs, application to scale-up of ELM system.

1. Introduction

The liquid membrane separation technique provides a potentially powerful technique for effecting a diverse number of separation operations. Compared to conventional polymeric membrane processes, liquid membrane processes have some advantages such as simple operation, high efficiency, extraction and stripping in one stage, larger interfacial area, higher selectivity to different phases due to their chemical nature, and scope of continuous operation. Because of these advantages, the liquid membrane technique has greater potential for recovery and removal of heavy metals and hydrocarbons from wastewater than conventional methods, which have lower separation efficiency and higher investment.

Liquid membranes have been applied as either supported (facilitated) type or emulsion type. The supported liquid membranes can be applied either as a thin sheet of supported liquid membrane or a hollow fiber supported liquid membrane. The thin sheet supported liquid membrane, which is the most simple in design, can be utilized for laboratory scale, but cannot easily be scaled up for industrial use because of capacity limitations. The equipment for emulsion liquid membrane (ELM) extractions can be batch, semi-continuous batch, or continuous. The major advantages of ELMs over the supported liquid membranes are the simple experimental system, and scale up. Many researchers have reported the application of liquid membranes for separation of many chemicals depends on the stability of liquid membranes, breaking the used ELMs(demulsification), and scale up in industrial use.

Applications of ELM technology are: 1)Extraction of metal ions: hydrometallurgical recovery of metal ions as well as removal of toxic metals from wastewater generated by various metallurgical industries, removing toxic substances from wastewater, 2)Biochemical and biomedical applications: (a) Blood oxygenation and artificial red blood cells: (b) Removal of toxins from the blood, and(c) Slow release of enzymes and drugs, 3) Separation of gas mixtures: CO_2 removal by bicarbonate ion as carrier for the preparation of enriched air, CO removal by cuprous chloride and tertamine macrocyclic ligand, 4) Liquid membranes (LM) as chemical reactors: LMs can also he used as chemical reactors for the reaction between two gases in a heterogeneous catalyst system where the reactants are encapsulated in an aqueous membrane containing catalyst. Some researchers have reported that the electric field acts only at the interface between droplets and the continuous phase rather than throughout the bulk of both phases. But the electric field can influence the emulsion droplets when heavy metals are encapsulated in the emulsion droplet. According to our experiments, approximately 70 percent of emulsion droplets in the extraction process are agglomerating along the iron shaft. That depends on the composition of the membrane phase. An emulsion droplet made with n-heptane is sensitive to the electric field but a droplet made with hexane is not. Teramoto and his coworkers [7] inserted the baffle plates to disperse the agglomerated emulsion liquid membranes but they did not report the effect of the baffle plates. There is no report on droplet breakage under the electrostatic field from the iron shaft of the agitator in metal ion separation. That may greatly affect on the stability of ELMs. Our research will focus on the effect of the electric field from the agitator shaft and design of a new system for copper extraction with recycled ELMs.

2. Review of previous work

The factors affecting the stability of emulsion droplets are divided into two groups:(1) forces such as mechanical and/or electrical force and (2) characteristics of emulsion droplets such as osmotic pressure. The parameters affecting the ELM extraction processes are voltage, frequency, temperature, degree of mixing, emulsion shape, and spacing of electrodes; density and viscosity of emulsion; interfacial tension, water drop size, size distribution, hold up, and surfactant concentration in oil phase; and electrolyte concentration in continuous phase [3].

When the osmotic pressure of the internal phase is higher than that of the external phase, the migration of water occurs from the internal phase to the external phase. This may directly cause a swelling of the vesicular structure of the internal phase, and may also contribute to various secondary phenomena such as thinning of the membrane phase, dilution of the solute in the internal phase, etc. The osmotic pressure can be neglected in all cases of when 1% w/w Tween 80 is present in the external phase. The water-permeation swelling increases with agitation time. Several mathematical descriptions of water transfer through ELM were reported. Matsumoto and Kohda [5] derived the following equation for measuring a trans-membrane volume-flux of water, which is given by

$$J_W = -L_P ART(g_2c_2 - g_1c_1) \tag{1}$$

where L_P is the hydrodynamic coefficient of the membrane, A is the area of the membrane, R the gas constant, T is temperature, g is the osmotic coefficient, c is the solute concentration, and the subscripts 1 and 2 refer to the less concentrated and the more concentrated solutions on the two sides of the membrane, respectively. The flux of water in moles per unit time is

$$\Phi_W \equiv J_W / \overline{V} \tag{2}$$

where $\overline{V}$ is the partial molar volume of water. From Equations (1) and (2)

$$\Phi_W \equiv -L_P ART(g_2c_2 - g_1c_1)/\overline{V} = -P_0 A(g_2c_2 - g_1c_1) \tag{3}$$

where P_0 is the osmotic permeability coefficient or water permeation coefficient of the membrane.

2.1 Breaking Emulsion in a Stirred Vessel

The rate of inter-phase mass transfer is affected by the physical, electrical, and chemical characteristics of the system and the mechanical features of the equipment. The former includes viscosities and densities of the phases, interfacial surface properties, diffusion coefficients, and chemical reaction coefficients. The latter includes the type and diameter of the impeller, vessel geometry, the flow rate of each phase, and the rotational speed of the impeller[1].

The flow in an agitated vessel cannot be treated analytically due to the presence of turbulent flow and the complexity of the governing equations of motion. At low vessel Reynolds number the flow is laminar. As $(N_{\mathrm{Re}})_T$ increases, turbulence spreads from the impeller to all parts of the vessel. The energy dissipation rate per unit mass throughout the vessel is not uniform. About 70% of the power was dissipated in the impeller stream and only 30% in the rest of the vessel was reported. Drew etc. [1] reported that for high tank Reynolds number, $(N_{\mathrm{Re}})_T > 10{,}000$, the average energy input rate of the impeller per unit mass of liquid, ε, is independent of the properties of the liquid and a function only of the vessel and impeller geometry. In this case

$$\bar{\varepsilon} = K' N^{*3} D_I^2 \tag{4}$$

where K' is a constant dependent only on the vessel and impeller geometries, N^* is the rotational speed of the impeller, and D_I is the impeller diameter.

2.2 Unsteady-State Mathematical Modeling

In this study, an unsteady-state mathematical model based on the advancing reaction front model, described by Matulevicius and Li [5], is proposed for facilitated carrier-meditated batch extraction of divalent copper (Cu ++). The model accounts for the internal-phase leakage effect due to membrane rupture [1]. Due to breakage of emulsion globules with time, some solute leaks, as does internal reagent from the reacted and unreacted parts of the emulsion globules, into the external phase. This model takes account the influence of leakage rate on the reaction front progress. The following assumptions were made for developing the present model.

1. The droplets inside the emulsion globules are immobile and uniformly distributed due to the considerable viscosity difference between the aqueous internal phase and the membrane phase and the presence of surfactant in the membrane phase.
2. The concentration gradient within the internal droplets can be neglected.
3. The system is monodisperse with globules of average size.
4. The main mechanism of leakage is the mechanical rupture of emulsion drops because the diffusion of internal reagent through the membrane is negligible.
5. At a given speed of agitation, the rate of change of internal-phase volume due to leakage is independent of time and proportional to the internal-phase volume.
6. Volume change of emulsion and external phase due to leakage can be neglected.

The rate of change of volume of the internal phase due to leakage is

$$\frac{dV_i}{dt} = -\psi V_i \tag{5}$$

V_i can be replaced by $\phi_t V$ because volume change of emulsion and external phase due to leakage can be neglected. ϕ_t is volume fraction of internal aqueous phase in the emulsion at any time t. Eq(5) becomes

$$\frac{d\phi_t}{dt} = -\psi\phi_t \tag{6}$$

Integrating Eq. (6), one obtains

$$\phi_t = \phi_{i0} e^{-\psi t} \tag{7}$$

The material balance of the solute ions contained in the external phase is

$$-V_e \frac{dC_e}{dt} = n \cdot 4\pi \cdot R_0^2 \cdot D_{eff} \cdot \frac{\partial C_m}{\partial r} \Big|_{r=R_0} - J_L \tag{8}$$

where J_L is volumetric leakage rate of solute from the reacted region of the globules $(R_0 \rangle r \rangle R_f)$ to the external phase.

$$J_L = \psi \cdot \phi_t \cdot n \cdot \tfrac{4}{3}\pi\left(R_0^3 - R_f^3\right) \cdot C_i \tag{9}$$

By substituting ϕ_t from Eq.(7) into Eq.(9), Eq.(10) is obtained:

$$J_L = \psi \cdot \phi_{i0} \cdot e^{-\psi t} \cdot n \cdot \tfrac{4}{3}\pi\left(R_0^3 - R_f^3\right) \cdot C_i \tag{10}$$

Simplifying the equation after substituting Eq.(10) into Eq.(8). Then Eq.(11) is obtained:

$$-\frac{dC_e}{dt} = \frac{3V_{tot}}{V_e}\left[\frac{D_{eff}}{R_0} \cdot \frac{\partial C_m}{\partial r} \Big|_{r=R_0} - \frac{\psi}{3}\phi_{i0} e^{-\psi t}\left(1 - R_f^3 \Big/ R_0^3\right) C_i\right] \tag{11}$$

The initial conditions for Eq.(11) is

$$C_0 = C_0^* \quad \text{at t} = 0 \tag{12}$$

The material balance of the solute in the reacted region of the globules $(R_0 \rangle r \rangle R_f)$ is

$$\frac{dC_m}{dt} = \frac{D_{eff}}{(1 - \phi_{i0})}\left[\frac{\partial^2 C_m}{\partial r^2} + \frac{2}{r}\frac{\partial C_m}{\partial r}\right] \tag{13}$$

The initial and boundary conditions of Eq.(13) are

$$C_m(0, r) = 0 \tag{14}$$

$$C_m(t, R_0) = C_m\big|_{r=R_0} = C_D \cdot C_e \tag{15}$$

$$C_m(t, R_f) = 0 \tag{16}$$

$$\frac{\partial C_m}{\partial r} \Big|_{r=R_f=0} = 0 \tag{17}$$

The rate of consumption of internal reagent is

$$-\frac{d}{dt}\left(\frac{4}{3}\pi R_f^3 \phi_t C_i\right) = m \cdot 4\pi \cdot R_f^2 \cdot D_{eff} \frac{\partial C_m}{\partial r} \Big|_{r=R_f} \tag{18}$$

Substituting ϕ_t from Eq.(8) into Eq.(18), then Eq.(19) is obtained:

$$-\frac{d}{dt}\left(\frac{4}{3}\pi R_f^3 \phi_{i0} e^{-\psi t} C_i\right) = m \cdot 4\pi \cdot R_f^2 \cdot D_{eff} \frac{\partial C_m}{\partial r} \Big|_{r=R_f} \tag{19}$$

or

$$-\frac{dR_f}{dt} = \frac{D_{eff}}{C_i \phi_{i0} e^{-\psi_t}} \frac{\partial C_m}{\partial r} \Big|_{r=R_f} - \frac{R_f \psi}{3} \tag{20}$$

The initial condition for this reaction is

$$R_f = R_0 \quad \text{at t} = 0 \tag{21}$$

2.3 Diffusivities in ELM

Three diffusivities are involved in transferring a solute through ELM. Lin and Long [4] reported correlation for calculating the solute diffusivity in an organic membrane phase, internal phase, and emulsion globules. First the solute diffusivity in the membrane phase is

$$D_m = 13.3 \times 10^{-8} \left(T^{1.47} \mu_m^e / V_C^{0.71} \right) \tag{22}$$

where $e = (10.2/V_C) - 0.791$, μ_m is the viscosity of the membrane phase which is obtained from experiments by viscometer, and V_C is the molar volume of the solute extractant complex at its normal boiling point. It can be evaluated from Schroeder rule. The diffusivity of solute in the internal phase is

$$D_i = [7.4 \times 10^{-8} \left(\Omega M_i^* \right)^{1/2} T] / \left(\mu_i V_A^{0.6} \right) \tag{23}$$

where Ω is the association factor of the solvent, equal to 2.6 for water, M_i^* is the average molecular weight of the internal phase, μ_i is the viscosity of the internal phase, and V_A is molar volume of solute at its normal boiling temperature.

The effective diffusivity in emulsion globules was given by Guha [2]. The correlation is following:

$$D_e = D_m \frac{(1 - \varphi_i)(1 + \varphi_i / \alpha_i)}{(1 - \varphi_i + \varphi_i / \alpha_i)} \tag{24}$$

in which D_m can be estimated from Eq.(22), α_i is partition coefficient of solute between the membrane and internal phase at equilibrium, and

$$\varphi_i = \frac{V_{io}}{V_{io} + V_{mo}} \tag{25}$$

2.4 Extraction Process

The following stoichiometric expression is assumed to describe the overall extraction reaction which occurs on either side of the membrane/solution interface:

$$Cu^{2+} + 2\overline{HR} \longleftrightarrow \overline{CuR_2} + 2H^+ \tag{26}$$

Stripping Process

$$\overline{CuR_2} + 2H^+ \longleftrightarrow 2\overline{HR} + Cu^{2+} \tag{27}$$

3. Experiment

Our experimental plan will be divided into three parts: (1)a basic screening experiment, (2) applying a hybrid system, and (3) scale-up. In part (1), the optimal conditions, which are amounts and types of membrane phase, surfactant, and carrier will be investigated. The most common shaft material for the agitator is iron. There are no reports of the effect of the shaft on heavy metal separation with emulsion liquid membranes. The stability of emulsion liquid membranes during the extraction processes will be investigated because the emulsion liquid membranes containing the heavy metals are agglomerating along the iron shaft of the agitator shown in Fig. 1.

4. Conclusion

The effect on the stability of ELMs from the iron shaft electric field must be considered as a parameter for separating the heavy metals with emulsion liquid membranes. This agglomeration will reduce the removal efficiency of liquid membranes, also it may increases the breakage of the emulsion liquid membranes. The advantages of our research are: reducing the amounts of ELMs for heavy metal separation with ELMs, applying for scale-up of ELM system.

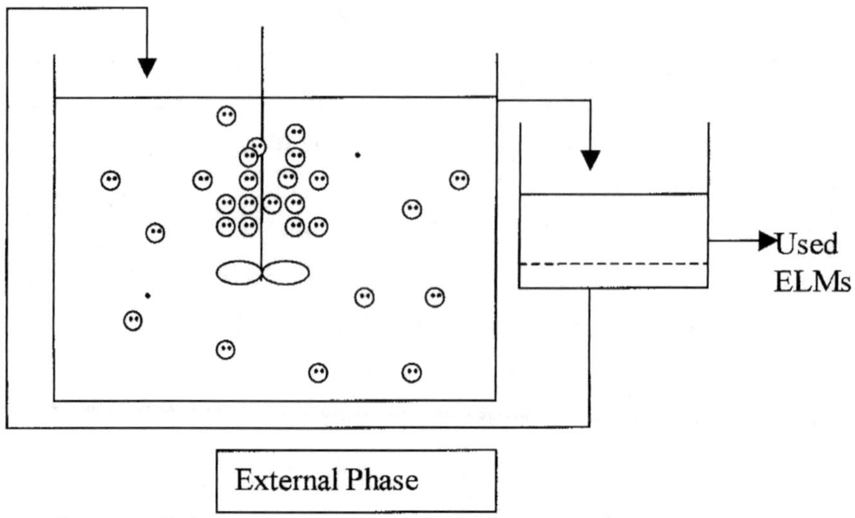

Figure 1. Schematic apparatus for this research.

References

1. Drew, T.B., G.R. Cokelet, J.W. Hoopers, and T. Vermeulen, Ed. 1981. Advances in chemical engineering. Academic press, New York.
2. Guha, A.K., C.H. Yun, R. Basu, and K.K. Sirkar. 1994. "Heavy metal removal and recovery by contained liquid membrane permeator," AIChE Journal, 40(7):1223-1237.
3. Kim, B., J. H. Moon, T. Sung, S. Yang, and J. Kim. 2002. "Demulsification of water-in-crude oil emulsions by a continuous electrostatic dehydrator," Separation science and technology, 37(6):1307-1320.
4. Lin, C. C. and R. L. Long. 1997. "Removal of nitric acid by emulsion liquid membrane:experimental results and model prediction," Journal of Membrane Science, 134:33-45.
5. Matsumoto, S. and M. Kohda. 1980. "The viscosity of W/O/W emulsions :an attempt to estimate the water permeation coefficient of the oil layer from the viscosity changes in diluted systems on aging under osmotic pressure gradients," Journal of Colloid and Interface Science, 73(1):13-20.
6. Matulevicius, E. S. and N. N. Li. 1975. "Facilitated transport through liquid membranes," Separation science and technology, 4(1):73-96.
7. Teramoto, M., T. Sakai, K. Yanagawa, and Y. Miyake. 1983. "Modeling of the permeation of copper through liquid surfactant membranes by continuous operations," Separation science and technology, 18(11):985-997.

Composite Pd and Pd/alloy Porous Stainless Steel Membranes for Fuel Cell Fuel Processor Applications

Yi Hua Ma, Ivan P. Mardilovich and Erik E. Engwall

Center for Inorganic Membrane Studies
Department of Chemical Engineering
Worcester Polytechnic Institute
Worcester, Massachusetts 01609

Prepared for Presentation at the 2003 AIChE Annual Meeting, Wednesday, November 19, 2003, Fuel Cell Membranes Session

Unpublished

Abstract

The separation of hydrogen from an integrated catalytic membrane reactor in a fuel cell fuel processor requires the utilization of gas separation processes that will provide high fluxes, high selectivity of separation and ability to operate at elevated temperatures. Inorganic membranes offer the required characteristics. Dense composite Pd and Pd/alloy membranes supported on a porous substrate, in particular, porous stainless steel (PSS), are especially suited for these applications because of their good high temperature thermal and mechanical stability, high separation factors and thin Pd layers for better permeation fluxes and lower membrane cost. There are several key issues that need to be addressed for the Pd composite membrane to be viable for industrial applications. They include intermetallic diffusion between elements in the substrate and the palladium layer, the difference in coefficients of thermal expansion and the pore size and pore size distribution of the substrate. In addition, the effects of gases, such as CO, CO_2 and possibly some sulfur containing compounds in the permeate stream, on the membrane stability also need to be considered. The unique method of in-situ controlled oxidation to produce an intermetallic diffusion barrier layer will be described in the presentation. Furthermore, the effects of CO and CO_2 on the stability of the membrane will be examined.

1. Introduction

The development of the Fuel Cell Electric Vehicle (FCEV) for low-emissions to comply with environmental regulations can provide a radical solution to electric powered vehicles. The FCEV carries fuel cells that utilize hydrogen as the fuel. Despite of the well-known fact that hydrogen is needed as the fuel to operate PEM fuels, the research and development of economical and efficient ways of generating hydrogen have only recently gained some momentum. For the foreseeable short term of hydrogen production, steam reforming of hydrocarbons including natural gas remains as one the most developed technologies for such an application. However, to obtain high methane conversion and high purity of hydrogen, high temperature operation, shift reactors and separators are required. A hydrogen membrane reformer is an attractive approach in that it combines hydrogen generation and separation in one single step. In such a unit, not only the conversion is increased but also hydrogen is separated from other reaction products, thereby, eliminating the requirement of additional shift reactors and hydrogen separator. Dense composite palladium and palladium/alloy porous stainless steel membranes are especially suited for this application due to their good thermal and mechanical stabillity, ease of module fabrication and process integration and high hydrogen flux. This presentation will discuss the synthesis of Pd and Pd/alloy membranes and examples will be given to demonstrate the advantage of a membrane reactor.

2. Composite Palladium and Palladium/alloy Porous Stainless Steel Membrane Synthesis

Currently, only foils, tubes or other bulk samples of palladium and palladium/alloy are commercially available. The industrial use of self supporting Pd membranes has been limited for economic reasons. Over the more recent past, research efforts have been focused on economically viable supported composite Pd membranes, which use a thinner dense Pd layer applied to a porous or otherwise permeable structural support. The major advantages of composite membranes are increased hydrogen flux and reduced consumption of noble metals. A number of porous substrates can be used as the support and various methods are available for depositing metals on the substrate. Our choices for the support and method of deposition are porous stainless steel (PSS) and electroless plating, respectively (Mardilovich et al., 1998; Ma et al., 2001). In the case of palladium alloy membranes, the challenges of applying a thin dense layer to a support are augmented by the difficulties of forming this layer from two or more metals and obtaining a uniform alloy throughout the layer.

The composite membrane synthesis mainly consists of pre-treatment of the porous stainless steel (PSS) support, surface activation and plating. Pre-treatment of the PSS support consists of: (1) Complete removal of foreign contaminants (grease, oil, dirt, corrosion products), followed by controlled oxidation at high temperatures ($\geq 400°C$) (Ma et al., 2001), (2) surface activation consisting of seeding the support surface with palladium nuclei, which during the electroless plating initiate an autocatalytic process of the reduction of a metastable Pd salt complex on the target surface; (3) Electroless plating with a plating solution typically consisting of $(Pd(NH_3)_4)Cl_2\ H_2O$, Na_2 EDTA, NH_4 OH

(28%), and H_2NNH_2 (1M) at pH around 10.4 and temperature of 60°C. The thickness of the Pd layer can be determined gravimetrically from the plated amount and the density of Pd (12.0 g/cm^3). In some cases the thickness can be estimated directly from the SEM picture. For the formation of Pd/alloy films, we apply alternative coating between Pd and the alloying element followed by high temperature annealing for alloy formation (Generally known as coating and diffusion treatment). Detailed description can be found in (Mardilovich et al., 1998; Ma et al., 2001; Ma et al., 2003a,b).

3. Hydrogen Permeation and Membrane Reactor

The permeation of hydrogen through Pd and Pd/alloy is by the solution and diffusion mechanism and is, in most cases, controlled by the diffusion of dissociated hydrogen molecules through the Pd bulk. The hydrogen flux is proportional to the difference of the square root of the pressures, known as the Sievert's law. Figures 1 and 2 show the flux as a function of the difference of the square root of pressure and the inverse of the membrane thickness, respectively. However, deviations from the Sievert's law existed in the literature due, possibly, to, among others, mass transfer resistances, surface reaction, defects in the membrane.

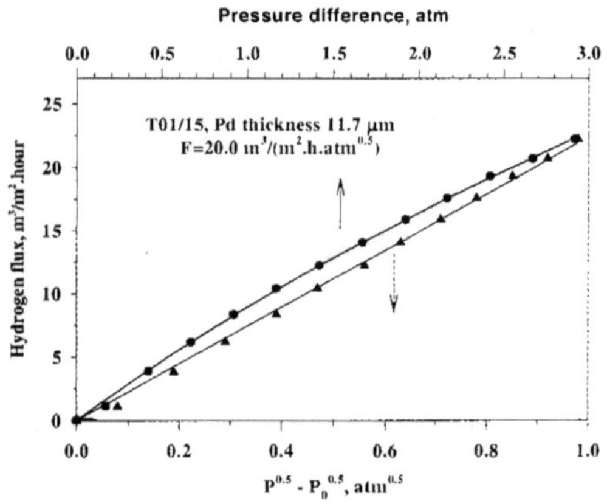

Figure 1. The hydrogen Flux as a function of the difference of the square root of the pressure (Sievert's law) at 350°C)

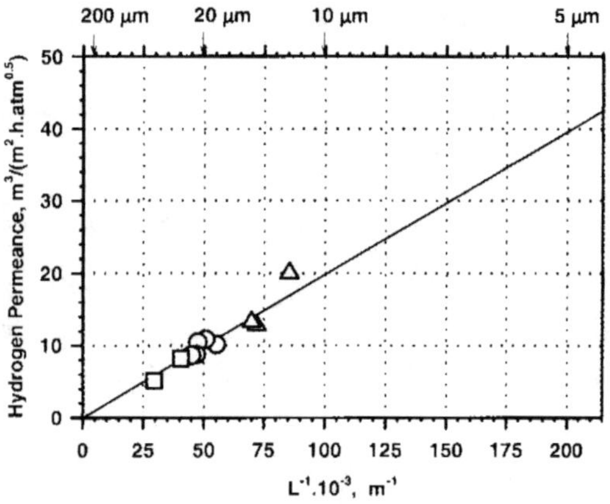

Figure 2. Hydrogen permeance versus reciprocal palladium thickness for a pressure difference of 1 atm and a temperature of 350 °C. The Pd layer thickness is indicated along the top. The grade of the supports were: **squares**-grade 0.5 μm; **circles**-grade 0.2 μm; **triangles**-grade 0.1μm.

The use of a membrane reactor for steam reforming was demonstrated by Lin and Rei (2000). They use one of our Pd membranes to carry out steam reforming of methanol for hydrogen production for direct fuel cell applications. The membrane reactor was packed with a copper-based catalyst for steam reforming while the gases exhausted from the membrane reactor were fed to a bed packed with Pd/Al_2O_3 oxidation catalyst. Since the membrane reactor and the oxidation bed were in a concentric arrangement, the required heat for steam reforming was directly supplied by the combustion of the gas mixture rejected from the membrane reactor. Their data showed that for hydrogen flux in the range of 1.36 – 2.54 m^3/m^2-h, the hydrogen recovery was in the range of 85 – 66%, respectively, with the exhaust from the system containing only water and CO_2.

4. Concluding Remark

Dense composite palladium, especially palladium/alloy, membranes will play an important and essential role in the production of pure hydrogen due to the expected increase in hydrogen demands in the future. The use of porous metal as substrates will facilitate the membrane module fabrication and process integration. The integration of Pd membrane reactor into a PEM fuel cell system will reduce the number of unit operations and allow hydrogen to be directly fed into the fuel cell for generating electricity. Of course, many technical issues, such as producing thin membranes with high separation factors and long term thermal and mechanical stability still need to be resolved. The use of palladium alloys to increase the hydrogen permeability and at the same time, to enhance the membrane's thermal stability should improve the process viability.

5. References

1. Mardilovich, P. P., Y. She and Y. H. Ma, "Defect- Free Palladium Membranes on Porous Stainless-Steel Support", *AIChE J*, **44**, 310(1998)
2. Ma, Y. H., P. P. Mardilovich and Y. She, "Hydrogen Gas-Extraction Module and Method of Fabrication", *US Patent 6,152,987*, 2000, Also, *European Patent No. 1 042 049*, 2003
3. Ma, Yi Hua, Mardilovich, Ivan P., Erik Engwall, "Thin Composite Palladium and Palladium/Alloy Membranes for Hydrogen Separation", *Annuals of the New York Academy of Sciences*, Vol. **984**, Advanced Membrane Technology, Eds., Norman N. Li, Enrico Drioli, W.S. Winston Ho and Glenn G. Lipscomb, 346(2003a)
4. Ma, Y. H., B. Ceylan Akis, M. Engin Ayturk, Federico Guazzone, Erik E. Engwall and Ivan P. Mardilovich, "Characterization of Intermetallic Diffusion Barrier and Alloy Formation for Pd/Cu and Pd/Ag Porous Stainless Steel Composite Membranes", *I&EC Research*, Submitted for publication(2003b)
5. Lin, Y. M. and M. H. Rei, "Effect of incipient removal of hydrogen through palladium membrane on the conversion of methane steam reforming: Experimental and Modeling", *Int. J. Hydrogen Energy*, **25**, 211(2000)

THIN SUPPORTED INORGANIC MEMBRANES FOR H_2 AND O_2 SEPARATION

Greg T. Quickel, Matthew L. Mottern, Di Yu, and Henk Verweij

Department of Materials Science & Engineering, Ohio State University, 2041 College Road, Columbus OH 43210-1178, USA

Internet: www.mse.eng.ohio-state.edu/fac_staff/faculty/verweij/

Prepared for presentation at AIChE 2003 Annual Meeting/November 19, 2003/Fuel Cell Membranes

August/2003

Unpublished

Abstract

Thin supported inorganic membranes show potential for industrial separation of H_2 from gas mixtures containing CO, CO_2, and H_2O *and* for separation of O_2 from air. Both applications are of interest to develop viable H_2 fuel processing technology for fuel cells. While several interesting membrane concepts for high-selectivity H_2 and O_2 separation are now well established, much work still needs to be done to improve the membrane's structural design, thermochemical stability, mechanical properties and, in particular for H_2, flux and selectivity. This paper will focus on state of the art and perspectives for thin supported amorphous silica (H_2) and perovskite (O_2) membranes.

Introduction

Permselective membranes are able to separate gases with energy requirements that approach the thermodynamic equilibrium limit. Two major separations, of interest for fuel cell technology are purification or separation from reaction mixtures of H_2 fuel, *and* separation of O_2 from air. Pure O_2 is used as a reactant in coal gasification or partial oxidation of hydrocarbons. The CO in partial oxidation product mixtures may react (shift) with H_2O to produce more H_2 and CO_2. After separation from CO_2 and H_2O, the H_2 may require final purification to bring unwanted CO at <10 ppm. H_2 separation can be done in a stand-alone membrane purification unit *or* in a membrane reactor. In the latter case H_2 is directly separated from mixtures with thermodynamically limited conversion to increase space yield and efficiency. Likewise O_2 separation can be done in a "passive" membrane unit or in a high temperature partial oxidation membrane reactor in which O_2 is introduced from pressurized air via an O_2-permselective membrane reactor wall. H_2-selective membranes are generally made as thin films on porous supports to provide sufficient strength. Depending on the detailed application, O_2-selective membranes can be applied *either* as stand-alone disks and tubes *or* as supported thin films. The rationale behind the combination of partial oxidation and water gas shift reaction is that heat produced in the exothermic first step can be used directly in the endothermic second step. As a consequence, application of membranes as indicated, is most beneficial if they can operate for extended periods at elevated temperatures.

Ceramic membranes offer perspectives of high flux-high selectivity operation at high temperatures in harsh environments for extended periods. The development of inorganic ceramic membranes began nearly 50 years ago for the separation of $^{235}UF_6$ from $^{238}UF_6$ with supported γ-alumina membranes. Attempts to broaden their application area in normal gas separation initially stalled because of the minimal Knudsen selectivity. This situation changed with the advent of dense membrane materials such as Pd, $La_{1-x}Sr_xCoO_{3-\delta}$ and YSZ-noble metal composites and micro-porous membranes around 1990. The ideal selectivity of dense membranes rests upon transport of atomic or ionic (H and O) species on interstitial and vacant sites, possibly accompanied by charge compensating electron transport. Micro-porous membranes consist of zeolites, carbon or amorphous silica with connected pores of sub 1 nm dimension acting as an adsorption site for small molecules. The size and adsorption characteristics of the micro-pores lead to selectivities many orders of magnitude more that Knudsen combined with reasonably high fluxes.

Amorphous silica membranes with ~0.5 nm pore Ø and ~50 nm thickness have been prepared by dip coating of supported γ-alumina membranes in a silica sol, followed by drying and calcining at 400-800°C [1,2]. Supported γ-alumina membranes with 3 nm pore Ø and ~1 μm thickness were prepared by dip-coating porous α-Al_2O_3 supports in a Boehmite sol, followed by drying and calcining at 600°C. 2 mm thick α-Al_2O_3 supports were made by colloidal filtration of well-dispersed 0.4 μm α-Al_2O_3 particles, followed by sintering of the filter compact [3]. Dense crystalline 7.5-20 μm thick perovskite $La_{0.5}Sr_{0.5}CoO_{3-\delta}$ membranes were made by Pulsed Laser Deposition (PLD) on top of macroporous supports of the same composition [4]. $La_{0.5}Sr_{0.5}CoO_{3-\delta}$ supports with 400 nm pore Ø, 2 mm thickness and sufficient surface quality were made by colloidal filtration of commercially available powder with an average particle size of 1 μm, followed by sintering at 1020°C, grinding and mechano-chemical polishing. In both cases the membrane side of the support must be smooth with a homogeneous porous microstructure such that it can be used to apply very thin membrane films without formation of pinholes due to surface defects ("sinkholes").

Process improvements in supported amorphous silica fabrication, especially the use of more homogenous supports and γ-alumina layers and clean-room techniques, occasionally resulted in membranes without detectable mesoscopic defects, and hence significantly improved transport properties. For these membranes the permselectivity, $\alpha^{s}_{l_1/l_2}$, provides a good indication of the separation performance for gases l_1 and l_2. $\alpha^{s}_{l_1/l_2}$ is the ratio of single gas permeances, f_{l_1} and f_{l_2}. f_l is defined as the ratio of flux, j_l, and pressure difference, Δp_l. Supported membranes calcined at 400°C and 30 nm thickness, showed $f_{H_2} = 2\times10^{-6}$ mole/(m^{2}sPa) and $\alpha^{s}_{H_2/CH_4} > 500$ at 200°C [5]. Molecules larger than CH_4 were completely blocked. Silica membranes calcined at 600°C showed no detectable CH_4 flux, with $f_{H_2} = 5\times10^{-7}$ mole/(m^{2}sPa) at 200°C. Silica membranes have $10 < \alpha^{s}_{H_2/CO_2} < 20$ and much higher values for $10 < \alpha^{s}_{H_2/CO} < 20$. They show a slight increase of f_{H_2} and a slight decrease of f_{CO_2} with increasing temperature so that H_2/CO_2 is expected to improve at high temperatures. Unfortunately little systematic data are available for temperatures >300°C.

Pinhole-free dense perovskite membranes show $\alpha^{s}_{O_2/l} = \infty$ in which l is any gas, other than O_2. These membranes have compositions like $La_xSr_{1-x}(Co_y,Fe_{1-y})O_{3-\delta}$. They are often studied as stand-alone disks or tubes. Contrary to what seems to be the case for microporous membranes, dense perovskite membranes exhibit a considerable surface transfer resistance for O_2 permeation, related to the kinetics of the $O_2 + 4e^- \rightleftarrows 2O^{2-}$ reaction in both directions. This has the consequence that these membranes have a critical thickness, Z^c, for which surface transfer resistance becomes equally important as bulk resistance. This has the consequence that not much can be gained by making the membranes much thinner than Z^c. For perovskite membranes, Z^c, strongly depends on temperature with typical values ranging from 2 mm at 300°C to 200 μm at 800°C [6]. It should be possible to improve surface transfer kinetics by dispersing thermochemically stable catalyst nano-particles on the surface but little progress has been shown for this approach. In [4] the dependence of j_{O_2} on thickness, Z, at 900°C is reported for $La_{0.5}Sr_{0.5}CoO_{3-\delta}$ with $p^{f}_{O_2} = 20$ kPa at the feed side and $p^{p}_{O_2} = 2$ kPa at the permeate side. It was found that j_{O_2} increases linearly with $1/Z$ from 10^{-2} mole/m^{2}s at Z = 2 mm to 0.2 mole/m^{2}s at Z = 0.1 mm where it leveled off. Diffusion-controlled $La_{0.5}Sr_{0.5}CoO_{3-\delta}$ permeation shows a strong Arrhenius-type of thermal activation, in this case with an effective activation energy of 122 kJ. The surface transfer resistance show a similar temperature dependence, be it with a lower activation energy such that at a given thickness, transport control may change with decreasing temperature from diffusion-controlled to surface-controlled.

Membrane durability tests at operating conditions that are relevant for H_2 fuel processing trains are scarce. Tests with bulk micro-porous silica [7] have shown that this material is likely to be stable at most of the conditions. In [8] it is indicated that hydrothermal stability of the amorphous silica might be improved further by making it hydrophobic by replacing terminal OH groups in the structure by CH_3 groups. Membrane tests up to now have been plagued by instability of the γ-alumina intermediate layer. In [9] it is shown that 200 nm Ø holes form in the γ-alumina after 23 hrs exposure at 475°C and $p_{H_2O} = 40$ kPa . These holes correspond with the macropores in the ceramic support directly underneath the γ-alumina layer. This problem could be solved by using supports, made by colloidal filtration and 120 nm Ø pores. However, with those supports, a more stringent, 100 hrs treatment at methane steam reforming conditions (SASRA: 600°C, in a 3:1 H_2O:CH_4 mixture at 2.5 MPa total pressure) led to significant blistering and delamination of the membranes [3]. Both phenomena were related to the fact that γ-alumina showed residual sinter activity during the aging tests. This sinter-activity could be suppressed by an addition of 6 mol-% La_2O_3. Blistering and delamination could be suppressed further by improving the adherence of the γ-alumina membrane to the α-alumina support by application of interlayer phosphate bonding. The combination of measures led to supported γ-alumina membranes that could be operated stable for 100 hrs at SASRA conditions [3]. Studies of the thermochemical stability of silica membranes, supported by such structures are underway.

Lifetime tests of unsupported perovskite membranes have focused on chemical reduction at the low p_{O_2} side and simple mechanical properties. The first can be addressed by replacing Co by Fe at the expense of j_{O_2} while the second is related to flaws in state-of-the-art ceramic powders and forming methods. Supported perovskite membranes react vigorously with normal α-Al_2O_3 supports and hence require a porous support structure with a chemical composition that is largely or completely the same as that of the membrane. This, in turn, has the consequence that support homogeneity is not as good as what can be obtained with α-Al_2O_3. The state of the art in α-Al_2O_3 support formation is much more advanced that that of perovskite powders, with their wide range of compositions adding to the problem. The scarce results reported [4] for thin supported perovskite membranes are plagued by pinhole formation due to macro-defects in the underlying supports. It was also found that at a thickness of 5 μm the membrane structure became instable due to surface tension effects and disintegrated spontaneously into the porous support structure.

For both supported silica and perovskite membranes the influence of support resistance becomes noticeable at the highest fluxes realized. This is related to the fact that the supports have to be made sufficiently fine porous to act as a bridge for the membrane layer and to avoid part of the stability problems as mentioned. While many interesting membrane concepts are explored, progress and applications are more and more hampered by reproducibility and life-time problems as mentioned. These problems are not so much caused by intrinsic membrane problems but rather by imperfections in the integrated supported membrane structures:

- Too dense and fine-porous structures lead to significant support resistance. This can be addressed by application of layer-wise or gradually coarsening structures.
- The occurrence of low-density agglomerates and other flaws in particles (powders) used for support and membrane preparation lead to "sinkholes" in the support and a membrane thickness larger than actually necessary to cover up the agglomerates. This can be addressed by clean room processing and classification and microsieving of the particles.
- Coarse irregular support surfaces lead to the need of relatively thick membrane to create a contiguous membrane layer. The support surface can be improved by machining and polishing or by using a forming method that leads to a smooth surface. Machining and polishing may lead to the creation of trapped debris and surface damage that lead to destructive rearrangement during life-time. Forming methods that lead to smooth surfaces are limited to colloidal deposition methods for which viable manufacturing is yet to be developed.

- The occurrence of abrupt interfaces between support and membrane layers will always carry the risk of delamination during life time, simply because of property differences and fluctuations in the ambient. This may be addressed eventually with graded structures that show a gradual transition in composition and microstructure in the direction, perpendicular to the membrane.
- Too much reactivity between subsequent layers may lead to disintegration while complete lack of reactivity may lead to weak interfacial bonding as shown for the γ-alumina/α-Al_2O_3 case.

These issues are nothing special in high temperature materials science but will require serious attention before any stable and operational membrane concepts can be introduced to the market. This made us decide to focus the membrane program at OSU at an integrated approach to high-definition particulate processing of complete supported membrane structures. The design of these structures will eventually be based on improved insights of overall transport, mechanical and thermochemical stability for a specific application environment. The next paragraph summarizes recent results with the synthesis supported γ-alumina membranes that are used for the preparation of supported silica membranes.

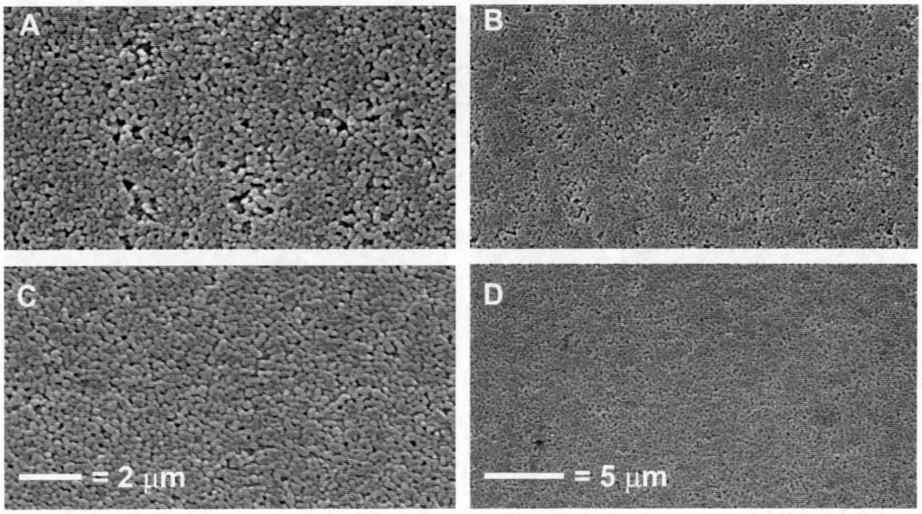

Figure 1: SEM of AKP30 support, using stabilization by APMA at 167 kg/m^2 (A and B) and HNO_3 at p_H=2 (C and D). The supports prepared with APMA show clear defects that will become "sinkholes" for the membrane to be supported.

Current membrane research at the Ohio State University

The colloidal filtration technique is chosen as the primary method for membrane substrate preparation, because this technique enables us to produce a homogenous porous ceramic green body with an optically smooth surface without machining and polishing. To demonstrate the level of control and definition, the procedure is summarized below:

- Two different α-alumina powders are investigated for substrates of different pore sizes: AKP-15 and AKP-30 with an average particle size of 400 and 800 nm, respectively. Both are high purity (99.9% α-Al_2O_3, Sumitomo Chemical Company) and have a narrow particle size distribution.
- Aqueous solutions of nitric acid and Darvan C (APMA) are used to prepare stable particle dispersions of the AKP powders. The nitric acid solution was found to be the most suitable dispersion medium for the AKP-30 substrate preparation. Typical support surface homogeneities obtained with difference dispersion media are shown in figure 1.
- Focused ultrasonic treatment produced by a Branson Model 102C ultrasonic probe is used for mechanical dispersion. A special designed dispersion beaker with cooling jacket is utilized to address colloidal instability due to heat dissipation.
- The dispersion agent is drained through a filtration manifold and the particles are blocked by an 800 nm pore Ø cellulose nitric membrane to form the green support body on top of it. This is done at class 100 clean room to prevent airborne particle surface contamination. Mechanical

vibrations, leading to standing wave patterns on the substrate surface must be avoided. Substrates are allowed to dry in the filtration manifold overnight, prior to release.

- Slight sintering at 850ºC for 10 hours is applied to achieve a maximum porous structure and sufficient substrate strength.

Figure 2. ~100 nm thick TEM FIB cross-section of macroporous α-Al_2O_3 support with two 1.5 μm thick γ-alumina layers, the 2nd is only partly shown.

A first result for thick 2×1.5 μm thick γ-alumina membranes, formed by colloidal filtration of 10 nm Boehmite particle precursor sols, followed by calcination at 600ºC are shown in figure 2. The γ-alumina layers show 100-300 nm low density areas (ellipse) due to agglomerates in the precursor γ-alumina. The dense packed support structure may cause transport constriction effects at the interface. Thinner layers may be realized by dilution of the sol and after removal of the (100-300 nm) agglomerates. Centrifugation of the precursor sol has shown the average particle sizes to decrease from 33 to 21 nm, as measured by dynamic laser scattering. A filtration sequence through 0.8, 0.45, and 0.22 μm cellulose nitric membranes has resulted in a visibly clearer sol. Dilution of that sol by 10 and 20 times has resulted in an gel layer before drying with an optical interference color that appears perfectly homogeneous. Structural studies and studies of deposition and membrane properties of amorphous silica layers, on such structures are underway.

References

1. R.J.R. Uhlhorn, M.H.B.J. Huis in 't Veld, K. Keizer and A.J. Burggraaf, "High permselectivities of microporous silica-modified γ-alumina membranes," *J. Mater. Sci. Lett.*, **8** [7] 1135-38 (1989).
2. R.M. de Vos and H. Verweij, "High-Selectivity, High-Flux Silica Membranes for Gas Separation," *Science*, **279** 1710-11 (1998).
3. A. Nijmeijer, H. Kruidhof, R. Bredesen and H. Verweij, "Preparation and Properies of Hydrothermally stable γ-alumina membranes," *J. Am. Ceram. Soc.*, **84** [1] 136-40 (2001).
4. L.M. van der Haar, Mixed-conducting perovskite membranes for oxygen separation — Towards the development of a supported thin film membrane, Thesis, University of Twente, Enschede, The Netherlands 2001*.
5. R.M. de Vos and H. Verweij, "Improved performance of silica membranes for gas separation," *J. Membrane Sci.*, **143** [1] 37-51 (1998).
6. M.W. den Otter, A study of oxygen transport in mixed conducting oxides using isotopic exchange and conductivity relaxation, Thesis, University of Twente, Enschede, The Netherlands 2000.
7. A. Nijmeijer, Hydrogen-selective silica Membranes for use in membrane steam reforming, Thesis, University of Twente, Enschede, The Netherlands 1999.
8. R.M. de Vos, W.F. Maier, H. Verweij, "Hydrophobic silica membranes for gas separation," *J. Membrane Sci.*, **158** [] 277-88 (1999).
9. R.M. de Vos, High selectivity, high flux silica membranes for gas separation — synthesis, transport and stability, Thesis, University of Twente, Enschede, The Netherlands 1999.

* Most of the University of Twente theses are currently available as pdf downloads on http://www.ub.utwente.nl/webdoc/docs/ct.shtml.

Modified Polybenzimadazole Type Membranes for PEM Fuel Cells

I.Cabasso , F. Johnson and Y. Yuan
The Michael Szwarc Polymer Research Institute
State University of New York – esf
Syracuse, New York 13210
e-mail: icabasso@syr.edu

INTRODUCTION

In polymer electrolyte membrane fuel cells (PEMFC), the polymer membrane, mostly ionomer, has a significant role. The membrane must have a number of characteristics, including high proton conductivity, very low permeability to gas, adequate hydrophilicity, and high durability in the harsh fuel cell environment. The latter calls for a polymer that can resist oxidation and reduction at high temperature in the presence of H_2 and O_2. Most of the polymer membrane electrolytes that have been used in PEM fuel cells are based on the sulphonic acid moiety, these include perfluorniated ionomers (such as Nafion®) and several hydrocarbon membranes, such as block copolymers of sulfonated polystyrene[1,2] and blend membranes of sulfonated poly(phenylene oxide) with poly(vinylidene fluoride)[3,4]. Since performance at high temperature is increasingly being required in many membrane applications, alternatives to the sulfonic acid group have been sought. While phosphoric acid fuel cells (PAFC) have been shown to endure high temperature application, grafting phosphoric acid groups on a polymer chain though the hydrolytic sensitive C-O-P bond is not feasible. Therefore, phosphonic acid polymer membranes, that are linked though the C-P- stable bond, have drawn attention[5,6].

A preliminary report form our laboratory[7,8] had indicated that partly hydrolyzed poly(phenylene oxide phosphonate ester) membrane, having phosphonic acid residues pendant to the polymer chain, can perform as a membrane electrolyte in PEM fuel cells and was tested at relatively low temperature and pressure.

The new type of proton exchange membranes-modified polybenzimadazole (PBI) have been developed for the PEM fuel cell in our laboratory[9]. PBIs are a class of high strength,

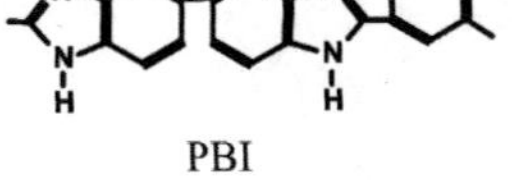

PBI

Poly-2,2'-(-m-phenylene)-5,5'-bibenzimidazole

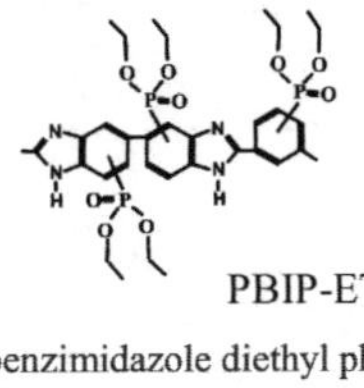

PBIP-ET

Polybenzimidazole diethyl phosphonate

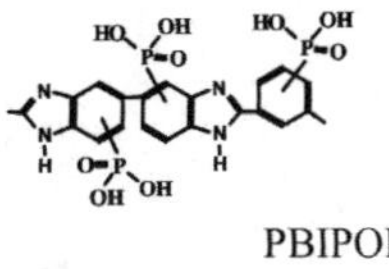

PBIPOH

Polybenzimidazole phosphonic acid

Scheme I

thermally stable polymers. Modified PBI type polymers have been functionalized with both diethyl phosphonyl, and phosphonic acid groups. Diethyl phosphonylation yield a alcohol soluble high temperature polymer; these PBIPs have been further hydrolyzed to yield phosphonic acid derivatives (PBIPOH), (see the scheme I). This PBIPOH membrane can be possesses high proton conductivity and good performance in the fuel cell testing at relative high temperature.

EXPERIMENTAL

Conductivity Conductivity of the membrane was measured by AC impedance analyzer. An EG&G 273A potentiostart with an oscillating signal generator and two phase analyzer was used to monitor real and imaginary impedance by controlling the AC current (1Hz to 100 kHz). The measurements were conducted at different temperatures with fully hydrated membranes.

Water Uptake The water uptake of protonated membrane was characterized by suspending the membrane in a control humidity chamber at constant temperature (22.0°C), or at 100% humidity under different temperature. Temperature and relative humidity were monitored within 0.1°C and 0.1%, respectively, using a Vaisala HMI/RH% probe. The membrane was allowed to equilibrate at each condition for at least seven days. The ionic charge density was obtained by a revere titration of protonated membrane with 0.1 N NaOH.

Fuel cell Testing The electrode were constructed in this laboratory[9]. Pt loading is ~0.2 mg/cm^2. The PBIPOH membranes were sandwiched between two electrodes and installed in the MEA without application of pressing at room temperature. Active MEA area is 5 cm^2. Fuel cell testing was conducted in a fuel cell test station (Fuel Cell Technologies Inc., Los Alamos) that provides humidification, temperature and pressure control feed gases (H_2/O_2), as well as to the MEA.

RESULTS AND DISCUSSION

Molecular weight, degree of substitution (DS) phosphonic acid groups and charge density (ICD) of PBIPs were determined by ^{1}H, ^{13}C-DEPT NMR, light scattering and titration. The results are show in table 1. a series of PBIP membranes with degree of substitution in the range of 0.3 up to 3.0 (per unit), and molecular weight, Mw, ~80,000 up to 140,000 have been produced. The charge density (ICD) of these membrane increase from 2.0 meq/g to 10.0 meq/g with DS increase from 0.3 to 3.0.

Table 1. Molecular weight (M_w) and Degree of Polymerization (DP) and charge density (ICD) of PBIs

Polymer	DS	M_w	dn/dc	DP	ICD (meqH$^+$/g)
PBI	-	70,800	0.114	230	-
PBIP	0.3	79,900	0.106	223	2.0
PBIP	3.0	142,000	0.132	228	9.4

Water uptake was characterized for PBIPOH (DS=3.0) membrane at various water activity and temperature. It can be seen (Figure 1) that the water uptake decreases with temperature increase and water activity decreases.

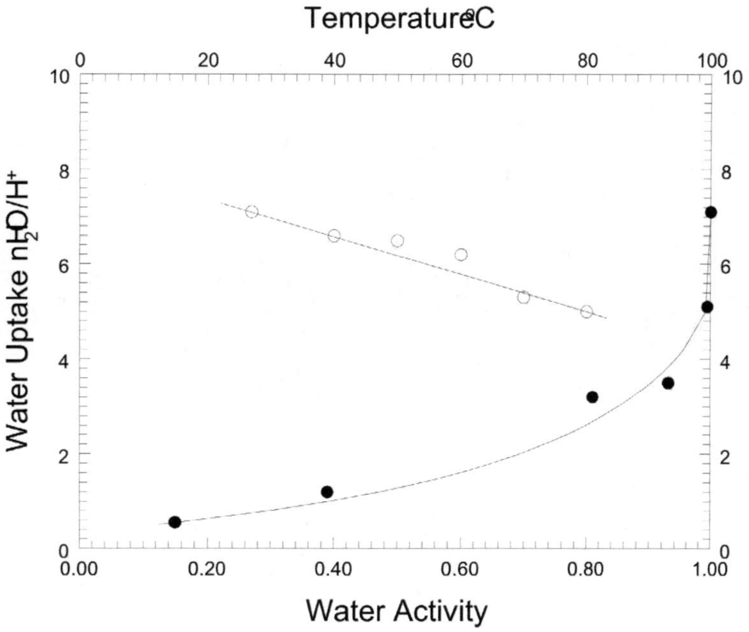

Figure 1 Water uptake n H_2O/H^+ of PBIPOH (ICD = 9.4 meq/g) at different temperature (o), and different humidity (•).

The maximum of water uptake of PBIPOH membrane is 7.0 H_2O molecules per charge. Compare with Nafion 112 (ICD=0.91) has ~21 H_2O molecules per charge, this value is low. However, the unique structure of PBIPOH contain 3 $-PO(OH)_2$ per unit. Therefore, there will be 42 water molecules per unit charges (each unit contains 6 charge).

The membranes display very high ionic conductivity that increases with the temperature and DS. Figure 2. shows the typical impedance spectrum of PBIPOH with various degrees of substitution and its blends with SPES, PBIS, and FSPES at 90°C. The proton conductivity can been calculated. With high DS=3, the PBIPOH exhibit very high productivity ~(0.5S/cm^2 at 95°C) compared to sulfonic acid membranes, though the phosphonic acid is much less acidic. The $-SO_3H$ membranes exhibit a pKa <<1, while phosphonic acid compounds commonly display pKa_1 in the range of 3-4. The acidity of the PBIPOH that has been used in here is slightly higher. From titration curve (figure 3.) shows the pKa_1 = 2.35 and pKa_2 = 6.95 for PBIPOH membrane. Standard oxidation testing of the membranes with H_2O_2 shows good oxidative stability.

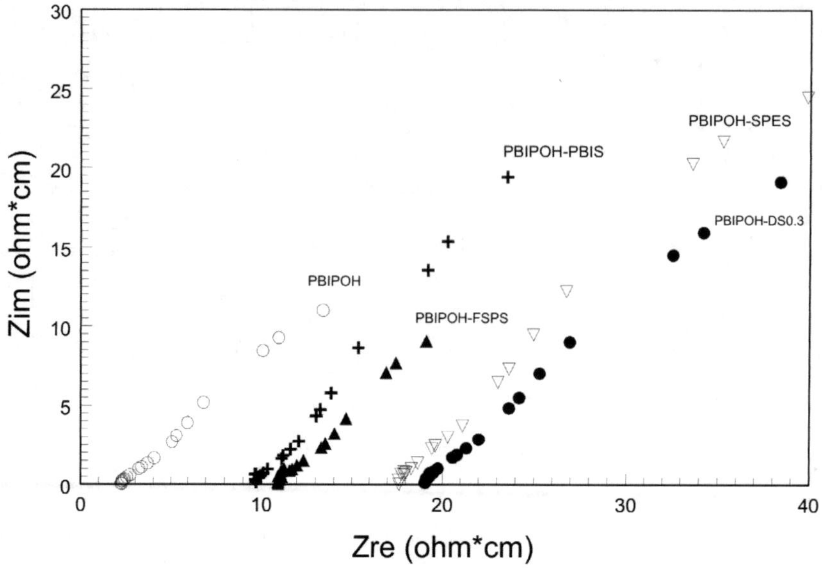

Figure 2. Impendence spectrum of PBIPOH and its blend membranes at 90°C.

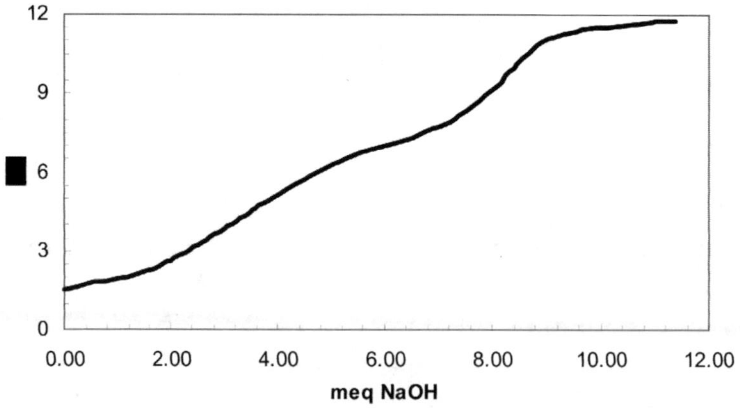

Figure 3. Typical titration curve of PBIPOH membrane.
ICD= 9.4 meq/g, and pKa_1 =2.35 and pKa_2 =6.95

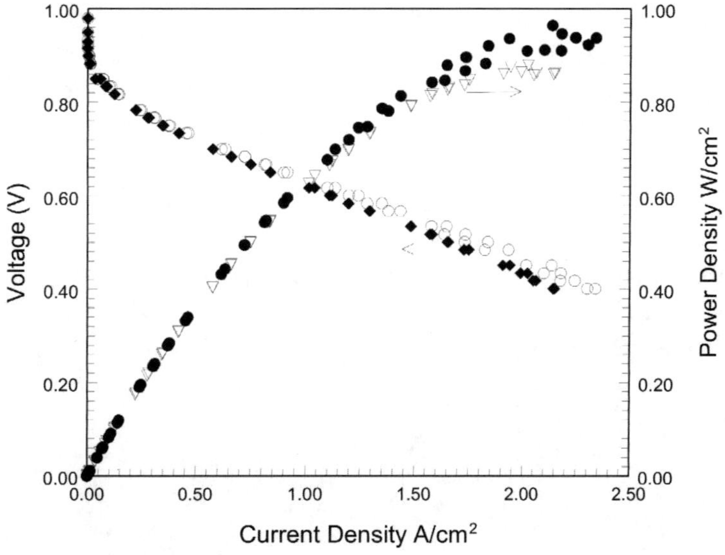

Figure 4. Performance of MEA using PBIPOH as membrane electrolyte (50 μm) at 80°C (•,o), and 100°C (∇,♦), 30 psi. The Pt loading on electrode is 0.2 mg/cm^2

PBIPOHs with a high degree of substitution are soluble in aqueous solution at pH>7, and can be cast into membranes for fuel cell applications. Figure 4 shows the cell potential and power density *vs.* current density of MEA using PBIPOH (DS=3) membrane as electrolyte at 80°C and 100°C, 30 psig. It is shown that current density of this MEA at 0.5 V has 2 A/cm^2 and maximum power density 0.98W/cm^2

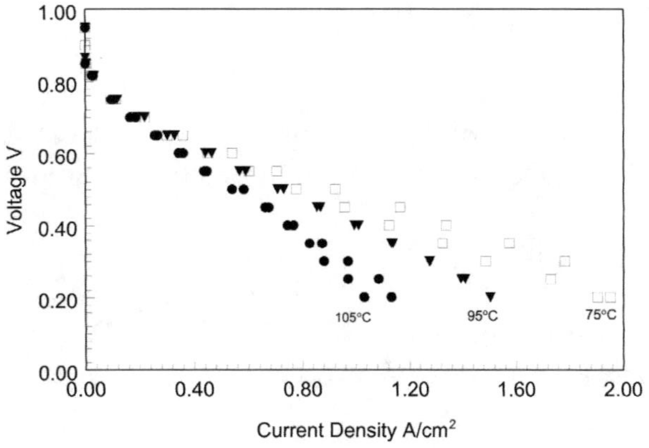

Figure 5. Performance of blend membrane (90% PBIPOH, 10%SPES) at 30 psi, with different temperature. H_2/O_2 as feed gases. The Pt loading on electrode is 0.2 mg/cm^2.

The phosphonic acid derivatives PBIP also can blend with sulfonated (-SO_3H) version of PBI, Polysulfone (PS), and poly(ether sulfone, PES) in different ratios[10]. After hydrolysis these blend membrane display very good performance in fuel cell testing. Figure 5 is performance of MEA using 90% PBIP, 10% SPES blend membrane as electrolyte at different temperature. No hysteresis is observed while obtaining the polarization curves even at high temperature (105°C), suggesting good water "management" in the MEA.

Short-term durability testing of the PBIPs at temperature range from 50°C to 120° with daily start-ups and shut down has been conducted for more then 500h without a reduction in the fuel cell performance. (figure 6.).

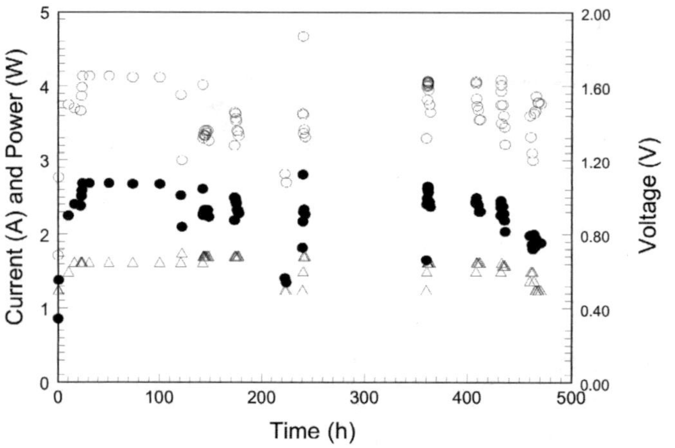

Figure 6. Durability of the PBIPOH membrane in the MEA operating at 95°C, 30 psi with H_2/O_2. The Pt loading on electrode is 0.2 mg/cm^2. Voltage (∇), Current (o), Power (•).

In conclusion, we have shown here that modified PBI, like phosphonylated PBIPOH and its blend membranes have high ionic conductivity and can operate at elevated temperature, with competitive performance when compared to sulfonic acid derivatives.

ACKNOWLEDGEMENT

We wish to acknowledge Ms. Elizabeth Loveless, student of Chem. Eng. supported by NSF (from the REU program in SUNY-ESF) this summer, for the help in producing some of the results.

REFERENCES

(1). Rider, J.N.; Serpico, J.M.; Ehrenberg, S.G.; Einset, A.G. *The Electrochemical Society Extended Abstracts,* **1994,** 95-2, 1070.

(2). Ehrenberg, S., Serpico, J.M., Wnek, G.E., Rider, J.N.; **1996,** U.S. Patent 5,679,482.

(3). Cabasso, I.; Yuan, Y.; Mittelsteadt, C.; **1999,** U.S. Patent 5,989,742.
(4). Cabasso, I.; Yuan, Y.; Mittelsteadt, C.; **2000,** U.S. Patent 6,103,41
(5). Cabasso, I.; Jagur-Grodzinski, J.; Vofsi, D., *J. Appl. Polym. Sci.* **1974,** 18,1969.
(6). Tran, C.H.; Cabasso,I., *J. Appl. Polym. Sci.* **1979,** 23, 2967.
(7). Xu, X and Cabasso, I., *Poly. Mat. Sci.*, **1993,** 68, 120 .
(8) Xu, X., Thesis. **1993**, State University of New York, ESF.
(9) Cabasso, I; Xu X.; Yuan, Y.; **1998**, US Patent 5,783,325.
(10) Cabasso, I; Johnson, F.; Yuan, Y.; US. Patent PCT 887-4166, **2001.**

Nafion-Acid Functionalized Zeolite Beta Nanocomposite Proton Exchange Membranes

Brett A. Holmberg, Yushan Yan, and Joe M. Norbeck

[1]University of California
Bournes Hall Rm 242
Riverside, CA 92521

[2]CE-CERT
1084 Columbia Ave
Riverside, CA 92507

Prepared for presentation at the AIChE Annual Meeting
November 20, 2003
T6 Topical Conference on Advanced Membrane-Based Separations
Session 76: Fuel Cell Membranes

October 2003

Unpublished

Abstract

Direct methanol fuel cells (DMFCs) have recently received a great deal of attention for their potential as compact, high efficiency power sources for portable and transportation applications (Wasmus and Kuver 1999). We have been working on Nafion-acid functionalized zeolite Beta nanocomposite membranes that take advantage of the molecular sieving and proton conductivity of acid functionalized zeolite Beta. By reducing methanol crossover without a significant loss in proton conductivity, our nanocomposite membranes have the potential to improve methanol oxidation efficiency and reduce cathode overpotentials. With the lower methanol crossover of our membranes, higher DMFC operation temperatures are more feasible, which could lead to improved DMFC power density.

Using a combination of two proven synthesis techniques, highly crystalline zeolite Beta nanocrystals were hydrothermally synthesized (Camblor et al. 1997) with phenethyl organic functional groups attached to the zeolitic inner pore structure and outer surface (Jones et al. 1998). Sulfonic acid groups were then added to the phenethyl side chains with a concentrated sulfuric acid treatment (Gilbert 1965). A suspension of our acid functionalized zeolite Beta was then added to a suspension of Nafion, and the mixture was cast into a membrane in a glass vessel.

A detailed analyses of the acid functionalized zeolite Beta nanocrystals will be given using powder X-ray diffraction (XRD) for crystallinity determination, Fourier Transform Infrared (FTIR) spectroscopy, Si^{29} NMR, and Fourier Transform Raman spectroscopy, with dynamic light scattering (DLS) for crystal size determination. Properties of the acid functionalized zeolite Beta nanocrystals will be demonstrated . Scanning electron microscopy (SEM) images for membrane morphology, fuel cell performance data, proton conductivity measurements using four-electrode impedance (Sone et al. 1996), and methanol permeability measurements (Tricoli 1998) from commercial Nafion 117, pure cast Nafion, and Nafion-acid functionalized zeolite Beta nanocomposite membranes will be presented.

References

Camblor, M. A., Corma, A., Mifsud, A., PerezPariente, J., and Valencia, S. (1997). "Synthesis of nanocrystalline zeolite Beta in the absence of alkali metal cations." Progress in Zeolite and Microporous Materials, Pts A-C, 105, 341-348.

Gilbert, E. E. (1965). Sulfonation and Related Reactions, Interscience Publisher, Santa Monica, CA.

Jones, C. W., Tsuji, K., and Davis, M. E. (1998). "Organic-functionalized molecular sieves as shape-selective catalysts." Nature, 393(6680), 52-54.

Sone, Y., Ekdunge, P., and Simonsson, D. (1996). "Proton conductivity of Nafion 117 as measured by a four-electrode AC impedance method." Journal of the Electrochemical Society, 143(4), 1254-1259.

Tricoli, V. (1998). "Proton and methanol transport in poly(perfluorosulfonote) membranes containing Cs+ and H+ cations." Journal of the Electrochemical Society, 145(11), 3798-3801.

Wasmus, S., and Kuver, A. (1999). "Methanol oxidation and direct methanol fuel cells: a selective review." Journal of Electroanalytical Chemistry, 461(1-2), 14-31.

Perfluorinated ionomers containing the sulfonimide acid function as PEMs for fuel cells

D. DesMarteau and S. Creager

Clemson University, Chemistry Department, Box 340973, Clemson, SC 29634-0973.

Abstract

The bis((perfluoroalkyl)sulfonyl)imide function possesses remarkable Brønsted acidity and high thermal stability. When incorporated into perfluorinated ionomers, the materials show excellent potential as PEMs for hydrogen – oxygen fuel cells. The many variations possible with the sulfonimide function allows the preparation of ionomers with novel polymer architectures as represented by the general formula

$$(CF_2CF_2)_nCF_2\underset{|}{C}F)$$
$$(\overset{|}{O}\text{-}CF_2\underset{|}{C}F)_zOCF_2CF_2SO_2[\overset{H}{N}SO_2(CF_2)_xSO_2]_y\overset{H}{N}SO_2R_f$$
$$\overset{|}{C}F_3$$

n=3-10; z=0,1; y=0,1; x=2,4,6,other; $R_f=C_nF_{2n+1}$, other.

The synthesis of the required vinyl ether monomers and their copolymerization by aqueous emulsion with tetrafluoroethylene and polymer properties will be reviewed. Single cell test data utilizing these materials in MEAs will be presented and compared with perfluorinated sulfonic acid ionomers.

1. Introduction

The challenge to develop PEM fuel cells for mobile power applications is formidable. For the automotive industry, it is now recognized that operation at >110°C is essential for a variety of system requirements. Many issues are as yet unresolved and the availability of a membrane material that can operate at a higher cell temperature without pressurization is a critical issue. The commercial PFSA (perfluorinated sulfonic acids) polymer materials are the leading candidates for use in PEM fuel cells. However it is nearly certain that these materials will not meet the requirements for a PEM fuel cell operating at >110°C using 1 atm air at the cathode. As the relative humidity of the system falls below 100%, the proton conductivity of the PFSA membrane decreases, the PFSA-containing catalyst layers in the MEA are adversely affected, and the performance of the fuel cell (single cell or stock) begins to decline. Of course this is a simplistic view of a complex electrochemical system, but these are certainly major factors in the decreased performance.

Many approaches are being tried to overcome the limitations of PFSA polymers. These range from modification of PFSA polymers with fillers, to new sulfonated aromatic polymers and acid doped polybenzmidizoles, to inorganic membrane materials. Our approach has been the development of new perfluorinated ionomers containing alternative acid functions. In this paper, our research on new polymer materials containing the sulfonimide acid function is briefly reviewed.

2. Novel Ionomers Containing the Sulfonimide Function

2.1 Background

In the 1980's, the Gas Research Institute initiated a major program to develop fluorinated electrolytes for use in phosacid fuel cells (PAFC). Under this program, the excellent electrochemical properties of bis((perfluoroalkyl)sulfonyl)imides were first demonstrated [1,2]. A large variety of molecular sulfonimides were prepared including multifunctional compounds [3,4]. After four years it became clear that the cost of molecular electrolytes of this type would be prohibitively expensive for PAFC-type fuel cells. However it was found that adding only 0.1% of certain difunctional compounds as an additive to a PAFC, gave a dramatic improvement in the performance, presumably due to enhanced oxygen reduction kinetics. This was first noted at International Fuel Cells [5] and confirmed by fundamental electrochemical studies [2]. An effort was begun in 1989 to develop perfluorinated ionomers containing the sulfonimide function (PFSI). The goal was a polymer system with long-term stability that could be incorporated in the catalyst layer of a PAFC. Subsequently, this effort was redirected towards developing new polymers for use as membrane materials in PEM fuel cells and in the catalyst layer of the MEA.

2.2. Preparation and Properties of PFSI Ionomers.

The preparation of a perfluorinated ionomer required a functionalized monomer that could be copolymerized with tetrafluoroethylene (TFE). The initial monomer chosen for development was $CF_2{=}CFOCF_2CF(CF_3)OCF_2CF_2SO_2N(Na)SO_2CF_3$ (**1**). This monomer was based on the PFSA monomer $CF_2{=}CFOCF_2(CF_3)OCF_2CF_2SO_2F$ (**2**) originally developed by DuPont and used in the commercial production of Nafion® [6]. However unlike the preparation of PFSA copolymers which are produced by emulsion polymerization of the water insoluble monomers **2** and TFE, monomer **1** is quite soluble in water. Furthermore, copolymerization with

TFE would lead directly to an ionomer material, whereas the copolymer of **2** with TFE is converted to an ionomer after melt processing of the copolymer. The major issue at the start of this research was whether or not monomer **1** could be effectively copolymerized with TFE to give a good polymer material. In fact, after some effort, excellent copolymers could be obtained. After some years, we determined that the good reactivity of **1** in an aqueous emulsion polymerization was due in part to the fact that **1** is an excellent surfactant.

The preparation of **1** as it is presently carried out is shown in Scheme 1. Initially we prepared **2** from TFE/SO_3 as taught by DuPont, but later we were able to obtain **2** from other sources.

$$\mathbf{2} \xrightarrow{Br_2} BrCF_2CFBrOCF_2CF(CF_3)OCF_2CF_2SO_2F\ (\mathbf{A})$$

$$(CF_3SO_2)O \xrightarrow{NH_3} \xrightarrow{NaOH} \xrightarrow{HMDS} CF_3SO_2N(Na)SiMe_3\ (\mathbf{B})$$

$$\mathbf{A} + \mathbf{B} \longrightarrow \xrightarrow{Zn} \mathbf{1}$$

Scheme 1

The conditions for a typical copolymerization of **1** with TFE in a 1L magnetically stirred autoclave are given in Table 1. Monomer **1** is added to the autoclave during polymerization using a high pressure syringe pump to maintain a nearly constant monomer concentration. The TFE is maintained at constant pressure (±5 psi) with the concentrations of **1** and TFE chosen for a given target equivalent weight. Copolymers having EWs in the range of 700-1500 have been produced. Most of the electrochemical testing of TFE copolymer of **1** has been carried out with an EW near 1200.

Table 1. Conditions for a Typical Polymerization of **1** with TFE.

Reactor	1000 mL	Average TFE drop rate (psi/hour)	9.80
Initial monomer in the autoclave (g)	4.64	TFE consumed (g)	6.3
$C_7F_{15}CO_2NH_4$ (g)	4.65	Added monomer (g)	4.64
$Na_2HPO_4 \cdot 7H_2O$	6.63	Added $Na_2HPO_4 \cdot 7H_2O$ (g)	0.74
NaH_2PO_4 (g)	3.36	Added NaH_2PO_4 (g)	0.38
$(NH_4)_2S_2O_8$ (g)	0.70	Addition volume (ml)	41.4
$NaHSO_3$ (g)	0.64	PTFE recovered after reaction (g)	0.12
H_2O (mL)	600	Polymer Weight (g)	12.12
Pressure of TFE (psia)	143-150	IEC (mgH/g)	0.83
Stirring speed (rpm)	212	EW	1170
Temperature (°C)	10-12	TFE in polymer (g)	6.33
Reaction time (hour)	5.5	Monomer in polymer (g)	5.78
Total TFE drop (psi)	54.1		

The properties of the PFSI copolymers (EW$\simeq$ 1200) have many similarities to Nafion® 1100. Their Tg is a few °C lower and their DSC is remarkably similar to that of Nafion® 1100. However the sulfonimide ionomer has much higher thermal stability in acid form and a typical TGA scan shows essentially no decomposition until nearly 400°C. This is in marked contrast to Nafion® which begins to decompose at near 150°C. Another significant difference is that sulfonimide ionomers typically show a larger water uptake in membrane form. To date, we have prepared membranes by solubilizing the polymer in DMF in acid form, followed by casting a film onto a polished stainless steel surface. The choice of DMF as a solvent resulted in the fortuitous formation of the $(CH_3)_2NH_2$- salts of the sulfonimide by reaction of the DMF with highly acidic sulfonimide. Subsequent annealing of the film in this salt form at 200-240°C gives a tough, plastic membrane material when converted back to the acid by boiling in concentrated nitric acid.

Some electrochemical and structural data on these membrane materials have been reported [7-17]. These data show that the PFSI ionomers are structurally similar to PFSA ionomers but there are differences in the polymer microstructure. Proton conductivity of sulfonimide membranes was in general comparable to or better than Nafion® 1100 at 100% RH and comparable EW values. At less than 100% RH the fall off in conductivity was faster for a 1200 EW PFSI and similar to Nafion® 1100 for a 1075 EW PFSI.

Single cell MEA testing of PFSI polymers, benchmarked against Nafion® 1135 MEAs under identical test conditions, has shown very encouraging results. Significantly higher power densities can be obtained at 80 and 110°C with a 1075 EW PFSI using solubilized Nafion® 1100 in the catalyst. Some preliminary results indicate that even higher performance is possible using the solubilized PFSI in the catalyst layer.

2.3. New Polymer architectures with PFSI polymers

The monofunctional pendant side chains in the PFSA and PFSI ionomers discussed above limit the degree of functionality (EW) that can be obtained, while still maintaining good mechanical properties of the polymer. In the case of sulfonimides, an alternative polymer structure is possible using multifunctional pendant side chains. In principle one can achieve lower equivalent weights while maintaining good mechanical properties using a difunctional monomer **3** (Fig. 1).

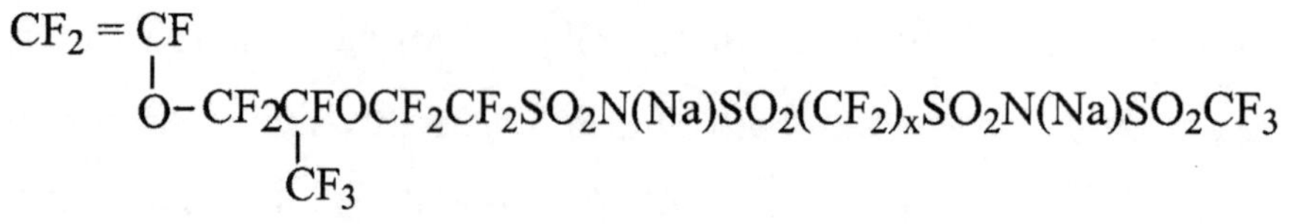

Figure 1. Difunctional PFSI monomer **3**

The synthetic route to **3** is summarized in Scheme 2.

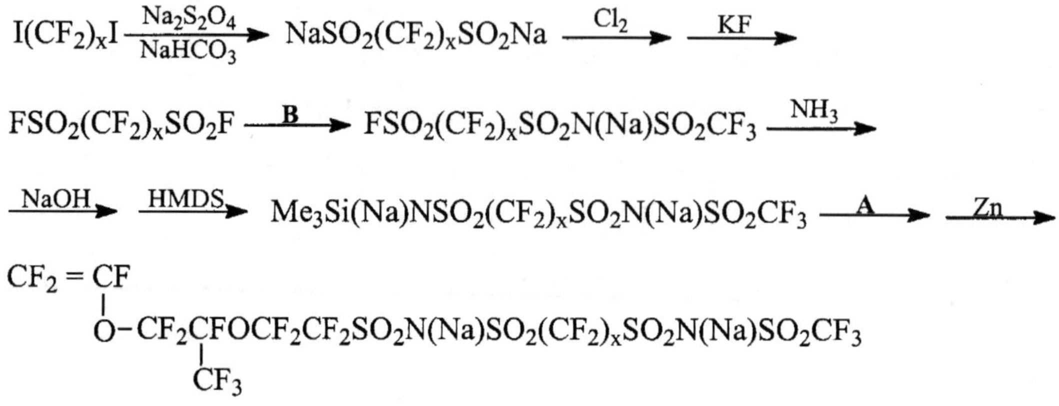

$(CF_2)_x = (CF_2)_4$ (**3a**), $(CF_2)_6$ (**3b**), $CF_2CF_2OCF_2CF_2$ (**3c**)

Scheme 2

Obviously the synthesis of monomer **3** is more complex than the monofunctional **1**, but **3a-c** can be obtained in excellent yields. The copolymerization of **3** with TFE has been carried out and good quality polymer materials can be obtained. Polymers of **3** have been obtained with EW values from 1100 to 2000. At an EW of 1100 for **3a**, the ratio of n (moles TFE/mol monomer) is 12.8, compared to n-5.2 for the copolymer of **1** at an EW=1100. Clearly the microstructure of the 2 polymers will differ considerably. At this point, we do not have enough information to understand this difference in any detail.

Polymers of **3** behave as a typical ionomer with properties similar to polymers of **1**. Conductivities provide some interesting insight into the potential of these polymers. Polymer **3a** with EW=1170 has higher conductivity (100% RH) than Nafion® 1100 (0.15 s/cm -vs- 0.0618 s/cm) but at 60% RH, Nafion® 1100 is more conductive. On the other hand polymer **3c** (EW 1220) exhibits higher conductivity than Nafion® 1100 from RH = 16-100%.

Detailed MEA testing of polymers **3** has not been carried out at this time. However, use of **3a** in catalyst layer with a Nafion® 1135 membrane showed considerable improvement in performance over the MEA using 1135 membrane and Nafion® 1100 in the catalyst layer.

3.3 Conclusions.

Sulfonimide ionomers clearly have excellent potential for use in PEM fuel cells. Unoptimized polymers in head to head comparison with Nafion® membranes show superior performance. The versatility of the sulfonimides in providing new polymer architecture has high potential to overcome some of the limitations of PFSA ionomers.

Acknowledgement

The support of this research by the Gas Research Institute, DOD-ARPA URI and DOE-SC EPSCoR is gratefully acknowledged.

References

1. M. Razaq, A. Razaq, E. Yeager, D. D. DesMarteau and S. Singh, "Bis ((trifluoromethyl)sulfonyl) Imide as an Alternate Electrolyte in Hydrogen-Oxygen Fuel Cells," *J. Applied Electrochem.*, 17 (1987) 1064.

2. M. Razaq, A. Razaq, E. Yeager, D. D. DesMarteau and S. Singh, "Perfluorosulfonimides as an Additive in Phosphoric Acid Fuel Cells," *J. Electrochem. Soc.*, 136 (1989) 385.

3. D. D. DesMarteau, S.-Z. Zhu, W. T. Pennington, Y. Gotoh and M. Witz, "Superacids of Nitrogen and Carbon," *J. Fluorine Chem.*, 45 (1989) 24.

4. D. D. DesMarteau, "Novel Perfluorinated Ionomers and Ionenes," *J. Fluorine Chem.*, 72 (1985) 203.

5. D. L. Maricle, International Fuel Cells, 1989.

6. A. Eisenberg and H. L. Yeager, "Perfluorinated Ionomer Membranes, ACS Symposium Series 180, American Chemical Society, Washington, D.C., 1982.

7. S. E. Creager, J. J. Sumner, R. D. Bailey, J.-J. Ma, W. T. Pennington and D. D. DesMarteau, "Effects of Ionomer Equivalent Weight and Crystallinity on Water Uptake and Proton Conductivity in Bis[(perfluoroalkyl)sulfonyl]imide-based Ionomers," *Electrochemical and Solid-State Letters*, 2 (1999) 434.

8. S. C. Savett, J. R. Atkins, C. R. Sides, J. L. Harris, B. H. Thomas, S. E. Creager, W. T. Pennington and D. D. DesMarteau, "A Comparison of bis[(perfloroalkyl)sulfonyl]imide Ionomers with Perfluorosulfonic Acid Ionomers for Applications in PEM Fuel Cell Technology," *Journal of the Electrochemical Society,* 149 (12) (2002) A1527.

9. J. R. Atkins, C. R. Sides, S. E. Creager, J. L. Harris, W. T. Pennington, B. H. Thomas and D. D. DesMarteau, "Effects of Equivalent Weight on Ionic Conductivity in Bis[(perfluoroalkyl)sulfonyl]imide Perfluorinated Ionomers," *Journal of New Materials for Electrochemical Systems,* 6 (1) (2003) 9.

10. D. D. DesMarteau, J.-J. Ma, M.-H. Tu, M.-T. Liu, B. Thomas and J. McClellan, "Synthesis and Properties of New Bis[(perfluoroalkyl)sulfonyl]imide Ionomers for Fuel Cells," *Polym. Mat. Science and Eng.,* 80 (1999) 598.

11. D. D. DesMarteau, S. E. Creager, S. Savett and B. Thomas, "New Bis[(perfluoroalkyl)sulfonyl]imide Ionomers for PEM Fuel Cells," *Polym. Mat. Science and Eng.,* 80 (1999) 600.

12. D. D. DesMarteau, "Perfluorinated Sulfonyl Monomers and Polymers and Membranes Made Therefrom", U. S. Pat. No. 5,463,005, Oct. 31, 1995.

13. S. E. Creager, J. J. Sumner, J. J. Ma and D. D. DesMarteau, "Proton Conductivity in Nafion 117 and in a Novel Bis((Perfluoroalkyl)sulfonyl)imide Ionomer Membrane," *Journal of the Electrochemical Society*, 145 (1998) 107.

14. B. Ma, C. E. Bunker, K. Simons, J.-T. Liu, J.-J. Ma, C. W. Martin, D. D. DesMarteau and Y.-P. Sun, "Steady-State and Time-Resolved Fluorescence Spectroscopic Probing of Microsructures and Properties of Perfluorinated Polyelectrolyte Membranes," *J. Electroanal. Chem.* 459 (1998) 15.

15. H. W. Rollins, T. Whiteside, G. J. Shafer, J.-J. Ma, M.-H. Tu, J.-T. Liu, D. D. DesMarteau and Y. P. Sun, "Nanoscale Metal Sulfides in Perfluorinated Ionomer Membranes," *J. Mater. Chem.*, 10 (2000) 2081.

16. C. E. Bunker, B. Ma, K. J. Simmons, H. W. Rollins, J.-T. Liu, J.-J. Ma, C. W. Martin, D. D. DesMarteau and Y.-P. Sun, "Fluorescence Spectroscopic Probing of Two Distinctive Micro-Environments in Perfluorinated Ionomer Membranes," *J. Photochem. Photobiol. A,* 129 (1999) 71.

17. H. W. Rollins, F. Lin, J. Johnson, J.-J. Ma, J.-T. Liu, M.-H. Tu, D. D. DesMarteau and Y.P.-Sun, "Nanoscale Cavities for Nanoparticles in Perfluorinated Ionomer Membranes," *Langmuir*, 16 (2000) 8031.

Functionalized Porous Semiconductors: Active Proton Exchange Membranes for Micro Fuel Cells

[1]Chang Lu, [2]Scott A. Gold, [2]Kuan-Lun Chu, and [2]Richard I. Masel

[1]Cornell University
G6, Clark Hall, Applied and Engineering Physics
Ithaca, NY 14853

[2]University of Illinois
Department of Chemical and Biomolecular Engineering
Urbana, IL 61801

Prepared for presentation at the AIChE Annual Meeting
November 20, 2003
T6 Topical Conference on Advanced Membrane-Based Separations
Session 76: Fuel Cell Membranes

October 2003

Unpublished

Abstract

Small proton exchange membrane (PEM) fuel cells are likely to displace batteries in many portable power applications. Presently, many investigators have been trying to incorporate fuel cells into small silicon devices, but so far there has been some difficulty in identifying proton-conducting membranes that bond well to silicon and are compatible with standard semiconductor processing. This paper demonstrates that functionalized nanoporous silicon membranes are promising electrolytes for micro fuel cells. The membranes can be fabricated using standard micromachining techniques. 30~50 micron thick nanoporous silicon membranes, with an average pore size of 10~15 nm exhibit proton conductivities comparable to those of Nafion. The membranes are easy to bond to silicon substrates and thermally stable to at least 160 C. There is no significant expansion and contraction upon exposure to water. A non-optimised prototype micro fuel cell, running on formic acid at 20 C and atmospheric pressure, gave an open circuit voltage of 0.73 V and a current density of 13.3 mA/cm2. These novel silicon electrolyte membranes provide unique opportunities for elegant and efficient micro fuel cell designs. The membranes will also be useful in sensors and other devices where proton conductivity is important to device operation.

Design and Performance of Limited Recharge Fuel Cell Vehicle

Galen J. Suppes
Chemical Engineering
University of Missouri-Columbia
Columbia, MO 65211

ABSTRACT

The plug-in recharge option for hybrid or electric cars provides opportunities to travel short distances sourced from grid electricity and longer distances sourced from liquid fuel that is reformed to hydrogen on the vehicle. The presentation evaluates the potential impact of a 20-mile plug-in recharge option for producing and storing hydrogen and oxygen on the vehicle. Equipment specifications and control strategies are discussed. The environmental advantages of such a system are tremendous and substantial replacement of imported petroleum with domestic electricity can be realized.

KEY WORDS

Plug-In, Fuel Cell, Charge, Electric, Hybrid

INTRODUCTION

Plug-In hybrid vehicles that can be recharged overnight to provide a limited-range of travel the subsequent day using the recharge. For conventional engine-battery hybrids this might translate to a slightly larger battery pack to allow 20-40 miles of transit rather than about 10—a batter charger would also be required. For a fuel cell hybrid vehicle, onboard electrolysis would be needed as well as a hydrogen storage tank.

A report of the Electrical Power Research Institute (see http://www.arb.ca.gov/msprog/zevprog/2003rule/1202wkshp/graham.pdf) suggests the following timeline for the feasibility of vehicles that can run on electrical recharge:

before 2000	Battery Electric Vehicles
2000	Hybrid Vehicles
2005	Plug-In Battery Recharge Hybrid
2008	Plug-In Hydrogen Recharge Hybrid
2015	Conventional Fuel Cells

Two distinct advantages of the Plug-In Hydrogen Recharge Hybrid are 1) feasibility at an earlier time than conventional fuel cell vehicles and 2) substantial less dependence on a hydrogen refueling infrastructure. Support of the Plug-In options is also growing with the CalCars group in California (see http://www.calcars.org)

BACKGROUND

Figure 1 illustrates two of many possible configurations for a plug-in fuel cell hybrid. In the hydrogen storage option, reversible fuel cells are used to generate hydrogen and oxygen which are stored in compressed tanks. This configuration can actually reduce the amount of secondary source battery storage that is necessary since stored pure hydrogen can power the fuel cells. Alternative to hydrogen storage, a larger battery pack can be used to provide the 20-mile recharge range.

In both the options of Figure 1, a liquid fuel is the primary energy source. Reformer technology converts the liquid fuel to hydrogen through sequential steps of reforming, water-gas-shift, and preferential oxidation of carbon monoxide remaining from the water-gas-shift. , The reformer technology is being developed for fuels ranging from natural gas, and methanol to gasoline and diesel. Development obstacles are many, including: 1) needs for ultra-low sulfur fuel,4 2) long startup times (significantly longer than the three seconds a typical vehicle owner is accustomed), 3) a hydrogen product that is diluted with nitrogen and carbon monoxide,7 and 4) development and optimization of the reformer system as well as all associated unit operations.

The plug-in fuel cell option addresses these challenges as follows:

- · Ultra-Low Sulfur Fuel – by substantially displacing liquid fuel consumption with grid electricity, more expensive fuels (like methanol, ethanol, and biodiesel) that are available today can be used. Refueling needs are also reduced, addressing potential problems with locating alternative fuel refueling stations.

- · Startup Times – because the vehicle would operate for several miles off the recharge, several minutes are available to start up the reformer system allowing this to be addressed with technology available within the timeframe of the plug-in fuel cell technology.

- · Dilution of Hydrogen – Diluting hydrogen reduces the fuel cell efficiency. This is less of a problem for a recharge vehicle where reliance on the liquid fuel and reformer is substantially displaced with grid electricity.

- · Needs for Development and Optimization – Less than optimal reformer systems would be acceptable since the vehicle will be less dependent upon the reformer hydrogen.

Also, the plug-in fuel cell hybrid would not require a hydrogen refueling infrastructure and could use compressed hydrogen and oxygen storage since the quantities of hydrogen would be far less than needed for a vehicle relying only on stored hydrogen for extended range travel (200-400 miles).

The major development obstacles of an extended-range fuel cell vehicle are overcome by the plug-in fuel cell hybrid that operates predominantly with the recharge option. This paper addresses design details of the plug-in fuel cell hybrid toward 1) identifying optimal recharge capacities, 2) quantifying the well-to-wheel efficiencies including the impact on greenhouse gas emissions, 3) comparing the costs of recharging batteries versus hydrogen versus ICE or reformed hydrogen and 4) comparing the battery recharge versus hydrogen recharge.

DISCUSSION

Optimal Recharge Capacities - Travel habits play an important role in fuel tank design for the plug-in fuel cell hybrid. Figure 2 shows the average daily travel habits in the U.S. About 47% of all miles put on passenger vehicles in a day are within the first 20 miles of travel by a vehicle. Furthermore, about 50% of the vehicles travel 20 miles or less per day, and these 20 miles are intensive in inner-city travel where fuel economy for conventional ICEs is poor and emissions have their greatest adverse impacts. A 20-mile recharge range would have an ample market and could displace up to 50% of petroleum fuel consumption. This impact could be extended by providing recharge options at employment locations for those workers who have commutes of over 10 miles each way.

Comparison of Fuel Costs - The fuel costs are about \$0.025 and \$0.032 per mile for gasoline ICE and hydrogen produced by grid (off-peak, \$0.056 per kW-hr). Recharging batteries using grid electricity would cost about \$0.021 per mile. These fuel costs are less than those of Table 1 because they do not include highway taxes and are based on a light-weight vehicle consistent with trends in hybrid vehicle technology. Continued escalation of petroleum prices would cause the gasoline prices to rapidly surpass the hydrogen electrolysis costs. In any case, the electricity costs are small in comparison to those costs summarized in Table 1 that are related to the depreciation, upkeep, and protection of the vehicle. For drive-by-wire technology, the reductions in vehicle costs (and depreciation) should more than compensate for the increased fuel costs.

Further details on this concept can be found in the following publication:

"Plug-in Fuel Cell Hybrids as Transition Technology to Hydrogen Infrastructure." G. J. Suppes*, S. Lopes, and C. W. Chiu. International Journal of Hydrogen Energy, Accepted for Publication, June 2003.

Table 1. Typical operating costs for vehicle. Ranges include low end cost of mid-sized car and upper end costs for luxury car and SUV[i].

	Vehicle Operating Cost Range (\$/mile)	Average Cost
Depreciation	22.5-31.8	49.2%
Insurance	6.9-10.5	14.3%
Financing	4.8-8.3	12.3%
Fuel	4.5-6.9	10.9%
Maintenance, Oil, Tires	4.7-5.3	9.1%
License & registration	1.4-3.2	4.2%

ACKNOWLEDGMENTS

Contributions from S. Lopes, and C. W. Chiu at The University of Missouri-Columbia are acknowledged.

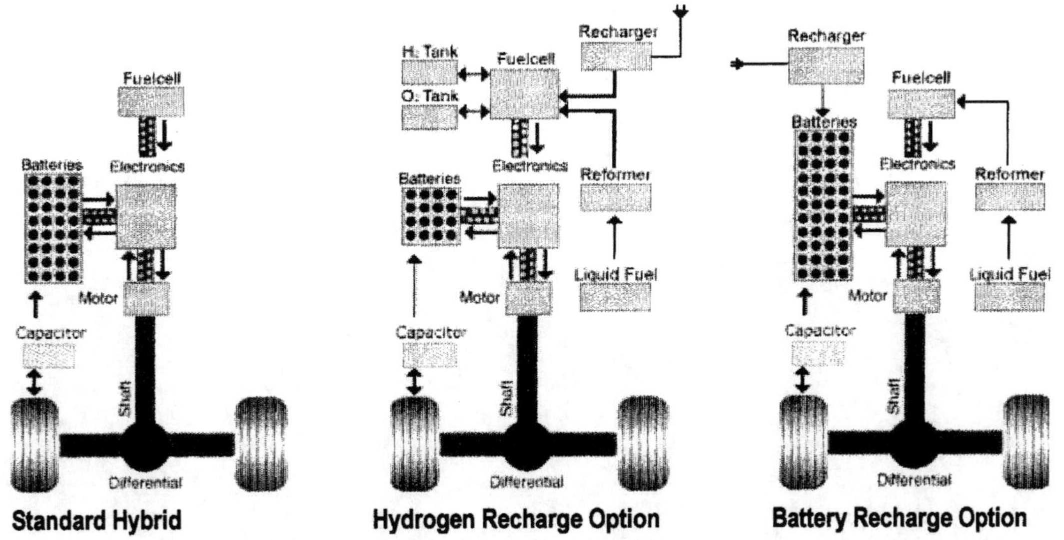

Figure 1. Configurations for hybrid, plug-in fuel cell hybrid with hydrogen recharge, and plug-in fuel cell hybrid with battery recharge.

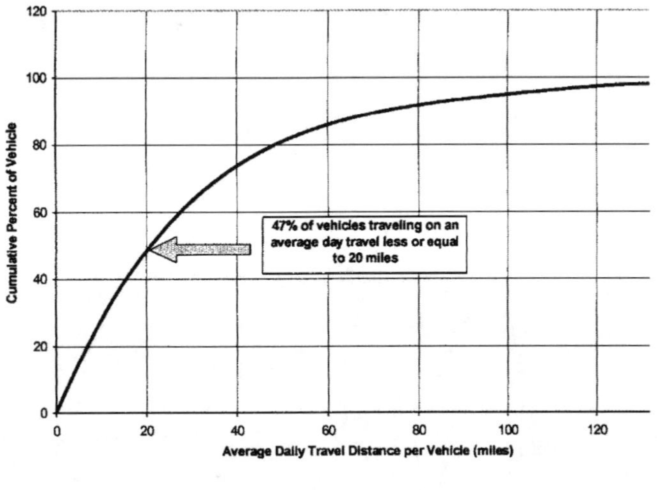

Figure 2. Pattern of average daily travel distance in U.S.A (Source- 1990 Nationwide Personal Transportation Survey).[i]

[i] Renyi, J., P. J. Schuurmans, "policy Implications of Hybrid Electric Vehicles-Final Report to NREL, National Renewable Energy Laboratory, Golden, CO, Report ACB-5-15337-01 (see www.ott.doe.gov/pdfs/nevcor.pdf), April 22, 1996.

REFERENCES

[i] Anonymous, Your Driving Costs, American Automobile Association, based on data from Runzheimer International (see http://www.vtpi.org/tdm/tdm82.htm). 1998.

Nonlinear Model Reduction of a Dynamic Two-dimensional Molten Carbonate Fuel Cell Model

Michael Mangold and Min Sheng

Max-Planck-Institut für Dynamik komplexer technischer Systeme
Sandtorstr. 1, 39106 Magdeburg, Germany
mangold@mpi-magdeburg.mpg.de

Key words: Molten carbonate fuel cell; dynamic simulation; reduced model; spatially distributed system; Karhunen-Loève decomposition; reduced state observer

Prepared for Presentation at the 2003 AIChE Annual Meeting,
San Francisco, CA, Nov. 16-21

September 2003
Unpublished

Abstract

A reduced nonlinear model of a planar molten carbonate fuel cell is presented. The model is derived from a spatially distributed dynamic model of the cell by applying the Karhunen Loève Galerkin procedure. The reduced model is of considerably lower order than the original one and requires much less computation time. The comparison between the two models shows that the reduced model can describe the dynamic of the temperature field with sufficient accuracy and has good extrapolation qualities with respect to changes in the model parameters. The usefulness of the reduced model for process control purposes is demonstrated by applying it in the framework of a state and parameter estimator.

1 Introduction

The molten carbonate fuel cell (MCFC) is a high-temperature fuel cell operated at 600°C - 700°C. Due to its high operation temperature, the MCFC offers advantages for the co-generation of heat and electricity. Furthermore, the high operation temperature enables internal reforming and makes the MCFC flexible with respect to the sorts of fuel that can be used. On the other hand, the operation temperatures and consequently the demands of the MCFC on thermal stability of the used materials are below those of solid oxide fuel cells[1]. Those properties make the MCFC an attractive candidate for decentralized power generation.

Currently, the development and operation of MCFCs as of other high temperature fuel cells is mainly based on experimental and empirical knowledge. However, an intuitive process understanding is difficult in this case, as an MCFC system is an highly integrated process whose behavior depends on numerous interactions between the electrochemical reaction steps, the internal reforming, mass transport processes, and the heat transfer inside a cell or stack. A deeper understanding of the physico-chemical processes in an MCFC can be obtained from mathematical models based on physical conservation principles. Model based process control and process design strategies can lead to a much better use of the fuel cells' capacities and increase the efficiency of the system, but require suitable dynamic process models. Such models should offer good extrapolation qualities on the one hand, but must also be solvable in the framework of a real-time process control environment or an optimization algorithm. Today, the majority of MCFC models available in open literature are steady state models and therefore not suitable for process control (a survey is given in [2]). Only very few detailed dynamic MCFC models have been published [3, 4, 5]. Those models consist of systems of algebraic and nonlinear partial differential equations in several space coordinates, which are too complex for many process control purposes. As a consequence, most control studies of the MCFC have to rely on rather simple dynamical models, black-box type input output models, or qualitative knowledge-based approaches [6, 7, 8, 9]. The purpose of this contribution is to derive a reduced nonlinear dynamic model of an MCFC that approximates a detailed spatially distributed model with good accuracy, but whose solution requires less computation time. In the next section, a detailed two-dimensional spatially distributed model of a planar MCFC with cross flow is presented briefly. This model is used as a starting-point and reference for the reduced model. A reduced set of model equations is developed by applying the Karhunen-Loève Galerkin to the reference model. The reduced model is validated in test simulations by comparison with the reference model. Based on the reduced model, a state observer is developed and tested in simulations.

2 Spatially Distributed Reference Model of the MCFC

A planar MCFC as shown in Fig. 1 is considered. The flows an anode and on cathode side are orthogonal to each other, thereby inducing a cross flow. At the anode, a reaction between H_2 and carbonate ions takes place that releases two electrons. The carbonate ions are produced at the cathode in a reaction of carbon dioxide and oxygen under consumption of two electrons. The transport of carbonate ions through the electrolyte closes the internal electrical circuit. The hydrogen consumed on the anode side is generated in a steam reforming reaction. In this work, a combination of external reforming in a pre-reformer and of direct internal reforming is considered. The pre-reformer is fed with steam and methane. The model of the process used here is based on the following assumptions:

- Spatial gradients of the concentrations and of the temperature are considered in y and in z direction.

- The main purpose of the model is to serve as a basis for the temperature control of the cell. As the dynamics of the temperature equations is much slower than that of the mass and charge balances, only the temperature equations is considered to be dynamic, the mass and charge balances are assumed to be at steady state.

- Temperature differences between the gases and the solid parts are neglected. This leads to a pseudo-homogeneous energy balance.

- The electrochemical reactions on anode and on cathode side are described by Butler-Volmer kinetics. For the reforming step, a kinetic expression from [10] is used.

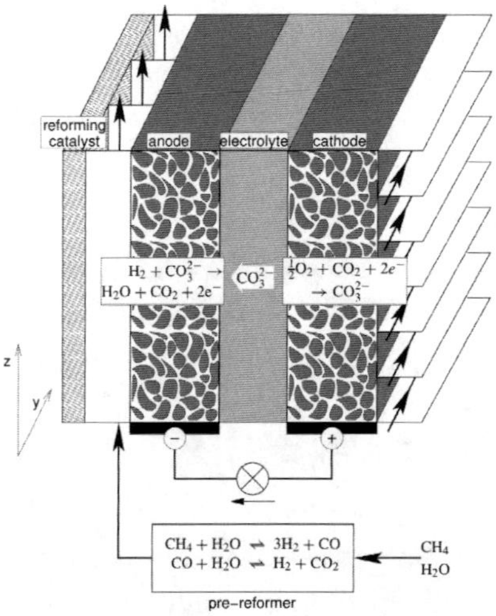

Figure 1: Basic structure of an MCFC with internal reforming

An energy balance of the system leads to a partial differential equation for the temperature that has the following structure:

$$
\begin{aligned}
0 \;=\; & (\rho c_P)d\frac{\partial T}{\partial t} - \frac{c_P^A \dot{n}^A}{L_y}\frac{\partial T}{\partial y} - \frac{c_P^C \dot{n}^C}{L_z}\frac{\partial T}{\partial z} + \lambda d\left(\frac{\partial^2 T}{\partial y^2} + \frac{\partial^2 T}{\partial z^2}\right) \\
& + i^A\left(\frac{(-\Delta h_R^A)}{\mathbb{R}T} - \Delta\Phi^A\right) + i^C\left(\frac{(\Delta h_R^C)}{\mathbb{R}T} - \Delta\Phi^C\right) + \frac{I_{Cell}}{L_y L_z}\Delta\Phi^M \\
& + (-\Delta h_R^R) r_R + (-\Delta h_R^W) r_w =: Res
\end{aligned}
\tag{1}
$$

The first line in Eq.1 contains terms caused by convective and dispersive heat transport. The terms in the second and in the third line are nonlinear sources caused by the chemical and electrochemical processes inside the fuel cell. The evaluation of those terms requires the solution of mass and charge balances which complete the reference model. A detailed description of the model will be given in [11].

3 Derivation of the reduced MCFC model

For model reduction of parabolic partial differential equations (PDEs) like (1), orthogonal projection methods have become a frequently used technique [12, 13, 14]. The basic idea is to represent the unknown variable, e.g. the temperature T , by an infinite sum of products of time dependent amplitude functions $T_i(t)$ and orthonormal, space dependent basis functions $\varphi_i(y,z)$:

$$T(t,y,z) = \sum_{i=1}^{\infty} T_i(t)\varphi_i(y,z), \tag{2}$$

A model reduction is achieved by approximating the infinite sum by a finite series

$$\tilde{T}(t,y,z) := \sum_{i=1}^{N^T} \tilde{T}_i(t)\varphi_i(y,z). \tag{3}$$

Clearly, the approximation (3) will not solve the PDE (1) exactly, but a nonzero residual will remain. In order to get conditions for the time functions $\tilde{T}_i(t)$, the projections of the residual Res in (1) onto the basis functions are required to vanish, i.e.

$$\int_y^{L_y}\int_z^{L_z} Res\,\varphi_i \stackrel{!}{=} 0, \qquad i = 1,\ldots,N^T. \tag{4}$$

This approach reduces the PDE (1) to N^T ordinary differential equations. For the model reduction of the MCFC model, not only the profile of the temperature, but also those of the molar fractions in the anode and in the cathode gas channels, as well as the profiles of the total molar flow rates have to be approximated by basis functions. The resulting reduced model is a low-order differential algebraic system of differential index one. The quality of the reduced model, i.e. its deviation from the original model, mainly depends on two factors. The first one is the number of terms considered in the approximations of the spatial profiles. The second is the choice of the basis functions. A good approximation of the complete model by a low order reduced model is achievable, if suitable problem-specific basis functions are chosen. One possibility would be to linearize the original spatially distributed system and to use the eigenfunctions of the linearized problem as basis functions [12]. In this contribution, another approach, the Karhunen-Loève decomposition method is applied, because by that method it is also possible to incorporate the nonlinear behavior of the system in the basis functions.

The Karhunen-Loève decomposition (K-L decomposition) was originally developed for the description of stochastic data[15]. By applying the K-L decomposition to a given stochastic field, eigenfunction are obtained that can reproduce the data with a certain accuracy and with a minimum number of degrees of freedom. For the solution of partial differential equation, the K-L decomposition method can be used to generate basis function for the Galerkin procedure[16]. The key issue in this approach is to determine suitable basis functions from the simulation results with the original model, taken at discrete time points, so-called snapshots. By the K-L decomposition, it is possible to extract the most typical or characteristic structure from these snapshots in the form of empirical eigenfunctions. In order to determine suitable basis functions for the MCFC model, the response of the complete model to a increase of the cell current and to a subsequent decrease to the original value is computed numerically by using the method of lines. The basis functions for the reduced model are computed from the transient solution of the complete model. For the temperature profile, between 1 and 5 basis functions are chosen. For the other variables of the reduced model, two basis functions for each gas are found to be sufficient.

4 Validation of the reduced model

Test simulations are made in order to validate the reduced model by comparison with the original model. Special emphasis is laid on the extrapolation qualities of the reduced model. The first test consists in the simulation of a randomly varying cell current. The results are shown in Fig.2. It is found that already the approximation of the temperature profile by a single basis function leads to a quite satisfactory behavior of the reduced model. The temperature error becomes very small, if 5 basis functions are used for the temperature. In all simulations shown in Fig. 2, the cell voltage of the reduced model matches the result of the complete model nearly perfectly. A second test is the response of the reduced model to a change of the steam-to-carbon ratio in the feed. This test is more challenging than the first one, because the basis functions for the reduced model were obtained from simulations with a fixed steam-to-carbon ratio. Nevertheless, the agreement between the reduced and the complete model is still very good, as can be seen in Fig. 3. The extrapolation qualities of the reduced model could be confirmed in further tests, where the kinetic parameters of the anodic and the cathodic reaction were varied. An example is given in Fig. 4, which contains steady state results for different kinetic parameters of the anodic reaction. Although the parameters are varied over a large interval, the error of the reduced model hardly changes. The K-L decomposition technique leads to a considerable reduction in terms of the order of the system as well as in terms of the computation time. This is illustrated for the test simulation in Fig. 2. After a spatial discretization, the complete model consists of about 12,000 equations. Its numerical solution requires about 43,000s of a CPU time on a PC. In comparison, the reduced model consists of 25 equations, if 5 temperature basis functions are used. Its numerical solution takes about 380s of CPU time on the same PC. The decrease of the computational time achieved by the model reduction is not quite as strong as the decrease of the order of the system. The reason is that the evaluation of the reduced model equations is more complicated as it requires a numerical quadrature.

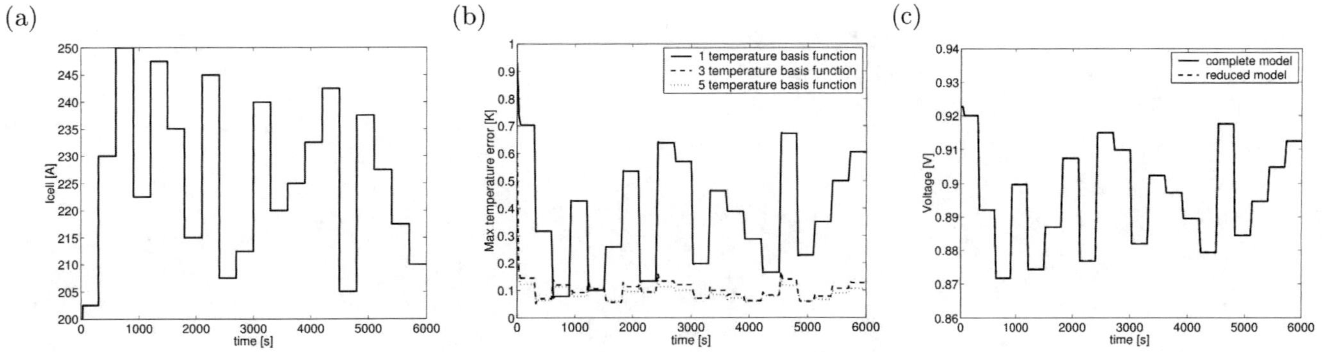

Figure 2: Validation of the reduced model by a test simulation with a randomly varying cell current; (a) cell current used as input signal; (b) maximum temperature error of the reduced model; (c) cell voltage of the reduced model and the reference model

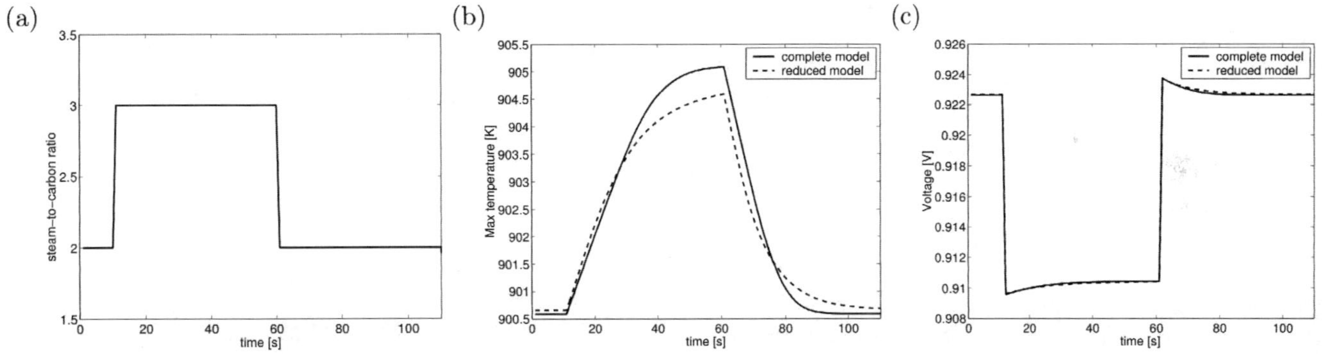

Figure 3: Validation of the reduced model by a test simulation with a randomly varying steam-to-carbon ratio; (a) ratio of the water and the methane concentration in the feed; (b) maximum temperature of the reduced model and of the reference model; (c) cell voltage of the reduced model and the reference model

5 Development of a state and parameter estimator based on the reduced model

In the following, the usefulness of the reduced model for process control tasks is demonstrated by applying it to a state and parameter estimation problem. Usually, the measuring information available from a fuel cell is rather limited. In most cases, temperature measurements can be taken only at a few points in the cell, and there is no information on the composition of the gases in the cell. A model based measuring system can provide additional

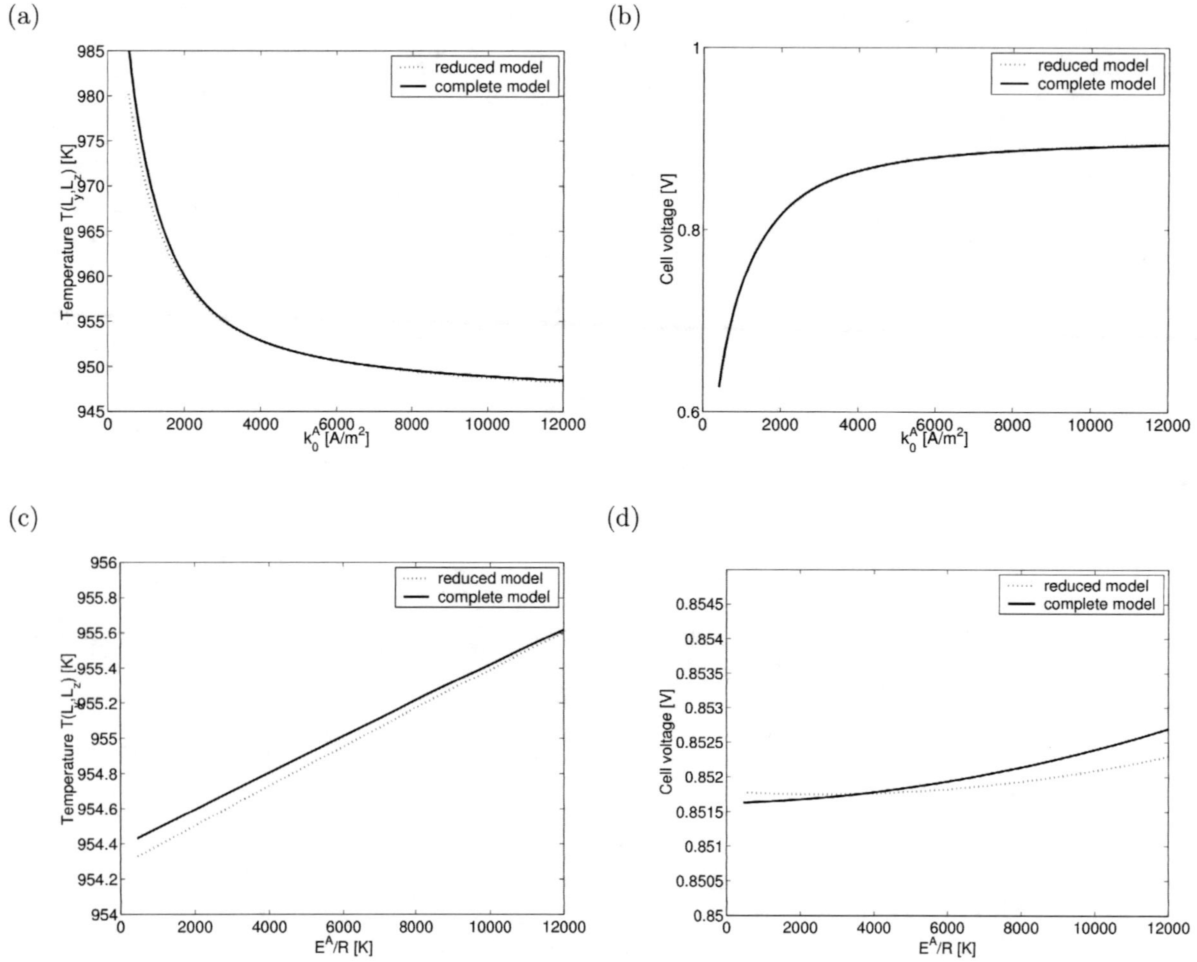

Figure 4: Validation of the reduced model by comparison of steady state results for different values of the anodic kinetic parameters; (a),(b) variation of the pre-exponential factor k_0^A of the anodic reaction; (c), (d) variation of the activation energy E^A of the anodic reaction

information on temperatures and concentrations that cannot be measured directly. In this work, the approach of a Luenberger state observer is used to design the model based measuring system. The reduced MCFC model serves as the simulator part in the observer. The observer gain matrix is obtained by solving an algebraic Riccati equation. A traditional Luenberger observer only corrects the states of a system, but does not correct wrong values of the model parameters. However, in the case of a fuel cell, the correction of model parameters, especially kinetic parameters, is necessary because of aging and degradation effects in the cell. Over the life cycle of a fuel cell, the activity of the catalysts decreases. Therefore, the values of kinetic parameters determined for a new cell may no longer be valid after some time of operation. In order to compensate for this effect in the observer, dummy equations for the kinetic parameters are added to the reduced model. The resulting observer is tested in simulations. Simulation data of the reduced model are used as measuring information for the observer. It is assumed that temperature measurements can be taken at three points in the cell, and that the cell voltage can be measured as well. In the simulation shown in Fig. 5, the observer model initially contains a wrong value for the pre-exponential factor of the anodic reaction k_0^A. Consequently, the steady state solution of the reduced model deviates strongly from the correct solution. This wrong steady state solution is taken as the initial condition of the observer model. At the beginning of the simulation, the observer mainly corrects the wrong temperature profile. Therefore, the temperature error decreases rapidly, whereas initially the error in the cell voltage is hardly reduced. In the second part of the simulation, the observer correction also affects the wrong parameter k_0A. This leads to an improvement of the estimate and to a very good match of observer state and reference states at the end of the simulation.

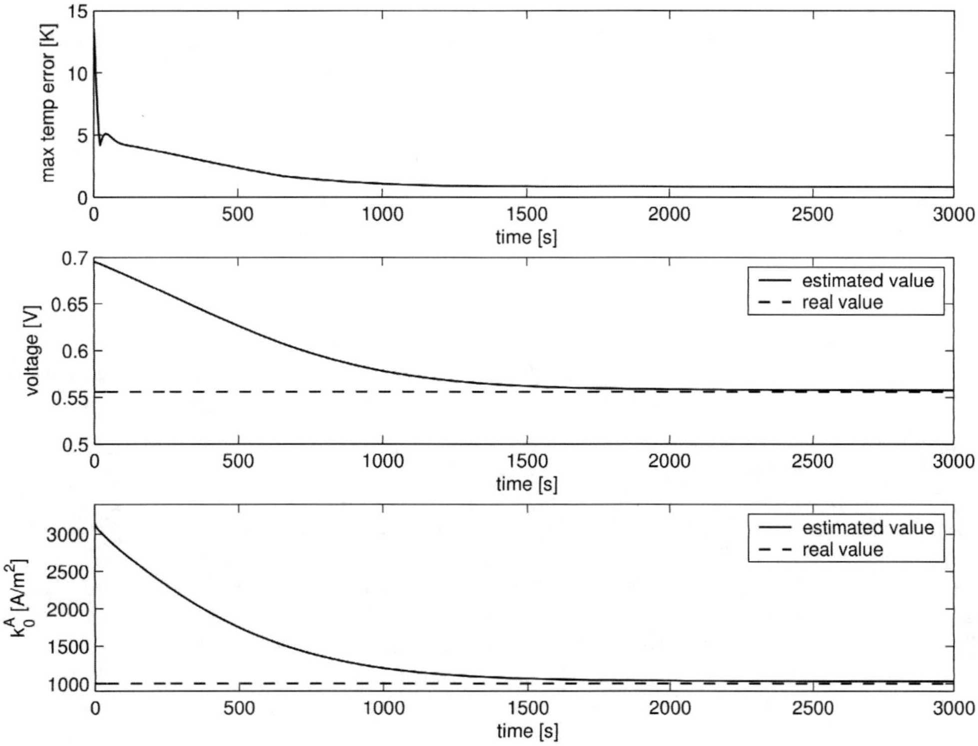

Figure 5: Simulation results of the state and parameter observer based on the reduced model; top diagram: maximum temperature difference between observer and reference system; middle diagram: estimated and reference cell voltage; lower diagram: estimated and reference kinetic parameter

6 Conclusions

A reduced model of an MCFC is obtained by applying the Karhunen-Loève Galerkin method to a two-dimensional spatially distributed cross-flow model of the cell. The basic idea of the method is to approximate the profiles of the spatially distributed variables by basis functions obtained from test simulations with a detailed reference model. For the MCFC model considered here, this technique proves to be successful. The reduced model produces results that are very close to those of the original model, but it reduces the computation time by a factor of more than 100. It may be regarded as a certain drawback of the K-L decomposition method, that the resulting eigenfunctions depend on the choice of the numerical test simulations or snapshots. However, the results of the previous section indicate satisfactory extrapolation qualities of the reduced model of the MCFC. That may be explained by the fact, that only the basis functions of the reduced model are obtained numerically. The reduced model itself is given in an analytical form and still contains the information on the physical correlations in the original model. Due to its properties, the reduced model is suitable for applications in the field of model based process control. An example is the state and parameter estimator developed here. The estimator is able to reconstruct the states of the cell, i.e. temperature, concentrations, and voltage, but also uncertain model parameters like kinetic coefficients with a good accuracy.

Acknowledgement

This work is supported by the German Federal Ministry of Education and Research under contract No. 03C0345B.

References

[1] L. Carrette, K.A. Friedrich, U. Stimming, *Fuel Cells* **2001**, 1, 5-32.

[2] J. Koh, B. Kang, H. Lim, *AIChE Journal* **2001**, 47, 1941-1956.

[3] W. He, Q. Chen, *Journal of Power Sources* **1998**, 73, 182-192.

[4] P. Heidebrecht, K. Sundmacher, *Chemical Engineering Science* **2003**, 58, 1029-1036.

[5] P. Heidebrecht, K. Sundmacher, *Fuel Cells* **2003**, 2, 166-180.

[6] M.D. Lukas, K.Y. Lee, H. Ghezel-Ayagh, *Control Engineering Practice* **2002**, 10, 197-206.

[7] B. Kang, J. Koh, H. Lim, *Journal of Power Sources* **2001**, 94, 51-62.

[8] C. Shen, G. Cao, X. Zhu, X. Sun, *Journal of Process Control* **2002**, 12, 831-839.

[9] X. Sun, G. Cao, X. Zhu, *J. Intell. Robot. Syst.* **2001**, 31, 299-316.

[10] K. Hou, R. Hughes, *Chemical Engineering Journal* **2001**, 82, 311-328.

[11] M. Mangold, M. Sheng, *Fuel Cells* **2003**, submitted.

[12] K. Hoo, D. Zheng, *Chem. Engng. Sci.* **2001**, 56, 6683-6710.

[13] J. Baker, P. Christofides, *Int. J. Control* **2000**, 73, 439-456.

[14] J. A. Atwell, B.B. King, *Math. Comp. Modelling* **2001**, 33, 1-19.

[15] Loève, M., *Probability Theory.* Van Nostrand, Princeton, NJ, **1955**.

[16] H.M. Park, D.H. Cho, *Chemical Engineering Science*, **1996**, 51, 81-98

169e

ON-BOARD HYDROGEN PRODUCTION FOR FUEL CELL VEHICLES VIA A MEMBRANE SEPARATOR AND AN EXTERNALLY FIRED METHANOL REFORMER.

By

H.W. Parker, T.T. Maxwell, R. Aksoy, J.J. Halmari, S.P. Mathakari, and R. Wang,

Departments of Chemical and Mechanical Engineering

Texas Tech University

Lubbock TX 79409

Prepared for presentation at the Fuel Processing Session I,

National AIChE Meeting, San Francisco, Nov 16-21, 2003

Abstract

Many options exist for on-board hydrogen production from liquid fuels for fuel cell powered vehicles. This paper reports on design and construction of an on-board system for hydrogen production from methanol. Methanol reforming is accomplished using an externally fired catalytic reactor. Carbon monoxide, separated from the reactor effluent by a membrane separator, is the reactor fuel. The reactor is operated at 1950 kPa to supply gas to the membrane separator at the maximum design pressure, and at 500 C, to minimize the amount of methanol remaining the reactor product. The methanol feed rate will be 0.32 l/min, sufficient to supply a 10 kW fuel cell package, but it can easily be expanded to larger sizes.

The membrane separator is an off-the-shelf, polymer-based model, and it is not expected to reduce the carbon monoxide to a sufficiently low level, 50 ppm, to be acceptable by proton exchange membrane fuel cells, usually considered for vehicular power, even with two stages of membrane separation. For this reason it was necessary to include a selective oxidation reactor to remove the carbon monoxide contaminants.

An adiabatic energy balance indicates about 30% excess energy is available from burning the separated carbon monoxide as fuel to the reformer. Much of this energy will be lost to the atmosphere from the rather small equipment. The opportunity exists to add small amounts of water to the methanol fuel to utilize the water gas shift reaction to produce additional hydrogen, if excess energy were available from the separated carbon monoxide.

Introduction

Fuel cell vehicles are popularly regarded as the most likely alternative fuel technology to replace internal combustion engine vehicles in coming decades. The low emissions and high energy efficiency associated with fuel cells suggest that fuel cells can have a strong impact on the transportation industry. There are serious space and safety considerations facing fuel cell vehicles that use on-board hydrogen storage. Although high performance storage vessels have been developed, concern for vehicle safety remains strong. This difficulty of on-board high-pressure hydrogen is part of the motivation behind the development of an on-board methanol reforming system.

Process Description

A flow diagram for the process is shown in Figure 1. Pure liquid methanol (0.32 l/min) is fed to methanol vaporizer through a chemical feed reciprocating pump at 27 C and 1950 kPa. The reforming reaction takes place at a relatively high temperature, 500 C, to minimize the equilibrium methanol concentration, as follows:

$$\mathbf{CH_3OH \quad \rightarrow \quad CO + 2H_2 \quad} - Q_1$$

The mixture is passed through a cooler where it is cooled down to 60 C. The product is then passed through a membrane separator where it splits into two streams viz. non-permeate and permeate. The permeate consists of hydrogen rich stream with around 8 mole % of CO. The non-permeate (1950 kPa) consists of CO rich stream with about 25 mole % of Hydrogen.

The reformer system is a unit which consists of a furnace shell for burning, with air, the carbon monoxide rich stream to vaporize the methanol and then reform it to carbon monoxide and hydrogen. The catalyst to be used here is alumina pellets. The tubing is arranged in a cross flow pattern so as to utilize the heat liberated by the exothermic combustion reaction in the shell housing.

$$\mathbf{CO + \tfrac{1}{2}\, O_2 \rightarrow CO_2 \quad + \mathit{Q_2}}$$

$$\mathbf{H_2 + \tfrac{1}{2}\, O_2 \rightarrow H_2O \quad + \mathit{Q_3}}$$

An adiabatic energy balance indicates about 30% excess energy is available from burning the separated carbon monoxide as fuel to the reformer.

The permeate stream is sent to a preferential oxidation reactor where the CO is selectively oxidized to CO_2. The PROX reactor is essentially a hydrogen purifier as the CO is detriment to the fuel cell platinum catalyst. Only about 10 ppm of CO is permissible for the 'safe' operation of fuel cell. The suitable catalyst for this reaction is Ru/alumina which has a high selectivity for CO oxidation.[1] **(Reference)** The O2/CO ratio is also a major design factor for the PROX reactor as any excess oxygen available for hydrogen oxidation will result in reducing the fuel cell efficiency.[2] **(Reference)** The reaction takes place in the PROX reactor at 150 C and atmospheric pressure as follows:

$$\mathbf{CO + \tfrac{1}{2}\, O_2 \rightarrow CO_2}$$

After the PROX reactor and a subsequent cooler which cools the product down to 80C, the hydrogen is fed to the Proton Exchange Membrane type Fuel Cell. The carbon dioxide, nitrogen and traces of oxygen are harmless to the Fuel Cell.

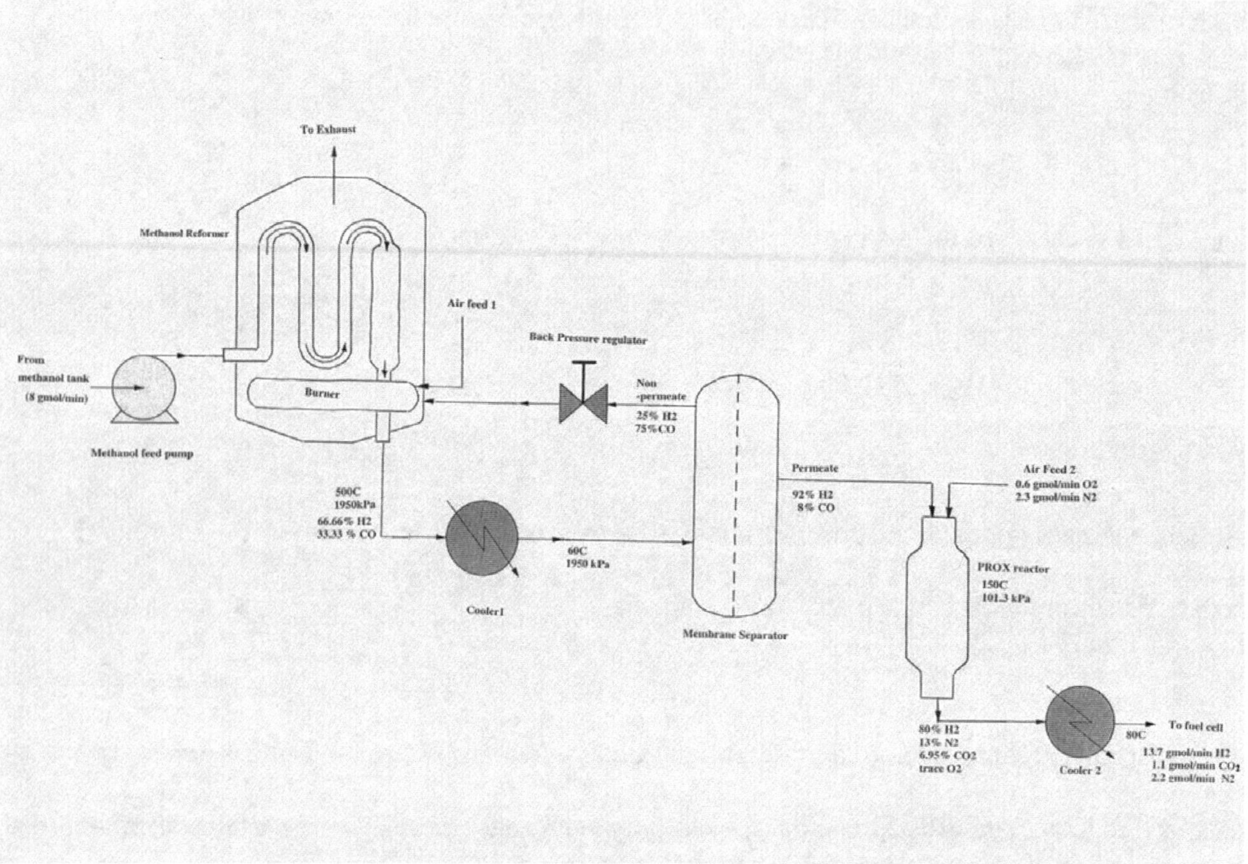

Flow sheet for On-board Hydrogen Generation System (Figure 1)

The key features to our design are the Membrane Separator and the externally fired Methanol Reformer.

Membrane Separator:

The membrane uses asymmetric hollow fiber membrane technology to separate gas components based on the principle of selective permeation.

Each gas has a characteristic permeation rate, which is a function of its ability to dissolve and diffuse through a membrane. Hydrogen, helium, and water vapor are "fast" gases and selectively diffused through the membrane wall while most of the “slow” gases such as carbon monoxide, nitrogen, methane, and heavier hydrocarbons remain behind. Some gases such as carbon dioxide and oxygen permeate at an intermediate rate between the fast and slow gases listed above.[3]

The driving force for the separation is the difference between the partial pressure of hydrogen gas on the inside of the hollow fiber and that on the outside.

In the PRISM membrane separator which we are using, feed gas will be introduced into the end port and flows down the inside of the hollow fibers. "Fast" gases, some of the intermediate gases, and a small amount of "slow" gases, will pass through the membrane wall to the outside of the fibers to be collected at low pressure as the permeate stream and exit out of a side port. Most of the "slow" gases and a small amount of the "fast" gases will continue to travel through the fiber until they reach the other end of the membrane separator where they shall exit the end port as the non-permeate.

The feed gas in this application may contain methanol, which in high concentrations can damage epoxy components of the membrane bundle. The membrane separator should not be exposed to feed gas with more than 20% relative saturation of methanol.[4] Special precautions should be taken for cold start-ups and during shutdowns to prevent this limit from being exceeded within the membrane separator.

Membrane separator permeates more gas when operated at a higher pressure and temperature, and thus a higher rate of hydrogen recovery. The membrane separator should be operated at a temperature that is, at a minimum, 6°C greater than the dew point of the feed gas or higher when certain contaminants are present. This is necessary to ensure that water vapor, oil vapor, and other contaminants in the feed gas will not condense on the membrane.[5]

At higher purity of fast gas in the feed, the membrane will subsequently permeate more gas thus giving a higher purity. The pressure drop from feed to non-permeate varies with operating pressure, operating temperature, feed rate, and membrane separator length. Pressure and temperature affect the volumetric flow through the membrane separator, and the greater the flow through the membrane separator, the greater is the pressure drop. The pressure drop on the non-permeate side should be typically 10 psig or less.

The proposed design of the externally fired methanol reformer

The planned methanol reformer will be a vertical 20 mm stainless steel preheater, 2 m long with liquid methanol entering the top. The bottom end of the tube will attach to a 25 mm reactor section filled with alumina catalyst support. This reformer will be enclosed by a 120 mm diameter tube lined with high temperature fiber insulation. A burner and air blower will supply very hot combustion gases to heat the reformer.

The methanol reformer will consist of the outer shell with a burner where the combustion of CO will take place, the tubing in which the preheating and vaporization of methanol is to occur followed by the reactor packed with alumina catalyst where the actual endothermic reforming reaction will take place. The flue gases and the methanol will flow in cross flow pattern for better heat exchange. The tubing will consist of the preheater vaporizer followed by the reactor where methanol reforming will take place. Propane gas will be used for testing and startup operations, although on a vehicle carbon monoxide should be stored from previous operation of the reformer.

Present status

At the closing date for submitting "records of presentation" to AIChE, August 2003, most of the equipment items are on hand and detailed design of the reformer/furnace is in progress.

Conclusions

Design and construction of a methanol reformer presented in this paper has provided several students in mechanical and chemical engineering hands on experience regarding anticipated future transportation power sources. The anticipated high permeation rate of carbon monoxide through the membrane separator is a major constraint on the proposed system.

Acknowledgments

The research on which this paper is based was funded by the State of Texas through the Advanced Technology Program. The authors also appreciate the support and facilities made available by the Advanced Vehicle Engineering Laboratory at Texas Tech University.

References

1. M.J. Kahlich, H.A. Gasteiger, R.J. Behm. *J. Catal. 182 (1999) 430-440*
2. 'Selective Catalytic Oxidation of CO in H2 for fuel Cell Applications', O. Korotkikh, R. Farrauto, *Engelhard Corporation.*
3. Installation & Operating Instructions, *Air Products and Chemicals, Inc.*
4. Installation & Operating Instructions, *Air Products and Chemicals, Inc.*
5. Installation & Operating Instructions, *Air Products and Chemicals, Inc.*

Development and Application of System Level Dynamic Models for Stationary Fuel Cell Power Plants

Mithun Kamat, UTC Fuel Cells, LLC (UTCFC)*
Jonas Eborn, United Technologies Research Center (UTRC)

Prepared for Presentation at AICHE Annual Fall Meeting, Nov 16-21 2003.
TI003: Fuel Processing Session I: Modeling and System Integration

* Corresponding author: Email: Mithun.Kamat@utcfuelcells.com

Abstract

Fuel Cell based power plants are truly heterogeneous systems involving mechanical, chemical, thermal and electrical systems. The power plant sub-systems include the Fuel Processing System (FPS) that converts readily available hydrocarbons to hydrogen, the Cell Stack Assembly (CSA) that utilizes air (oxygen) and hydrogen to produce electricity, the Thermal Management System (TMS), and the Power Conditioning System (PCS). The constant interactions between the sub-systems typically result in complex dynamic behavior during normal 'on-load' operations as well as the start-stop cycles. As such, highly integrated systems design and control paradigms are required to ensure safe and reliable operation during the entire cycle.

In a joint effort between UTC Fuel Cells, UTCFC, and United Technologies Research Center, UTRC, system level dynamic models are developed and deployed for the design and analysis of PEM based fuel cell power systems [Eborn, et.al, 2003]. The process models include all major process subsystems, FPS, CSA, TMS, and the fuel and air delivery systems. The models are lumped but physics-based and capture the dynamic behavior of the respective components. The controls models include the continuous control algorithms as well as the supervisory sequences.

This article provides an overview of the modeling effort undertaken by a team of UTRC and UTCFC engineers over a period of two years. The system level modeling team has made demonstrated gains toward the development of model based fuel cell system design capability in stationary applications. Two notable applications within the context of fuel cell program life cycle, systems dynamic analysis, and the conceptualization and verification of controls architecture are discussed and some simulation results are presented. Further, we discuss the lessons learned in the development of reusable model libraries constructed in the open-code language Modelica, and in the integration of a system level dynamic model. The system level model typically is on the order of around 500 states and 20000 variables. Maintaining the balance between the level of detail (model fidelity) and the execution speed is a very critical issue that is typical of the scale of the problem.

1. Introduction

The design of fuel cell systems poses a major challenge in systems engineering. There are several strongly interacting subsystems in a fuel cell system that includes a broad range of engineering disciplines. Figure 1 shows a typical schematic of fuel cell based power system.

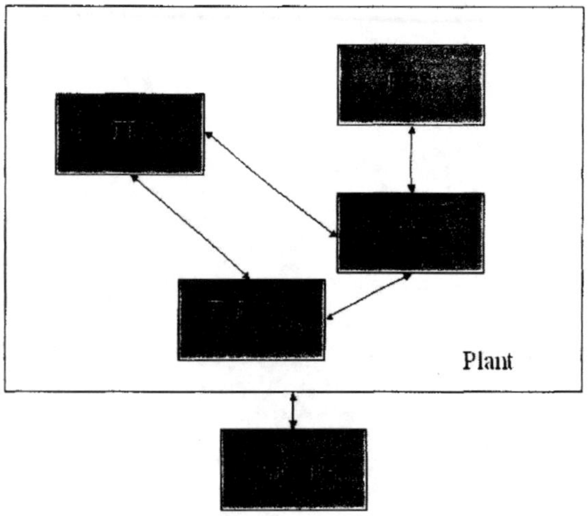

Figure 1: Schematic of a typical fuel cell system

The Proton Exchange Membrane (PEM) fuel cell stack (CSA) electrochemically convert oxygen and hydrogen into electrical power with water as a by-product. The direct current from the CSA is inverted into alternating current for the power grid through a Power Conditioning System (PCS). The water produced in the fuel cell is distributed through a Thermal Management System (TMS) that manages the heat integration of the power plant and is optimized to increase overall system efficiency. The oxygen for the fuel cell is normally supplied through an air delivery system. However, unlike air that is readily available, hydrogen is processed onboard through a Fuel Processing System (FPS) due to the lack of current energy distribution infrastructure. The FPS converts readily available hydrocarbons (such as natural gas) into hydrogen. We avoid here a detailed discussion of the specific components in the FPS. Put in a nutshell, FPS consists of a train of reactors and heat exchangers that manage the exit hydrogen quality and heat integration. Finally, the power plant includes a digital control system (Controller) that enables the power plant to undergo optimal start-stop procedures and transient load operations while maintaining overall system objectives of switchability, controllability and flexibility.

2. System Level Dynamic Modeling of Fuel Cell power systems

As discussed in the previous section, fuel cell systems are truly heterogeneous systems with strong interactions between mechanical, chemical, thermal and electrical components and subsystems. The fundamental goal of the UTRC and UTCFC joint effort of System Level Dynamic Modeling (SLDM) was to establish model-based design capability enabling rapid design and prototyping of fuel cell power plants. Consider the V-diagram of a product specification, design and deployment process in Figure 2 [Sage, 1996]. We can broadly classify the development process into four phases, concept design, preliminary design, detailed design, and validation and deployment. System level models find applications in each of these phases.

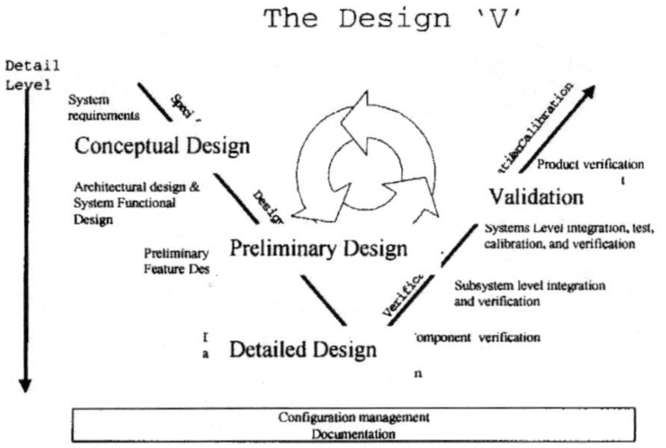

Figure 2: V-Diagram illustrating product specification, design and deployment.

During concept design, steady state models can assist the evaluation of different system configurations that meet the overall requirements. The steady state design provides a range of feasible operating points with flow, pressure drops and other design data necessary for the dynamic models. During the preliminary and detailed design phase, the SLDM models assist in evaluating dynamic specifications as well as system transient performance. The models are also used to verify the controls architecture. During the validation and deployment phase, test data are made available to the dynamic models. The tasks thereon are two-staged. The dynamic models are first validated using component test data, and system level simulations are then exercised to validate the final design. In some cases, validated system models may be used to debug test stand issues. The ultimate goal of the SLDM program is to support all the phases of product development including conceptual system design and integration to final controlled system design enabling rapid prototype design. Clearly, the modeling objectives are different during the different phases of product development. Our aim is to develop a library of dynamic models that enable rapid reconstruction and reconfiguration of system level models.

3. Dynamic model

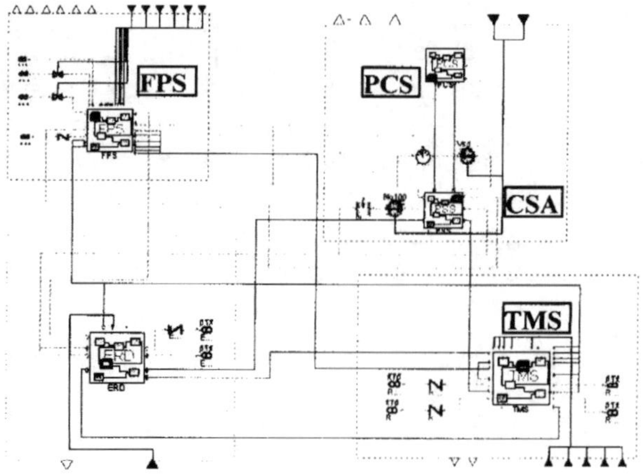

Figure 3: Fuel cell system level model in Dymola, including FPS, CSA, PCS and TMS.

Figure 3 shows a snapshot of the system level dynamic model built for fuel cell program at UTCFC for stationary applications. As seen in the figure, the system model contains all the major subsystems such as FPS, CSA, TMS, and PCS. All the unit operations and balance of plant component models are library elements constructed in the modeling language, Modelica [Modelica website]. The interconnections between different subsystems are physical streams, which contain information such as flow, composition, temperature, density, etc. On the outside of the system are triangular input and output connections, carrying signals to and from the control system. The thermodynamic properties are calculated based upon ideal gas laws, which is an adequate assumption for the operating range of the application. These properties are calculated using an external library called ThermoFluid [Tummescheit, 2002], which is a free library available in public domain. It is important to note that the system model is constructed from components using iconic "drag & drop" approach. However, simulation of such highly heterogeneous system with more than 500 states and several thousand equations is a non-trivial numerical issue for the modeling tool. We discuss this aspect in more detail in Section 4. Please note that the control system (continuous control loops and sequential logic) is implemented in Matlab (Simulink/Stateflow). A closed loop model is built in simulink with the plant model being interfaced to Simulink via S-function block. In the following section, we briefly discuss the key applications of system model in fuel cell program life cycle at UTCFC.

3.1. Dynamic analysis

Dynamic models find their primary application in the system analysis of fuel cell power plants. Understanding the dynamic behavior of the power plant is key to successful design of the prototype. Dynamic requirements for fuel cell systems are written taking into account several key factors, including safety, reliability, and flexibility. System level dynamic analysis may reveal components limiting transient performance, non-minimum phase characteristics, and dominant dynamics in the system, and key interactions between systems. Load following capability is one area of dynamic performance evaluation of the design. Dynamic models are also used to predict the emissions level during load transients. These analyses typically result in requirements documents that may affect design (e.g. hardware requirements), and/or controls architecture.

3.2. Controls Verification

The control system is designed with several objectives, mainly rapid load tracking, sustained performance and robustness at all operating points, while ensuring safe operation. Fuel Cell power plant control system is a challenging problem simply due to the vastly different time scales of the different subsystems. Furthermore, stringent plant wide control objectives result due to highly integrated system design. Depending upon the development stage, dynamic models can provide both qualitative as well as quantitative understanding of the control system behavior, thus ensuring rapid design as well as risk reduction. During the conceptual and preliminary design phase, the physics based dynamic models can provide a good qualitative understanding of the system behavior and can be used to verify the performance of the continuous as well as sequential control structure. Figure 4 shows a typical simulation of fuel cell power plant. The system model is exercised from start-up conditions, to on-load operation where the system undergoes and up & down transient, followed by a normal shutdown. The plot shows the stack current, process airflow and fuel flow at different controller states. The controller states are sequential logic that step up the power plant at different conditions such as start-up, shutdown. During on-load operation (state 150), the continuous controllers are in automatic mode. Such simulations typically may highlight any abnormalities in the system flows, pressures or sequencing logic. Please note that the stack current and process flow values are not shown in order to not reveal any proprietary data. Furthermore, the control system performance may be evaluated to obtain specifications on measurement uncertainty by including randomly generated error (or predetermined offset) in the sensor measurements in the model.

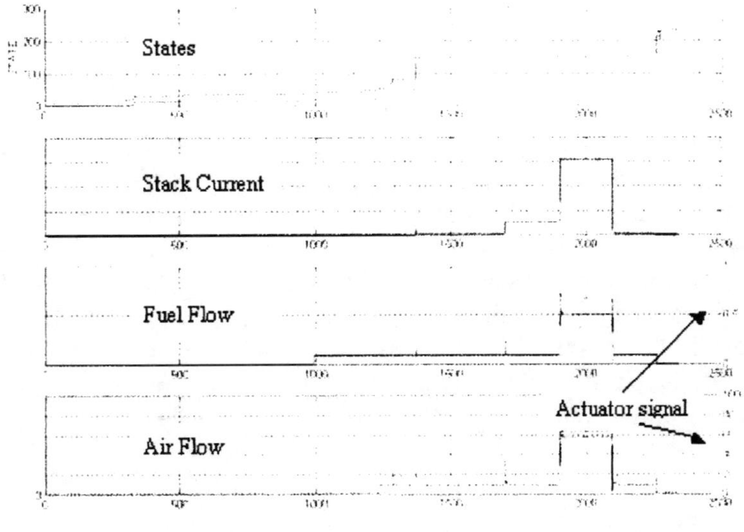

Figure 4: Sample results of an entire load cycle simulation (includes start-up, normal on-load operation, and shutdown).

During the final design phase, measurement and hardware data is usually available. These data can be used to validate the dynamic models, which can then be used in closed loop simulations to determine controller parameters. Hardware troubleshooting is another area where dynamic models can help, for instance, evaluating the performance with different signal to noise ratios.

Our current efforts are also focused on model based controls design, wherein linear plant models may be used in evaluating different advanced controls concepts such as multivariable controller. Further, linear models generated from dynamic models or reduced order models may be used in Hardware-in-the-loop (HIL) testing of the controller hardware and software. HIL testing is currently being implemented as standard work at UTCFC for new systems design.

4. Lessons learned

The SLDM program has highlighted the importance of model library management. Several team members work concurrently on different models, which puts requirements on both version control and quality management. Model quality has three important aspects, robustness, accuracy, and documentation. To ensure minimum turnback time, and smooth functioning, the model development process involves the following:

- Model version control, e.g. using a tool such as CVS, concurrent version system.
- Model documentation that included model equations as well as key references.
- Peer code reviews
- Pre-determined success criteria: the model should pass standard tests to ensure physical soundness and robustness. These issues are discussed in more detail below.
- Standard naming conventions for components, measurements, etc. that are consistent with the conventions used in P&I diagrams.
- Parameter management through library scripts.

Model robustness is an important issue that deserves special mention here. The initial versions of the model libraries suffered from problems with robustness, slow or non-convergence, solver crashes, etc. This was partly because the development effort involved newly developed tools in a new application. It was also however, due to the complexity of the system models. Models on the order of 700 states and 20000 equations, together with the nonlinear highly interacting flow network result in several initialization issues that are prominent in the simulations of startup or shutdown cycles, where the system operates in zero-flow or back flow conditions, and the temperatures and compositions are outside the nominal range of operation. Issues of especially common occurrence are chattering and singular Jacobians. While chattering is usually caused by flow reversals or phase switches, singular Jacobians are often the result of semi-empirical expressions for catalytic reaction rates and non-linear voltage characteristics of the fuel cell. In subsequent versions of the model library, these problems have been avoided through several different measures.

- Implementation of the code in a style such as to avoid singularities, e.g. by guarding division by zero. Although validity range of such expressions are still limited, robust operating range may be extended.

- Avoid unnecessary discrete events caused by 'if' statements, by using specific constructs such as Modelica keyword noEvent.
- Benchmark testing of the component models by rigorous Taguchi-style testing [Taguchi, & Yokoyama, 1993]. The model is exercised through a wide range of predetermined extreme operating conditions (that extend outside the expected operating envelope) to ensure model robustness. Any potential model problems are thus revealed early and enable rapid integration of system level models.
- Avoid unnecessary complexity in the models that may increase the model order. In other words, use the simplest possible models and add complexity only when fidelity requirements call for it. Although the issues are not observed in the component model testing, they are especially prevalent in large complex system models. Adhering to Occam's rule and keeping model order to a minimum avoids potential problems.

5. Conclusions

A Dynamic model library has been developed that supports the design and development of fuel cell programs at UTCFC. System level dynamic models are built using the library components, ensuring the right balance between the speed of execution and the level of detail. The system level dynamic models find applications spanning the entire development cycle, from conceptual system design and integration to final controlled system design. The current work is focusing on developing multiple fidelity model libraries while ensuring robustness over the entire operating envelope of the system. The ultimate goal is to have a suite of validated model libraries together with a simulation environment that enables rapid construction and reconfiguration of system level models. Ongoing work is also focused on the usage of dynamic models in model based controls design and analysis. An important aspect in this work is the ability of the models to interface with analysis software for control system design.

6. References

- A. P. Sage, et.al, "Handbook of Systems Engineering and Management".
- G. Taguchi, Y. Yokoyama, Taguchi Methods Design of Experiments. ASI Press, 1993.
- H. Tummescheit, "Design and Implementation of Object-Oriented Model Libraries Using Modelica", PhD thesis ISRN LUFTD@/TFRT--1063--SE, August 2002, Department of Automatic Control, Lund Institute of Technology, Lund, Sweden.
- J. Eborn, L. Pedersen, C. Haugstetter, and S. Ghosh, "System Level Dynamic Modeling of Fuel Cell Power Plants", In *Conference Proceedings of 2003 American Controls Conference*, ACC'03. June 2003, Denver, Colorado.
- Modelica Association: http://www.modelica.org

Acknowledgements

The SLDM activity has truly been a team effort of engineers from UTRC and UTCFC. The authors wish to acknowledge the contributions of the team members. We would especially like to thank Tim Crowley, Yiping Wang, Josh Isom, Chris McCready, and Keshava Kumar at UTCFC, and Subbarao Varigonda, Lars Pedersen, Chris Haugstetter, Thordur Runolfsson, and Shubhro Ghosh at UTRC.

H_2S AND CO REMOVAL FOR PEM H_2-O_2 FUEL CELLS USING ULTRA-HIGH CONTACTING EFFICIENCY MICROFIBROUS ENTRAPPED CATALYSTS AND SORBENTS

Bong-Kyu Chang, Yong Lu, and Bruce J. Tatarchuk

Center for Microfibrous Materials Manufacturing (CM^3)
Department of Chemical Engineering
Auburn University
Auburn, AL 36849-5127

Prepared for Presentation at the 2003 Annual Meeting, San Francisco, CA, Nov. 16-21, Fuel Processing Session II: Catalysis and Kinetics (Session #170)

November 2003

Unpublished

Abstract

An important processing step in the reformation of commercial and logistical fuels for PEM fuel cells is the removal of trace levels of H_2S and CO. The goal of this work is to develop an effective and compact "reactive fuel filter" operating at the stack temperature but capable of protecting the poison intolerant nature of the MEAs.

In this study, a new class of microfibrous media is described which permits entrapment of fine grains of supported ZnO sorbents and PROX-CO catalysts from 10 to 200 μm in diameter. The structure of the microfibrous carrier facilitates improved heterogeneous contacting while the small particle size reduces intraparticle transport limitations and increases external surface area. Since reactive particles are supported within a highly porous matrix at volume fractions well below the tap density, normal Ergun-type relationships involving pressure drop versus particle size for packed-beds are favorably altered. Proper design and materials selection permits concentrations of H_2S and CO to be reduced from 20,000 ppmv to less than 0.1 ppmv within layers a few millimeters in thickness at residence times at or below 150 ms at temperatures from 298 to 473 K.

To accomplish the above noted requirement, a two-layer design of media is prepared approximately 2 mm in thickness. The first layer consists of supported ZnO sorbents. The second layer consists of a supported PROX catalyst. Various catalysts have been screened for potential application including Pt, Au, Cu-containing catalysts. Investigations have been conducted from: 298 to 473 K, 0.1 to 20,000 ppmv H_2S/CO, and 0 to 30 % H_2O.

Integral and differential reactor data/modeling are described, as well as sorbent/catalyst characterization, and overall system performance.

1. Introduction

During the last several decades, there has been a significant increase in the research on fuel cells since it can offer a cleaner alternative for conventional internal combustion engines. Due to both kinetic and thermodynamic constrains of hydrocarbons (e.g.: methanol) reforming, however, significant concentration of CO are produced from the hydrogen producing catalytic reformers. And there is a limit, unfortunately, regarding the thermodynamic conversion of the water-gas shift reaction (WGSR), so that a CO concentration which is sufficiently low to be acceptable by the PEMFC system cannot be achieved using WGSR only. The reformate coming from the fuel processing section contains 50~65 vol% H_2, 25~40 vol% CO_2, 2~5 vol% H_2O, and 1 vol% CO. The fuel cell performance significantly deteriorates if the feed stream contains even trace amount of CO (more than 20 ppm) since the Pt or Pt-Ru anodes are poisoned by CO, which strongly chemisorbs on the active sites, thus blocking the sites where the dissociation/oxidation of H_2 takes place. Among the many approaches to reduce CO concentration in the reformed gas mixture, selective catalytic oxidation of CO to CO_2 has been found to be the most effective way to remove the trace amount of CO from hydrogen.

In our previous study[1], over 150 catalysts were evaluated for selective oxidation of CO in model reformate in the temperature range of from room temperature to 250°C. Based on the amount of catalyst used, Pt/γ-Al_2O_3, promoted Pt-M/γ-Al_2O_3, Au/α-Fe_2O_3, and CuO-CeO_2, showed complete CO conversion activity in model reformate in integral reactor conditions. In the present study, catalytic activities of the four catalysts are investigated in differential reactor conditions and reveal that the promoted Pt-M/γ-Al_2O_3 catalyst is a superior candidate compared to other catalysts. The promoted Pt-M/γ-Al_2O_3 catalyst is, then, examined in the presence of excess water for practical PEM fuel cell applications and the result indicates that the promoted Pt-M/γ-Al_2O_3 catalyst retains its highest activity even in the presence of 30vol% water in the feed gas.

1.1. Microfiber-based composite materials using wet lay paper-making/sintering process

Microstructured materials have potential for enhanced mass and heat transfer compared to typical catalyst particulates used in industrial processes. A new-patented class of composites materials made by a wet lay paper- making/sintering process can incorporate particles as small as 10 μm into a micrometal fiber matrix (Cahela and Tatarchuk, 2001).

Enhanced heat transfer due to the larger surface to volume ratio of small particles and the micrometal matrix results in more nearly isothermal conditions in fixed bed reactors minimizing hot spots, resulting in higher selectivity, and allowing processing to be done with smaller safety factors (Cahela and Tatarchuk, 2001).

The wet lay paper-making/sintering process allows continuous adjustment of void volume from upwards of 98% down to values similar to packed beds of particles (Marrion, Cahela, Ahn, Tatarchuk, 1994). This allows the composite catalyst/adsorbent material to be optimized to the appropriate void volume for different types of applications. For example of these microfibrous composite carriers, SEM images of 150-250 μm γ-Al_2O_3 support particles entrapped in a mirometal fiber matrix of 4 and 8 μm nickel fibers are shown in figure 2. An SEM image of a sintered mesh of 4 μm nickel fibers is shown in figure 1.

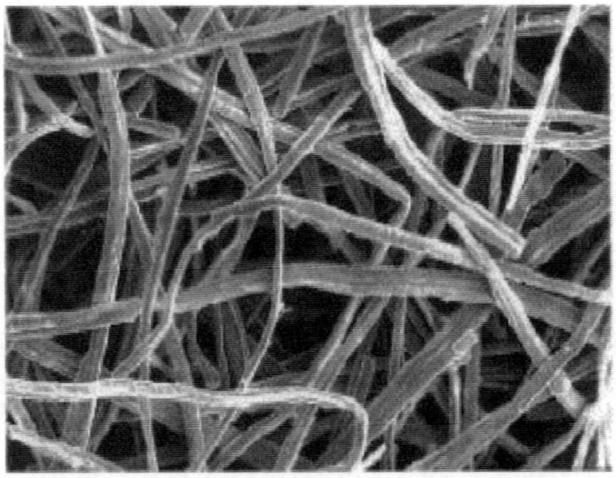

Figure 1. SEM of mesh of 4 μm nickel fibers (Cahela and Tatarchuk, 2001)

Figure 2. SEM images of 150-250 μm γ-Al_2O_3 support particles entrapped in a micrometal fiber matrix of 4 and 8 μm nickel fibers

2. Experimental

Catalysts are prepared by conventional co-precipitation or incipient-wetness method followed by drying and calcination processes.

Differential reactor experiments are carried out using a 4mm I.D. quartz tube reactor. 0.01g (10mg) of catalysts are diluted with 0.1g of inert support material (α-Al_2O_3) to make the reactor isothermal. The particle size is 150-250 μm (60-100 mesh). Dry feed gas consists of 1% CO, 1% O_2, 20.6% CO_2, 40% H_2, and balance N_2 and feed flow rate is controlled by gas mass flow controller. The total feed flow rate is changed from 50 to 200 (ml/min) to produce the differential reactor conditions. The temperature range investigated in the differential reactor study is 100 to 180°C.

The reaction products were analyzed with gas chromatograph (TCD-GC);

* Conversion of CO to CO_2 (%) = $(C_{CO}^{in} - C_{CO}^{out}) / C_{CO}^{in} \times 100\%$

* Selectivity for CO oxidation over H_2 oxidation (%)

= O_2 consumption by CO oxidation / Total O_2 consumption × 100%

= $\frac{1}{2}(C_{CO}^{in} - C_{CO}^{out}) / (C_{O2}^{in} - C_{O2}^{out}) \times 100\%$

3. Results and Discussion

(1) Integral reactor

In our previous study[1], over 150 catalysts were evaluated for selective oxidation of CO in H_2-rich gas in the temperature range of from room temperature to 250 °C.

Among the catalysts investigated Pt/γ-Al_2O_3, promoted Pt-M/γ-Al_2O_3, Au/α-Fe_2O_3, and CuO-CeO_2, showed extremely high CO conversion activity in model reformate in integral reactor conditions based on the amount of catalyst loaded. Figure 3 shows the results of the integral reactor study.

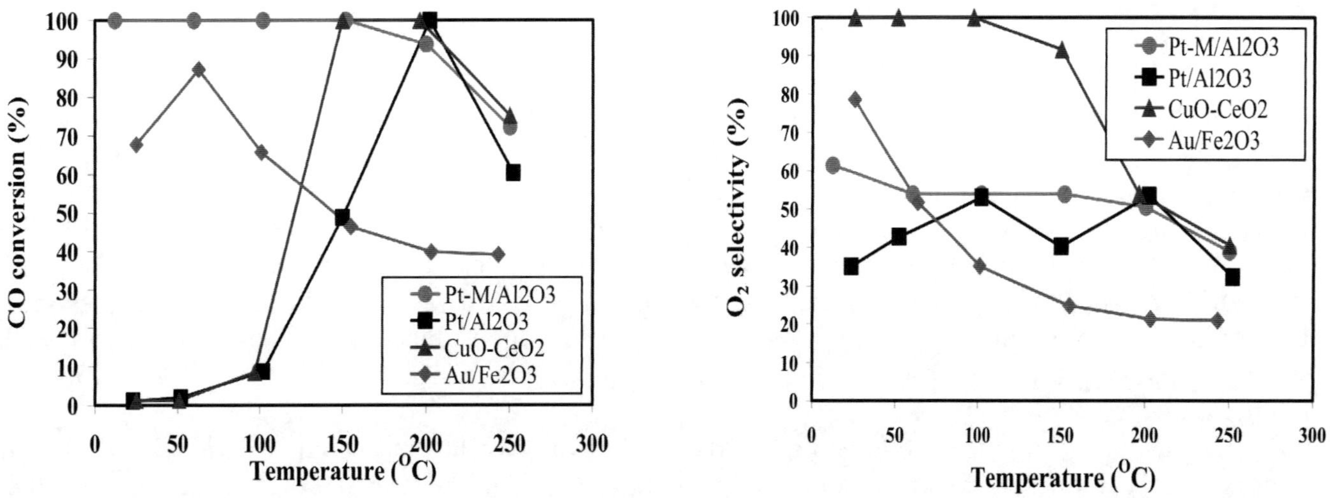

Figure 3. Comparison of the performance of Pt-M/Al_2O_3 (1.1wt%Pt), Pt/Al_2O_3 (1.1wt%Pt), Au/α-Fe_2O_3 (5at%Au), and CuO-CeO_2 (5.7% Cu, Cu:Ce = 0.143:1) in the presence of CO_2 and H_2O. *(1.1% CO, 1.1% O_2, 20.6 % CO_2, 40% H_2, 2.9% H_2O, and balance N_2 in the reactant feed, 0.1g catalyst, total flow rate 100 ml/min)*

And table 1 summarizes the comparison of the four catalysts in integral reactor in terms of active reaction temperature range, CO conversion, and O_2 selectivity.

Table 1. Comparison of the performance of Pt-M/Al_2O_3 (1.1wt%Pt), Pt/Al_2O_3 (1.1wt%Pt), Au/α-Fe_2O_3 (5at%Au), and CuO-CeO_2 (5.7% Cu, Cu:Ce = 0.143:1) in the presence of CO_2 and H_2O. (*1.1% CO, 1.1% O_2, 20.6 % CO_2, 40% H_2, 2.9% H_2O, and balance N_2 in the reactant feed, 0.1g catalyst, total flow rate 100 ml/min*)

Catalyst	Temp. (°C)	CO conversion (%)	O_2 selectivity (%)
Pt-M/Al_2O_3	RT – 200	90 – 100	50 – 60
Pt/Al_2O_3	200	80 – 100	20 – 50
Au/α-Fe_2O_3	RT – 50	65 – 95	50 – 80
CuO-CeO_2	120 – 200	80 – 100	50 – 95

(2) Integration of H_2S and PROX CO utilizing microfibrous entrapped Pt-M/Al_2O_3 catalyst and ZnO sorbent

In this study, a new class of microfibrous media permits entrapment of fine grains of supported ZnO sorbents and PROX-CO catalysts from 10 to 200 μm in diameter. The structure of the microfibrous carrier facilitates improved heterogeneous contacting while the small particle size reduces intraparticle transport limitations and increases external surface area. Since reactive particles are supported within a highly porous matrix at volume fractions well below the tap density, normal Ergun-type relationships involving pressure drop versus particle size for packed-beds are favorably altered. Proper design and materials selection permits concentrations of H_2S and CO to be reduced from 20,000 ppmv to less than 0.1 ppmv within layers a few millimeters in thickness at residence times at or below 150 ms at temperatures from 298 to 473 K.

To accomplish the above noted requirement, a two-layer design of media is prepared approximately 2 mm in thickness and shown in figure 4. The first layer consists of supported ZnO sorbents. The second layer consists of a supported PROX catalyst.

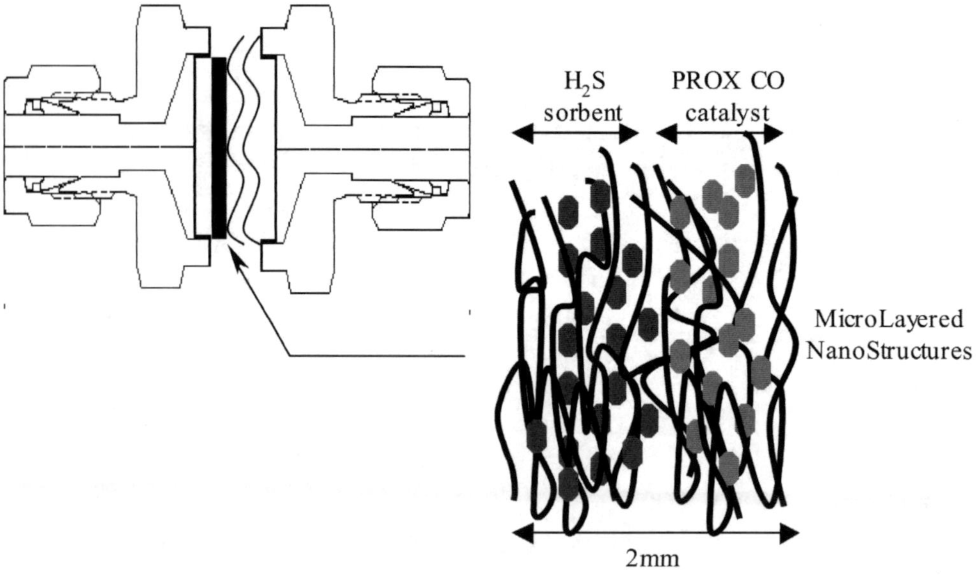

Figure 4: Integrated H_2S and PROX systems in a few mm thickness utilizing microfibrous entrapped catalysts and sorbents.

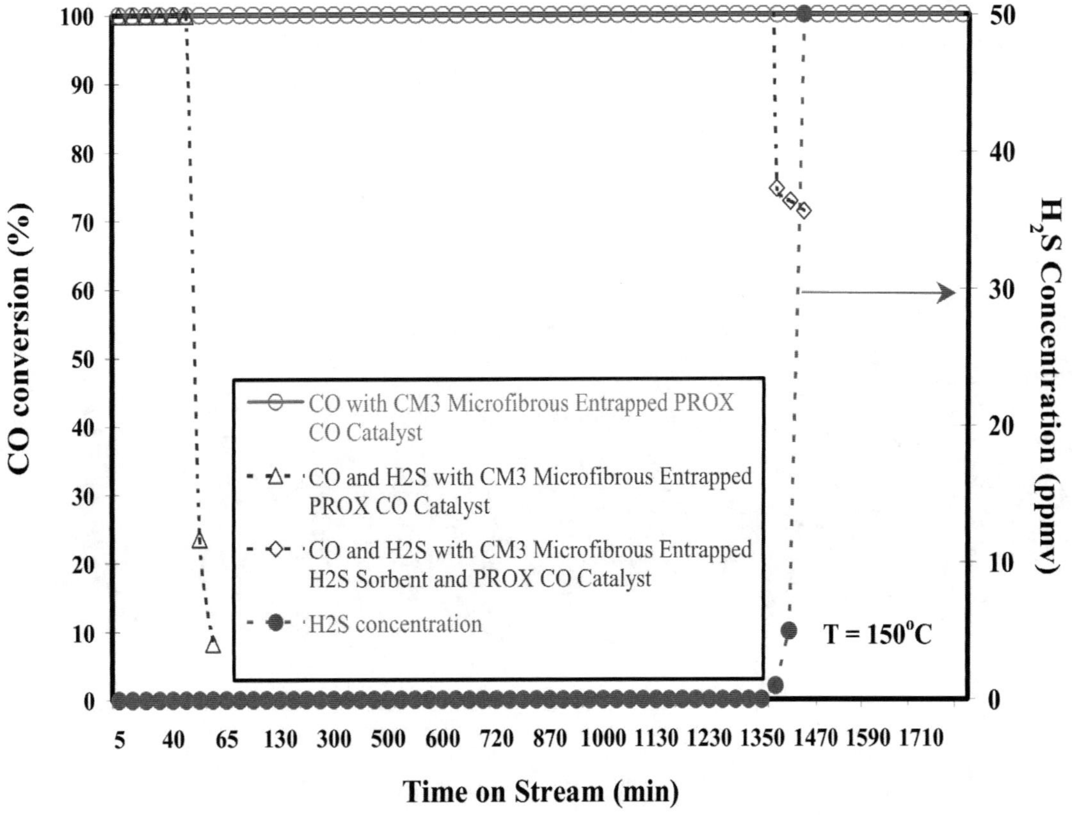

Figure 5: CO conversion and H_2S breakthrough curve over microfibrous entrapped Pt-M/Al_2O_3 for PROX CO from model reformats in the presence of 30% H_2O and 50 ppmv H_2S (when used); *dry gas 1% CO, 1%O_2, 20% CO_2, 40% H_2, and balance N_2, GHSV = 47,000 hr^{-1}, CO:O_2 = 1:1, reduced in H_2 at 250°C for about 2hrs, 2.5wt% Pt, M:Al = 0.01:1, T = 150°C*

Figure 5 shows CO conversion and H_2S concentration over microfibrous entrapped Pt-M/Al_2O_3 (2.5wt% Pt, M:Al = 0.01:1) at 150°C prepared by conventional impregnation method. CO conversion activity of 100% was maintained at least 30 hours when CO presents only in the inlet. CO conversion activity, however, dropped dramatically at around 3 hours after 50 ppmv H_2S was introduced together in the inlet with CO. Finally, H_2S sorbent was placed on top of the microfibrous entrapped PROX CO catalysts and the integrated H_2S and PROX CO removal unit in one reactor was investigated in the presence of both 1% CO and 50 ppmv H_2S in the feed stream. As it can be seen clearly, complete CO conversion activity was recovered for approximately 24 hours before CO was detected in the outlet. H_2S started being detected at around same time with CO, and the breakthrough time for the H_2S over the integrated system was about 24 hours.

4. Conclusions

Pt/γ-Al_2O_3, promoted Pt-M/γ-Al_2O_3, Au/α-Fe_2O_3, and CuO-CeO_2 have demonstrated promising catalytic activities for selective CO oxidation in integral reactor conditions in the temperature range from 298 to 523K with up to 30% H_2O. Catalytic activities of these four systems have been further evaluated in differential reactor studies in the absence of heat and mass transfer limitations. Specific activities occur in the order: promoted Pt-M/γ-Al_2O_3 > CuO-CeO_2 > Au/α-Fe_2O_3 > Pt/γ-Al_2O_3. Comparison of specific activities and selectivities over the temperature range from 298 to 473K indicate that promoted Pt-M/γ-Al_2O_3 is the best candidate for operation with up to 30% H_2O from 298 to 373K.

5. References

1. Chen. L.; Chang. B-K.; Lu. Y.; Yang. W.; and Tatarchuk. B. J. Prepr. Pap. - *Am. Chem. Soc., Div. Fuel Chem.*, **2002**, *47* (2), 609.

PREFERENTIAL OXIDATION IN MICROCHANNEL REACTORS FOR SCALABLE FUEL PROCESSING SYSTEMS

X. Ouyang,[1] P. Ho,[2] H. Chen,[1] W.C. Shin,[1] L. Bednarova,[1] W. Lee,[1] R.S. Besser[1],
S. Pau,[3] C.S. Pai,[3] J.A. Taylor,[3] W.M. Mansfield[3]

[1] Stevens Institute of Technology
Castle Point on Hudson
Hoboken, NJ 07030

[2] Reaction Design, Inc.
6440 Lusk Boulevard, Suite D-209
San Diego, CA 92121

[3] New Jersey Nanotechnology Consortium
Lucent Technologies
Murray Hill, NJ 07974

Prepared for Presentation at AIChE 2003 Annual Meeting, Nov., 2003, Fuel Processing Session III

Abstract

Microchannel reactors have demonstrated value for fuel-processing unit operations of both portable and larger scale fuel cell systems. As a key component, a preferential oxidation (PrOx) reactor to mitigate CO from a reformate stream was successfully demonstrated through the use of silicon micromachined reactors with novel Pt/Al_2O_3 thin-film wall catalysts. Productivity of the reactors, as measured by space-time-yield, is on par with the best reactors published to date. To increase understanding of the reaction mechanisms, CHEMKIN© simulations were used, which effectively predicted reactor conversion and selectivity. The combined experimental/simulation work serves as a model approach for microchannel reactor realization for hydrocarbon fuel processing applications.

1. INTRODUCTION

The use of hydrocarbon fuels for the production of hydrogen for fuel cells is inevitable for exploitation of PEM fuel cell technology [1]. Although appropriate infrastructure may ultimately be created which can provide hydrogen gas to automotive, residential, and stationary installations, it is likely that an ongoing segment will exist for the conversion of hydrocarbon sources into hydrogen-bearing feed streams, especially for portable applications. Hydrocarbon reforming results in the presence of CO in the product stream as a consequence of reaction thermodynamics [2]. With water-gas-shift reaction (WGS), the stream still contains ~1% CO, which is nevertheless sufficient to poison the fuel cell catalysts, rendering them ineffective and causing cell performance to severely degrade [1,3]. Consequently, a deep CO removal process, which eliminates CO to below 10 ppm, is imperative in order to support sustainable fuel cell function. Preferential oxidation of CO (PrOx) is currently the most promising and efficient approach compared to other routes, such as palladium-based membrane purification and catalytic methanation [4].

This research focuses on the use of silicon microchannel reactors. Microchannel architecture is quite efficient for a number of reasons, especially in regard to the excellent transport properties arising from microscale geometry. With gas phase systems at low to moderate temperatures, mass transport effects are essentially negligible and hence intrinsic kinetics dominate. Similarly, heat transport resistances are diminishingly small, such that thermal equilibration is rapid and isothermality generally holds. Microchannel-based systems also have the advantage of compactness, which is especially suitable for portable applications (1-50 W). Microchannel reactors are not limited, however, to compact applications. Rather, microchannel systems can be scaled in a straightforward way by replication of basic units [5]. Once the basic reactor unit is characterized, a scaled reactor behaves in a predictable manner since critical reaction geometry is unaltered.

As a catalyst for PrOx, Pt/Al_2O_3 is one of the most promising catalysts because of its high conversion and stability at moderate temperature (~200°C). This catalyst system has been known for some time as Engelhard was awarded a patent on it in 1963 [6]. More recently, Kahlich, et al. [7] observed a maximum CO conversion at 200 °C with a selectivity of ~40% on Pt/Al_2O_3. Effort was also made to deduce a simplified kinetic model and to seek the effect of surface hydroxyl species (OH(s)). Gorte, et al. [8] discovered a relationship between CO desorption and Pt particle size, i.e.,

CO desorption rate rises with decreasing Pt particle size, which in turn influences CO reaction rate. Ajmera, et al. [9] demonstrated a differential packed-bed Si microreactor for CO oxidation as a useful tool for simple kinetic rate law deduction.

The approach we have explored for PrOx on Pt/Al_2O_3 is illustrated in Fig. 1. The goal was to pursue understanding of detailed reaction mechanisms through advanced elementary kinetic modeling which is based on experimental feedback. Ultimately, we will apply this understanding to design a PrOx reactor for portable power applications.

Silicon MEMS technology was used to produce reactors possessing microscale geometry with high aspect-ratio for characterization [10,11,12]. Sol-gel synthesis techniques were adopted to create novel catalyst deposition methods for producing metallic Pt catalysts on porous, thin-film Al_2O_3 supports. Reaction characterization of the microreactor was performed in a microkinetic array test rig wherein microreactors can be characterized rapidly and efficiently. In addition to the rapid parallel testing scheme, the approach is efficient in that it occupies a relatively small footprint in the lab, and consumes less chemicals and electricity.

Initial modeling to design the laboratory reactor was performed using lumped empirical rate laws, while an advanced model later formulated was based on the CHEMKIN© reaction simulator. Finally, after using experimental data to refine the CHEMKIN© model, the latter provided insight into the rate-limiting steps within the overall reaction mechanism.

2. EXPERIMENTAL

2.1 Microreactors, Microkinetic Array, and Reaction Characterization

Silicon was adopted as the microreactor material because of its mature fabrication infrastructure, low cost and mass production capability. Forty microreactors were manufactured on each Si wafer (p-type <100> 8 inch) by silicon bulk micromachining techniques. Photolithography and deep reactive ion etching (DRIE) by inductively coupled plasma (ICP) were the major techniques applied [12]. Each reactor consisted of dual gas inlets, a pre-mixer, and a single reaction channel with an outlet (Fig. 1). The microreactor had a channel that was 4.5 cm long, 500 μm wide and 470 μm deep. Pt/Al_2O_3 catalyst was deposited in the microchannel before the silicon chip was sealed with a Pyrex™ cover by anodic bonding (AB).

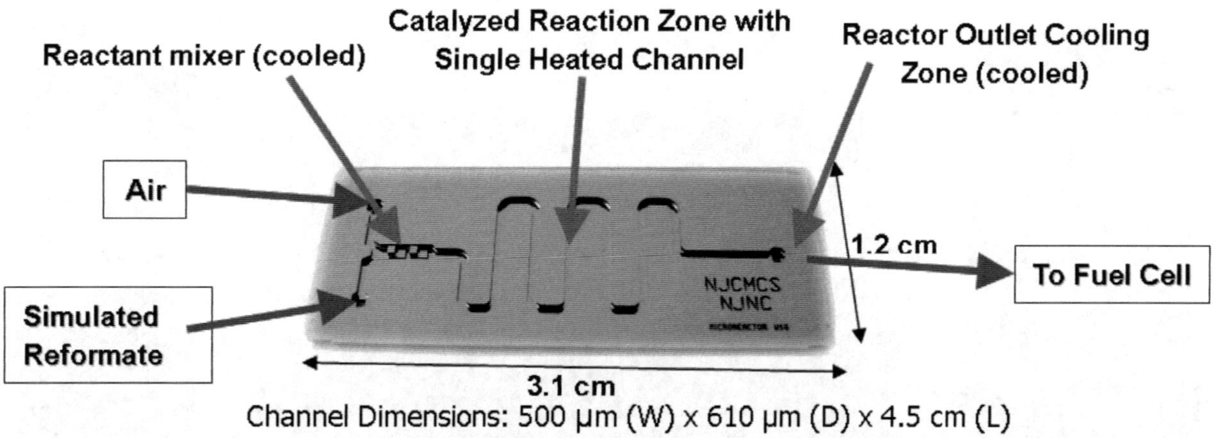

FIG. 1. Si microreactor used in this research, with a pre-mixer, catalyzed reaction zone and an outlet cooling zone for reaction quenching.

Alumina support was prepared from aluminum isopropoxide (Aldrich, 99.99+%) [13,14] by adding it to deionized water at 85 °C to start the hydrolysis reaction resulting in a sol. After 45 minutes under continuous stirring, HNO_3 was added. Platinum metal was introduced by using a solution of $H_2PtCl_6.xH_2O$ (Aldrich, 99.9+%) in 1,3-butanediol (Aldrich, 99%, anhydrous) at room temperature. The final platinum content in the catalyst was 2 wt%.

The catalyst was deposited into the microreactor by infiltrating the microchannel with liquid precursor into the open channel, then drying in air at 120 °C for 12 hours and calcined at 500 °C for 2 hours to make a solid thin-film, adherent to the walls of the microchannel.

The crystal structure of the alumina support was studied by XRD using a Siemens D5000 diffractometer. A scanning electron microscope (LEO 982) was used to examine the thin-film catalyst inside the microchannel. Support surface area and pore size were determined from BET measurement (Quantachrome AUTOSORB-1). Platinum particle size was determined using Chemisorption (Quantachrome CHEMBET-3000) and was supported by TEM bright field images (Philips CM20 FEG TEM/STEM).

SEM images of the microchannel cross section showed uniform thin-film layers on each of the three walls (Fig. 2), with the layer thickness ranging from ~2 nm to ~15 nm depending on the number of coating layers. The 2~15 nm thickness resulted in catalyst weights of 0.15~1.5 mg per reactor. However, thicker layers on the side walls than on the bottom of the microchannel were observed. BET results showed 480 m^2/g surface area with 2-3.7 nm pore distribution and Pt metal particle size between 2.5~8.0 nm with ~20% dispersion.

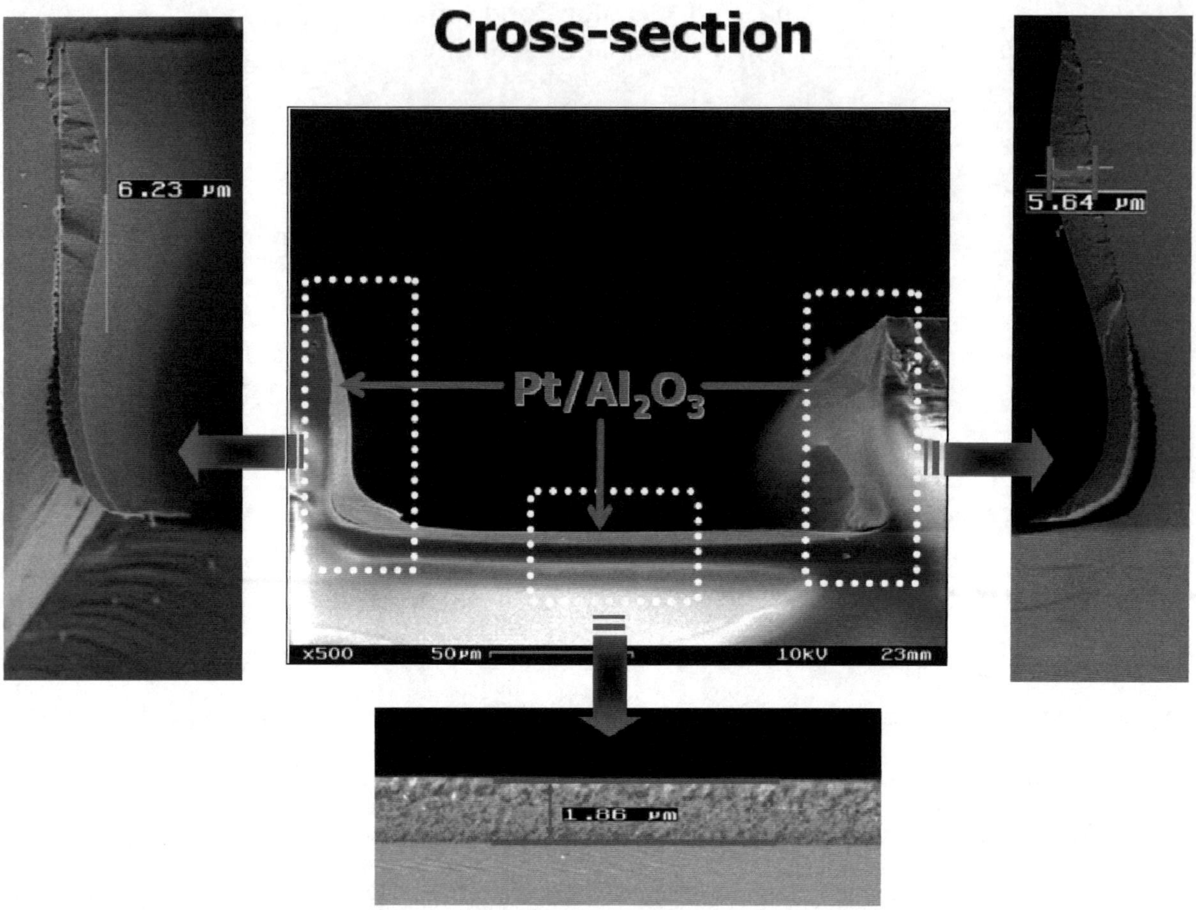

FIG. 2. SEM photo of cross-section of the microchannel with thin-film coating. Thickness on the sidewalls was ~6 μm while on the bottom was ~2μm. The delamination at the corner of the channel was an artifact of sample preparation for SEM examination.

The microkinetic array permits the simultaneous testing of four microreactors which can possess independently defined operating conditions of reactant flow rate, pressure, and temperature. In addition to the rapid parallel testing scheme, the approach is efficient in that it occupies a relatively small footprint in the lab, uses shared analytical instruments (MS and GC), and is completely automated. A detailed description of this apparatus is found elsewhere [15]. In order to begin testing, the reactor was evacuated at room temperature for 30 minutes followed by heating (10°C/min) in pure H_2 (5 Ncm^3/min) to 450 °C and kept at this temperature for 4 hours, then allowed to cool to room temperature. Feed was introduced and temperature was stepwise increased at 10 °C/min from 20 °C to 100 °C, 120 °C, 140 °C, 160 °C, 180 °C, 200 °C, 240 °C, 260 °C, 280 °C and 300 °C. The temperature was kept at each set point for 40-60 min. The feed consisted of simulated reformate (1.7% CO, 68.0% H_2, 21.0% CO_2 and N_2 as balance) and air (78.1% N_2 and 21.9% O_2). The flow of the reformate was kept at 5 Ncm^3/min and dry air at 0.5 Ncm^3/min ($\lambda = 2.57$) to accommodate the initial reactor design for a 0.5 W_e fuel cell. The products were analyzed by gas chromatograph (Varian CP-4900) and mass spectrometer (Stanford Research Systems QMS-200).

2.2 CHEMKIN© Simulation

CHEMKIN©, an integrated software package for modeling complex chemical kinetic processes which originated at Sandia National Laboratory, was used to understand the intrinsic reaction mechanism and to provide insight into the processes occurring in a reactor for quantities that can be difficult to measure. It contains reactor models targeted towards simulation of specific geometries or idealized flow conditions. In CHEMKIN©, it is not possible to perform simulation for boundary-layer flow along the full length of microchannel (which prevails in our devices) because of numerical difficulties that arise when CO is substantially depleted and hydrogen oxidation predominates. Although the plug flow condition is less appropriate than a boundary-layer flow or full CFD simulation in this case, such calculations are computationally much more tractable in CHEMKIN©. In order to justify the use of the CHEMKIN© turbulent flow model in a laminar-flow-dominated microchannel, the CHEMKIN© "CRESLAF" module, which can simulate laminar, chemically reacting, boundary-layer flow in cylindrical or planar channels, was used for comparison to plug-flow solutions under conditions where converged solutions could be achieved. Contour plots of CO mole fraction across the cross-section of the microchannel from CRESLAF simulation gave essentially flat profiles indicating nearly plug-flow behavior. Moreover, profiles of CO mole fraction as a function of distance along the microchannel showed that the boundary-layer flow at centerline and edge were nearly indistinguishable and agreed well with the plug-flow results (Fig. 3). Because there was almost no radial variation in concentration for boundary-layer flow, the plug-flow model for further simulation was adopted.

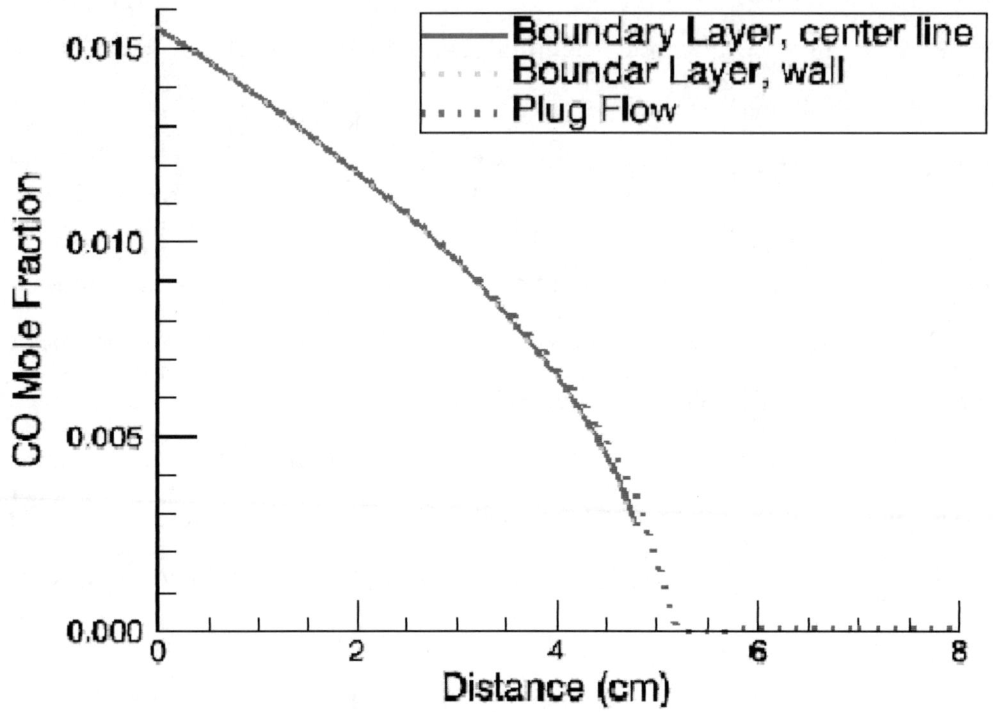

FIG. 3. Results of simulations with a "CRESLAF" module for boundary layer flow compared with the CHEMKIN plug flow model agreed well at both edge and center of the channel.

Detailed microkinetic models with a set of reaction pathways and rate coefficients for each pathway were developed using elementary reaction data for partial oxidation of C_1 and C_2 hydrocarbons on Pt [16]. The published microkinetic mechanism by Zerkle, et al. [16] has been reduced by omitting about 400 gas-phase reactions, and eliminating the hydrocarbon chemistry. This yielded a mechanism with 8 gas-phase species (O_2, CO, CO_2, H_2, H_2O, O, H, OH), 8 surface species (Pt(s), O(s), H(s), OH(s), H_2O(s), C(s), CO(s), CO_2 (s)), and a reaction network of 28 surface reactions (Table 1).

1. H_2+Pt(s)+Pt(s) → H(s)+H(s)	15. OH(s)+Pt(s) → H(s)+O(s)
2. O_2+Pt(s)+Pt(s) → O(s)+O(s)	16. H_2O(s)+Pt(s) → H(s)+OH(s)
3. H_2O+Pt(s) → H_2O(s)	17. OH(s)+OH(s) → H_2O(s)+O(s)
4. CO_2+Pt(s) → CO_2(s)	18. H_2O(s)+O(s) → OH(s)+OH(s)
5. CO+Pt(s) → CO(s)	19. H+Pt(s) → H(s)
6. CO(s) → CO+Pt(s)	20. H(s) → H+Pt(s)
7. CO_2(s) → CO_2+Pt(s)	21. O+Pt(s) → O(s)
8. C(s)+O(s) → CO(s)+Pt(s)	22. O(s) → O+Pt(s)
9. CO(s)+Pt(s) → C(s)+O(s)	23. OH+Pt(s) → OH(s)
10. CO(s)+O(s) → CO_2(s)+Pt(s)	24. OH(s) → OH+Pt(s)
11. CO_2(s)+Pt(s) → CO(s)+O(s)	25. H(s)+H(s) → Pt(s)+Pt(s)+H_2
12. CO(s)+OH(s) → CO_2(s)+H(s)	26. O(s)+O(s) → Pt(s)+Pt(s)+O_2
13. CO_2(s)+H(s) → CO(s)+OH(s)	27. H_2O(s) → H_2O+Pt(s)
14. H(s)+O(s) → OH(s)+Pt(s)	28. H(s)+OH(s) → H_2O(s)+Pt(s)

Table 1. Detailed microkinetic models with a set of reaction pathways, rate coefficients for each reaction pathway were developed using elementary reaction data for partial oxidation of C_1 and C_2 hydrocarbons on Pt.

3. RESULTS AND DISCUSSION

3.1 Preliminary Microchannel Design

At the outset, design of the microchannel was accomplished with an analytical model using a lumped LHHW rate expression for CO oxidation [17], as indicated in Equation 1. Channel dimensions (height, width and length), catalyst parameters (catalyst thickness and density), flow rates and temperatures were used in the simulation. Results showed that for a $500\mu m \times 470\mu m$ channel with literature catalyst parameters, 4.5 cm was sufficient to fully convert CO at 200°C. The reactor built from this initial design was used to acquire thin-film catalyst and reactor productivity data

$$R(c,T) = \frac{k_p^0 \exp(-\frac{Ea}{RT})CcoCo_2}{(1 + K_a^0 \exp(\frac{\Delta H}{RT})Cco)^2} \quad [1]$$

3.2 Productivity of PrOx Microreactor

The productivity of Pt/Al_2O_3 thin-film catalytic microreactors were evaluated with space-time-yield (STY). STY is defined here as desired product molar flow per unit time and precious metal weight (Equation 2). We calculated STY for other PrOx reactor systems with Pt/Al_2O_3 catalyst in the literature and compared them to our microreactors.

$$STY = \frac{\dot{n}_{CO_2}}{W_{Pt}} \quad [2]$$

As shown in Table 2, productivity of the Si microreactor with thin-film Pt/Al_2O_3 catalyst is at least an order of magnitude higher than the other more conventional systems. Since the dimension of the smallest reactor approaches microscale geometries, it is not immediately apparent why such discrepancy should exist, although catalyst utilization variation is possible. This interesting anomaly is a topic for future research.

Reference	**STY (s-1)**	**Reactor**	**Catalyst System**
This work	**0.126**	Microreactor	2 wt% Pt/Alumina
Kahlich, et al.	0.008	Quartz Tube Packed-Bed	5 wt% Pt/alumina
Son, et al.	0.009	Quartz Tube Packed-Bed	5 wt% Pt/Alumina
Avgouropoulos, et al.	0.018	Quartz Tube Packed-Bed	5 wt% Pt/alumina
Oh, et al.	0.004	SS Packed-Bed	0.5 wt% Pt/alumina
Manasilp, et al.	0.013	Packed-Bed	2 wt% Pt/alumina

Table 2. PrOx reaction productivity of the thin-film catalytic microreactor was compared with other PrOx systems in literature: with the calculation of space time yield (STY), our microreactor demonstrated superior reaction productivity.

3.3 Reaction Analysis with CHEMKIN©

Simulations were performed with the same catalyst (1.5 mg, 2 wt% Pt/Al_2O_3) and reactor geometries (as shown in Fig. 1) as used in the experiments. Revision and refinement of the kinetic model were done with iterations of comparison between simulation and experimental results.

The consistency of the simulation was checked by comparing the predicted conversion behavior with experiment, as indicated in Fig. 4. These results indicate nearly perfect agreement in the light-off curves for CO, O_2 and H_2 within the same temperature range. With some confidence in the model's ability to simulate the reactor, we examined details of the mechanism in the gas and surface phases.

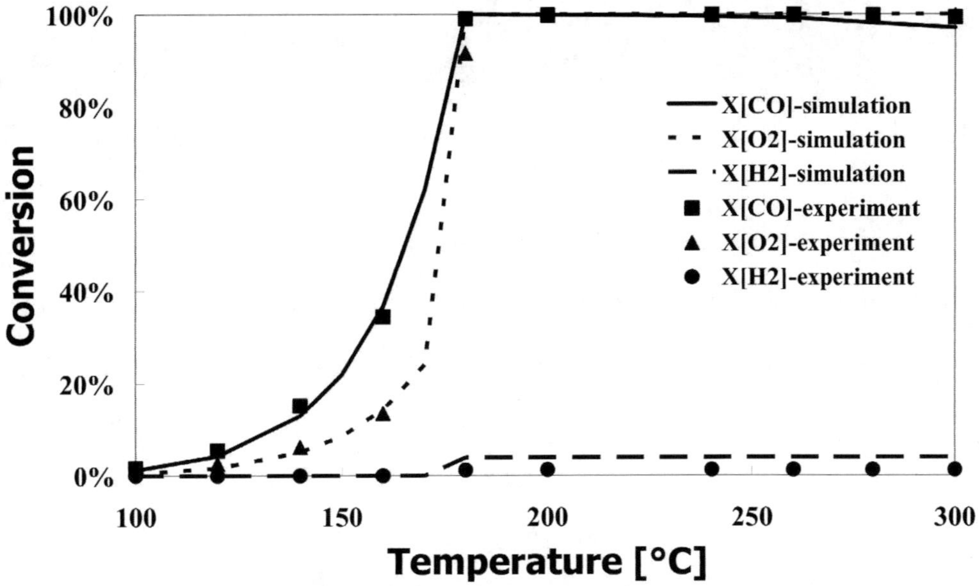

FIG. 4. CHEMKIN© simulation agreed well with experimental results for X_{CO}, X_{O2} and X_{H2}.

An important aspect of the reaction mechanism scheme involves the rates of surface reactions leading to CO conversion. In Fig. 5, the model indicated that the major route for CO conversion is through the reaction between adsorbed CO and adsorbed OH (reaction 12 in Table 1). The rate of this reaction is 10 orders higher than for the reaction between CO(s) and O(s) (reaction 10). OH(s) is formed by reaction of H(s) and O(s) (reaction 14), which in turn are formed by dissociative adsorption of H_2 and O_2, respectively (reactions 1 and 2).

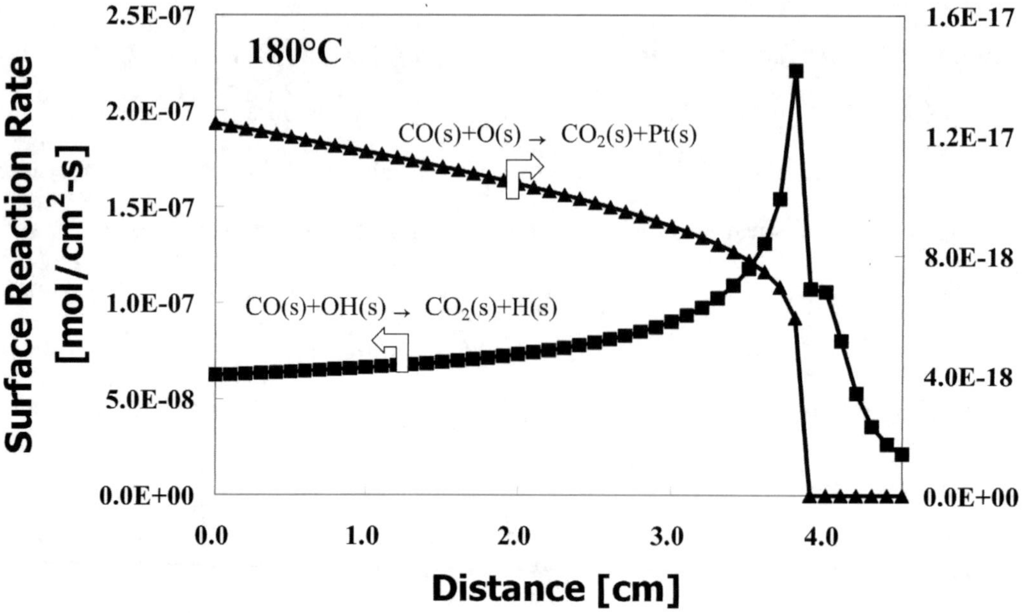

FIG. 5. Reaction rates of two surface reactions to convert CO to CO_2: rate of surface reaction CO(s)+OH(s) → CO_2(s)+H(s) was 10 orders of magnitude higher than CO(s)+O(s) → CO_2(s)+Pt(s).

Fig. 6 shows the predicted mole fractions for the most important gas-phase species as a function of position along the reactor length at 180°C. This plot shows that CO and O_2 are steadily consumed and CO_2 gradually produced up to a distance of 3.8 cm along the microchannel. However, the sudden simultaneous decrease of H_2 and O_2 within 3.8 to 4.0 cm indicates that H_2 oxidation is only activated after essentially all of the CO is converted. Further insight is obtained through an observation of surface species along the microchannel, as provided in Fig. 7. CO(s) and H(s) dominate on the surface with relatively small amount of O(s) adsorbed up to 3.8 cm along the channel. However, there is an abrupt decrease of the surface CO(s) and an increase in available surface Pt site density beyond 3.8 cm. The active surface is mostly covered by CO(s) up to 3.8 cm. At 3.8 cm, CO(s) concentration decreases drastically, which opens up Pt sites (Pt(s)) for O adsorption. These O(s) in turn oxidize more CO(s) and further expedite the Pt(s) concentration increase. As a result of the rapid Pt(s) increase, O(s) concentration rose to 10 times the original level (at 3.7 cm) and both CO(s) and H(s) dropped as a result of oxidation by O(s). At further distance, gas phase O_2 was depleted (4.0 cm) and a significant amount of H_2O was formed due to these surface reactions, as indicated in Fig. 6. The hydrogen oxidation was limited by the amount of O_2 present, implying that excess O_2 can severely reduce product selectivity.

After this point in the microchannel (beyond 4.0 cm), the gas species approach a new steady state reflecting the absence of gas-phase CO or O_2 (Fig. 6). On the catalyst surface, however, CO(s) coverage increases due to re-adsorption and surface dissociation of CO_2, which lead to the reverse water-gas-shift reaction (reverse WGS). This effect of additional reactor length, which serves to decrease overall CO conversion, is most noticeable at higher reaction temperatures (Fig. 8). This demonstrated that the operating temperature, microchannel length, and loading of the thin-film catalyst are important for consideration in suppressing undesired side reactions. Carbon formation was insignificant during simulation at all temperatures, suggesting that it is not a major catalyst deactivation mechanism.

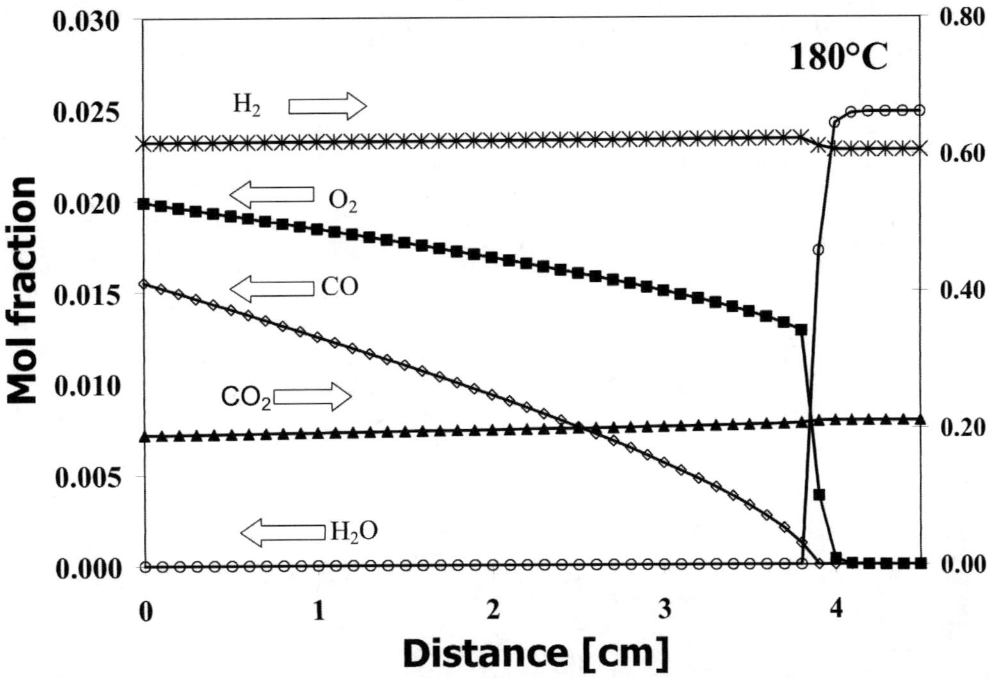

FIG. 6. CHEMKIN© simulation: mole fractions of important gas species along the microchannel at 180°C.

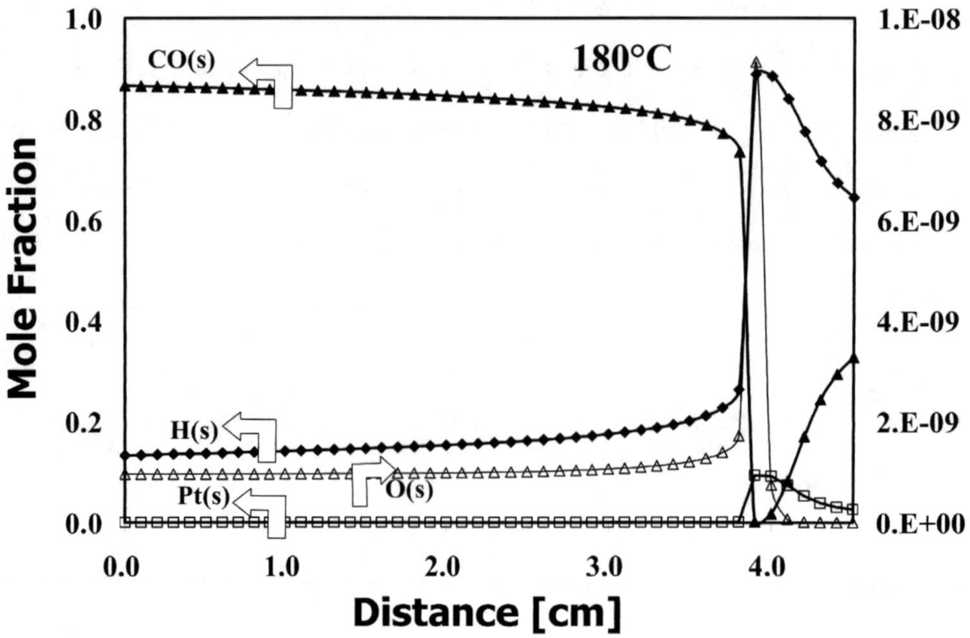

FIG. 7. CHEMKIN© simulation: mole fractions of surface species along the microchannel.

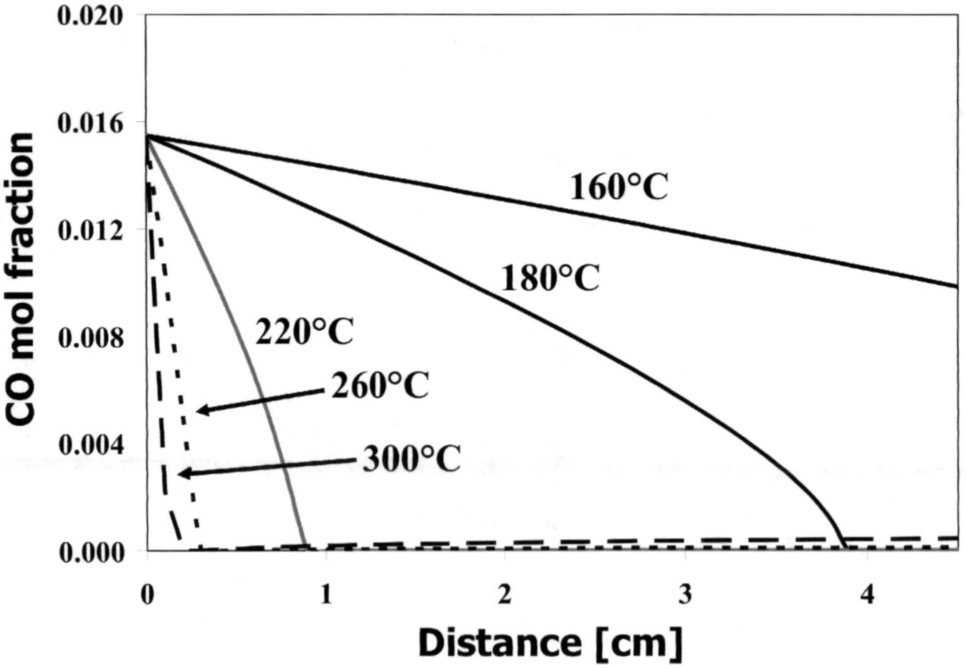

FIG. 8. CHEMKIN© simulation: mole fractions of surface species along the microchannel.

4. CONCLUSIONS

As demonstrated by these results, microchannel reactors deploying thin-film wall catalyst coatings are a competitive option for PrOx reactors for portable or scaled hydrogen processing systems for fuel cells. Productivity factors achieved were equal to or better than the best values for PrOx reactors reported in the literature. The combination of excellent transport behavior and catalyst efficiency is no doubt responsible, with further research being motivated to understand how these factors contribute. In addition, the methodology of combining CHEMKIN© simulation with a rapid microreactor characterization scheme serves as a model for an efficient, rational basis for understanding and design of microchannel reactor systems.

References

1. C. Song, "Fuel Processing for Low-Temperature and High-Temperature Fuel Cells Challenges, and Opportunities for Sustainable Development in the 21st Century," *Catalysis Today* **77** (2002) 17–49.

2. T.V. Choudhary, D.W. Goodman, "CO-Free Fuel Processing for Fuel Cell Applications," *Catalysis Today* **77** (2002) 65-78.

3. S.H. Oh, R.M. Sinkevitch, "Carbon Monoxide Removal from Hydrogen-Rich Fuel Cell Fedstreams by Selective Catalytic Oxidation," *Journal of Catalysis* **142** (1993) 254-262.

4. Akkarat Manasilp, Erdogan Gulari, "Selective Oxidation over Pt/alumina Catalysts for Fuel Cell Applications," *Applied Catalysis B: Environmental* **37** (2002) 17-25.

5. Ch. Wille, V. Autze, H. Kim, U. Nickel, S. Oberbeck, "Progress in Transferring Microreactors from Lab into Production – an Example in the Field of Pigments Technology," *Proceedings of IMRET6 – 6th International Conference on Microreaction Technology* (2002).

6. M. Brown, A. Green, US Patent 3,216,783 (1965).

7. M.J. Kahlich, et al., "Kinetics of the Selective CO Oxidation in H_2-Rich Gas on Pt/Al_2O_3," *Journal of Catalysis* **171,** (1997) 93–105.

8. G.S. Zafiris, et al., „CO Oxidation on Pt/α-Al2O3(0001): Evidence for Structure Sensitivity," *Journal of Catalysis* **140** (1993) 418-423.

9. S.K. Ajmera, C. Delattre, M.A. Schmidt, K.F. Jensen, "Microfabricated Differential Reactor for Heterogeneous Gas Phase Catalyst Testing," *Journal of Catalysis* **209** (2002) 401-402.

10. H. Surangalikar, X. Ouyang, R.S. Besser., "Experimental Study of Hydrocarbon Hydrogenation and Dehydrogenation Reactions in Silicon Microfabricated Reactors of Two Different Geometries," *Chemical Engineering Journal* **93** (2003) 217-224.

11. R.S. Besser, X. Ouyang, H. Surangalikar, "Hydrocarbon Hydrogenation and Dehydrogenation Reactions in Microfabricated Catalytic Reactors," *Chemical Engineering Science* **58** (2003) 19-26.

12. R.S. Besser, W.C. Shin, "Deep Reactive Ion Etching Characteristics of a Micromachined Chemical Reactor," *Journal of Vacuum Science and T*echnology B, Microelectronics and Nanometer Structures **21(2)** (2003) 912

13. S. Zhao, R.S. Besser, "Selective Deposition of Supported Platinum Catalyst for Hydrogenation in a Micromachined Reactor," *Proceedings of the 6th International Conference on Microreaction Technology (IMRET VI)*, March 10-14, 2002, New Orleans, LA, American Institute of Chemical Engineers, (2002), pp. 289-296.

14. A. Manasilp, E. Gulari, "Selective CO Oxidation over Pt/alumina Catalysts for Fuel Cell Applications," *Applied Catalysis B: Environmental* **37** (2002) 17–25.

15. X. Ouyang, R.S. Besser, "Development of a Microreactor-Based Parallel Catalyst Analysis System for Synthesis Gas Conversion," *Catalysis Today* **84** (2003) 33-41.

16. D. K. Zerkle, M.D. Allendorf, M. Wolf, O. Deutschmann, "Understanding Homogeneous and Heterogeneous Contributions to the Patinum-Catalyzed Partial Oxidation of Ethane in a Short-Contact-Time Reactor," *Journal of Catalysis* **196** (2000) 18-39.

17. R.H. Venderbosch, W. Prins, W.P.M. van Swaajj, "Patinum Catalyzed Oxidation of Carbon Monoxide as a Model Reaction in Mass Transfer Measurement," *Chemical Engineering Science* **53** 19 (1998) 3355-3366.

A Modeling Study of CO_2-Selective Water-Gas-Shift Membrane Reactor for Fuel Cell

ABSTRACT

Water gas shift (WGS) reaction is critical to hydrogen purification for fuel cells. Since the WGS reaction is reversible, the reaction is not efficient, resulting in a high concentration of unconverted CO (about 1%) in the H2 product and a bulky, heavy reactor. Using a CO_2-selective membrane reactor shifts the reaction towards the product side, which enhances the conversion of CO and increases the purity of the H_2 product at a high pressure. Also air can be used as the sweep gas to remove the permeate, CO_2, on the low-pressure side of the membrane to have a high driving force for the separation. In this study, the reaction and transport process in the countercurrent WGS membrane reactor was simulated with a one-dimensional non-isothermal model using the recent experimental data of membrane selectivity and permeability, and the influences of several system parameters were investigated.

In this modeling study, the synthesis gases with different CO concentrations from autothermal reforming of gasoline with air were used as the feed gas, while heated air was used as the sweep gas. The reaction rate equation for the Cu/ZnO catalyst from literature was incorporated into the model. The modeling results have shown that a CO concentration of less than 10 ppm, a H_2 recovery of greater than 97%, and a H_2 concentration of greater than 54% (on the dry basis) are achievable. As the CO_2/H_2 selectivity increased, the recovery of H_2 increased, without affecting the membrane area requirement and the low CO attainment. Increasing sweep-to-feed flow rate ratio enhanced the permeation driving force but decreased the feed side temperature and thus the reaction rate, resulting in a net effect balanced between them. When an improved catalyst with higher activity was applied, the membrane area requirement reduced significantly. Higher membrane permeability also resulted in the reduction of the membrane area. These results have shown that both the reaction rate and the membrane permeability play an important role on the overall reactor behavior. In addition, it is expected with the advancement of the high temperature proton-exchange-membrane fuel cell (120 – 150°C) that the constraint for CO concentration can be relaxed to about 50 ppm in the near future. Therefore, the membrane area could be reduced significantly based on the modeling results.

KEY WORDS

CO_2-selective membrane, Membrane reactor, WGS reaction, Fuel cell

INTRODUCTION

Fuel cell is an efficient energy conversion device for both transportation and stationary power generation. It converts chemical energy directly to electrical energy, which is about two times efficiency of an internal combustion engine in terms of gas mileage. On the other hand, fuel cell is also an environmental friendly device; water is the only emission when hydrogen is the fuel [1, 2].

In the applications of fuel cell, one of the key problems is the purity of hydrogen, as the most preferable fuel, before feeding it into the fuel cell. For example, CO concentration should be less than 10 ppm to avoid poisoning platinum, which works as the electrocatalyst in polymer electrolyte membrane fuel cells (PEMFCs) [3]. The Water gas shift (WGS) reaction is an important industrial reaction to convert CO to CO_2 and H_2. However, as an exothermic and reversible reaction, a bulky, heavy reactor is needed to achieve such a low CO concentration with the traditional fixed bed reactor. With the continuous removal of CO_2, CO_2-selective WGS membrane reactor is a promising approach to enhance

CO conversion and increase the purity of H_2 under relatively low temperature. Also, the compact, lightweight membrane reactor will be very beneficial to the application of fuel cell in automobiles.

BACKGROUND

WGS reaction is a reversible, exothermic reaction that is thermodynamically unfavorable at elevated temperatures, which result in a high concentration of unconverted CO (~1%) in the final hydrogen product.

$$CO(g) + H_2O(g) \Leftrightarrow CO_2(g) + H_2(g) \qquad \Delta H_r = -41.1 \quad KJ/mol \tag{1}$$

As one of two main types of commercialized membrane modules, the hollow fiber membrane module has shown excellent mass transfer performance because of its high specific membrane area (about 3000 ft^2/ft^3 for gas separation). In this work, the WGS membrane reactor was configured to be a hollow fiber membrane module with catalyst particles packed inside the fibers. It is assumed that the hollow fiber module was composed of CO_2-selective facilitated transport polymeric membrane, and that CO_2 and H_2 were the only two gases permeating through the membrane. The CO_2 permeability of this membrane was in the range from 1000 to 8000 Barrers, and the CO_2/H_2 selectivity, expressed in Equation (2) [4], was in the range from 10 to 80.

$$\alpha = \frac{y_{CO_2}/y_{H_2}}{x_{CO_2}/x_{H_2}} \tag{2}$$

The catalyst packed was the commercial Cu/ZnO catalyst for lower-temperature WGS reaction. The reaction kinetics reported in the literature was applied in the model [5, 6]. The feed gas was the syngas from the autothermal reforming, with the composition of 1% CO, 9.5% H_2O, 41% H_2, 15% CO_2, and 33.5% N_2. Heated air was used as the sweep gas.

MODEL DESCRIPTION

A one-dimensional non-isothermal model was developed to simulate the countercurrent hollow fiber WGS membrane reactor. According to the schematic diagram of the hollow fiber WGS membrane reactor shown in Figure 1, the molar and energy balances were performed on both lumen and shell sides, respectively.

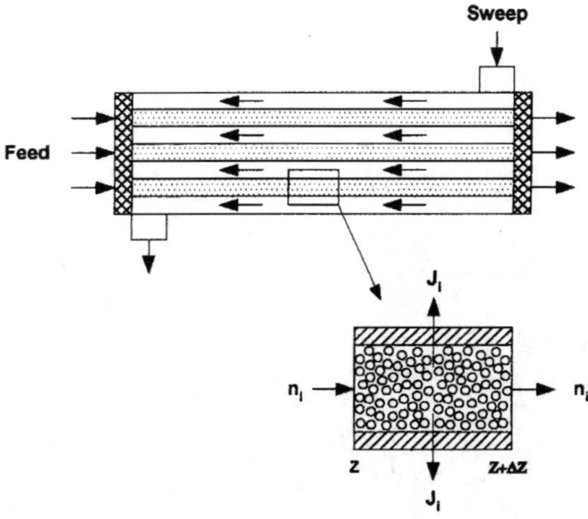

Figure 1. Schematic diagram of hollow fiber water gas shift membrane reactor.

MOLAR BALANCE

Lumen side

Based on the volume element from z to z + Δz, the molar balance for gas species i is expressed as,

In - Out + Generation = Accumulation

or

$$n_{fi}\big|_z - n_{fi}\big|_{z+\Delta z} - \pi\, d\, \Delta z\, J_i + \frac{1}{4}\pi\, d^2\, \Delta z\, r_i = 0 \tag{3}$$

where n_i, J_i and r_i are the molar flow rate, permeation flux and reaction rate of species i, respectively.

$$J_i = P_i \frac{\Delta p_i}{\ell} \tag{4}$$

$$r_i = 1.0\times10^3 \exp\left(13.39 - \frac{5557}{T_f}\right)\frac{\rho_b P_f}{n_t R T_f} n_{CO}\left(1 - \frac{n_{fH_2} n_{fCO_2}}{K_{eq} n_{fCO} n_{fH_2O}}\right) \tag{5}$$

where $K_{eq} = \exp\left(-4.33 + \dfrac{4577.8}{T_f}\right)$ (6)

Dividing both sides of equation (3) by Δz and taking the limit as Δz → 0 gives

$$\frac{dn_{fi}}{dz} = \frac{1}{4}\pi d^2 r_i - \pi d J_i \qquad (7)$$

Shell side

Similarly, the molar balance for shell side is carried out and the resulting equation is

$$\frac{dn_{si}}{dz} = -\pi d J_i \qquad (8)$$

ENERGY BALANCE

Lumen side

$$\frac{d\Sigma(n_{fi}\, c_{pfi} T_f)}{dz} = \Delta H_r \frac{1}{4}\pi d^2 r_i - \pi d U_i \Delta T \qquad (9)$$

Where U_i is the overall heat transfer coefficient, which contains both heat conduction through the membrane and energy carried by permeating gases. Due to the small dimension of the lumen, a uniform temperature at z is assumed.

Shell side

$$\frac{d\Sigma(n_{si}\, c_{psi} T_s)}{dz} = -\pi d U_i \Delta T \qquad (10)$$

The boundary conditions of above differential equations are listed as follows:

At z = 0,

$$T_f = 140^0 C,\ n_{f,CO} = 0.01\, n_{t0},\ n_{f,H_2O} = 0.095\, n_{t0},$$
$$n_{f,H_2} = 0.41\, n_{t0},\ n_{f,CO_2} = 0.15\, n_{t0}$$

At z = L,

$$T_p = 140^0 C,\ n_{s,CO} = 0,\ n_{s,H_2O} = 0,$$
$$n_{s,H_2} = 5\times10^{-7}\, n_{t0}\,\gamma,\ n_{s,CO_2} = 370\times10^{-6} n_{t0}\,\gamma$$

Where n_{t0} is the feed mole flow rate, and γ is the sweep-to-feed molar flow rate ratio. In this work, n_{t0} is 1 mol/s, which can provide sufficient H_2 molar flow rate to generate a power of 50 kW for a five-passenger car via the fuel cell.

The bvp4c solver in Matlab® was used to solve the above differential equations of the boundary value problem with given boundary conditions. This solver approximates the solution as a series of cubic polynomials on each subinterval of the mesh, and the values at both ends and the midpoint of each subinterval are satisfied. And then, the resulting nonlinear algebraic equations are solved iteratively by linearization.

A reference case was chosen at CO_2/H_2 selectivity of 40, CO_2 permeability of 4000 Barrers, feed and sweep pressures of 3 and 1 atm, respectively, inlet sweep-to-feed molar flow rate ratio of 1, membrane thickness of 5 μm, and 52,500 hollow fibers (length of 61 cm and inner diameter of 0.1 cm). Based on this case, the effects of CO_2/H_2 selectivity, sweep-to-feed flow rate ratio, catalyst activity, and permeability on the reactor behavior were then investigated.

RESULTS AND DISCUSSION

REFERENCE CASE

Figures 2 and 3 show the profiles of the feed-side mole fraction of CO, CO_2 and H_2 along the length of the countercurrent membrane reactors, respectively. The calculation results demonstrate that this membrane reactor could reduce CO concentration from 1% to 9.82 ppm as the removal of almost all of CO_2 and enhance H_2 concentration from 45.30% to 54.95% (on the dry basis). In this case, H_2 recovery was 97.38%.

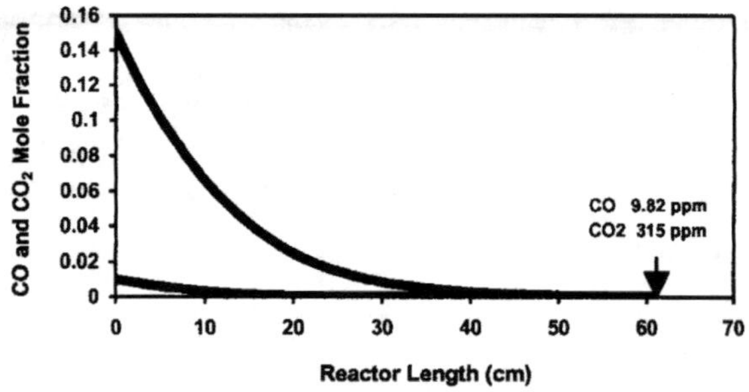

Figure 2. CO and CO_2 mole fraction profiles along the length of membrane reactor.

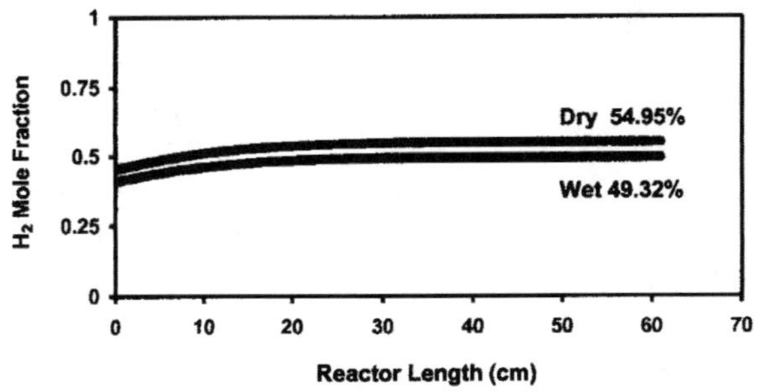

Figure 3. H_2 mole fraction profile along the length of membrane reactor.

In addition, it is expected with the advancement of the high temperature proton-exchange-membrane fuel cell (120 – 150°C) that the constraint for CO concentration can be relaxed to about 50 ppm in the near future. Therefore, the membrane area could be reduced significantly to 39,000 based on the modeling study.

EFFECTS OF CO_2/H_2 SELECTIVITY

Figures 4 and 5 present the effects of CO_2/H_2 selectivity on the exit CO concentration and H_2 recovery, respectively. Exit CO concentration is slightly increased as the CO_2/H_2 selectivity increases. This is due to the fact that the higher selectivity means reduced H_2 permeability, and the lower H_2 permeation rate is unfavorable to the WGS reaction rate. The H_2 recovery increases from 89.85% to 98.68% as CO_2/H_2 selectivity is increased from 10 to 80, which indicates that the higher selectivity decreases the H_2 loss because of the reduction in H_2 permeation through the membrane.

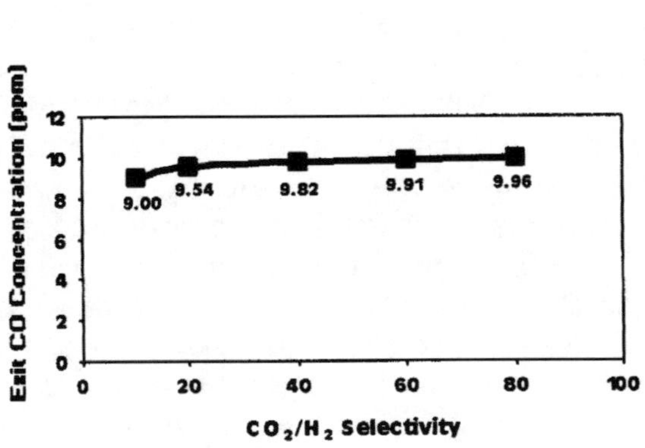

Figure 4. Exit CO concentration vs. CO_2/H_2 selectivity.

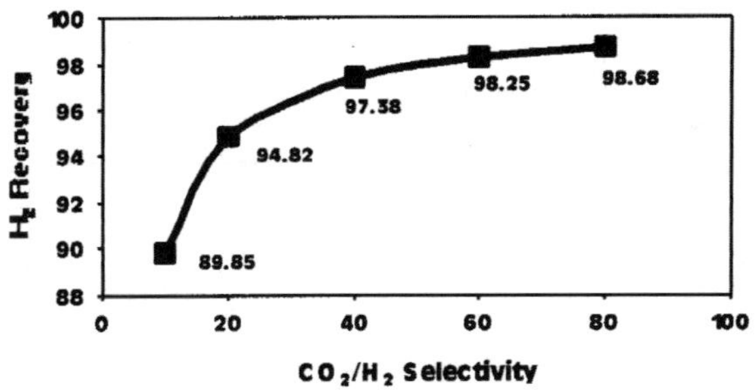

Figure 5. H_2 Recovery vs. CO_2/H_2 selectivity.

EFFECTS OF SWEEP-TO-FEED RATIO

Figures 6 and 7 show the effects of inlet sweep-to-feed flow rate ratio on the exit CO concentration and H_2 recovery, respectively. Increasing sweep-to-feed ratio reduces exit CO concentration firstly and then increases it slightly after minimum value, whereas sweep-to-feed ratio does not have a significant effect on the H_2 recovery. The higher sweep-to-feed flow rate ratio causes the lower CO_2 concentration on the sweep side and then the higher CO_2 permeation driving force. However, it will also enhance heat transfer and then decrease the feed side temperature, which would be unfavorable to WGS reaction. Therefore, an optimal value exists as the result of the tradeoff between the effects on CO_2 permeation rate and WGS reaction rate.

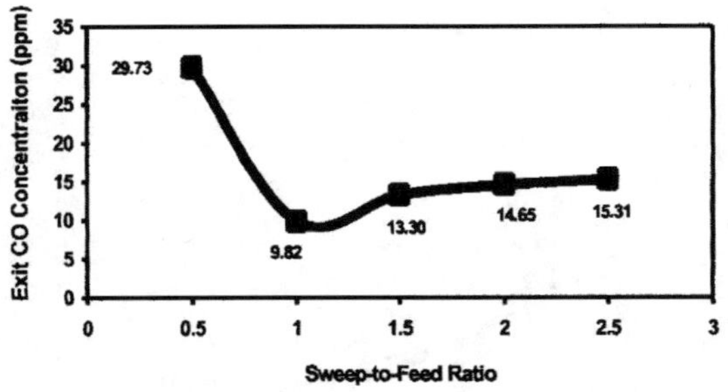

Figure 6. Exit CO concentration vs. sweep-to-feed ratio.

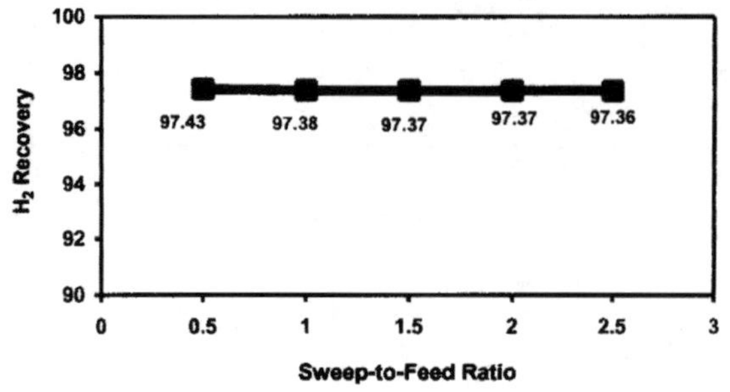

Figure 7. H_2 recovery vs. CO_2/H_2 selectivity.

EFFECT OF CO_2 PERMEABILITY

The membrane areas required for CO feed exit mole fraction < 10 ppm were calculated with five different CO_2 permeabilities ranging from 1000 to 8000 Barrers. Figure 8 demonstrates that the membrane area drops quickly as permeability increases from 1000 Barrers to 6000 Barrers. After that, it reaches an asymptotic value gradually. Increasing CO_2 permeability increases the CO_2 permeation rate and enhances CO_2 removal, which will shift the WGS reaction towards the product side. After the permeability exceeds certain value, the overall system becomes reaction control, the influence of the permeability become less significant.

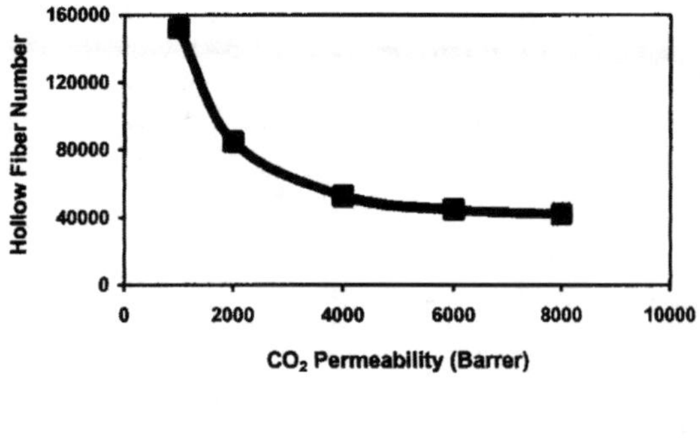

Figure 8. Hollow fiber number vs. CO_2 permeability.

EFFECT OF CATALYST ACTIVITY

The effect of catalyst activity was studied by apply several different reaction kinetics to the modeling based on the Cu/ZnO kinetics in the literature. As can be seen in Figure 9, increasing catalyst activity reduces the membrane area significantly. The higher catalyst activity makes the reaction rate faster. This promotes the CO_2 permeation rate because of the higher CO_2 partial pressure on the feed side and thus the higher driving force. It is expected the catalyst with higher activity and larger specific area would be developed in the near future.

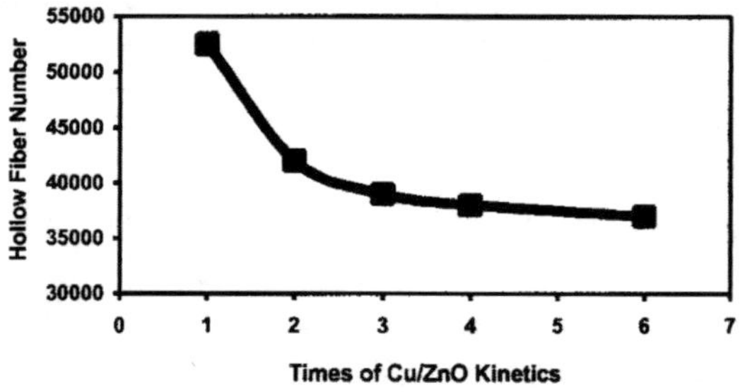

Figure 9. Hollow fiber number vs. catalyst activity.

SUMMARY

The CO_2-selective WGS membrane reactor was proposed to reduce CO concentration and obtain high purity H_2 from autothermal reforming syngas. A one-dimensional non-isothermal model was developed to simulate the hollow fiber membrane reactor. Based on the modeling results, a CO concentration of less than 10 ppm, a H_2 recovery of greater than 97%, and a H_2 concentration of greater than 54% (on the dry basis) are achievable. The modeling study shows that both the reaction rate and the membrane permeability play an important role on the overall reactor performance. As the CO_2/H_2 selectivity increased, the recovery of H_2 increased, without affecting the membrane area requirement and the low CO attainment. Increasing sweep-to-feed flow rate ratio enhanced the permeation driving force but decreased the feed side temperature and thus the reaction rate, resulting in a net effect balanced between them. Higher membrane permeability resulted in the reduction of the membrane area. When an improved catalyst with higher activity was applied, the membrane area requirement reduced significantly.

ACKNOWLEDGMENTS

The first author would like to thank Louei El-Azzami for his helpful discussions on this work. This work was supported by U S Department of Energy and H2fuel, LLC.

REFERENCES

1. G. J. K. Acres, "Recent advances in fuel cell technology and its applications", J. Power Sources, 100 (1-2), 60 (2001).
2. Brian M. Barnett, and W. Peter Teagan, "The role of fuel cells in our energy future," J. Power Sources, 37 (1), 15 (1992).
3. S. Ahmed and M. Krumpelt, "Hydrogen from hydrocarbon fuels for fuel cells", Int. J. Hydrogen Energy, 26 (4), 291 (2001).
4. W. S. W. Ho and K. K. Sirkar, eds., Membrane Handbook, Chapman and Hall, New York, 26 (1992).
5. J. Moe, Design of water-gas shift reactors, Chem. Eng. Progr., 58 (3), 33 (1962).
6. R. L. Keiski, O. Desponds, Y.-F. Chang and G. A. Somorjai, "Kinetics of the water-gas shift reaction over several alkane activation and water-gas shift catalysts", Appl. Catal. A: General, 101 (2), 317 (1993).

CONTACT

Jin Huang, PhD student, Department of Chemical Engineering, The Ohio State University.

Email: huangj@che.eng.ohio-state.edu

W.S. Winston Ho, University Scholar Professor, Chemical and Materials Science and Engineering, Center for Materials Research, The Ohio State University.

Email: ho@che.eng.ohio-state.edu

DEFINITIONS

c_p: Heat capacity (J/mol /K)

d: Hollow fiber diameter (cm)

ΔHr: Heat of reaction (J/mol)

J: Permeation flux (mol/cm^2/s)

K_{eq}: Reaction equilibrium constant (atm^{-2})

ℓ: Membrane thickness (cm)

L: Length of reactor (cm)

n: Molar flow rate (mol/s)

p: Pressure (atm)

P: Permeability (Barrer)

r: Volumetric reaction rate (mol/cm^3/s)

R: Ideal gas constant (atm cm^3/mol/K)

T: Temperature (^{0}C)

U_l: Overall heat transfer coefficient (W/cm^2/K)

x: Feed side molar fraction

y: Sweep side molar fraction

z: Axial position along length of reactor (cm)

GREEK LETTERS

α: Selectivity

γ: Sweep-to-feed molar flow rate ratio

ρ_b: Catalyst bulk density (g/cm^3)

SUBSCRIPTS

0: Initial

f: Feed side

i: Species

s: Sweep side

t: Total

Optimization of Autothermal Reactor for Maximum Hydrogen Production

Bijan F. Hagh
ES Logic, LLC
4231 E. Regency Avenue
Orange, CA. 92867, USA

Prepared for Presentation at the 2003 AIChE Annual Meeting, San Francisco, CA., Nov. 19

November, 2003

Unpublished

Abstract

Efficient generation of hydrogen is an important enabling technology for commercialization of fuel cells for homes and cars of the future. A methodology for optimization of an autothermal reactor with respect to system parameters is described. Furthermore, a framework for theoretical interpretation of reforming reactions is developed using an atomic balance approach, which is utilized to determine the reforming reaction space and maximum monolith temperature for a methane reformer.

Keywords: Hydrogen, Reforming, Fuel Processing, Optimization

Introduction

Hydrogen is increasingly recognized as the efficient and sustainable fuel of the future and there is a growing need for distributed generation of hydrogen for industrial, transportation and residential markets. Hydrogen is the preferred fuel for powering fuel cells in homes and cars and a hydrogen refueling infrastructure will be needed in the near future.

There are a number of major processes for hydrogen generation, but in actuality these technologies are points along a spectrum that spans combustion through steam reforming [1]. Figure 1 identifies the primary chemical reaction and the thermal requirements for each of these processes. Combustion is defined as the process of combining oxygen (O_2) with a hydrocarbon to form stoichiometric quantities of carbon dioxide and water as products. For a methane fuel, oxygen to carbon molar ratio of 2 provides sufficient oxygen to convert all the carbon and hydrogen in the fuel into carbon dioxide (CO_2) and water (H_2O) respectively. Partial oxidation (POX) is the process in which sub-stoichiometric amount of oxygen is added to the process, resulting in carbon monoxide and hydrogen products instead. Autothermal reforming (ATR) combines partial oxidation and steam reforming (SR) in a single process. The steam to carbon (S/C) ratio is the amount of H_2O being added to the reaction relative to the amount of carbon in the fuel on a molar basis. Not shown in Figure 1, another reaction that simultaneously occurs in ATR is water gas shift (WGS) reaction, which is slightly exothermic:

$$CO + H_2O = CO_2 + H_2 \qquad \Delta H = -41.2 \text{ kJ/mole}$$

Finally, fuel decomposition is simply a reforming process in which no O_2 or H_2O is added, and therefore, the products are simply H_2 and carbon or soot.

Figure 1 shows that combustion and POX reactions are exothermic or produce heat, while SR and fuel decomposition reactions are endothermic or consume heat. Typically, ATR reactions are considered to be thermally self-sustaining, and therefore, do not produce or consume external thermal energy. Since both fuel decomposition and SR processes are endothermic, heat must be generated external to the reformer process. Catalysts are commonly used to enhance the reaction rate of the reforming processes at lower temperatures.

Independent parameters that affect the performance of an ATR reactor are inlet feed temperature, S/C, O_2/C Feed, and pressure. Dependent parameters include reformate outlet temperature, conversion and reactor heat loss. For methane reforming, the molar ratio of H_2 produced in the ATR reactor to methane feed (H_2/CH_4 Feed) is a key performance parameter which multiplied by the ratio of Low Heating Values (LHV) of H_2 to CH_4 results in the ATR fuel processor efficiency. H_2/CH_4 Feed can alternatively be viewed as H_2 yield of the ATR reforming process. Optimization of an ATR reactor for maximum H_2 yield is not a trivial matter and requires complex and iterative calculations. A number of studies have been published in the past where maximum H_2 yield has been reported for limited ATR reactor conditions [2,3,4]. These findings do not shed light on the overall optimization of the ATR H_2 yield in the reforming spectrum spanning from combustion to SR.

In this study, a framework for theoretical interpretation of reforming reactions is developed using an atomic balance approach. ATR feed is assumed to be pure methane but the approach can also be applied to other hydrocarbon feeds. This theoretical framework is used to determine the boundaries of possible solutions for data analysis (i.e., Reforming reaction space) and maximum monolith temperature. Finally, H_2 yield is determined based on energy and material balance coupled with chemical equilibrium requirements. The assumptions inherent in the ATR energy and material balance are as follows:

- Complete consumption of O_2 without formation of carbon soot, which is a reasonable assumption at a S/C level of 1.5 or above [2,5].
- SR and WGS reactions to be at equilibrium at ATR outlet temperature. This is again a reasonable assumption based on previous studies [5].
- Adiabatic reaction or a known Heat Loss

Reforming Reaction Space

ATR reaction stoichiometrics can be represented as:

$$CH_4 + a\,O_2 + b\,H_2O = c\,CO + d\,CO_2 + e\,H_2$$

Where,

$a = O_{2,\ Feed}/(CO+CO_2)_{Reformate} = (O_2/CH_4\ Feed)/X$
$b = H_2O_{Reacted}/(CO+CO_2)_{Reformate}$
$c = CO_{Reformate}/(CO+CO_2)_{Reformate} = (CO/CH_4\ Feed)/X$
$d = CO_{2,Reformate}/(CO+CO_2)_{Reformate} = (CO_2/CH_4\ Feed)/X$
$e = H_{2,Reformate}/(CO+CO_2)_{Reformate} = (H_2/CH_4\ Feed)/X$
$X = CH_4$ Conversion $= (CO+CO_2)_{Reformate}/(CO+CO_2+CH_4)_{Reformate}$

The above assumes that all O_2 is consumed in the ATR without formation of carbon soot. Based on atomic balance for O, C and H, it can be shown that:

$b = e-2 = (H_2/CH_4\ Feed)/X - 2$
$c = 2-2a-b$
$d = 2a+b-1$

If c=0 or $CO/CO_2=0$,

$e = 4-2a$
$b = 2-2a$ or $a = 1-0.5b$
$d = 1$

The above means that a plotted versus b and e versus a will be straight lines. The upper solid lines in Figures 2 and 3 represent the case of $CO/CO_2=0$.

If d=0 or $CO/CO_2=\infty$,

$e = 3-2a$
$b = 1-2a$ or $a = 0.5-0.5b$
$c = 1$

The lower solid lines in Figures 2 and 3 represent the case of $CO/CO_2=\infty$. In general, it can be shown that for a given CO/CO_2 in the reformate:

$e = A-2a$

Where, $A = [3\ (CO/CO_2)+4]/[(CO/CO_2)+1]$
and, $3 \leq A \leq 4$

Maximum Monolith Temperature

The atomic balance approach can also be applied to determine the maximum monolith temperature. Metal catalysts supported on corderite Monoliths are commonly used in ATR reactors but corderite does not dissipate heat easily. This is a relevant problem since the exothermic POX has a faster kinetic rate than endothermic SR reaction, which can result in the formation of a hot spot near the entrance of a monolithic ATR reactor. Maximum temperature should approximately occur where b=0 (i.e., no water consumed in the ATR reaction). If b=0, we would have:

$e = 2$
$c = 2-2a$
$d = 2a-1$

Which means that:

$H_2/(CO+CO_2) = 2$
$CO/CO_2 = (2-2a)/(2a-1)$

The above conditions can be input into the ATR energy and material balance to calculate the maximum monolith temperature and conversion.

Discussion

As can be seen in Figures 2 and 3, a = 0, b = -2 and lines depicting CO/CO_2=0 and $CO/CO_2=\infty$, constitute the possible boundaries of the ATR reaction space in the full range of combustion to steam methane reforming. The ATR reaction space is independent of conversion, S/C, pressure, temperature, heat loss and assumption of equilibrium. The line a=0 represents SR reaction where oxygen does not take part. The line b=-2 represents combustion where 2 moles of water are produced per more of methane reacted.

Figure 2 shows that at given b, a higher inlet temperature will require lower a. This makes sense since at higher inlet temperatures less oxygen for exothermic oxidation step is required. The adiabatic reactor operating lines in Figures 2, 3 and 4 are developed based on S/C of 6 and reactor pressure of 97 psig (770 kPa).

Figure 3 shows that regardless of the inlet temperatures, higher a will result in lower e. This makes sense since as a goes up b comes down and hence hydrogen production is lowered. Figure 3 is helpful for data analysis since the relationship between hydrogen production and oxygen consumption is unidirectional, the reaction space is clearly identifiable, and x and y coordinates can easily be calculated from raw data.

In contrast, Figure 4 shows that at lower inlet temperatures, $H_2/(CH_4$ feed) will go through a maximum with respect to $O_2/(CH_4$ feed). This maximum moves higher as inlet temperature increases and disappears around 1,200 °C (i.e., steam methane reforming reactor inlet conditions). The difference between Figures 3 and 4 is conversion of methane. For example, at 500 °C at low oxygen levels the conversion and hydrogen production is low. At higher oxygen levels, hydrogen production steadily climbs as the energetics of the ATR reaction improves. However, at conversion of 97%, hydrogen production experiences a maximum yield.

On the rhetorical question of whether a SR or ATR fuel processor is more efficient [3,4], the answer depends on the inlet reactor conditions. At high inlet temperatures (i.e., 1,200 °C), SR reaction will always yield more hydrogen than ATR, however, as the inlet temperature drops, introduction of oxygen in the reactor results in a higher hydrogen yield in the ATR. It is important to note that for a typical ATR-based hydrogen generation system, where air is the source of oxygen, introduction of nitrogen diluent may result in a lower recovery in the downstream hydrogen purification and separation system resulting in a lower overall system efficiency for an ATR compared to SR.

Effect of S/C, Pressure and Heat Loss

Figures 5, 6 and 7 show the effects of S/C, pressure and heat loss on H_2 yield for an ATR reactor. Higher S/C generally yields more hydrogen with the maximum yield shifting to a lower (O_2/CH_4 Feed) at given inlet temperature. Higher pressure and heat loss have inhibitory effect on the $H_2/(CH_4$ Feed) with maximum H2 yield shifting to higher O2/(CH4 Feed) values at given inlet temperature.
Table 1 summarizes the maximum H_2 yields and provides the reactor inlet and outlet temperatures for the reforming data shown in Figures 4 to 7. For non-adiabatic reactors, heat loss as a percentage of the feed enthalpy is also provided in Table 1.

Maximum H_2 Yield

Figure 8 shows that maximum H_2 increases with decreasing levels of oxygen in the reactor with pressure showing an inhibitory effect. The beneficial effects of increasing inlet feed temperature and S/C on maximum H_2 yield are shown in Figures 9 and 10 respectively. Finally, Figure 11 shows that the rate of change of maximum H_2 yield with respect to inlet feed temperature is dependent on S/C.
It is important to note that in fuel processing systems where ATR feed is pre-heated with hot reformate, inlet temperature and S/C may not be truly independent parameters and a higher inlet temperature may require lower S/C. Higher inlet temperatures may also require specialized construction materials to insure reformer life.

Conclusion

From the simulation data presented in this paper, a reactor optimization strategy can be deduced; operate at the highest possible S/C and inlet feed temperature while minimizing heat loss and reactor pressure. At these conditions, the H_2 yield can be maximized with respect to O_2/(CH_4 feed) by multiple solutions of ATR energy and material balance coupled with reaction equilibrium conditions. The data and conclusions of this paper are applicable to any ATR regardless of size and are independent of feed flowrates.

References

[1] Farrauto R, Cuzens J, Hwang S, Swamy D, Korotkikh O. Catalytic Auto-Thermal Fuel Processing for Fuel Cells. Proceedings of AIChE Spring Meeting in Atlanta, Georgia, March 2000.

[2] Chan, SH, Wang, HM. Thermodynamic Analysis of Natural-gas Fuel Processing for Fuel Cell Applications. International Journal of Hydrogen Energy 2000; 25: 441-449.

[3] Doss E, Kumar R, Ahluwalia RK, Krumpelt M. Fuel Processors for Automotive Fuel Cell Systems: a Parametric Analysis, Journal of Power Sources 2001; 102: 1-15.

[4] Ahmed S, Krumpelt M. Hydrogen from Hydrocarbon Fuels for Fuel Cells. International Journal of Hydrogen Energy 2001; 26: 291-301.

[5] Hochmuth JK. Catalytic Partial Oxidation of Methane over a monolithic Supported Catalyst. Applied Catalysis B: Environmental 1992; 1: 89-100.

Figure 1: Reforming Technologies Spectrum

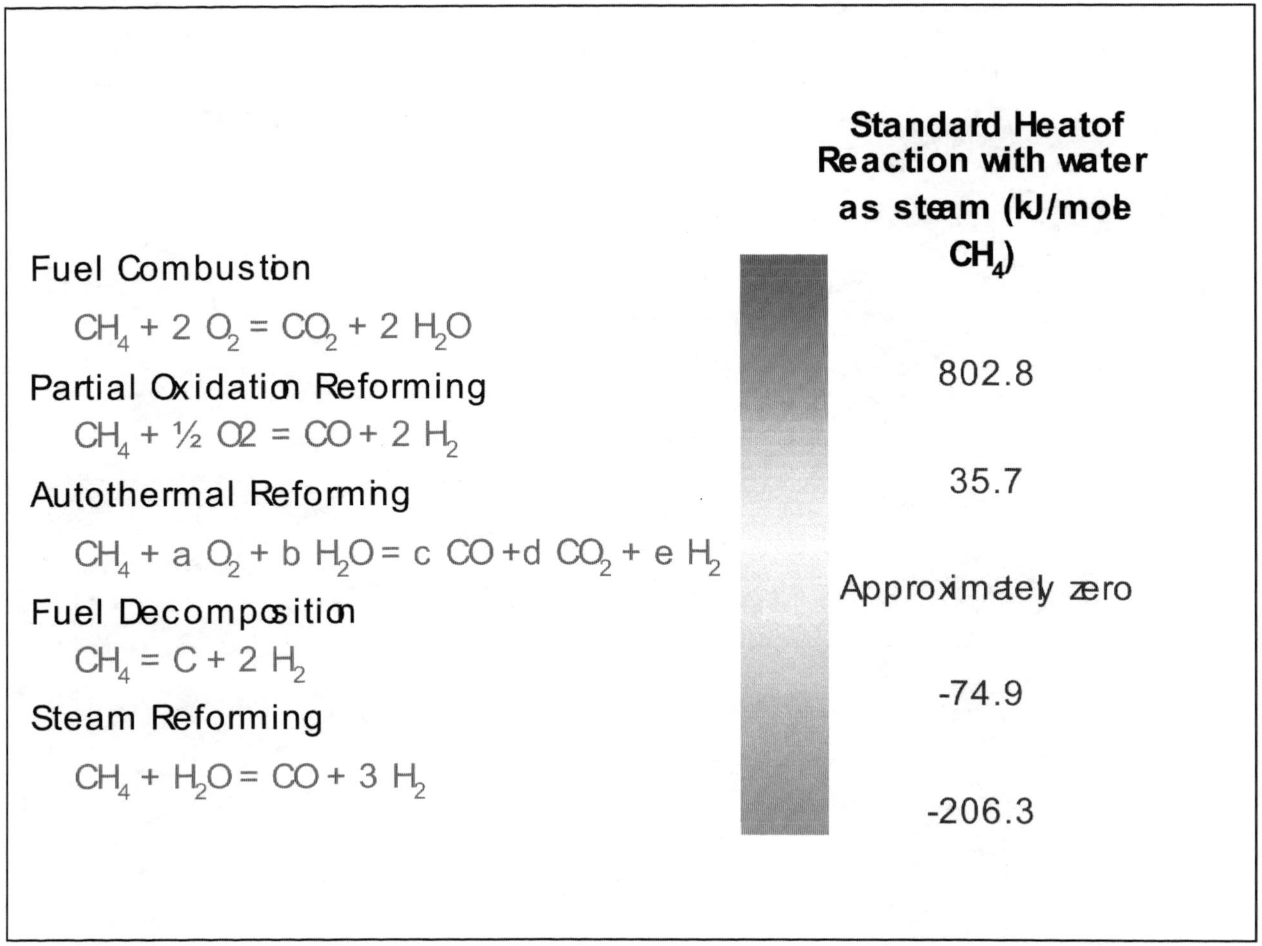

Figure 2: Reforming Reaction Space at S/C = 6 and P = 97 psig (770 kPa)

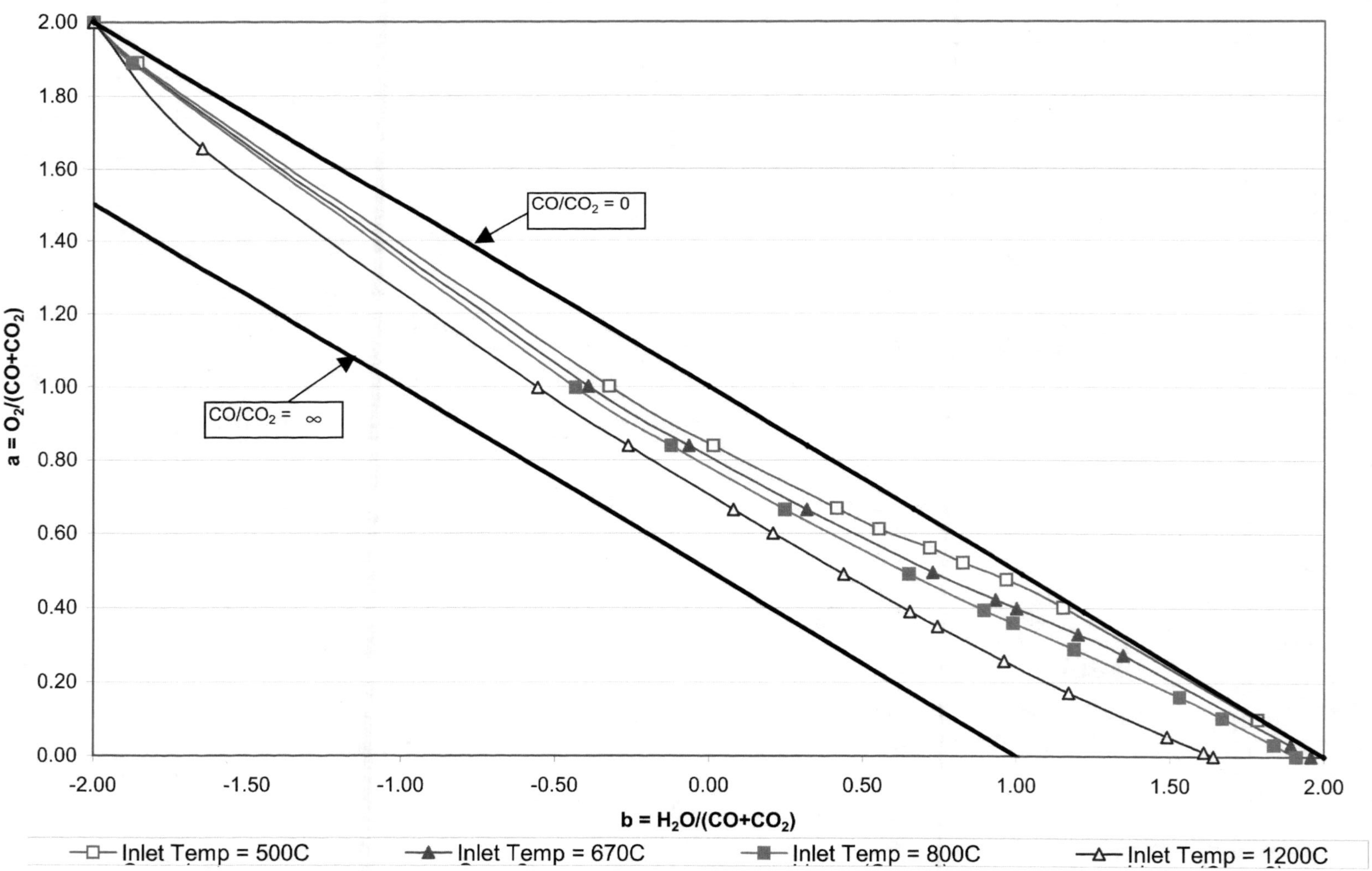

Figure 3: Reforming Reaction Space at S/C = 6 and P = 97 psig (770 kPa)

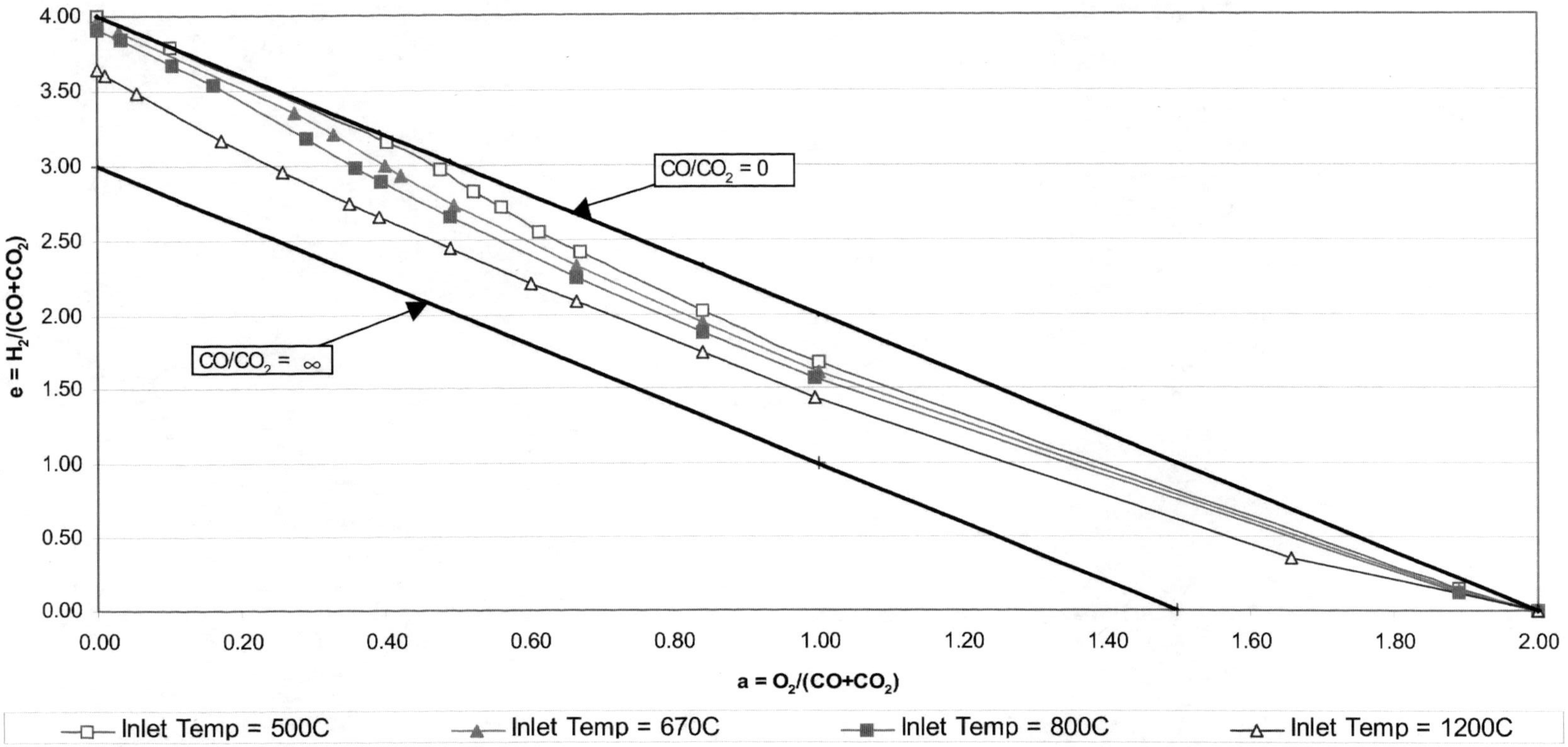

Figure 4: Effect of Inlet Temperature on Hydrogen Yield at S/C = 6 and P = 97 psig (770 kPa)

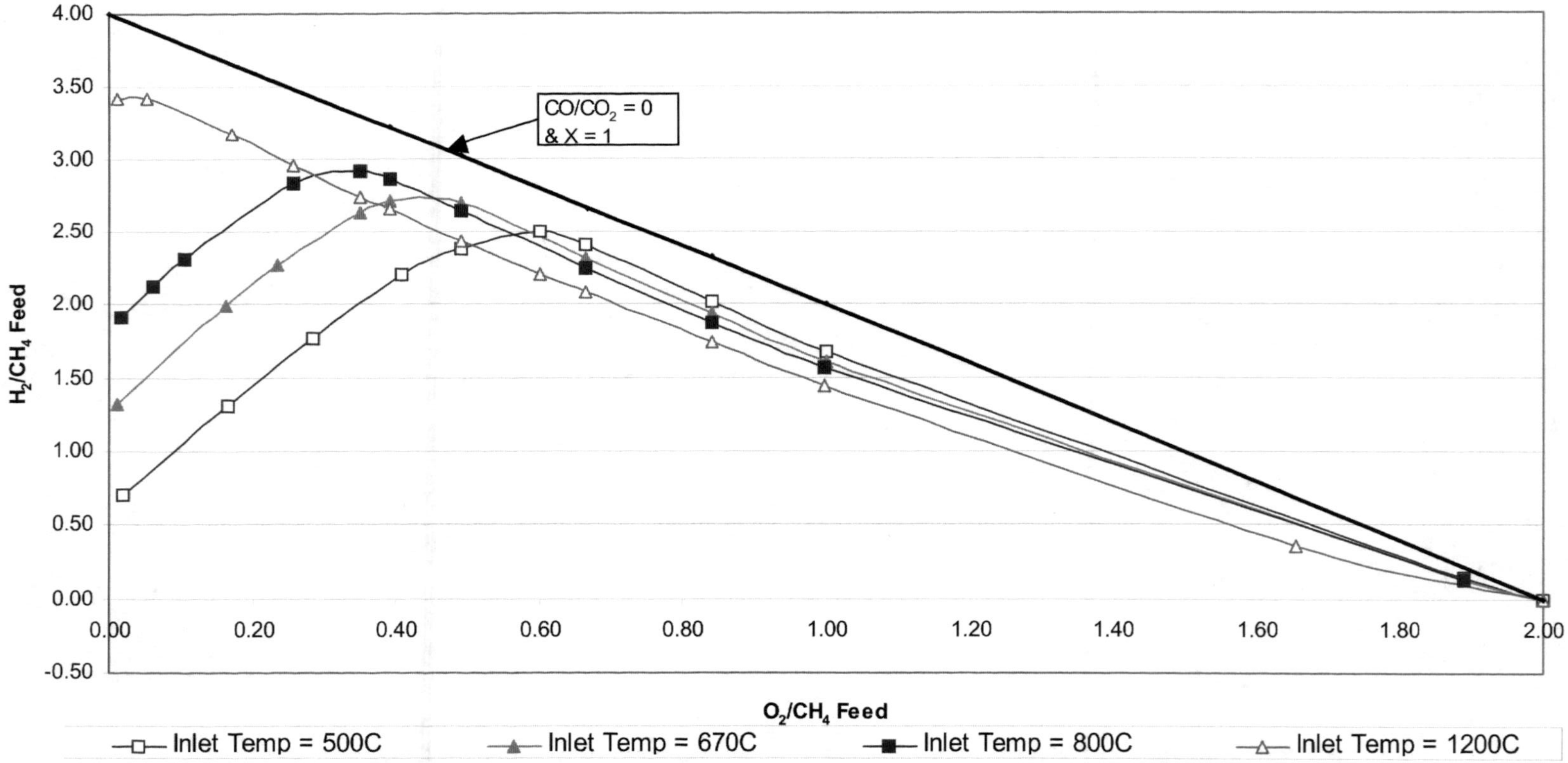

Figure 5: Effect of Pressure on Hydrogen Yield at S/C = 3

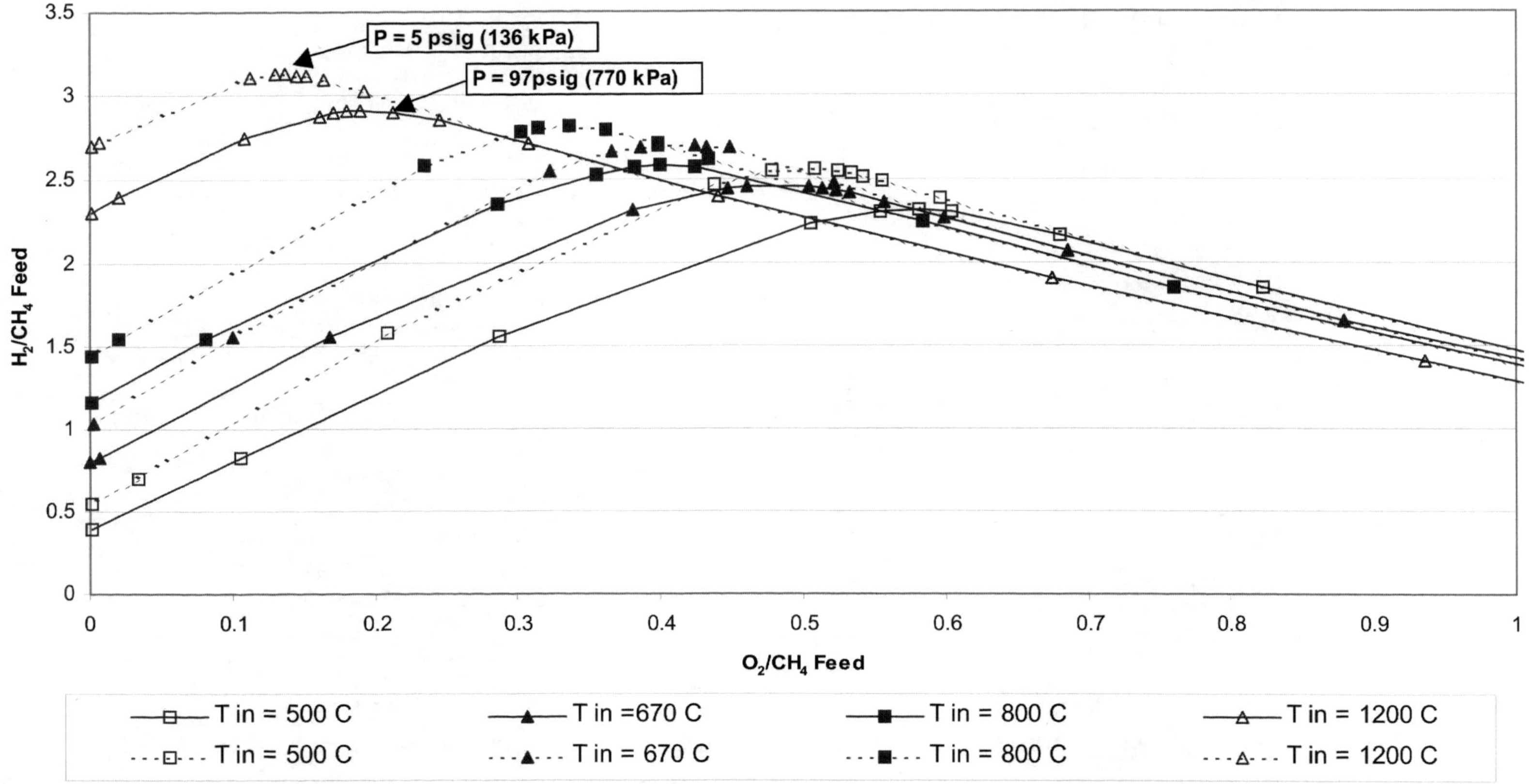

Figure 6: Effect of S/C on Hydrogen Yield at P = 97 psig (770 kPa)

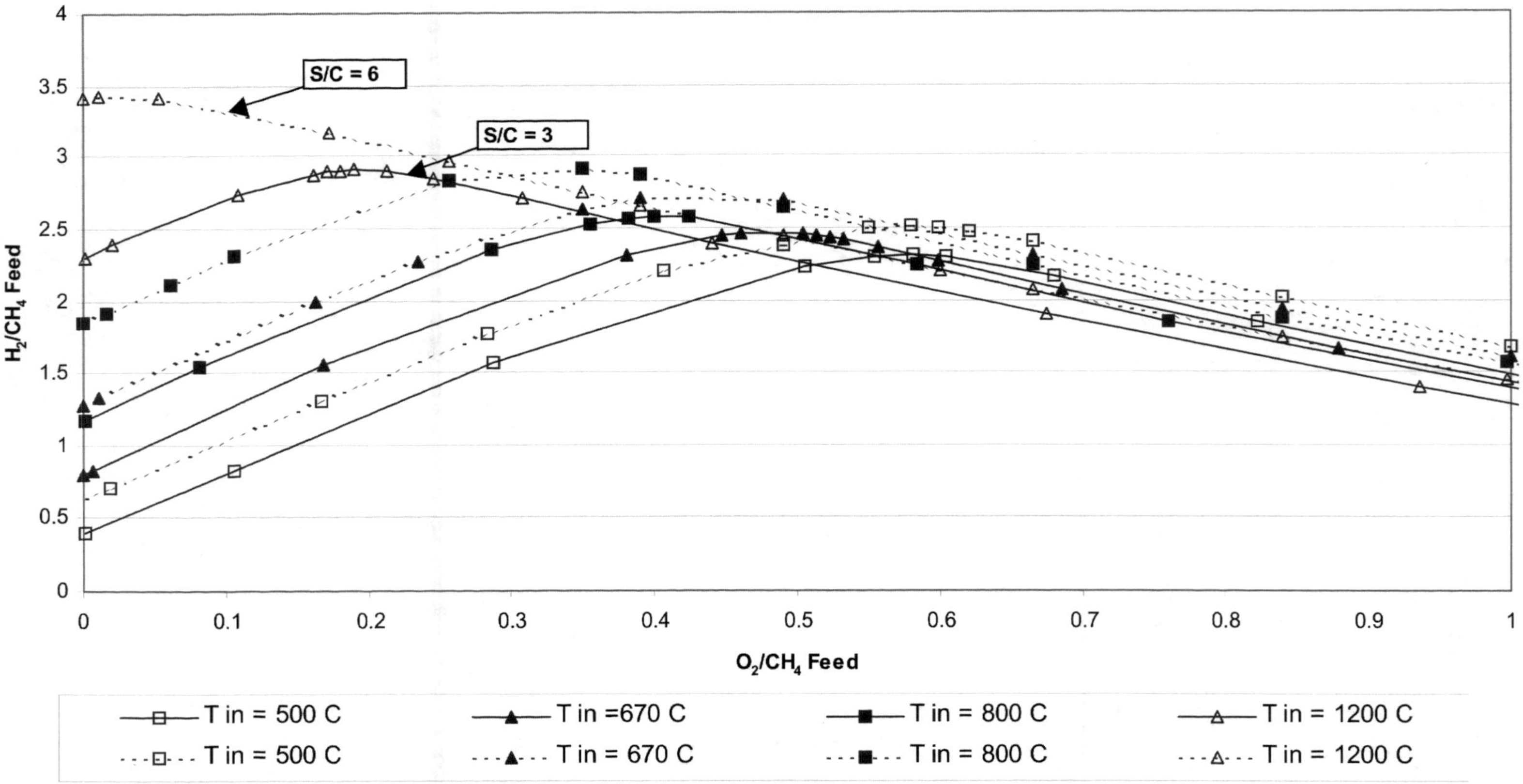

Figure 7: Effect of Heat Loss on Hydrogen Yield at S/C = 3 and P = 97 psig (770 kPa)

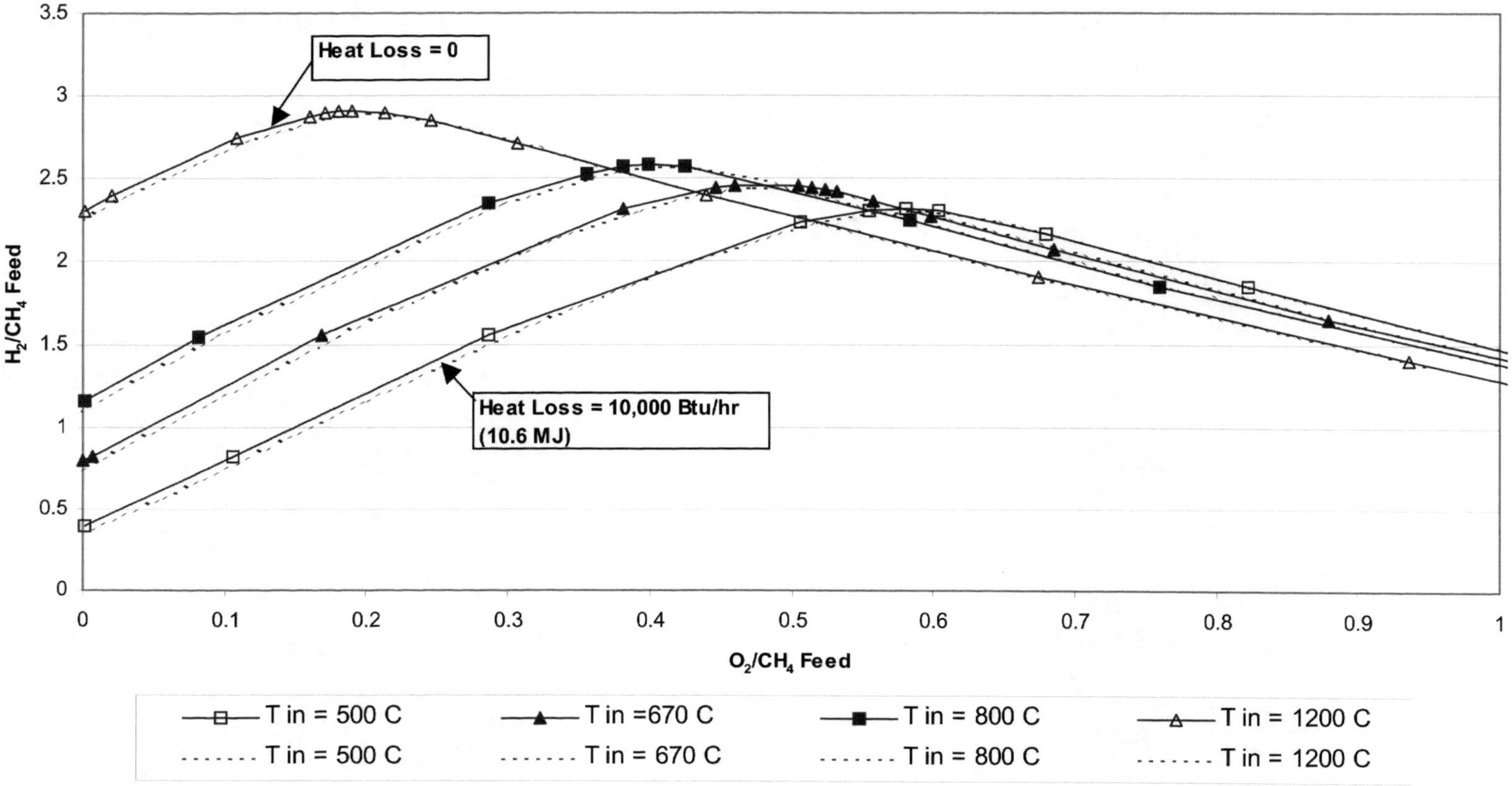

Figure 8: Effect of Pressure on Max. Hydrogen Yield at S/C = 3

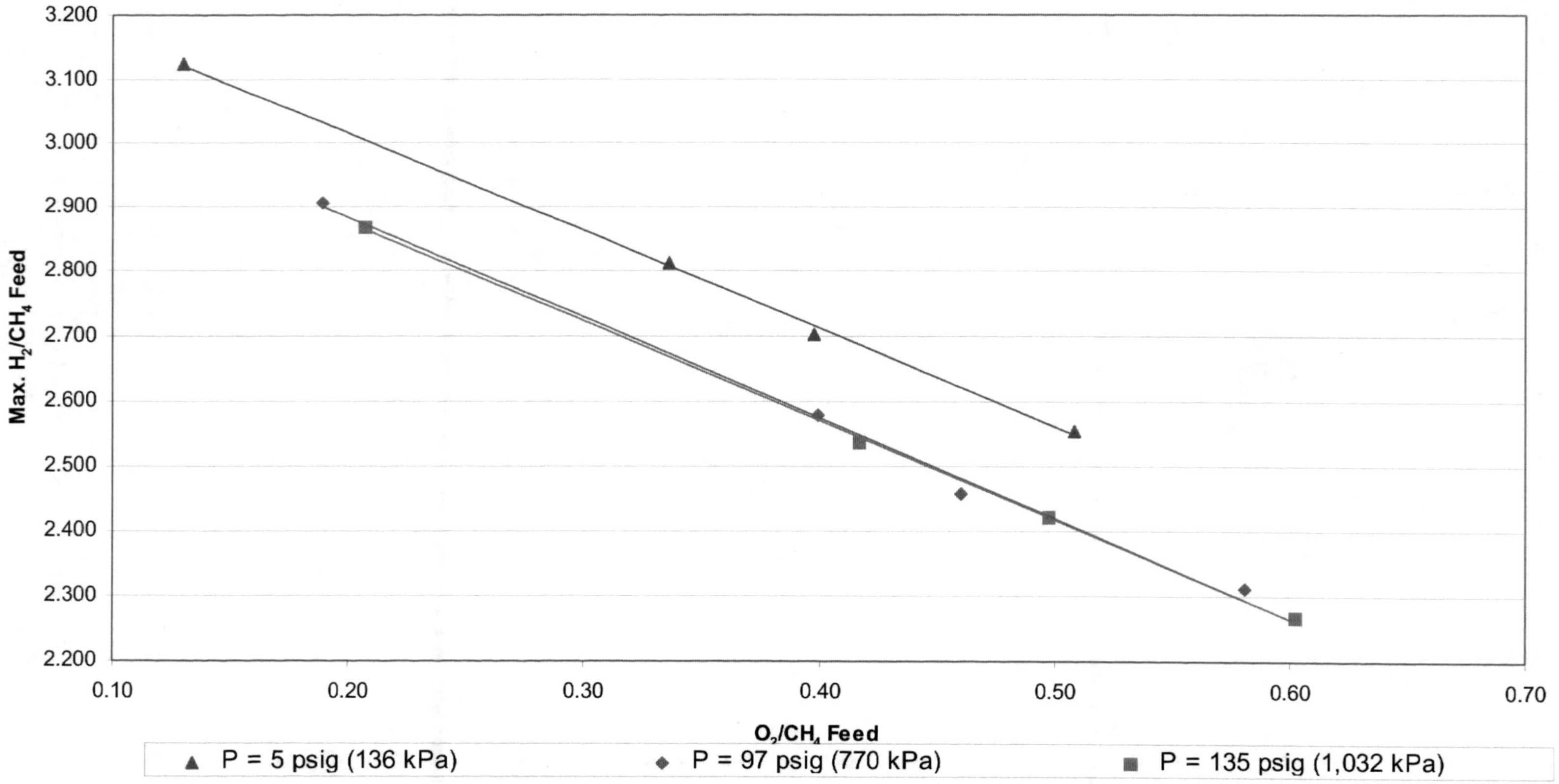

Figure 9: Effect of Inlet Temperature on Max. Hydrogen Yield at S/C = 3

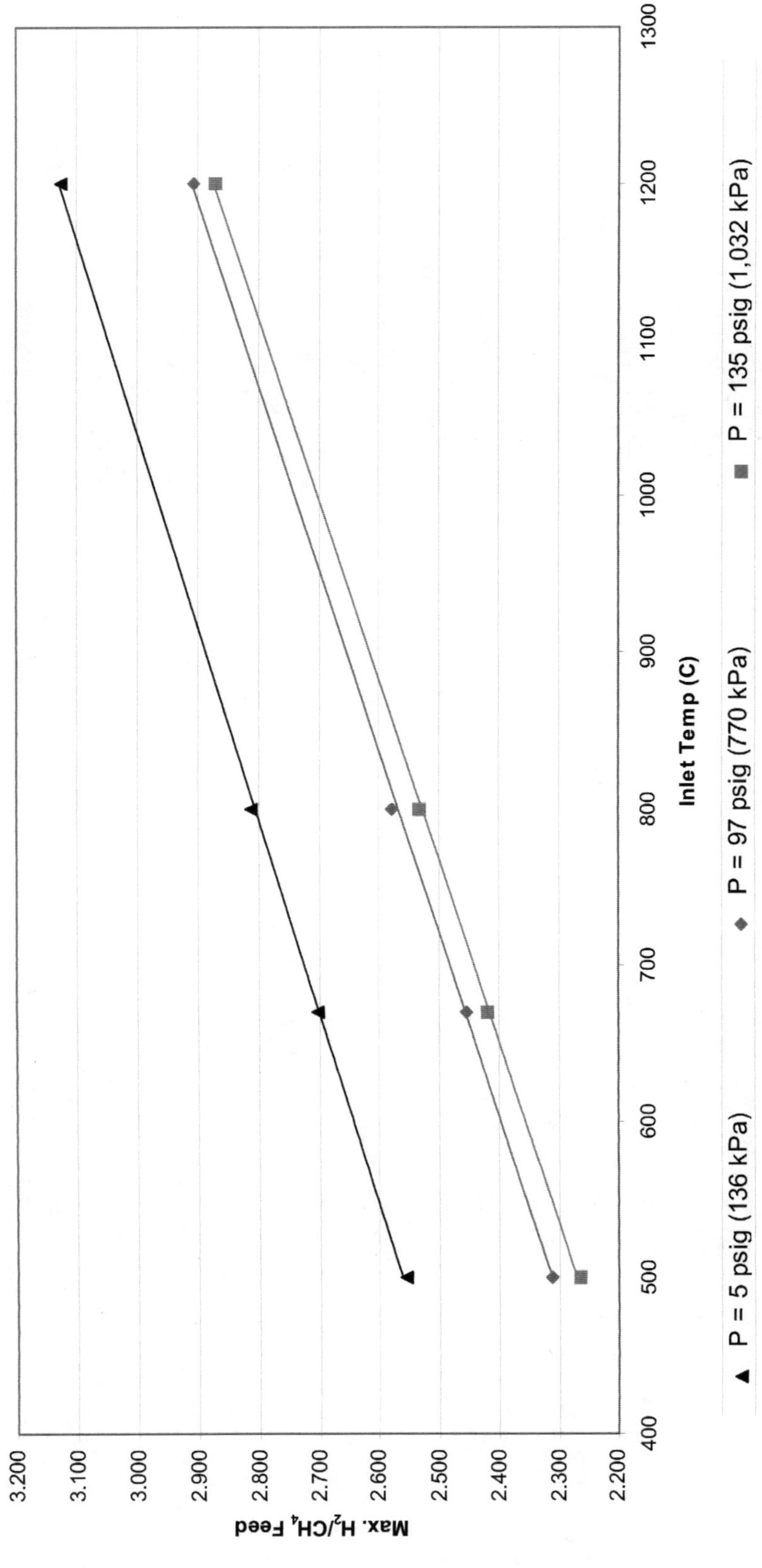

Figure 10: Effect of S/C on Max. Hydrogen Yield at P = 97 psig (770 kPa)

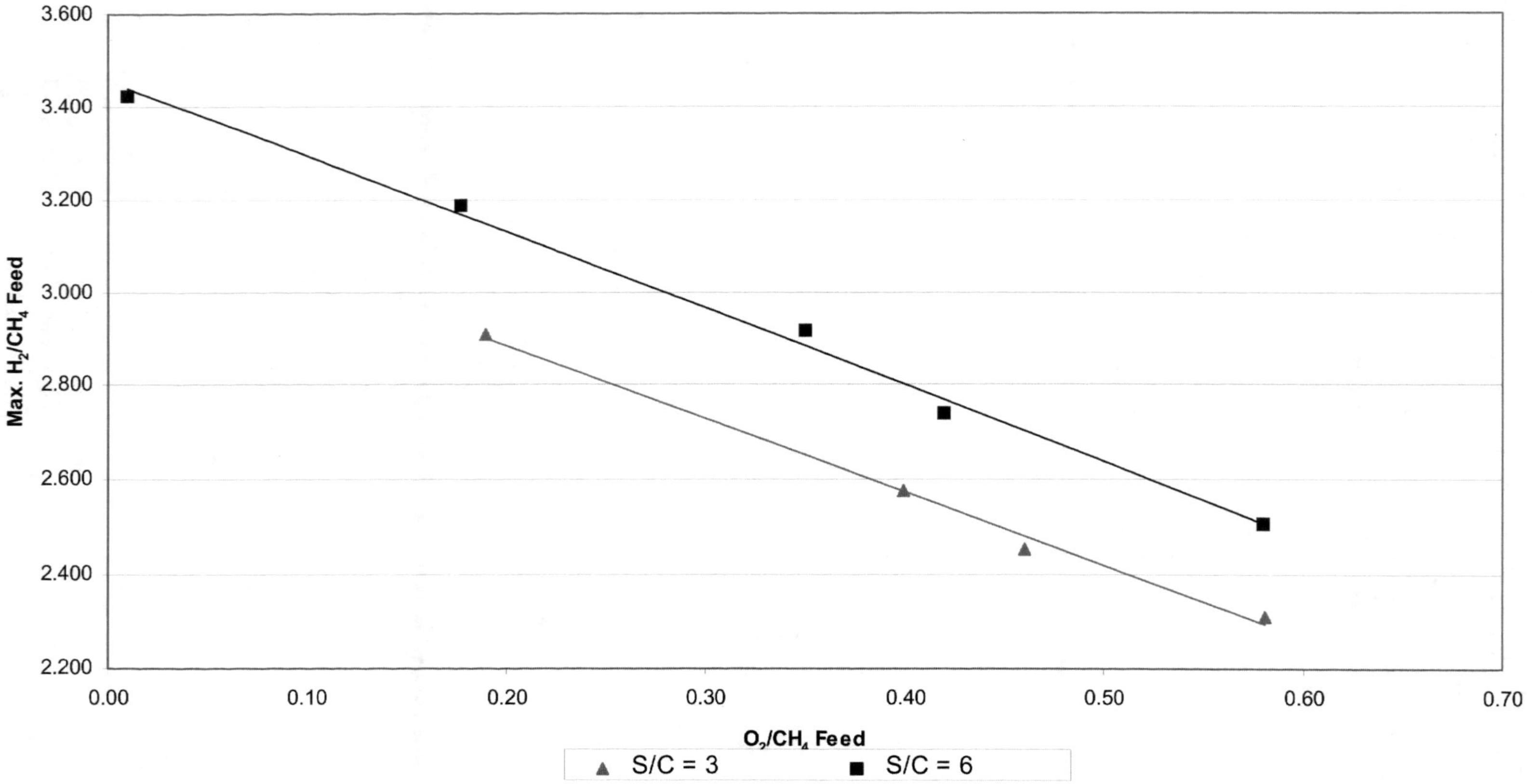

Figure 11: Effect of Inlet Temperature on Max. Hydrogen Yield at P = 97 psig (770 kPa)

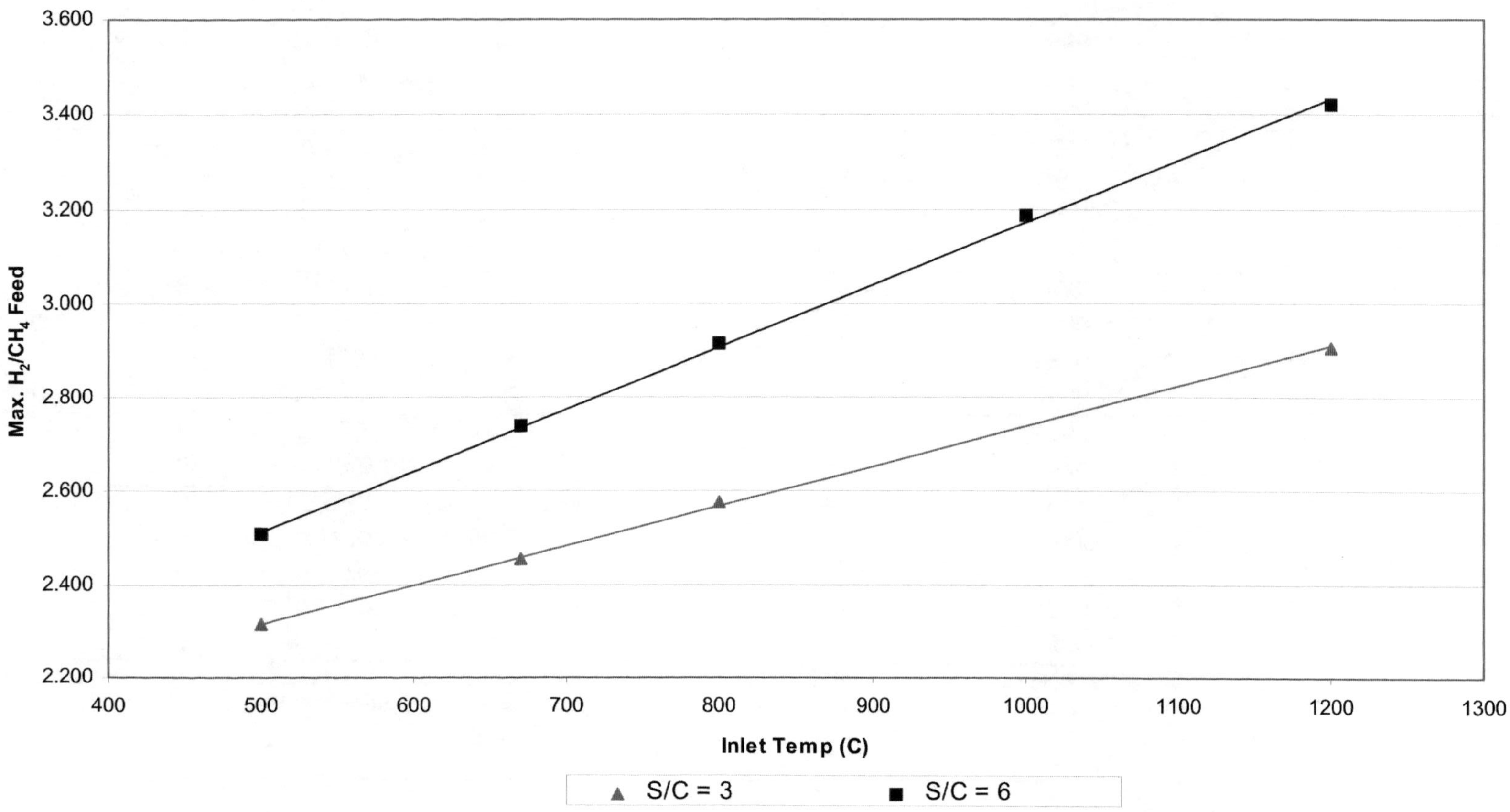

Table 1 - Maximum H_2 Yield at Various Reactor Conditions

O_2/CH_4 Feed	S/C	Pressure psig (kPa)	Heat Loss Btu/hr (MJ/hr)	Heat Loss/Feed Enthalpy	Inlet Temperature (C)	Outlet Temperature (C)	H_2/CH_4 Feed
0.58	6	97 (770)	0	0	500	688	2.507
0.42	6	97 (770)	0	0	670	692	2.739
0.35	6	97 (770)	0	0	800	734	2.915
0.18	6	97 (770)	0	0	1000	737	3.185
0.01	6	97 (770)	0	0	1200	746	3.423
0.35	6	5 (136)	0	0	670	608	3.013
0.25	6	5 (136)	0	0	800	621	3.191
0.51	3	5 (136)	0	0	500	650	2.555
0.40	3	5 (136)	0	0	670	650	2.700
0.34	3	5 (136)	0	0	800	670	2.813
0.13	3	5 (136)	0	0	1200	700	3.124
0.60	3	135 (770)	0	0	500	775	2.265
0.50	3	135 (770)	0	0	670	790	2.419
0.42	3	135 (770)	0	0	800	800	2.535
0.21	3	135 (770)	0	0	1200	850	2.869
0.58	3	97 (770)	0	0	500	750	2.312
0.46	3	97 (770)	0	0	670	750	2.456
0.40	3	97 (770)	0	0	800	775	2.577
0.19	3	97 (770)	0	0	1200	820	2.907
0.59	3	97 (770)	10,000 (10.6)	3.34%	500	750	2.298
0.47	3	97 (770)	10,000 (10.6)	2.50%	670	750	2.444
0.41	3	97 (770)	10,000 (10.6)	2.07%	800	775	2.565
0.20	3	97 (770)	10,000 (10.6)	1.42%	1200	825	2.897

Compression Moldable Composite Bipolar Plates for Fuel Cells

Jianhua Huang and Donald G. Baird
Department of Chemical Engineering, Virginia Tech
Blacksburg, VA 24061-0211

ABSTRACT

A new technology is developed to produce economical bipolar plates with high electrical conductivity and mechanical properties. The composite consisting of graphite particles, thermoplastic fibers and glass fibers is generated by means of a wet-lay process to yield highly formable sheets. The sheets together with additional graphite particles are then stacked and compression molded to form bipolar plates with gas flow channels and other features. The plates containing 65 wt-% graphite have a bulk conductivity of over 200 S/cm, well exceeding the DOE target (100 S/cm) for composite bipolar plates. This value of conductivity appears to be the highest of all polymer composite plates with the same or similar graphite loadings, reaching the range of carbon/carbon composite bipolar plates (200~300 S/cm, Oak Ridge National Laboratory). In addition, the plates have flexural and tensile strengths higher than any other polymer composites with the same graphite content. Because the plates can be generated without high temperature pyrolisis and chemical vapor infiltration processes, they can be manufactured at much less cost compared to the carbon/carbon composite plates.

KEY WORDS

fuel cell, bipolar plate, conductive material, graphite filled composite, compression molding

INTRODUCTION

The significant and growing interest in fuel cells for stationary power and transportation has been demonstrated by the attention these technologies are receiving from both government and industry especially the automotive sectors. As one of the key and costly components of PEM fuel cells, bipolar plates must have high electrical conductivity, sufficient mechanical integrity, corrosion resistance, low gas permeability, and low-cost in both materials and processing when applied to the civil market [1]. Historically bipolar plate materials have been metals coated with corrosion resistant layers or graphite with a seal treatment (to lower the gas permeability). In both cases, the bipolar plates require extensive machining and post processing, resulting in hardware costs far more expensive than the costs for the raw materials alone. To date, the costs of bipolar plates (~ \$10/plate with 400 cm^2) dominate the stack costs. Unless economical bipolar plates are developed, PEM fuel cells cannot be applied to the civilian markets and compete with established power technology [2].

To replace graphite bipolar plates and lower the cost, a variety of composite bipolar plates have been developed. Most of them were made by compression molding of polymer matrix (thermoplastic or thermoset resins) filled with conductive particles (such as graphite powders) or fibers [3-6]. Because most polymers have extremely low electronic conductivity, excessive conductive fillers have to be incorporated, and it is difficult to get high conductivity and sufficient mechanical properties at the same time. To solve this problem, Oak Ridge National Laboratory (ORNL) recently developed carbon/carbon composite bipolar plates [2, 7]. The manufacture process consists of multiple steps, including the production of carbon fiber/phenolic resin preforms (by slurry-molding process) followed by compression molding, and the pyrolysis and densification on surface by a chemical vapor infiltration (CVI) process. The plates have high conductivity and the technology has been transferred to Porvair Fuel Cell Technology for scale up [8]. However, this process seems to be too complicated and costly to be applied in civil applications like automotive transportations.

Recently another technology for making conductive materials attracted our interest. Tucker et al [9] made wet-lay sheet materials (mats consisting of graphite particles, thermoplastic fibers and glass or carbon fibers) using a slurry-making process and used them in compression molding to form composite plaques with excellent conductivity and mechanical properties. While the work was not targeted at fuel cell bipolar plates, it is possible that this kind of structure helps to achieve highly conductive, tough yet economical bipolar plates. The goal of the present work is to explore ways to increase the electrical conductivity of the materials further (from below 100 S/cm to as high as possible) while keeping the mechanical integrity of the composite as required for bipolar plates. The electric and mechanical properties of new material will be evaluated and compared with other composite bipolar plates. We will also evaluate the formability of the composites to see whether the bipolar plates with gas flow channels can be molded. This feature is very important for the composite materials to be used as bipolar plates of low cost.

EXPERIMENTAL

Wet-lay sheets made from 50 wt% graphite particles, 40 wt% polyester fibers and 10 wt% glass fibers were donated by DuPont. We note that the polyester may not be an idea matrix for application in fuel cell environment, which however will not affect the present work for testing the concept. If the result turns out to be satisfactory, there would be several measures that could be used to solve this problem. For example, the PET fiber may be replaced by thermotropic liquid crystalline polymer (TLCP) fibers. TLCPs are known as excellent matrices for fuel cell application and can also be used for making the wet-lay sheet materials[9]. It is also possible to mold a laminate bipolar plate with chemical resistant layers (such as mixture of fluoropolymer and graphite powders) covering the PET based material. The graphite powder that was added in the compression molding step was TIMREX provided by Timcal America Inc.

The sheet materials were cut according to the mold size and stacked together with additional graphite powders in the mold. The assembly was placed in a hydraulic press and pressed at 277°C and 900~1500 psi for 10 minutes. Then the platen heaters were turned off and the mold was allowed to cool. The pressure was maintained until the mold temperature reached 200°C. The pressure was then allowed to drop as mold temperature deceased further. When the mold temperature reached 30°C, the platens were opened and the assembly was removed from the press. The flat plaque or bipolar plate with gas flow channels was then removed from the mold.

The bulk conductivities (in-plane) were measured using the van der Pauw method according to ASTM Standard F76-86. The typical size of the specimens is 25.4mm in diameter and 1~2 mm in thickness. The sheet resistance, R_S, was obtained from the two measured characteristic resistances R_A and R_B by numerically solving the van der Pauw equation:

$$\exp(-\pi R_A/R_S) + \exp(-\pi R_B/R_S) = 1 \qquad (1)$$

The resistivity ρ is given by $\rho = R_S d$, where d is the thickness of the specimen. The volume conductivity $\sigma = 1/\rho$.

The tensile and flexural (three-point bending) tests were performed at room temperature (23°C) on an Instron 4204 tester in accordance with ASTM D638 and D790 standards, respectively. The specimen sizes were of L(Length)×W(Width) = 76.2×7.7 mm for the tensile test, and L×W=76.2×12.7 mm for the flexural test. The thickness of the samples was about 2 mm.

RESULTS AND DISCUSSION

In Fig.1 are presented the bulk conductivities (in-plane) of compression-molded plaques using wet-lay sheets (wet-lay material) or wet-lay sheets plus additional graphite (wet/dry lay material). It was noted that the graphite content of the wet/dry lay materials could reach 85 wt%, which is much higher than what the wet-lay sheets could hold [9]. As the graphite content increases, a significant increase in bulk conductivity was observed. Because the composites with graphite higher than 75% have apparently poor mechanical properties, further research was focused on the material with 65% graphite.

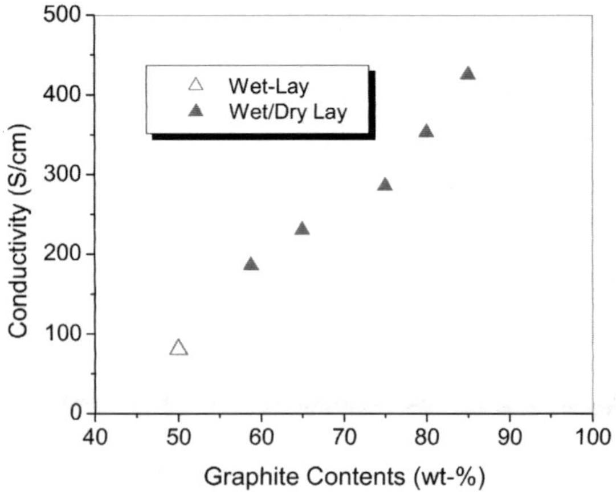

Figure 1. Conductivity of wet-lay and wet/dry lay composite materials.

The bulk conductivities (in-plane) of wet/dry (W/D) lay material and other state-of-the-art composite materials for bipolar plates are listed in Table 1. The plates containing 65 wt % graphite have a bulk conductivity of over 200 S/cm, well exceeding the DOE target (100 S/cm) for composite bipolar plates [1]. This value of conductivity is also the highest of all polymer composites with the same or similar graphite loadings,

Table 1. Property comparison for composite bipolar plates.

Company/Patent	Resin	Fillers, Wt %	Conductivity (S/cm)	Strength (MPa)***	
				Flexural	Tensile
US 6,248,467 (LANL)	Vinyl Ester	68% Graphite	60	29.6	23.4
Commercial*	Unknown	Unknown	105	20.7	19.3
Premix Inc.*	Vinyl Ester	68% Graphite	85	28.2	24.1
BMC Inc.*	Vinyl Ester	69% Graphite 2% Cotton 1% C Black	30	37.9	26.2
Plug Power**	Vinyl Ester	68% Graphite	55	40.0	26.2
Plug Power Target**			85	58.6	41.3
US 4,214,969 (GE)	Fluoropolymer	74% Graphite	119	35.1-37.2	/
US 4,339,322 (GE)	Fluoropolymer	74%(Graphite, Carbon Fiber)	109	42.7	/
US 5,942,347 (GTI)	Phenolic Resin	77.5% Graphite	53	/	/
US 6,171,720 (ORNL)	Phenolic Resin (Precursor)	Carbon Fiber	200-300	N/A	/
This Work (VT)	Polyester	65% Graphite 7% Glass Fiber	230	53.0(2.35)	36.5(2.06)

* From Ref[6].
** From Ref[1].
*** Data in brackets represent the standard deviations.

reaching the range of carbon/carbon composite bipolar plates (200~300 S/cm) developed by the Oak Ridge National Laboratory [2] and Porvair Fuel Cell Technology [8]. We note that our technology involves no pyrolysis and CVI processes which represent over 70% of total cost of C/C plates [2], and it should be possible to manufacture the plates at much less cost compared to the carbon/carbon plates.

In addition to the electrical conductivity, the bipolar plates should also have adequate mechanical properties to be applied in the fuel cell stacks. However, for polymer composites doped with conductive particles or fibers, it is difficult to get high conductivity and sufficient mechanical properties at the same time. In Table 1 is also listed the mechanical properties of W/D lay material and other composite plates. The flexural and tensile strengths of W/D lay composite are 53.0 MPa and 36.5 MPa, respectively, both are the highest in all polymer composite plates with the same or similar graphite loadings. In comparison with the present and target values of the Plug Power Inc. (see Table 1), our result represents significant progress toward the goals for commercialization. It is noted that Besmann et al. [2] reported flexural strength of 175 MPa for their carbon/carbon plates. However, because the property was obtained by means of a biaxial flexure test, not the standard three-point flexure as defined by ASTM D790, it is difficult to compare their results with other strength data listed in Table 1.

It is apparent that the mechanical properties of the wet/dry lay material have a close relation with the structure of the wet-lay sheet materials. It has been shown that compression molded wet-lay material has excellent mechanical properties [9]. This is believed to be the result of the unique structure of the wet-lay sheets, including the interaction of the reinforcing fibers and the layered structures formed in the slurry-making process. In Figs. 2 and 3 are presented the mechanical properties of wet/dry lay materials as compared to the wet-lay materials (containing 50 wt% graphite). As graphite content increases from 50 wt% to 65 wt%, the modulus of the material increases, which can be attributed to the addition of graphite powder which has a modulus significantly higher than that of the matrix (polyester). In contrast, the strength and maximum strain decrease in both tensile and flexural tests. This may also be attributed to the addition of graphite powder that acts like defects in the polymer matrix when the composite undergoes a tensile or flexural test. The loss of tensile or flexural strength in W/D lay composites may also be caused by the decrease of glass fiber contents after more graphite (in addition to that contained in the wet-lay materials) is added. In fact, W/D lay material containing 65% graphite has only 7 wt% glass, which is less than original wet-lay material (10 wt%) by 30%. Nevertheless, as can be seen from Figs. 2 and 3, W/D lay composites still have 89% tensile strength and 84% flexural strength relative to the wet-lay materials. That is, the W/D lay materials containing 65% graphite still retain the good mechanical properties offered by the reinforcing fibers and layered structures in the wet-lay sheets.

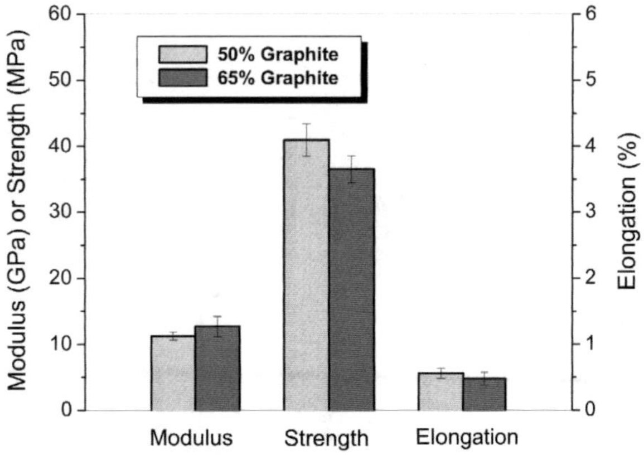

Figure 2. Tensile properties of compression molded wet-lay and wet/dry lay composites.

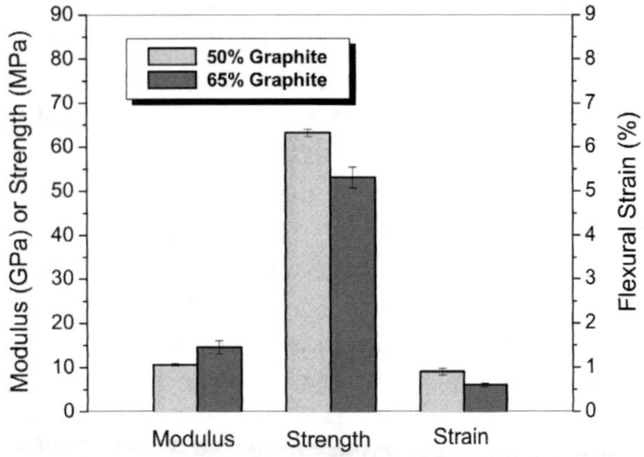

Figure 3. Flexural properties of compression molded wet-lay and wet/dry lay composites.

For all composite materials for bipolar plates, the formability of the material is very important if the gas flow channels are to be readily formed during a molding process.. This is because the flow channels in bipolar plates are numerous (i.e. densely spaced), narrow and relatively deep, e.g. 0.8 mm or 1/32 inch in width and depth. For our composite containing stacks of wet-lay mats, we are particularly concerned with whether the material is deformable enough to allow the formation of the channels and other features of bipolar plates as required during compression molding. If not, the flow channels would have to be machined and the advantages of polymer composites would vanish.

To evaluate the formability of the wet/dry lay composite, we first made a mold with a few straight grooves based on a standard bipolar plate design (7-channel, Los Alamos National Laboratory) and used it in the compression molding. The result showed that the composite has good formability, and the flow channels (simplified) formed are as good as other composite systems. On this basis, a standard mold was designed and fabricated for making the composite bipolar plates. In Fig.4 is presented the bipolar plate generated with this mold and wet/dry lay materials. It can be seen that the composites are highly formable and complicated flow channels and other features of the bipolar plates can be readily generated by compression molding. In our next work we will test the performance of new bipolar plates in PEM fuel cells. We will measure the through-plane conductivity of the material and use it together with the in-plane conductivity to predict the performance of the plate in actual fuel cells.

Figure 4. A compression molded composite bipolar plate with wet/dry lay materials.

SUMMARY

This work showed that a composite consisting of conductive graphite fillers and a thermoplastic matrix reinforced with glass fibers could be compression molded to form bipolar plates with high electrical conductivity and mechanical properties. The materials containing 65 wt % graphite have a bulk conductivity of over 200 S/cm. This value of conductivity is the highest of all polymer composites with the same or similar graphite loadings, reaching the range of carbon/carbon composite bipolar plates (200~300 S/cm) as reported by Oak Ridge

National Laboratory. The flexural and tensile strengths of the composite are 53.0 MPa and 36.5 MPa, respectively, which also appear to be the highest. Because the plates can be generated without high temperature pyrolisis (for carbonization) and chemical vapor infiltration (for densification) processes, it should be possible to manufacture them by compression molding at much less cost compared to the carbon/carbon plates under commercialization.

ACKNOWLEDGMENTS

The financial support of this work from US Department of Energy is gratefully acknowledged.

REFERENCES

1. J.G. Clulow et al., in "Fuel Cell Technology: Opportunities and Challenges", Topical Conference Proceedings, 2002 AIChE Spring National Meeting, March 10-14, 2002 New Orleans, LA. pp.417-425.
2. Besmann, T.M. et al., J. Electrochem. Soc., 147(11) 4083-4086 (2000).
3. Lawrance, US 4,214,969 (1980).
4. Balko et al., US 4,339,322 (1982).
5. Jousse, F et al., EUROMAT '99, Biannu. Meet. Fed. Eur. Mater. Soc. (FEMS) 13, 181-186 (2000).
6. Wilson et al., US 6,248,467 (2001).
7. Besmann et al., US 6,171,720 (2001).
8. D. Haack, R. Torre, K. Butcher, in "Fuel Cell Technology: Opportunities and Challenges", Topical Conference Proceedings, 2002 AIChE Spring National Meeting, March 10-14, 2002 New Orleans, LA. pp.454-459.
9. Tucker et al., US 5,614,312 (1997).

Modeling High-Temperature Electrocatalysts for Alcohol-Based Fuel Cells

Michael A. Rottmayer, Thomas L. Reitz
Air Force Research Laboratory
1950 Fifth St. Wright-Patterson, OH 45433

ABSTRACT

Fuel cell anode catalysts are an essential component for facilitating the direct conversion of chemical energy into electricity. There is an increasing interest in the direct alcohol/air proton exchange fuel cells (PEMFC), particularly for the development of unmanned air vehicles. To date, the development of direct alcohol fuel cells (DAFC) have primarily focused on using methanol, ignoring higher alcohols such as ethanol and propanols as potential fuel cell materials where the energy density increases substantially (Fig. 1).

Figure 1. Energy density comparison of higher alcohols to methanol.

Fuel	C_8H_{18}	Propanol	Ethanol	Methanol
Whr/kg	13,300	9,300	8,250	6,088

Ethanol is a renewable fuel readily available by fermentation of any sugar-containing plant material. In comparison to methanol, it is much less toxic and has nearly comparable kinetics at higher temperatures in an operational fuel cell environment. The disadvantage of ethanol as a fuel is the difficulty with electrochemically splitting the C-C bond, resulting in incomplete reaction products such as acetaldehyde and ethyl acetate. Electro-catalysts can be designed to promote electrochemical oxidation through the increase in catalytic activity, thus becoming a major factor in fuel cell efficiency.

Platinum is generally used in PEMFC due to its proven catalytic activity for hydrogen and methanol electro-oxidation at operating temperatures. In catalytic processes, the catalyst material is not consumed and can be recycled for future use. Platinum also offers high electrical conductivity and is stable at high temperatures. However, at PEMFC operating temperatures, platinum offers low catalytic activity for complete oxidation of higher alcohols resulting from high oxidation over-potential due to its CO/OH adsorbates. Various candidate electro-catalysts can be studied computationally in an attempt to enhance electrical output by increasing fuel oxidation efficiency. By employing platinum-based electro-catalysts, the exceptional electrical properties of platinum will be retained while allowing opportunities for supplementary metal(s) to minimize the poisoning effects.

Using Gaussian 98W, a molecular modeling and simulation software, studies began through model development of the pure ethanol molecule and its adsorption onto a platinum cluster. In order to control the number of variables involved in computational chemistry, precise structures were determined by bond distance, energy values, and vibrational spectra. Consistency with literature values was achieved. The ethanol infrared spectra obtained through the frequency calculations closely resemble literature Fourier Transform Infrared Spectroscopy data (1). Platinum cluster bond distances were gathered via post-structure optimization and determined to be within four percent of bulk crystal literature data (2).

Once confidence was reached with molecular development, the methanol oxidation scheme was studied. The direct methanol fuel cell (DMFC) has been studied extensively describing pure platinum catalytic activity with methanol (3). The objective of our investigation was to match computational structure and energy characteristics with cited experimental data. A density functional theory model was shown to accurately predict dissociative adsorption of methyl hydrogen through the platinum cluster. Bond distance and adsorption energy values calculated agree with cited experimental values to within five percent (4).

The ethanol oxidation scheme is being studied in order to understand the adsorption, reaction, and dissociation on the platinum cluster. Preliminary models show promising results with an initial adsorption through methylene carbon onto the platinum surface in agreement with identified adsorbates cited in literature (Fig. 3) (5).

Figure 3. Adsorbates in the ethanol oxidation reaction determined by in-situ IR spectroscopy.

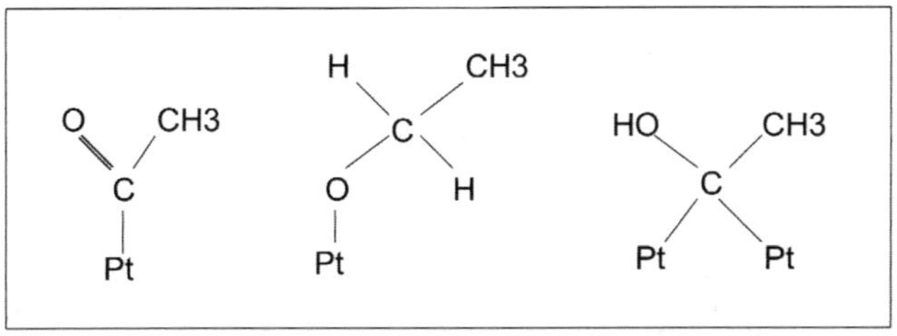

Transition state structures are being approximated in order to find the minimum energy pathway. This pathway, determined through the intrinsic reaction coordinate (IRC) analysis, traces the steepest descent path from the transition state to both reactants and products; activation energy is extrapolated from this procedure.

Based on the study of the ethanol oxidation mechanism and pathway on a pure platinum surface, the catalyst can be modified to examine the effects of additional metals. Alloy compounds will be computationally evaluated to determine optimal activity for the relaxation of the C-C bond.

KEY WORDS Electrocatalyst Design, Computational Modeling, Direct Alcohol Fuel Cell

REFERENCES

1. http://webbook.nist.gov/chemistry/.
2. Surf. Sci., **463**, 66, (2000).
3. Surf. Sci., **463**, 66, (2000).
4. Chem Phys., **177**, 399, (1999).
5. Electrocatalysis, Wiley-VCH, New York, (1998)

CONTACT

Michael A. Rottmayer, MRottmayer@aol.com.

Combinatorial exploration of ternary fuel cell electrocatalysts for DMFC anodes - a comparative study of the PtRuCo, PtRuNi, PtRuW systems

*Peter Strasser *, Qun Fan, Martin Devenney and W. Henry Weinberg*

Symyx Technologies, Inc., 3100 Central Expressway, CA 95051, USA

pstrasser@symyx.com

Prepared for Presentation at the AiChE meeting, Topical Session 173a,

November 21, 2003

Abstract

A high throughput electrochemical screening platform is presented that allows the combinatorial thin-film synthesis and parallel screening of 64 different electrocatalysts with respect to trends in the intrinsic electrocatalytic activity as a function of composition. Catalyst synthesis and screening were performed on a wafer supporting a 64-element electrode array with individual addressability (electrocatalyst library). The electrocatalyst library considered in this study included 6 evenly-spaced stoichiometries of each of the six ternary systems PtAB, A,B = Ru,W,Ni,Co. After sputter deposition synthesis, all 64 catalysts were screened in parallel for their methanol oxidation activity in acidic media, and their individual geometric and specific chronoamperometric current densities were monitored and evaluated against standard PtRu catalysts. The surface-area normalized activity data of the three PtRuX systems is considered in particular, and it suggest a general intrinsic activity trend PtRuCo > PtRuNi > PtRu > PtRuW ~ Pt. A discussion of these findings and their comparison to theoretical DFT-based predictions of ternary electrocatalysts is provided.

1. Introduction

Direct methanol fuel cells (DMFCs) are considered as one of the most promising alternative energy conversion devices for portable power applications. At the anode catalyst of a DMFC, methanol is oxidized to CO_2 and protons, while oxygen is reduced and water formed at the cathode side of an DMFC. Especially at low temperatures, severe challenges in terms of cell efficiency and power density arise due to a rapid deactivation and poisoning of the anode catalyst surface by adsorbed CO formed through the dissociative adsorption and subsequent H-stripping of methanol [1,2]

Initially, pure Pt was used as anode fuel cell catalyst. While it remained unclear how to suppress the adsorption and formation of surface-bound CO, remedies allowing the reduction of surface CO were found in the alloying of Pt with Ru, especially at the 50:50 stoichiometry [1,2]. This resulted in significant increases of the activity of the Pt-alloy anode catalysts and made Pt-Ru alloys probably the most intensively studied and most commonly employed anode electrocatalyst system for DMFC anodes. Still, much effort is dedicated to finding new anode catalysts with even more improved activity, with lower Pt content, yet acceptable stability. To achieve this, additional can be introduced into the binary PtRu catalyst resulting in multi-component catalyst systems. Due to the large number of possible third alloy elements, the experimental compositional space to be screened is tremendous. As a result of that, traditional electrocatalysts research characterized by a 'sample-by-sample testing' approach is very inefficient.

In the present paper we present a combinatorial high-throughput strategy for the discovery of three- and more component fuel cell alloy catalysts that are more active than the PtRu system. Combinatorial strategies basically consist of the fundamental idea of creating a more or less controlled large diversity of different materials from which the one with most suitable properties is selected. This 'diverse synthesis followed by selection' principle aims at accelerating the speed of chemical experimentation and of collection of chemical information [3-10]. High-throughput experimentation for materials discovery often requires the application of unconventional innovative synthesis or screening techniques for a particular materials property of primary interest such as catalytic activity. Furthermore, individual samples are typically of smaller dimensions compared to traditional testing. As a result of this, combinatorial high throughput materials discovery methods are best considered tools that very rapidly provide useful trends in the property-parameter space of materials.

We have used such an approach to generate and compare activity trend maps of a number of ternary PtM1M2 systems with an emphasis on a comparison of PtRuM, M=Ni,Co,W. The screening trends suggest that the PtRuCo ternary system contains a low-Pt, active, yet unexpectedly stable region in compositional space.

2. Experimental Methods

Experimental Combinatorial and High-throughput Electrocatalysis

Combinatorial catalyst synthesis and high-throughput electrochemical screening are achieved by a proprietary array of 64 individually addressable, circular electrode pads fabricated using lithographic techniques on an insulating 3" quartz wafer [11] (Figure 1). Each pad is approximately 1.7 mm in diameter (~0.02 cm^2 area) [8-11]. The electrode pads are providing ohmic contacts with the catalytic electrode materials to be deposited thereon, interfacing the materials library and a multichannel potentiostat/galvanostat (Arbin Instruments Inc.). The material of the 64 supporting electrodes was chosen to be Titanium, which is chemically inert and non-corrosive under the chosen experimental conditions.

For the rapid, parallel synthesis of the 64-element metal alloy electrocatalyst library, automated rf-magnetron vacuum sputtering was employed. The individual constituents of each catalyst composition were deposited onto the 64-element electrode array using contact masks (see Fig. 2a) [8-10]. A proprietary moving-shutter technique allowed the controlled deposition of thin-film gradients ranging from a few Angstroms to several micrometers [11]. Compositional variations in the thin-films were achieved either by sputtering parallel and orthogonal thickness gradients (see Fig. 2b) or else by sputtering multiple non-gradient metal slabs of varying thickness. Typical individual gradient sputtering steps cover a thickness range from about 5 to 15 Å (Fig. 2b, left). Sputtering of superlattice structures as shown on the right of Fig. 2b eventually led to the buildup of thin-film thicknesses of up to several thousand Ångstroms. The deposition conditions were chosen such that the multi-element superlattice structures underwent in-situ atomic mixing of their individual materials components [11]. X-ray diffraction (XRD) characterization of the films was performed using a Philips X-ray diffraction system (Cu Kα, 40 kV, 50 mA, GADDS for NT). Auger spectroscopic investigations of the alloy surfaces of sputtered, non-annealed thin-films showed that the surface compositions were very close to the bulk composition. This observation was similar to results on sputter-cleaned, arc-melted bulk catalysts [12].

For electrochemical characterizations, the array of electrocatalysts was combined with a cylindrical cell body such that all 64 electrocatalysts were facing upwards and were exposed to the electrolyte. All contact pads at the edge of the wafer were isolated from the electrolyte using an O-ring seal. A Pt-mesh of about the same circular area as the cross section of the cylindrical cell body was placed in parallel to the quartz wafer at a distance of about 5 cm and served as the common counter electrode. A Hg/Hg_2SO_4 electrode was used as reference electrode. The electrode was sitting in a glass compartment with Luggin-Haber capillary, which was filled with base electrolyte. The capillary tip of the reference electrode compartment was placed between the working electrode array and the counter electrode mesh. The distances between the capillary tip of the reference compartment and the working electrodes at the edge and at the center of the array ranged from about 3 to 4 cm. This distance was chosen to be large compared to the distance between individual working electrodes (about 3 mm) in order to minimize differences in the uncompensated ohmic resistance between catalysts. Voltage drops across the electrolyte between the reference electrode tip and individual working electrodes were considered negligible given the conductivity and the magnitude of the measured currents. The cell body was further equipped with gas inlet channels for purging the electrolyte with Argon, a gas-tight lid as well as a conventional liquid-seal gas outlet. All potentials in this study will be referred to with respect to the reversible hydrogen scale (RHE) scale with an assumed potential difference of +0.65 V between the two scales.

The anode catalysts were electrochemically pretreated using repeated voltammetric cycling of all 64 catalysts between 0 – 0.7 V/RHE in 0.5 M H_2SO_4 at a rate of 20 mV/s and 200 mV/s. The pretreatment was stopped as soon as a time-stable voltammogram was obtained for most compositions. Some alloy compositions appeared to continue to change their base voltammetry, especially at large overpotentials, which is likely to indicate corrosive degradation and limited stability of those compositions; that same compositions also exhibit very large hydrogen adsorption integrals. The hydrogen adsorption charge of each catalyst over a potential range of 0.05 – 0.35 V / RHE on the slow voltammograms was used as an estimate of its active surface area using a hydrogen electrosorption pseudocapacity of 210 C/cm^2. Subsequently, the catalyst array was evaluated with respect to its methanol oxidation activity in a 1 M methanol, 0.5 M sulphuric acid electrolyte. Parallel linear potential scans and parallel chronoamperometric screening (+0.45 V / RHE, 5 minutes) of all catalysts yielded current density-time curves. All experiments were performed at ambient pressure and room temperature.

3. Results

3.1 XRD characterization of a binary thin-film library

The combinatorial thin-film alloy synthesis method used in this study is a serial deposition technique that critically relies on the complete in-situ mixing of the atoms of the individual metal layers. To demonstrate the validity of our alloy synthesis approach with respect to atomic mixing, but also to demonstrate the materials diversity accessible in thin-film format, we chose to synthesize and structurally characterize a simple binary Pt-Fe library. Pt-Fe is a well-known binary alloy system exhibiting both substitutional solid solution phases as well as intermetallic compounds. Figure 3a illustrates the 64-element library design of the PtFe binary wafer: Going from left to right in each row, the composition of individual electrocatalysts changes from pure Pt to pure Fe in seven evenly spaced compositional steps, whereas the rows are identical to each other. Design redundancies are generally undesirable for an efficient screening of a materials diversity, yet may help enhance the precision of the materials trend analysis Specifically, from left to right, the design contains the individual stoichiometries Pt, Pt86Fe14, Pt71Fe29, Pt57Fe43, Pt42Fe58, Pt28Fe72, Pt14Fe86, and Fe. After sputter deposition of the 64 catalysts according to Fig. 3a, XRD analysis was performed on each individual electrocatalyst spot along rows. Fig. 3b and Fig. 3c show the resulting diffraction patterns within any given row for an un-annealed (that is as deposited) and an annealed library, respectively. Annealing was carried out at 400 °C under vacuum for 12 hours. In Fig 3b, the bottom most profile indicates the familiar three diffraction peaks (from left to right, (111) at ~40 2θ, indicated by the black dot, (200) at ~46 2θ, and (220) at ~67.5 2θ) of the pure Pt catalyst. The relative intensity differences of the three peaks compared to the materials database (see three vertical lines at bottom) suggest a preferred orientation of the Pt catalyst. With increasing Fe content, the prominent (111) peak position shifts gradually to the right indicating the contraction of the lattice units due to substitution of Pt with Fe in the fcc lattice. Finally, the topmost profile indicates the typical line profile of a pure Fe phase. The smaller peak to the left of the (111) peak is attributable to the Ti electrode array supporting the binary PtFe electrocatalysts. The gradual transition of the (111) alloy peak, together with the absence of any pure Pt or Fe phases on alloy catalysts evidences that complete atomic mixing is achieved under our deposition conditions. The as-deposited profiles of Fig. 3b are consistent with substitutional solid solutions of PtFe (disordered alloys) over the entire range of molar ratios. In contrast, Fig. 3c, shows the diffracting patterns after annealing of the binary library. Now, new diffraction lines indicate the formation of intermetallic phases (ordered alloys) of Pt_3Fe (third profile form bottom, black balls) and PtFe (fourth, fifth, sixth profile from bottom, black squares) on catalyst spots that come closest in stoichiometry. Fe-rich catalysts (see second profile from top) also show the formation of a PtFe phase with the concurrent presence of a pure Fe phase as expected based on their stoichiometry.

The sputter deposition method used in this study results in well-mixed Pt alloys, and allows access to both ordered and disordered alloys depending on post-synthesis processing.

3.2 Combinatorial Synthesis and high-throughput screening of a diverse Pt-alloy electrocatalyst library

Pt, Ru, Co, Ni and W were chosen as variables spanning the compositional space screened in this study, with Pt being a constituent in all catalysts (Pt-alloy catalyst concept). This combination of compositional space and constraints resulted in a total of six ternary systems (PtRuCo, PtRuNi, PtRuW, PtCoNi, PtCoW, and PtNiW) to be synthesized and screened on the 64 –element wafer. Fig. 4 illustrates the typical degree of sampling of each ternary system. Six ternary compositions were synthesized for each ternary system, in particular the three 20:20:60 and the three 40:40:20 compositions. The design also included a pure Pt catalyst, and a number of binary PtRu catalysts that were used as standard catalysts against which the activites of ternary alloys were evaluated. A repeat sample of the Pt:Ru 6:4 catalyst provided a measure of the experimental error of the experimental method.

The electrocatalytic activity of the 64-element PtRuWCoNi library was evaluated in high-throughput screening mode using anodic potential sweep and chronoamperometric potential step methods. Fig. 5 shows a chronoamperometric measurement of the most active catalysts at +450 mV / RHE over 5 minutes. While the methanol oxidation currents of most catalysts remained below absolute currents of 0.1 μA (0.005 mA/cm^2 geometric current density), a few composition exhibited very high sustained activity. The chronoamperometric profiles of all catalysts follows the familiar pattern of rapidly decaying capacitive current contributions followed by a slow asymptotic approach of a stationary-state activity. Fig. 6a and Fig. 6b report the final geometric current densities $J = I_{alloy} / A_{geometric}$ after 5 minutes of each individual channel over their channel numbers. The high-activity channels are labeled with their chemical compositions. The pure Pt standard catalyst was deposited on channel 56. The observed geometric current density observed for the Pt standard catalyst compares well with those reported on sputter-cleaned smooth alloy catalysts: At +450 mV/ RHE, the pure Pt catalyst exhibited an initial geometric current density of about 2.5 μA/cm^2 which dropped to 0.5 μA/cm^2 after 5 minutes. In comparison, in Fig 7c of ref. [14], a study performed in 0.5 M methanol, a sputter-cleaned Pt electrode showed an initial geometric current density of about 3 μA/cm^2.

As commonly known, alloying Pt with Ru increases the catalytic activity strongly [1,2,12] The variation between the two Pt:Ru 6:4 spots (channels 55 and 64) yields a rough measure of the experimental error of the screening method; from that, it is obvious that some of the 64 catalysts show large, statistically significant activity improvements over the PtRu compositions. As an additional cross-check of the high-throughput method, channel 63 (Fig. 6b) reports the activity of a Ru-rich PtRu alloy catalyst (Pt:Ru 20:80). As expected, high Ru contents are detrimental to the methanol oxidation activity leading to activities lower than Pt itself.

Based on the data in Fig. 6a and 6b, the following composition-activity relationships within the 64 electrocatalysts can be identified: First, most W-containing, gray-shaded compositions show very low to no current activity (marked as 'W-containing compositions' in Fig. 6a). A closer look at the compositions of these catalysts reveals that their combined molar fractions of all three base metals ($x_W + x_{Co} + x_{Ni}$) roughly equals 0.5 – 0.6. Second, there is a group of very active W-containing, Ru-free compositions with a combined base-metal molar fraction ($x_W + x_{Co} + x_{Ni}$) of about 0.8. Due to their high base metal content, these compositions undergo strong surface corrosion and surface roughening during the pretreatment resulting in high current densities due to their increased surface area; this is confirmed by their much larger H-adsorption integrals in base electrolyte compared to the corrosion-stable Pt or PtRu catalysts. Finally, there is a group of W-free electrocatalysts (black dots in Fig. 6a, see the labeled compositions in Fig. 6b) that are, with one exception, clustered closer around the PtRu reference activity. Among them, Ru-rich compositions exhibit a lower activity than the PtRu standards (see Pt 20 Ru 60 M 20 and Pt 20 Ru 40 M 40 with M = Co,Ni, channels 15,31,58,60). For PtRuNi, as the Pt content is increased at constant Ru, the activity rises to levels comparable to the PtRu standard (channel 28,30,32) [13]. In case of PtRuCo, however, the Co-rich Pt:Ru:Co = 1:1:3 exhibits very high activity (channel 10, $Pt_{20}Co_{60}Ru_{20}$), and as Pt replaces Co (channels 9,16) the activity drops stepwise to levels comparable to PtRu.

In order to rank and prioritize alloy electrocatalyst with respect to their *intrinsic catalytic activity*, electrochemical current data is typically normalized to the real surface area of the catalysts rather than to their geometric surface area. Fig. 6c reports the surface-area-normalized electrochemical activity of the most active alloy catalyst compositions. These activities were obtained dividing the absolute measured currents I_{alloy} of each catalyst by its real Pt surface area as determined electrochemically using the respective hydrogen adsorption integral. The order of activity of the most active electrocatalyst has obviously changed based on their electrochemically determined surface area. Analysis of the real surface area of Pt:Ru:Co 20:20:60 (channel 10) reveals no significant increase of its surface area compared to the PtRu standards. This suggests a favorable degree of stability and gives rise to a significant intrinsic activity improvement for this compositions. This is in contrast to the most active W-containing, Ru-free compositions that appear to corrode more severely and show reduced specific activity after correction for their surface area; while a few W-containing catalysts still maintain activities comparable or slightly better than the standards, their longer-term stability remains questionable.

Fig. 6d summarizes the basic observed trends of this experimental study: The Pt-area-normalized activities of the most active PtNiRu and PtCoRu compositions were normalized to that of pure Pt and plotted on a logarithmic scale. The physical meaning of the quantity plotted on the y axis correspond to a Butler-Volmer exponent, that is a potential-dependent activation energy difference. The more positive this values is, the higher is the intrinsic active of the corresponding composition. Fig. 6d highlights the catalytic activity gains of ternary alloys starting from pure Pt followed by PtRu, PtRuNi and PtRuCo compositions [13] where PtRuW compositions range between Pt and PtRu.

4. Discussion

Combinatorial high-throughput screening strategies were applied to screen the activity-compositional space of some ternary Pt alloys.

The experimental thin-film screening aimed at identifying *activity trends as function of composition* among six ternary metal alloys, specificially between PtRuM, M = Ni,Co,W, with respect to the methanol electrooxidation. These trends can then be used to identify suitable alloy compositions that can serve as starting points of further optimizations.

High-throughput screening methods often involve simplifications of the realistic materials environment by separating off parameters that are not considered significant for an improvement of intrinsic materials properties. In the present case of anode fuel cell catalysts, the influence of the supporting carbon, of the conductive polymer and other parameters associated with a typical gas-diffusion electrode were neglected by using a sputter-thin-film catalyst approach. Sputtered thin-films, unlike alloy electro-codeposition methods, allow a more precise control of the alloy stoichiometry, and are typically dense and very smooth so that the initial real electrode surface area of each catalyst is essentially equal to the geometric area of the supporting electrode pad. This facilitates a direct comparison of catalytic activities of different alloys based on geometric current activity only, provided a similar Pt molar fraction and no significant change of the surface area during testing. In case of alloys that are not corrosion-stable during the screening, a correction for changes in their surface area is crucial for a fair comparison between the activities of electrocatalysts. In terms of its intrinsic surface-area-based activity a PtRuCo ternary catalysts clearly outperformed the PtRu standard catalysts, followed by compositions of the PtRuNi ternary catalyst system. This general order of activity agrees favorably with a recent study of ternary methanol electrocatalysts electrodeposited onto organic polymer films where PtRuCo outperformed PtRuNi followed by PtRu catalysts [14,15].

In the present study, the best performing Co-containing ternary catalyst at channel 10 ($Pt_{20}Co_{60}Ru_{20}$) also appeared fairly corrosion-stable judged by its relatively small surface area. Given its low Pt content, this finding is surprising and will now deserve further investigation in the vicinity of this ternary stoichiometry.

The mechanism of the experimental activity enhancement remains speculative. Traditionally, catalytic enhancements are explained based on optimized surface atomic geometries for adsorbed species (*geometric/ensemble effect*) [2,12], optimized electronic interaction effects between a modified surface and an adsorbate (*ligand effect*), or the presence of two or more different surface sites with synergistic functionalities (*bifunctional / multifunctional effect*) [2]. None of the three mechanisms can be disregarded at this moment in the present case. It is known, however, that at potentials below 0.5 V / RHE, the methanol oxidation on PtRu binary alloys is rate limited by the availability of free surface sites for methanol adsorption and dehydrogenation [2]. The number of free surface sites could be increased either by an enhanced oxidative removal of adsorbed CO by oxygenated species adsorbed on the base metal component (*enhanced* bifunctional effect of a ternary alloy) or else by decreasing the level of CO poisoning of the surface by changing its CO adsorption energy (ligand effect of a ternary alloy). The ligand effect would make the formation of the adsorbed intermediate CO less favorable, while the formation of a reactive intermediate could be preferred. While both mechanisms rely on the availability of oxygenated species (adsorbed water or OH) in the neighborhood of methanol adsorption sites and, in this sense, both are bifunctional, the ligand effect explicitly relies on the combined presence of Ru and Co in the optimal stoichiometric ratio. In fact, neither Pt 20 Ru 80 nor Pt 20 Co 80 shows a methanol oxidation activity similar to Pt 20 Co 60 Ru 20. In fact, a recent theoretical DFT study on the CO tolerance of a variety of ternary Pt alloys [10], reproduced the activity trends found in the present study and lends further support for a predominant ligand effect of the third alloyed element as explanation for the activity enhancement.

References

[1] J. Lipkowski and P.N. Ross, eds, *Electrocatalysis,* Wiley-VCH, New York, **1998**

[2] N.M. Markovic and P.N. Ross *Surface Science Reports* **2002**, 45 117

[3] S. Miertus and G. Fassina *Combinatorial Chemistry and Technology*, Marcel Dekker, New York, **1999**

[4] X. Xiang, X. Sun, G. Briceno, Y. Lou, K. Wang, H. Chang, W. Wallace-Friedman, S. Chen and P. Schultz *Science* **1995**, 268, 1738

[5] B. Jandeleit, D.J. Schaefer, T.S. Powers, H. Turner and W.H. Weinberg *Angewandte Chemie (Int. Edition)* **1999**, 38, 2495

[6] B.Gurau, R. Viswanathan, R. Liu, T.J. Lafrenz, K.L. Ley, E.S. Smotkin and S. Sarangapani J. Phys. Chem. B **1998**, 102, 9997;

[7] E. Reddington, A. Sapienza, B. Gurau, R. Viswanathan, S. Sarangapani, E.S. Smotkin, T.E. Mallouk, *Science* **1998,** 280, 1735

[8] P. Strasser, S. Gorer, and M. Devenney, *Proceedings of the International Symposium on Fuel Cells for Vehicles*, 41st Battery Symposium, The Electrochemical Society of Japan, Nagoya **2000**, p.153;

[9] P. Strasser, S. Gorer, and M. Devenney, in *Direct Methanol Fuel Cells***,** S.R. Narayanan,S. Gottefeld, T. Zawodzinski, eds., Proceedings Volume 2001-4, The Electrochemical Society, New Jersey, **2001**, p. 191

[10] P. Strasser, Q. Fan, M. Devenney, and W.H. Weinberg, P. Liu, and J.K. Norskov *J. Phys. Chem. B,* October 2003, in press

[11] US Patent No's 6,045,671, 6,364,956, 5,985,356, 6,371,640, 6,187,164, 6,468,806, 5,985,356 and 6,004,617. Additional patents pending

[12] H. Gasteiger, N. Markovic, P.N. Ross and E. J. Cairns, *J.Phys.Chem.* **1993**, 97, 12020

[13] US Patent No's 6,517,965 Additional patents pending; US Patent application No. 2002/0146614, pending

[14] C. Lamy, A. Lima V. LeRHun, F. Delime, C. Coutanceau and J.-M. Leger J. Power Sources **2002**, 105, 283

[15] A. Lima, C. Coutanceau, J.-M. Leger and C. Lamy, *J.Appl.Electrochem.* **2001**, 31, 379

Figure Captions

Figure 1: A 64-element addressable electrode array on a 3'' quartz wafer is used as support for the synthesis of chemically-diverse electrocatalyst libraries.

Figure 2: a) A Rf magnetron sputter technique with automated target selection as well as automated moving shutters and physical shadow masking is employed for deposition of the electrocatalyst libraries. b) Sequential gradient sputtering of very thin material slabs lead to in-situ thin-film alloy formation on the contact spots. The width of the slabs on the left extends over the entire width of the electrode array. Superlattice deposition in z direction results in the buildup of thicker films of the desired stoichiometric gradient.

Figure 3: a) Design of a 64-element electrocatalyst library of Pt-Fe binary alloys. The square and the 64 round spots represents the wafer substrate and the location of individual electrocatalyst alloys, respectively. The pie-chart character of each catalyst represents its chemical stoichiometry ranging, from left to right, from pure Pt to pure Fe. An eightfold redundancy is built into the design along columns. During synthesis, this library design will be mapped onto the real physical electrode array wafer. The design was made using Symyx proprietary Library Studio® Software package.

b) Sequence of eight XRD profiles taken along the third row of the library given in a) *before post-synthesis processing*. The black dots indicate diffraction from the dominant (111) plane. The profiles of the alloy compositions are consistent with substitutional solid solution structures with fcc structure.

c) Sequence of eight XRD profiles taken along the third row of the library given in a) *after post-synthesis processing* (400 °C, 12h, vacuum). Intermetallic phase formation is observed. The black dots indicate diffraction peaks from the Pt_3Fe phase, while black squares indicate the formation of a PtFe phase.

Figure 4: Example of a ternary metal-alloy compositional space sampled and screened. Black dots indicate individual stoichiometries synthesized in the electrocatalyst library. Emphasis was placed on the 6 ternary stoichiometries Pt:A:B (20:20:60, 40:40:20 and all permutations thereof). The other 5 ternary systems were sampled as shown.

Figure 5: Parallel chronoamperometric screening of an electrocatalyst library. Shown are the measured current-time curves of the seven most active compositions after a potential step (t=10 s) from open circuit potential (OCP) to +450 mV / RHE. During the inital 10-second period (not shown on x axis), the OCP was measured. Shown are channels 2,3,4,10,19,42,43; compare Fig. 6a. Stoichiometric information is given for the three most active compositions. Conditions: 1 M Methanol, 0.5 M H_2SO_4, room temperature, E = +450 mV / RHE.

Figure 6: a) Screening results of the chronoamperometric screen in Fig. 5. Dots indicate the final chronoamperometric geometric current density of each channel. The gray coloring of the dots encode the W content of the electrocatalysts (White – high, black – low). The composition of a few active channels are indicated. b) Blow-up of the low-activity portion of a) with all W-containing alloy catalysts removed. Indicated are the ternary alloy catalysts of interest. PtRu binary as well as the pure Pt spot serve as standards. Gray-coding indicated the catalysts Ni content (White=high, black=low). c) Pt surface-area normalized electrochemical activity (area-specific activity) of the most active alloys. The Pt surface-area was determined electrochemically using the hydrogen adsorption integral of each individual catalyst. d) Summary of *activity trends* in Fig. c). Plotted are the area-specific activities of Fig. c) after normalization with repect to pure Pt. The logarithmic value represents relative kinetic improvement. Obvious are the gains in activity going from pure Pt, to PtRu, PtRuNi and PtRuCo compositions.

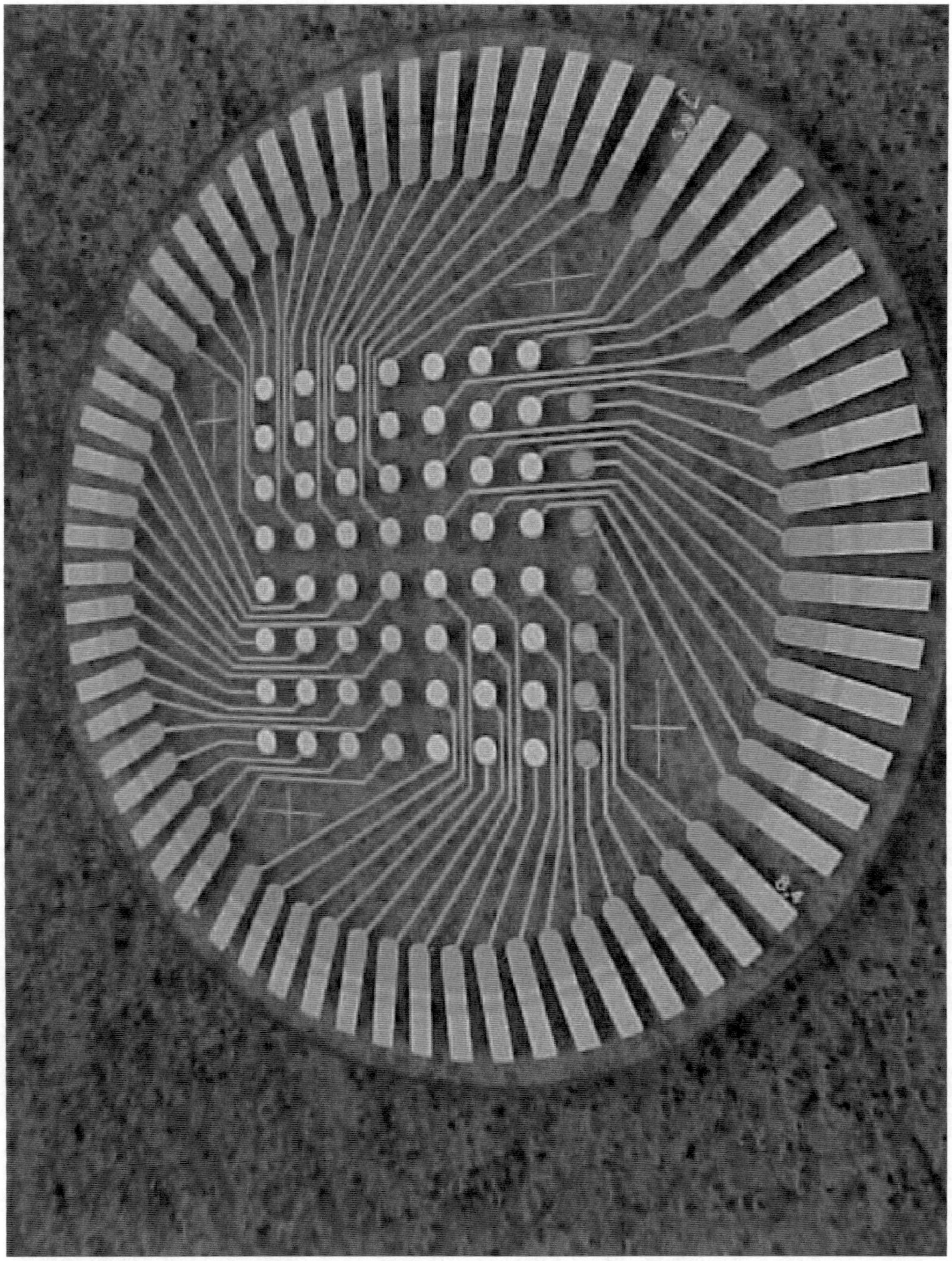

Figure 1

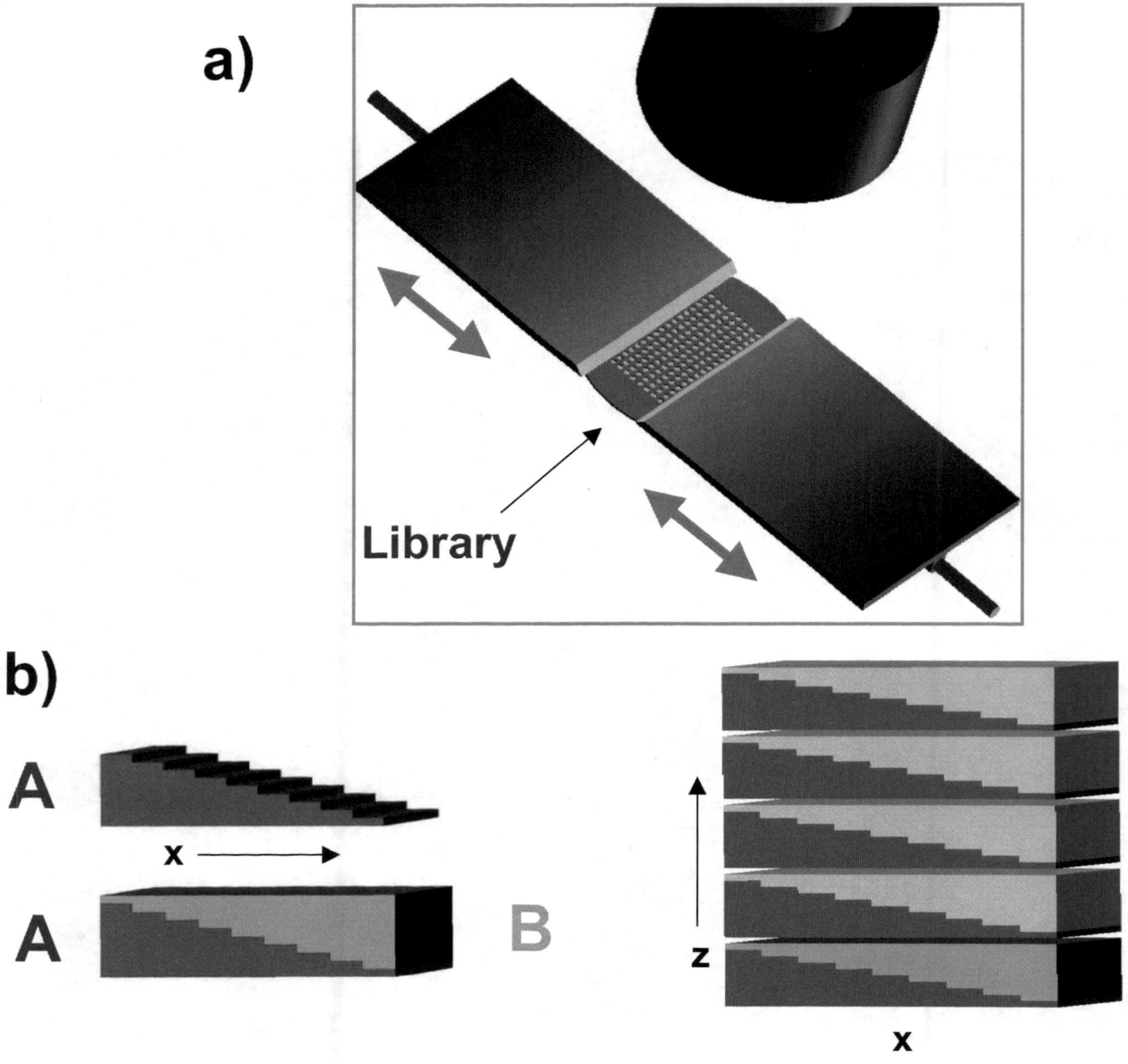

Figure 2

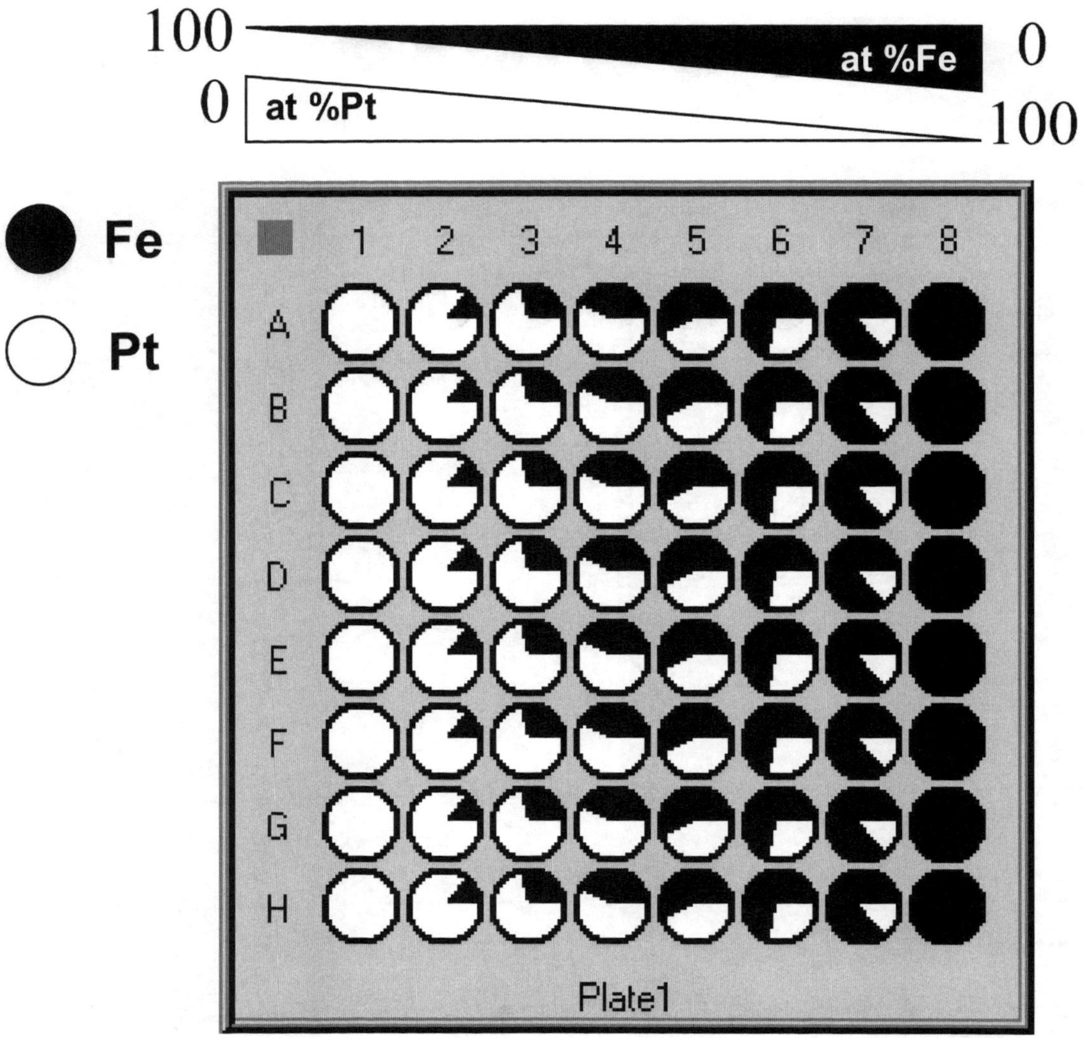

Figure 3a

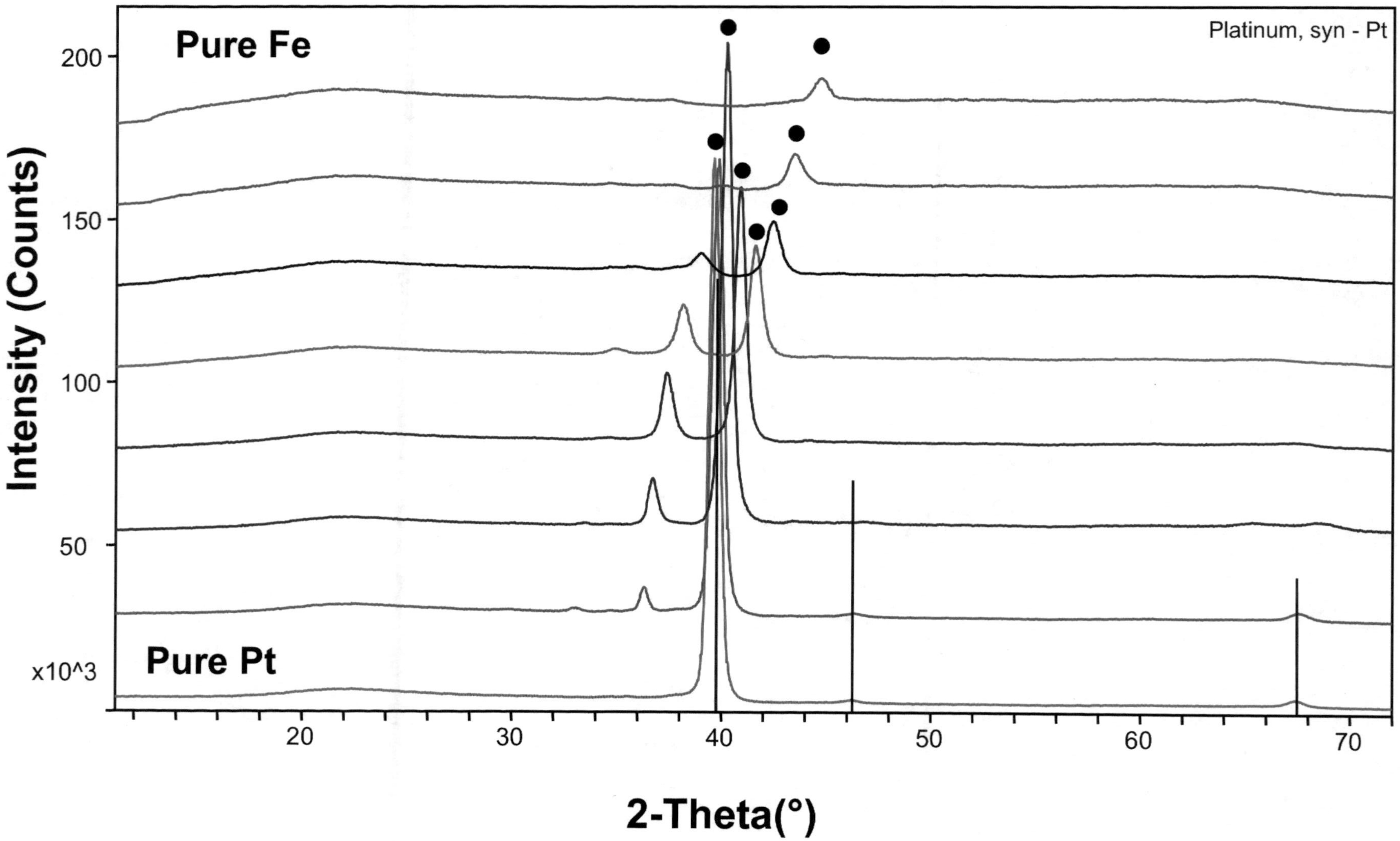
Pure Fe
Platinum, syn - Pt
Pure Pt
Intensity (Counts)
200
150
100
50
x10^3
20
30
40
50
60
70
2-Theta(°)

Figure 3b

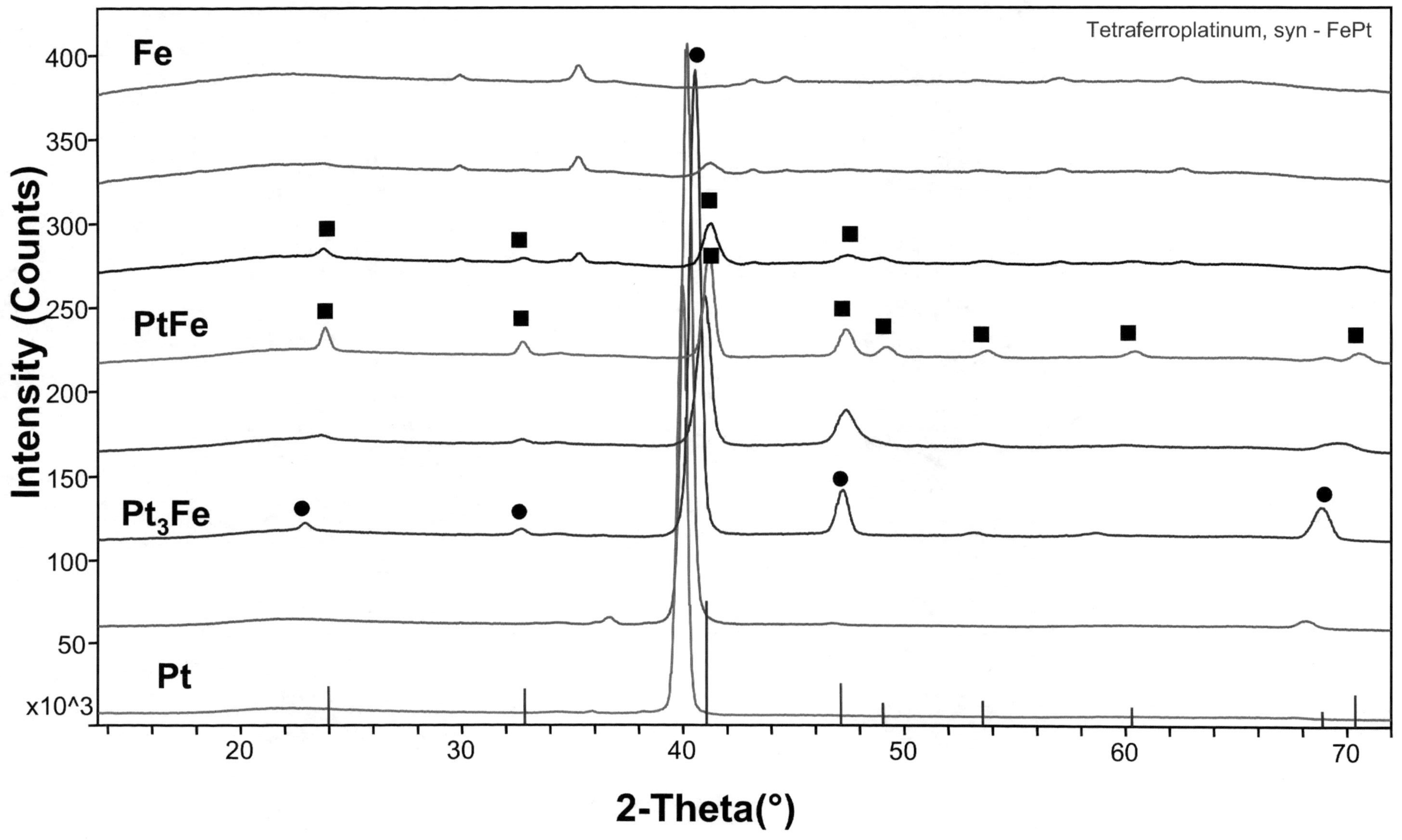

Tetraferroplatinum, syn - FePt
Fe
PtFe
Pt3Fe
Pt
Intensity (Counts)
x10^3
400
350
300
250
200
150
100
50
20
30
40
50
60
70
2-Theta(°)

Figure 3c

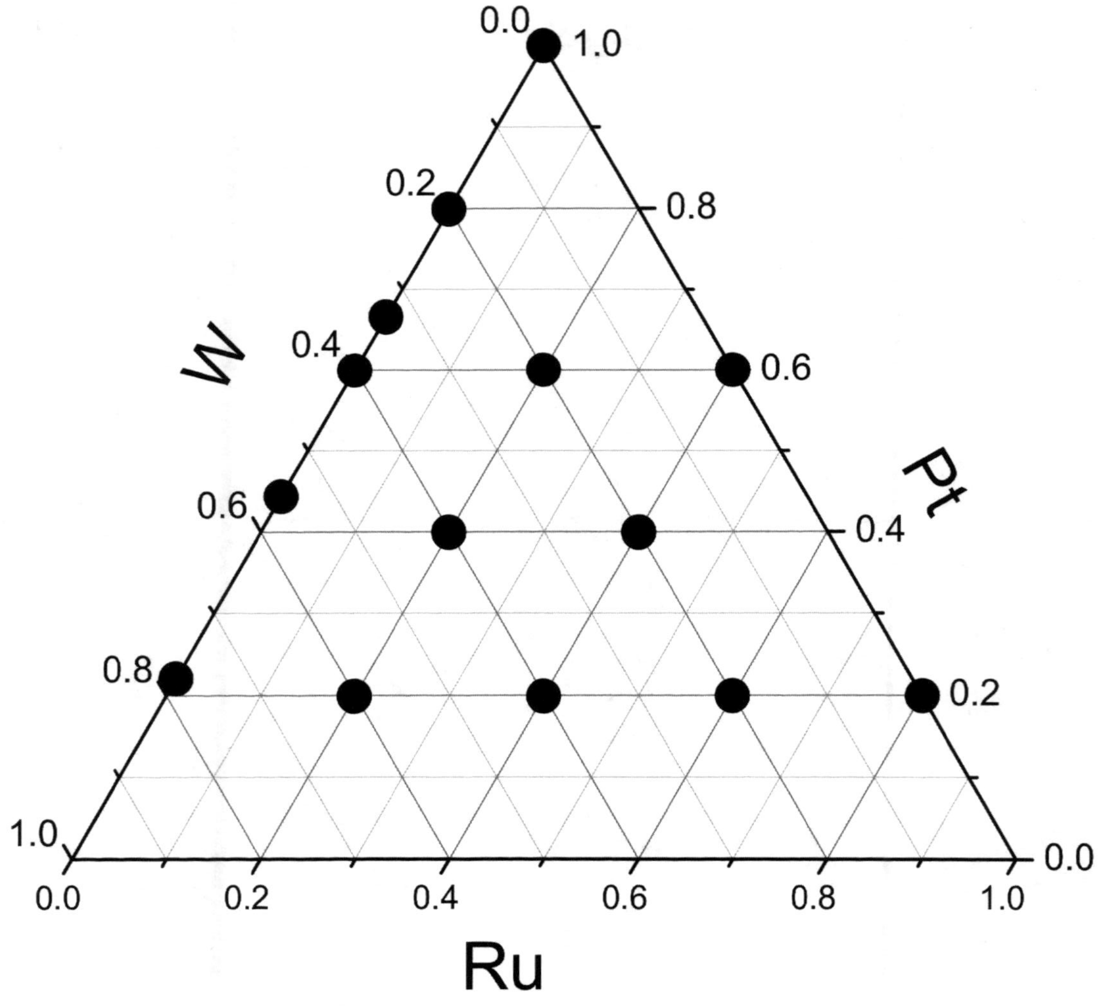

Figure 4

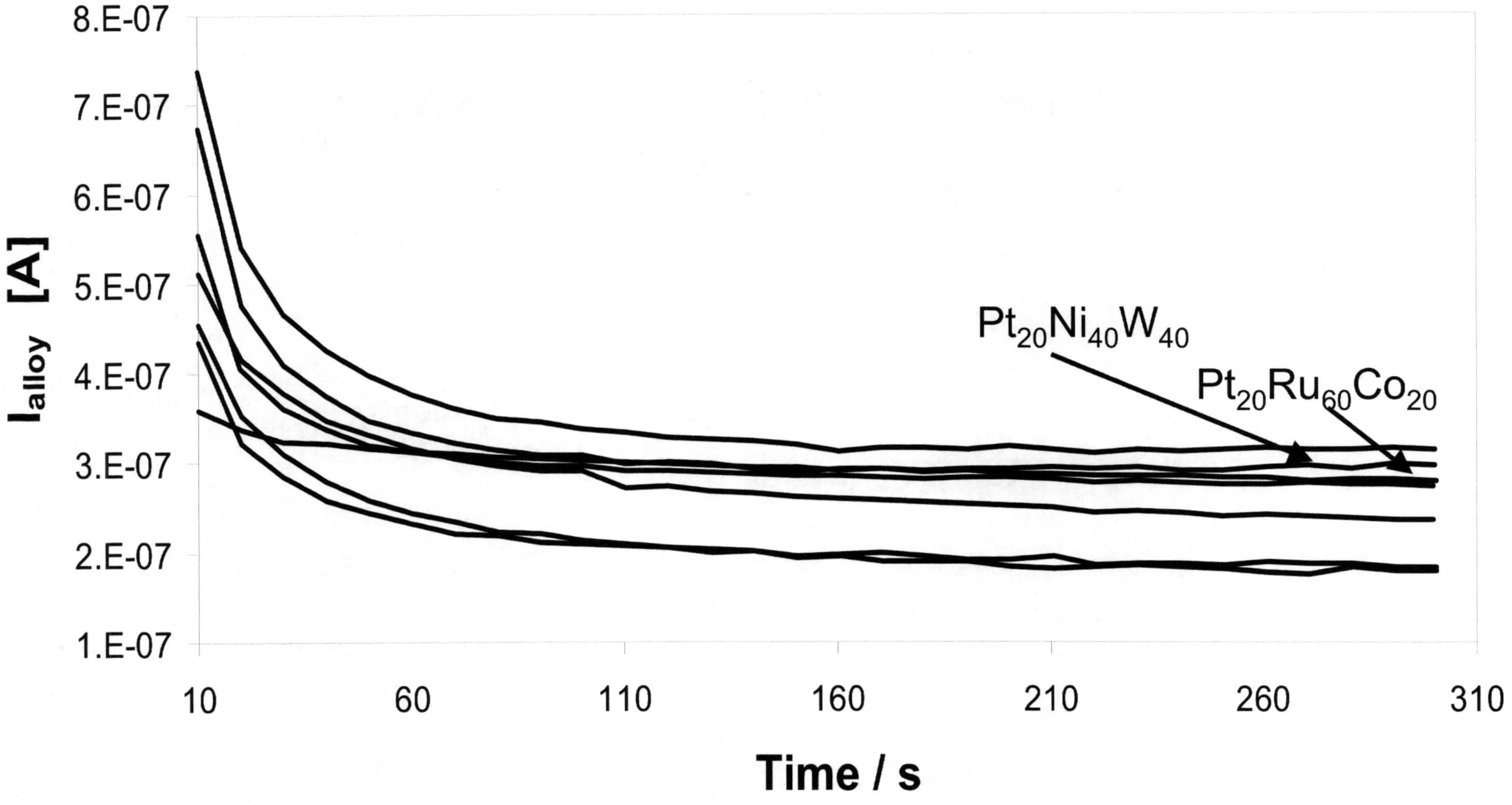
8.E-07
7.E-07
6.E-07
5.E-07
4.E-07
3.E-07
2.E-07
1.E-07
I_alloy [A]
10
60
110
160
210
260
310
Time / s
Pt20Ni40W40
Pt20Ru60Co20

Figure 5

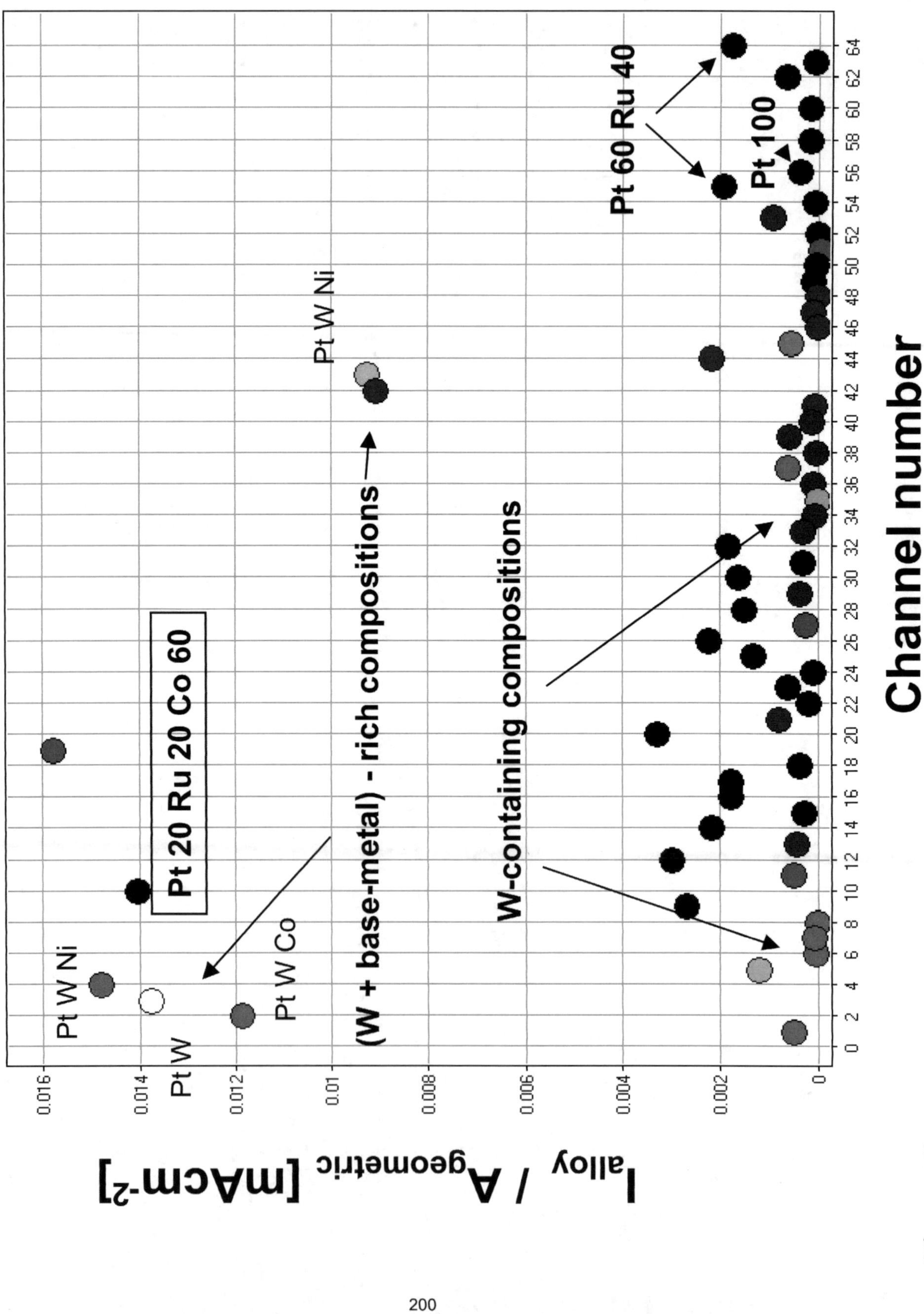

Pt W Ni
Pt W
Pt W Co
Pt 20 Ru 20 Co 60
(W + base-metal) - rich compositions
Pt W Ni
W-containing compositions
Pt 60 Ru 40
Pt 100
Channel number
I_alloy / A_geometric [mAcm^-2]
0.016
0.014
0.012
0.01
0.008
0.006
0.004
0.002
0
0 2 4 6 8 10 12 14 16 18 20 22 24 26 28 30 32 34 36 38 40 42 44 46 48 50 52 54 56 58 60 62 64

Figure 6a

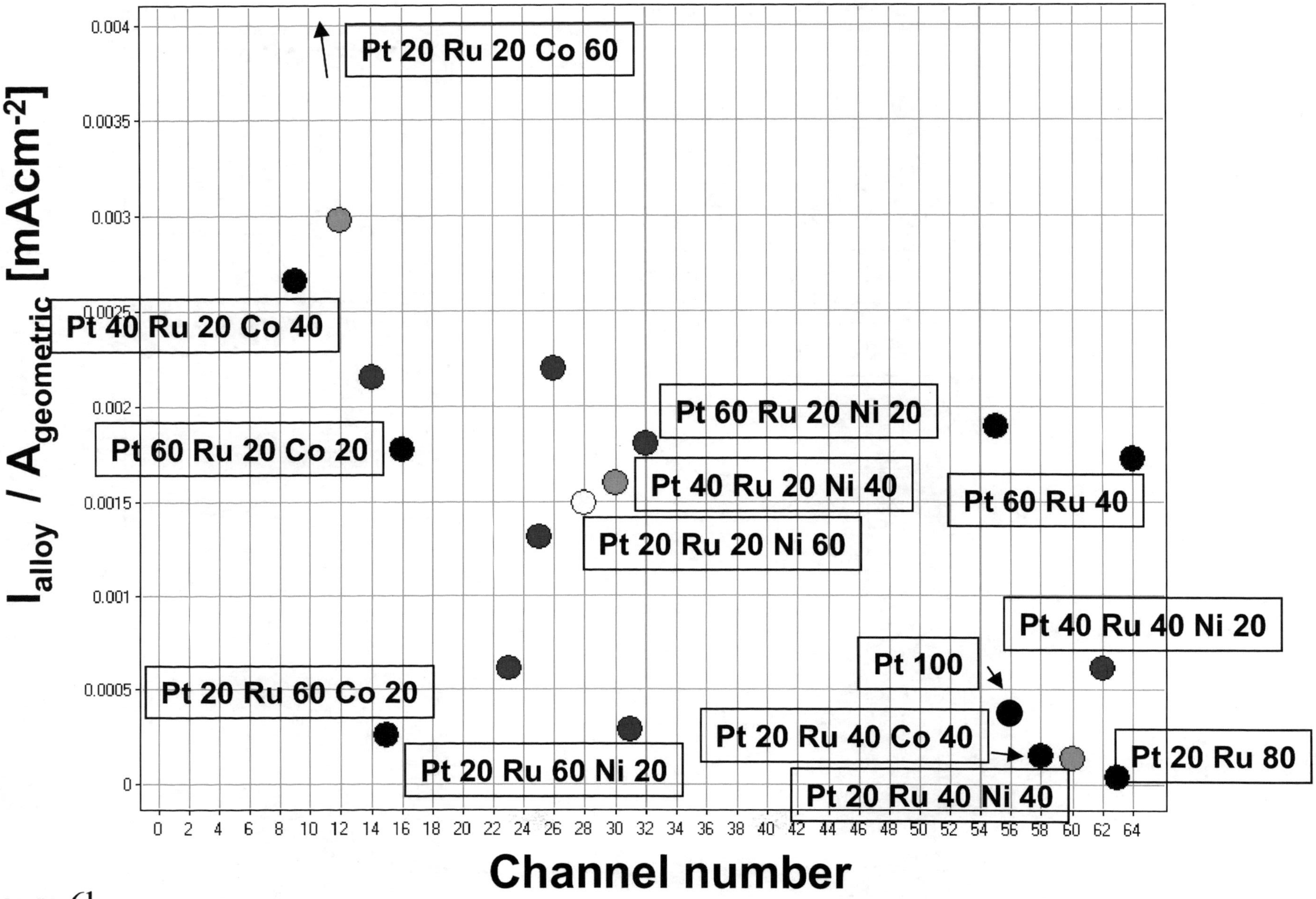
Pt 20 Ru 20 Co 60
Pt 40 Ru 20 Co 40
Pt 60 Ru 20 Co 20
Pt 20 Ru 60 Co 20
Pt 20 Ru 60 Ni 20
Pt 60 Ru 20 Ni 20
Pt 40 Ru 20 Ni 40
Pt 20 Ru 20 Ni 60
Pt 60 Ru 40
Pt 40 Ru 40 Ni 20
Pt 100
Pt 20 Ru 40 Co 40
Pt 20 Ru 80
Pt 20 Ru 40 Ni 40
0.004
0.0035
0.003
0.0025
0.002
0.0015
0.001
0.0005
0
0 2 4 6 8 10 12 14 16 18 20 22 24 26 28 30 32 34 36 38 40 42 44 46 48 50 52 54 56 58 60 62 64
I alloy / A geometric [mAcm-2]
Channel number

Figure 6b

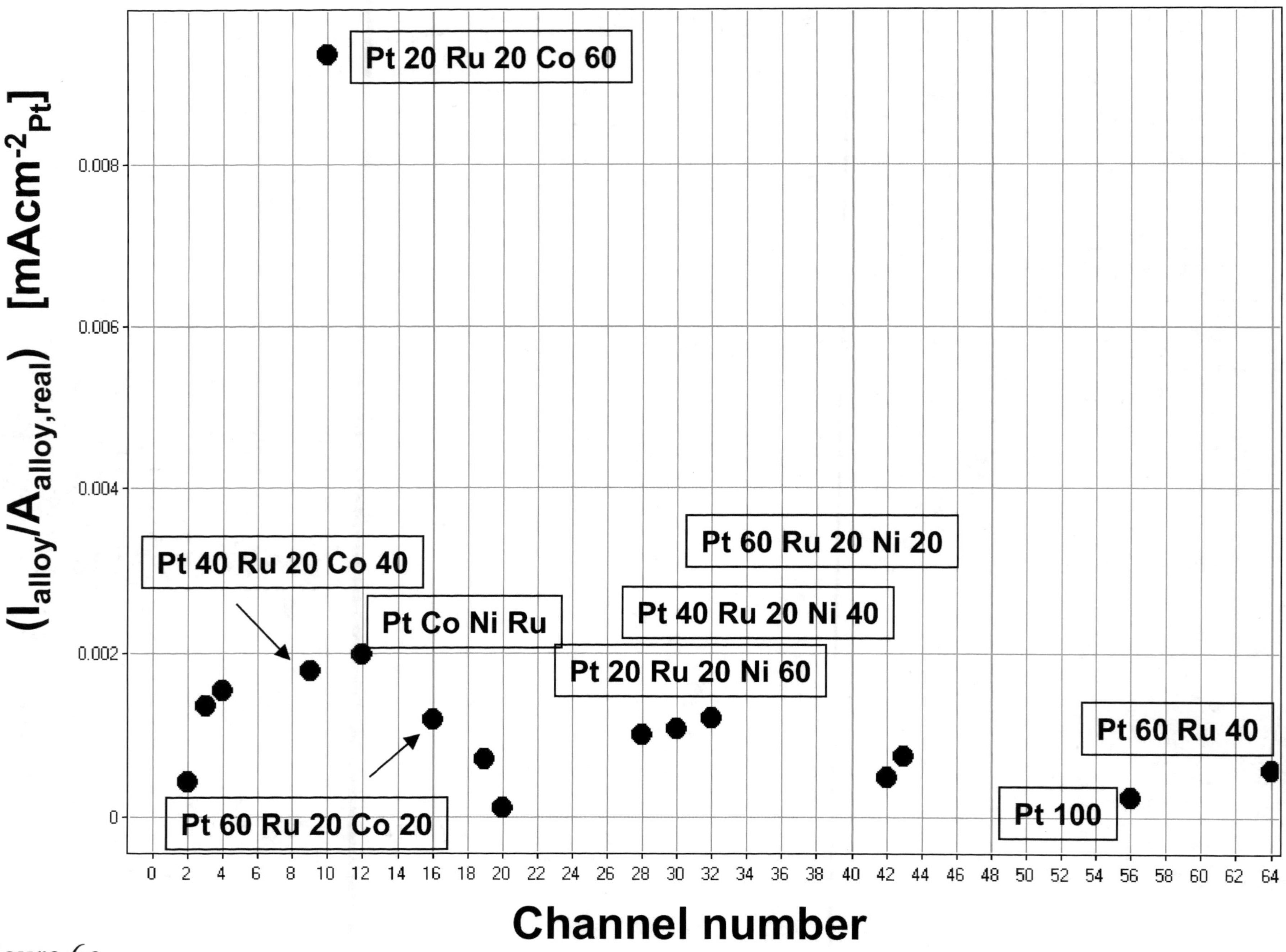

Pt 20 Ru 20 Co 60
Pt 40 Ru 20 Co 40
Pt Co Ni Ru
Pt 60 Ru 20 Co 20
Pt 60 Ru 20 Ni 20
Pt 40 Ru 20 Ni 40
Pt 20 Ru 20 Ni 60
Pt 60 Ru 40
Pt 100
0.008
0.006
0.004
0.002
0
0 2 4 6 8 10 12 14 16 18 20 22 24 26 28 30 32 34 36 38 40 42 44 46 48 50 52 54 56 58 60 62 64
(Ialloy/Aalloy,real) [mAcm-2Pt]
Channel number

Figure 6c

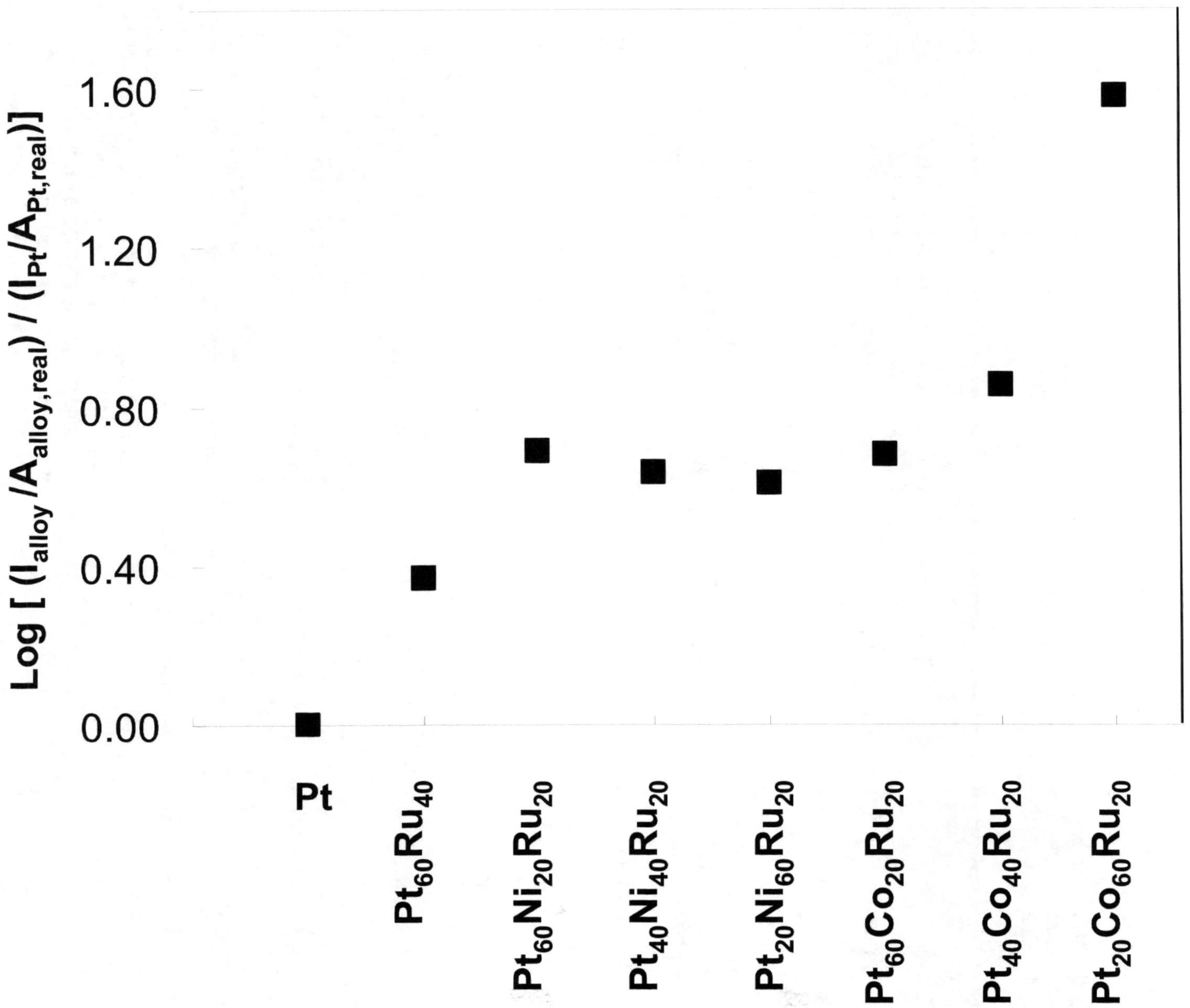
Log [(I_alloy /A_alloy,real) / (I_Pt/A_Pt,real)]
1.60
1.20
0.80
0.40
0.00
Pt
Pt60Ru40
Pt60Ni20Ru20
Pt40Ni40Ru20
Pt20Ni60Ru20
Pt60Co20Ru20
Pt40Co40Ru20
Pt20Co60Ru20

Figure 6d

Characterization of Carbonized Charcoals for Fuel Anodes

Kazuhiro Mochidzuki
Institute of Industrial Science, University of Tokyo
1-8 Yayoi-cho, Inage-ku, Chiba 263-0022, Japan

Katsuaki Tadokoro, Michael J. Antal, Jr.
Hawaii Natural Energy Institute, University of Hawaii
1680 East-West Road, Honolulu, HI 96822, USA

Maria Toth, Gábor Várhegyi
Hungarian Academy of Science
P. O. Box 17, Budapest 1525, Hungary

ABSTRACT

Because coal does not conduct electricity and graphite is costly and inert, little attention has been given to the development of a fuel cell that utilizes a consumable carbon anode (fuel anode) to generate power. On the other hand, carbonized charcoals show low electrical resistivity and high reactivity. Also, they are competitive in price with fossil fuels. Carbonized charcoals are promising materials for the fuel anodes. In this paper we show that a packed bed of carbonized charcoal particles subject to a compressive pressure (ca. 8 MPa) can be a good electrical conductor. The seven order of magnitude decrease in the electrical resistivity of charcoal with increasing heat treatment temperature (HTT) was observed. Low electrical resistivities (< 0.2Ω-cm) were manifest by many different charcoals after carbonization at a HTT of 950 °C. Furthermore, elemental analysis, XRD and FTIR measurements were carried out to characterize the carbonized charcoals.

KEY WORDS

Carbon-fueled fuel cell, carbonized charcoal, consumable carbon anode.

INTRODUCTION

While it is well known that hydrogen fuel cells are recognized to be among the promising technologies to generate electric power, little attention has been given to carbon-fueled fuel cells. However, carbon fuel cells (or batteries) played an important role in the history of fuel cell research. Sixteen years after Grove[1] demonstrated an "electrolyte gas cell" that gave an open circuit EMF of about 1 V between H_2 and O_2, Becquerel attempted to build a fuel cell that consumed coal[2, 3]. At that era a number of researches on carbon fuel cell were performed and contributed to the accumulation of fundamental knowledge, though no one was able to build a practical system. Interest in carbon fuel cells resurfaced during the 1970's, when the Stanford Research Institute (SRI) attempted to develop a coal based fuel cell that employed molten lead at temperatures of 500 to 900 °C [4, 5]. This work was abandoned in 1981. Recently, Scientific Applications and Research Associates[6] reported progress in further developing the SRI concept. At Stanford University Gur and Huggins[7] demonstrated a high temperature (725 to 955 °C) fuel cell that employed stabilized zirconia as a solid electrolyte and a graphite anode. To the best of our knowledge, these recent developments have not led to the demonstration of a practical carbon fuel cell. Summarizing the status of carbon fuel cells, Bockris and Srinivasen[3] concluded that carbon fuel cells are impractical because (i) coal is not an electrical conductor, and (ii) graphite is too scarce and expensive to be used as a fuel. Thus the history of carbon fuel cell research suggests that the chief obstacles to the development of a biocarbon fuel cell are the electrical conductivity of the biocarbon, its cost, and its reactivity.

On the other hand, it has been known for centuries that biocarbons can possess very high electrical conductivities. In 1810 carbonized charcoal electrodes were used in an arc lamp, and in 1830 carbonized charcoal was used as an electrode for primary batteries. These electrodes were made from powdered charcoal or coke bonded with sugar syrup or coal tar, pressed and carbonized.[8] Recently, Coutinho, Luengo, and their co-workers[9-11] reported extensive studies of biocarbon electrodes manufactured from charcoal particles bonded together by wood tar and subsequently carbonized. The measured electrical resistivity of the electrode fell to 10^{-2} Ω-cm for carbonization at temperatures above 900 °C. This suggests that carbonized charcoal could be used to fabricate the consumable anode of a carbon fuel cell. Furthermore, we remark that charcoal can be produced from biomass inexpensively in yields that approach the theoretical limit.[12-15] Carbonized charcoal is also easy to store, and an established infrastructure exists to deliver charcoal to consumers worldwide. Moreover, carbonized charcoal - unlike graphite - is extremely reactive[14, 16-19]. For these reasons we have a keen interest in the development of biocarbon fuel cells.

PREPARATION OF CARBONIZED CARCOALS

A tubular furnace, which can reach temperatures as high as 1200 °C, was employed to carbonize the high-yield charcoal samples. A quartz tube (122 cm long, 1.99 cm I.D.) was placed within the furnace and used to carbonize a measured amount of charcoal in a ceramic boat. By locating type K thermocouples in the boat, it was possible to accurately measure the heat treatment temperature (HTT) of the charcoal. Nitrogen gas was delivered at 1.0 L/min to the tube during the carbonization process, thereby ensuring that the carbonization process, including the heat-up and cool-down steps, occurred in an inert environment. In a few cases, charcoals were carbonized within a closed crucible in a muffle furnace. This procedure was employed in our earlier work[13], and is able to carbonize larger quantities of charcoal more conveniently than the tubular furnace, though some weaknesses accompany the use of a muffle furnace (e.g., the reported temperature does not agree with that of the carbon within the crucible).

As shown in **Table 1**, the mass loss of the macshell charcoal increased monotonically with increasing HTT, except for the 1050 °C sample that was produced from a sample of flash carbonization charcoal with much lower volatile matter content. The carbonization process enlarged the surface area and pore volume of the charcoals. The values of surface area and pore volume of the macnut shell charcoal increased with the increasing HTT, except the sample treated at 650 °C. Values of the H/C and O/C ratios of the macshell carbons (see **Table 2**) decreased monotonically, except for the 950 °C sample. Of the remaining carbons listed in Tables 1 and 2, the leucaena wood carbon resembles the kukui nut shell carbon (except for their respective ash contents). A comparison of the 950 °C carbons reveals a considerable range in the values of each of the properties listed in Tables 1 and 2.

Table 1. Carbonized charcoal employed in this work.

Feed	HTT [a] [°C]	Burnoff [wt%]	S_{BET} [b] [m²/g]	V_p [c] [cc/g]
Macnut shell	650	21	216	0.135
	750	22	169	0.115
	850	30	194	0.132
	950	32	244	0.164
	1050	22	NA	NA
Kukui nut shell	950	NA	22	NA
Coconut husk	950	NA	437	NA
Leucaena wood	950	NA	310	NA

a) Heat treatment temperature, b) BET surface area by nitrogen adsorption, c) total pore volume.

Table 2. H/C and O/C Ratios[a] for the Carbonized Charcoals Employed in This Work.

Feed	HTT [°C]	H/C	O/C
Macnut shell	650	0.2969	0.0571
	750	0.2141	0.0494
	850	0.1178	0.0472
	950	0.1192	0.0504
	1050	0.0463	0.0223
Kukui nut shell	950	0.0693	0.0365
Coconut husk	950	0.1008	0.0740
Leucaena wood	950	0.0595	0.0325

a) Results on a mole basis.

ELECTRIAL RESISTIVITY

The electrical resistivity of the biocarbon samples was determined by a two-probe packed-bed technique at room temperature (*ca.* 20 °C). **Figure 1** shows the experimental set-up. The Nickel electrodes at the top and bottom of the 1.9 cm diameter packed bed - contained in an alumina tube - enable measurement of the electrical resistance of the bed. The packed bed is compressed by the upper electrode, which is forced against the bed by a pneumatic piston. Note that the electrodes are insulated from the apparatus by Teflon and alumina.

The results of the resistivity measurements of carbonized macshell charcoals are shown in **Figure 2**. This figure displays the effect of increasing compressive pressure on the resistivity of each of these carbons, which decreases by more than five orders of magnitude with increasing HTT from 650 to 1050 °C. As displayed in this figure, the

1050 °C macshell carbon's resistivity is about twice that of the graphite powder. Because these values approach the limit of detectability of our equipment, the exact ratio may differ somewhat from a factor of two. Under a compressive pressure of 7.8 MPa the apparent bulk density of the packed bed decreases from 0.70 to 0.58 g/cm^3 as the HTT increases from 650 to 950 °C. This decrease is due to increased porosity of the carbon at higher HTT. **Table 3** summarizes the results of the resistivity tests. Incidentally, the higher heating value of these carbonized charcoals was about 35 MJ/kg. This value is somewhat lower than those of natural gas, propane and crude oil, but much higher than those of most coals. If we used the carbonized charcoals in compressed packed beds shown in Table 3, the volumetric energy densities of the samples range from 15 to 29 GJ/m^3 and can be compeered with those of gasoline (32 GJ/m^3), ethanol (21 GJ/m^3), LNG (23 GJ/m^3) and other conventional fuels. When it is recognized that these carbon beds retain their bulk density after an initial compaction, the dense carbon packed beds are attractive energy carriers.

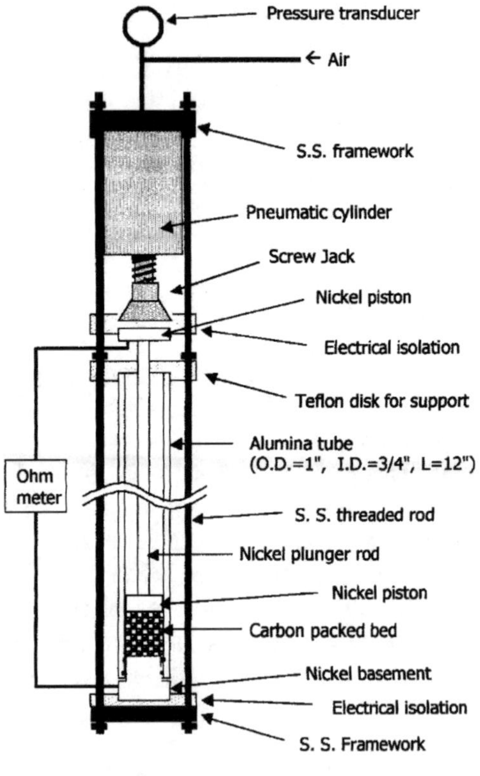

Figure 1. Apparatus employed to measure the electrical resistivity of a packed bed of carbonized charcoal.

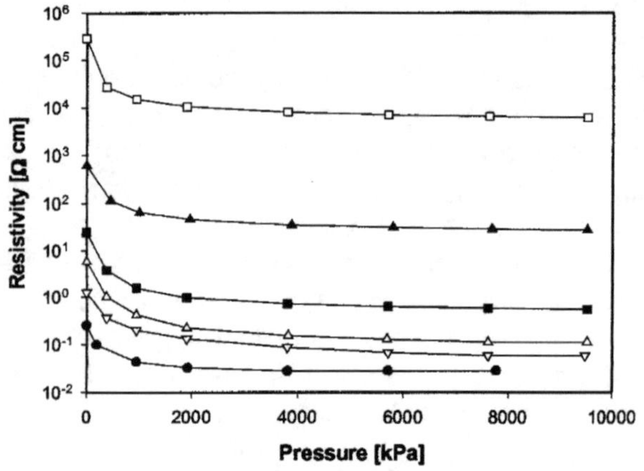

Figure 2. Resistivity versus compressive pressure for packed beds of 20/40 mesh macnut shell charcoals carbonized at 650, 750, 850, 950, and 1050 °C, and graphite powder (for comparison). Symbols (□: 650 °C, ▲: 750 °C, ■: 850 °C, △: 950 °C, ▽: 1050 °C, ●: graphite).

Table 3. Summary of the resistivity and bulk density of packed beds of carbonized charcoal samples under a compressive force of ca. 7.6 MPa.

Feed	HTT [°C]	ρ [a] [Ω-cm]	d [b] [g/cc]
Macnut shell	650	6.6 x10^3	0.70
	750	2.9 x10^1	0.65
	850	5.8 x10^{-1}	0.65
	950	1.1 x10^{-1}	0.58
	1050	5.9 x10^{-2}	0.61
Kukui nut shell	950	1.8 x10^{-1}	0.82
Coconut husk	950	1.8 x10^{-1}	0.46
Leucaena wood	950	1.6 x10^{-1}	0.44

a) Resistivity, b) bulk density of packed bed.

FTIR AND XRD ANALYSES

To learn about surface functional groups, FTIR spectra were measured by a FTIR spectrometer system (Perkin-Elmer 1710); including a DTGS detector, DRIFT accessory and microsampling. The specimens were examined in powder form, using KBr as reference material. The spectra were recorded from 4400 to 400 cm^{-1} by averaging 100 scans at 4 cm^{-1} resolution. Also, X-ray diffraction (XRD) measurements were carried out for phase and crystal structure identification with a Philips PW1710 diffractometer using CuK$_\alpha$ radiation and a graphite monochromator (45 kV, 35 mA, divergence 1°). Average crystallite size (L_{002}) and deformation (lattice strain, ε) were determined from single line profile Fourier (Voigt method) analysis.

As shown in **Figure 3**, the macshell charcoal has an alkyl aromatic structure with many oxygen containing (C-O-H, C=O, C-O-C) functional groups that give rise to characteristic bands in the IR spectrum.[27] During carbonization these functional groups are destroyed. At 650 °C the O-H groups (3700~2000 cm^{-1}), the aliphatic C-H groups (3000~2800 cm^{-1}) and in a great part the C=O groups (~1700 cm^{-1}) decompose, and condensed aromatic structures form with characteristic C-H out-of-plane bending modes (three main band components) between 900 and 700 cm^{-1}. At 750 °C most of the aromatic C-H groups were lost. A weak C-H absorption remains at about 880 cm^{-1}, which is assigned to lone H-atoms at the edges of the condensed aromatic sheets.[29] A weak, broad band with some sub-bands can be observed between 1700 and 1000 cm^{-1} due to the skeletal stretching and bending modes of aromatic structures containing residual O and/or N heteroatoms. At 850 °C the intensity of this weak, broad feature further diminishes, and so does the aromatic C-H (~880 cm^{-1}) band, as well. With increasing carbonization temperature between 650 and 850 °C the overall intensity of the infrared spectrum decreases along with the loss of functional groups. The baseline in the spectra of the carbonized charcoals is shifted upward. This baseline shift (increasing diffuse absorption) is assigned to low energy electron excitations of condensed aromatic structures. It is a well-known phenomenon in carbonized coal spectra.[29] All of the carbonized samples obtained from macshell, leucaena wood, coconut husk and kukui nut shell charcoals with HTT of 950 °C show spectra similar to each other as well as to the spectra of 850 °C carbonized macshell charcoal and graphite.

Figure 4 displays XRD patterns of macshell charcoals carbonized at increasing temperatures. The XRD patterns underwent almost no visible change with increasing HTT above 750 °C. **Table 3** demonstrates this observation. The scattering domains of the macshell carbons contained only 2-3 layers in the 002 direction. Small increase in L_{002} indicates some increase in ordering as a function of HTT in the macshell experiments. The estimated aromaticity γ_a of the domains did not significantly increase with HTT above 750 °C. At a HTT ≥950 °C the FTIR spectra and electrical resistivities were similar to those of graphite particles. On the other hand, the XRD pattern of the carbonized charcoals underwent virtually no change at increasing HTT from 750 to 950 °C. Evidently the 3 orders of magnitude decrease in electrical resistivity (with increasing HTT from 750 to 950 °C) is not associated with any change in the crystal structure of the samples. Furthermore, the XRD patterns do not resemble those of graphite or turbostratic carbons. These facts cause us to visualize carbonized charcoals to be a macromolecular, cross-linked, three dimensional, aromatic structures replete with conjugation and π bonds that facilitate the conduction of electrons. It can be concluded that the carbonized charcoals have electro-conductive amorphous structures. This property makes the carbonized charcoals reactive whereas graphite is inert. As described in the previous section, the carbonized charcoals are porous. The large surface area provided by the porous structure provides may contribute to the enhancement of reactivity when the carbonized charcoals are used as the fuel anodes. The electrochemical reactivity of the samples is under investigated. We are currently testing a carbon-fueled fuel cell that employs the carbonized charcoal packed bed described herein as its consumable anode.

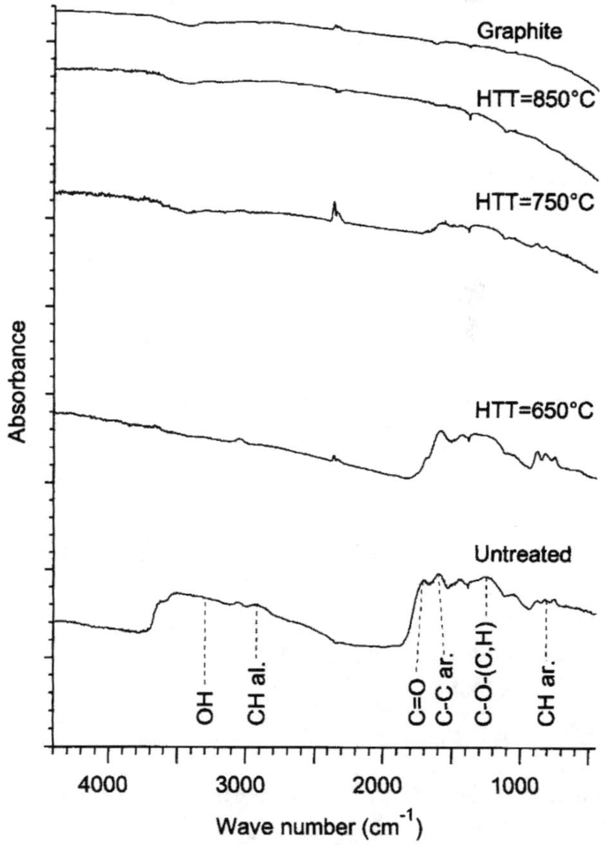

Figure 3. Effect of heat treatment temperature on the FTIR spectra of macadamia nutshell char. The spectrum of the synthetic graphic is shown for comparison. (Each curve is drawn one unit higher than the curve beneath it.)

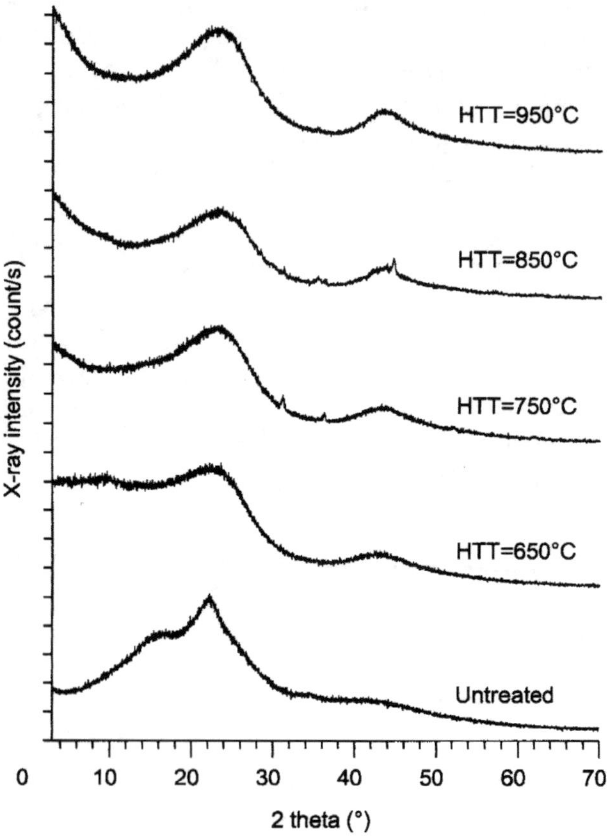

Figure 4. Effect of the heat treatment temperature (HTT) on the XRD spectra of the macadamia nutshell carbon.

Table 3. Carbonized charcoal employed in this work.

Feed	HTT [°C]	L_{002} [a] [Å]	*d*(002) [b] [m²/g]	γ_a [c]
Macnut shell	650	6	4.01	0.74
	750	7	3.98	0.79
	850	7	3.87	0.96
	950	8	3.79	0.97
	1050	8	3.76	0.99
Kukui nut shell	950	7	3.85	0.99
Coconut husk	950	6	3.89	0.99
Leucaena wood	950	6	3.83	0.99
Synthetic graphite		233	3.35	1.00
Natural graphite		918	3.36	1.00

a) Average crystal size determined from single-line profile Fourier Analysis, b) distance between crystallites planes in the 002 direction, c) aromaticity factor.

SUMMARY

Under compressive pressure a compact packed bed of carbonized charcoal can have an electrical resistivity comparable to that of a compact packed bed of graphite. A compact carbonized charcoal packed bed can have a very large surface area and high reactivity as well as an energy density comparable to conventional fuels. Furthermore, the wholesale price of biomass charcoal is competitive with gasoline and natural gas. In light of the above facts, we conclude that compact packed beds of carbonized charcoal have promising applications as electrodes and consumable anodes in carbon-fueled fuel cells and batteries.

ACKNOWLEDGMENTS

We thank Lloyd Paredes, Brent Shimizu, Caera McNally, Carolyn Wallace, and Kurt Verheyden (all of UH) for assistance with the experimental work, Prof. Jim Brewbaker (UH) and Dana Gray (Oils of Aloha) for biomass samples, and John Wayte for his skillful fabrication of the apparatus displayed in Figure 1. Research at the University of Hawaii (UH) was supported by the Office of Naval Research (contract #N00014-01-1-0928), the Coral Industries Endowment of UH, and the Osaka Gas Co. G. Várhegyi is grateful for OTKA grant T 037705 that supported his participation in the work.

REFERENCES

1. Vielstich, W. Fuel Cells.; Wiley-Interscience: London, 1965.
2. Williams, K. R., Ed., An Introduction to Fuel Cells., Elsevier Publishing Co.: Amsterdam, 1966.
3. Bockris, J. O. M.; Srinivasan, S. Fuel Cells: Their Electrochemistry.; McGraw-Hill Book Co.: New York, 1969.
4. Anbar, M.; McMillen, D. F.; Weaver, R. D.; Jorgensen, P. J. Method and Apparatus for Electrochemical Generation of Power from Carbonaceous Fuels. U.S.A. Patent 3,970,474, 1976.
5. Anbar, M. Methods and Apparatus for the Pollution-Free Generation of Electrochemical Energy. USA Patent 3,741,809, 1973.
6. Pesavento, P., personal communication, 1998.
7. Gur, T. M.; Huggins, R. A. Direct Electrochemical Conversion of Carbon to Electrical Energy in a High Temperature Fuel Cell. J. Electrochem. Soc. 1992, 139, L95.
8. Ford, A. R.; Greenhalgh, E. Industrial Applications of Carbon and Graphite. In Modern Aspects of Graphite Technology; L. C. F. Blackman, Eds.; Academic Press: London, 1970; p 258.
9. Coutinho, A. R.; Luengo, C. A. Preparing and Characterizing Electrode Grade Carbons from Eucalyptus Pyrolysis Products. In Advances in Thermochemical Biomass Conversion; A. V. Bridgwater, Eds.; Blackie Academic & Professional: London, 1993; p 1230.

10. Coutinho, A. R.; Luengo, C. A. Mass Balance of Biocarbon Electrodes Obtained by Experimental Bench Production. In Developments in Thermochemical Biomass Conversion; A. V. Bridgwater and D. G. B. Boocock, Eds.; Blackie Academic & Professional: London, 1997; p 305.
11. Coutinho, A. R.; Rocha, J. D.; Luengo, C. A. Preparing and characterizing biocarbon electrodes. Fuel Processing Technology 2000, 67, 93.
12. Antal, M. J.; Croiset, E.; Dai, X. F.; DeAlmeida, C.; Mok, W. S. L.; Norberg, N.; Richard, J. R.; Majthoub, M. A. High-Yield Biomass Charcoal. Energy Fuels 1996, 10, 652.
13. Antal, M. J.; Allen, S. G.; Dai, X.; Shimizu, B.; Tam, M. S.; Gronli, M. G. Attainment of the theoretical yield of carbon from biomass. Ind. Eng. Chem. Res. 2000, 39, 4024.
14. Antal, M. J.; Gronli, M. G. The Art, Science, and Technology of Charcoal Production. Ind. Eng. Chem. Res. 2003, 42, 1919.
15. Antal, M. J.; Mochidzuki, K.; Paredes, L. S. Flash Carbonization of Biomass. Ind. Eng. Chem. Res. 2003, in press.,
16. Dai, X.; Antal, M. J., Jr. Synthesis of a High-Yield Activated Carbon by Air Gasification of Macadamia Nut Shell Charcoal. Ind. Eng. Chem. Res. 1999, 38, 3386.
17. Tam, M. S.; Antal, M. J., Jr. Preparation of Activated Carbons from Macadamia Nut Shell and Coconut Shell by Air Activation. Ind. Eng. Chem. Res. 1999, 38, 4268.
18. Conesa, J. A.; Sakurai, M.; Antal, M. J. Synthesis of a High-Yield Activated Carbon by Oxygen Gasification of Macadamia Nut Shell Charcoal in Hot, Liquid Water. Carbon 2000, 38, 839.
19. Tam, M. S.; Antal, M. J.; Jakab, E.; Varhegyi, G. Activated Carbon from Macadamia Nut Shell by Air Oxidation in Boiling Water. Ind. Eng. Chem. Res. 2001, 40, 578.

CONTACT

Michael J. Antal, Jr.: email: mantal@hawaii.edu,
URL: www.hnei.hawaii.edu.

Effects of Convection through Gas Diffusion Layer to Limiting Current in PEMFC Operating Between 80°C and 120°C

Minkmas V. Williams, H. Russell Kunz, James M. Fenton
Chemical Engineering Department, University of Connecticut, Storrs, CT 06269
Tel: 860-486-3606, Fax: 860-486-2959, E-mail: minkmas@engr.uconn.edu

ABSTRACT

In proton-exchange membrane fuel cells (PEMFC), gas diffusion layers (GDL) serve as current collectors that allow ready access of fuel and oxidant to the anode and the cathode catalyst surfaces, respectively. Three gas diffusion layers with distinctively different gas permeability have been used to study the effects of convection through gas diffusion layer to the cathode limiting current using air and 4%oxygen/nitrogen as oxidant gases. The study has been done at three operating conditions: 80°C/75%R.H. cathode inlet, 100°C/70%R.H. cathode inlet, and 120°C/35%R.H. cathode inlet to evaluate the influence of cell temperature, oxygen mole fraction, and relative humidity on the limiting current. The graphite flow field used was a conventional single-serpentine pattern where diffusion is usually assumed the main transport phenomenon through the GDL. For the three GDLs, increasing dry gas flow rate increased the limiting current. The increased limiting current at higher dry cathode flow rate could be due to higher average oxygen mole fraction improving diffusion or more convection. Effects of increasing average oxygen mole fraction as the dry gas flow rate increases were separated out by analyzing the limiting current from air and 4%O_2/N_2 together and by analyzing the water and mass balance at each limiting current condition, giving results verifying higher limiting current at higher dry gas flow rate due to increased convection alone. Cell relative humidity has shown to be another significant parameter affecting the limiting current besides average oxygen mole fraction due to diffusion and dry gas flow rate due to convection, by reducing oxygen transport through the ionomer thin film of the cathode catalyst layer as the relative humidity decreases. Comparison of the three GDLs showed that a GDL with higher gas permeability in the micro-porous layer has higher limiting current from more convection accentuating the significance of high gas permeability as criteria for the design of effective GDL. Convection has proven to be significant to the limiting current of hydrogen/air PEMFC even in a conventional flow field pattern as what used.

Manipulation of Bioparticles with Electrode Polarization Effects

Adrienne R. Minerick*, Ronghui Zhou, and Hsueh-Chia Chang
University of Notre Dame
Department of Chemical and Biomolecular Engineering
Notre Dame, IN 46556
*Current address: Mississippi State University
Dave C. Swalm School of Chemical Engineering
Mississippi State, MS 39762

Paul Takhistov
Rutgers University
Department of Food Science
New Brunswick, NJ 08901

ABSTRACT

Dielectrophoretic (nonuniform AC fields) tools have been gaining popularity for use in microseparations. This work examines the motion of a suspension of human red blood cells in high-frequency AC fields. Rapid cell repulsion velocities from the polarized electrode surface are seen in the first few moments of AC field application and are tabulated for parallel and orthogonal electrode configurations. During the first few minutes, cellular polarization influences cell - cell interactions such that RBC pearl chains form along field lines due to attraction between polarized cells. This is followed by classical negative dielectrophoretic motion of the chains across field lines to regions of weaker field. The resonant frequency of electrode polarization and cell repulsion occurs at around 1 MHz. The interaction of the double-layer polarization mechanism of both the cell and the electrode is a distinct and potentially important bioparticle manipulation tool. This presentation record is a brief introduction; the full study is published in a special Miniaturization III issue of Electrophoresis [1].

KEY WORDS

Red blood cells, Erythrocytes

Microdevices

Dielectrophoresis

Electrode Polarization

INTRODUCTION

As electrokinetic tools are gaining popularity, especially in microdevices, the need for understanding fundamental forces on the micro scale becomes greater. AC electrokinetics have been useful in the separation and movement of a variety of cells [2]. It is typically employed as dielectrophoresis (DEP) where electrically polarized particles are induced to move by a non-homogeneous AC electric field. In addition to the usual advantages of using electric fields in micro-devices, AC fields are superior to DC ones because electrode bubble generation is minimized at high frequencies. In this talk, we propose a new AC electrokinetic technique for microfluidic separation of bioparticles that does not require an inhomogeneous field and with much higher mobilities than traditional DEP.

BACKGROUND

Nonlinear dielectrophoresis differs considerably from the DC electrokinetics used in many biological applications. DC electrokinetics are capable of differentiating between cells based on net surface charges and size. The bioparticles have, for the most part, a uniform Debye layer around them. However, nonlinear AC dielectrophoresis is capable of differentiating based on the conductivity of the boundary, interior, and constituents of the cell. DEP can also distinguish cellular capacitance and polarizability of the Debye layer. Nonuniform AC fields have been used for a variety of applications in the last 20 years from electroporation to electrofusion, but the possibilities of separating and selectively manipulating cells based on conductivity, capacitance, and polarizability is only beginning to be explored.

Classical dielectrophoretic theory classifies particle polarization mechanisms to determine their DEP motions and dielectric dispersion behavior, based on their conductivity with respect to the medium. During each half cycle in an uniform field (parallel electrodes), a more conducting particle will draw the current flux lines into it causing an induced dipole that is aligned with the field [2]. A less conducting particle will cause field lines to diverge around it and cause a depletion of ions at the particle poles creating an induced dipole aligned against the applied field. In both cases, the dipole orientation will flip with the field in the next AC half cycle. The result is that no net electrical forces act on the particles (and no motion is observed) over longer time periods in a uniform AC field [2]. However, in non-uniform fields, the current flux lines are not uniform through or around the cell and a net polarization can form. The resulting net dielectric forces can act on the polarizable particle and cause motion either toward high field density or low field density regions.

The DEP force acting on the particle is dependent on the field gradient as well as the local field, which directly impacts the dipole moment. When the particle is more conductive than the medium, the biased dipole moment causes a force toward regions of high electrical field. One should note that field density is greatest at points of high curvature on electrodes and attraction to these regions is known as positive DEP. When a particle is less conductive than the medium, the force from the dipole moment is toward regions of low field density. This behavior is known as negative DEP. Field density is lower at flat electrodes or far from electrode surfaces [3].

SUMMARY

At frequencies in the 10^3-10^6 Hz range, the cells polarize periodically in time at the same frequency as the AC field even with a spatially homogeneous field. This polarization is classically attributed to Maxwell-Wagner interface / membrane relaxations modeled by an RC circuit [4]. In other work within our group, we have shown that this time-periodic polarization is due to charging of the internal and external double layers of the cell. Experimental observations indicate that platinum electrodes also develop a polarized layer, presumably via similar mechanisms [5]. With specific polarizations of the electrode and bioparticle determined by the bulk and interior conductivity, the interaction of the polarized layers can result in a strong repulsive force. We present a series of experiments in homogeneous and inhomogeneous fields to quantify the repulsion between bioparticles and platinum electrodes and to compare it to DEP forces.

The cell repulsion from the platinum electrodes due to electrode polarization interacting with cell membrane polarizations are observed to be the strongest force acting on the particles in the first few seconds of field application. It is stronger than DEP due to inhomogeneous fields by a factor of ten in usual micro-channel geometries. Repulsion velocities are quantified as a function of blood age and species. In a 0.1 S/m isotonic phosphate buffer saline medium, we observe the largest cell mobilities at an optimal frequency of 1 MHz, corresponding to the inverse diffusion time across the double layer of the cell and of the electrode's polarized layer. This suggests that the dielectric responses of both particles and electrodes are mostly determined by normal electro-migration of ions from the bulk to their interfaces. The corresponding velocity approaches 50 microns per second at the optimum frequency. Sensitivity to RBC age and species suggests that the surface proteins and membrane ion channels can affect the capacitance of the interface to accommodate the ions from the bulk. Such surface ion accumulation and polarization mechanisms are different from the classical dielectric theories [1]. The data presented could help develop new non-DEP methods for manipulating and separating bioparticles based on their interface charges.

ACKNOWLEDGMENTS

This work was supported by NASA grant NAG5-10503, the Clare Booth Luce Fellowship Fund, and the Center for Environmental Science & Engineering Bayer Predoctoral Fellowship Fund. The authors would like to thank Kenneth Olson's lab for providing trout blood samples and the Center for Transgene Research at Notre Dame for providing mouse blood samples. Appreciation is also extended to Agnes Ostafin for use of the video microscope setup.

REFERENCES

1. Minerick, A.R., R. Zhou, P. Takhistov, and H. -C. Chang. "Manipulation and Characterization of Red Blood Cells with AC Fields in Micro-

Devices" [accepted - *Electrophoresis*: Special Issue Miniaturization III].
2. Pethig, R. "Dielectrophoresis: Using Inhomogeneous AC Electrical Fields to Separate and Manipulate Cells" *Crit. Rev. Biotechnol.* 16(4): 331-348. 1996.
3. Gascoyne, P.R.C., and Vykoukal, J., "Particle Separation by Dielectrophoresis" *Electrophoresis.* 2002, 23, 1973-1983.
4. Martinsen, O.G., Grimnes, S., and Schwan, H.P. "Interface Phenomena and Dielectric Properties of Biological Tissue" *Encyclopedia of Surface and Colloid Science.* Marcel Dekker, Inc. 2002.
5. Onaral, B. and Schwan, H.P. "Linear and nonlinear properties of platinum electrode polarization II: time domain analysis" *Med. & Biol. Eng. & Comput.* 21, 210-216. 1983.

CONTACT

For further information, please contact Adrienne Minerick at minerick@che.msstate.edu.

DEFINITIONS

Dielectrophoresis: electrically polarized particles are induced to move by a nonuniform AC electric field.

SIMULATION OF ELECTROKINETIC MICROFLUIDICS IN COLLOIDAL SYSTEMS

Dzmitry Hlushkou[1], Drona Kandhai[2], Andreas Seidel-Morgenstern[1], and Ulrich Tallarek[3]

[1]Max-Planck-Institute for the Dynamics of Complex Technical Systems

Germany, Magdeburg, 39106, Sandtorstrasse 1

[2]Kramers Laboratorium voor Fysische Technology,

Faculty of Applied Science, Delft University of Technology

The Netherlands, Delft, 2628 BW, Prins Bernhardlaan 6

[3]Institut für Verfahrenstechnik, Otto-von-Guericke-Universität Magdeburg

Germany, Magdeburg, 39106, Universitätsplatz 2

Key words: electrokinetics, electroosmosis, microfluidics, computer simulation

Prepared for presentation at the 2003 Annual Meeting, San Francisco, CA, Nov. 16-21

July, 2003

Unpublished

Abstract

A numerical scheme for simulating electrokinetic microfluids in systems with an arbitrary geometry is developed. This scheme is based on the numerical solution of the coupled Poisson, Nernst-Planck, and Navier-Stokes equations. While traditional finite difference methods are used to resolve the first two problems, the lattice-Boltzmann method is applied to the latter. The developed numerical scheme has been applied to computer simulations of electroosmotic flow through a simple-cubic array of micro-sized hard spheres. Volumetric electroosmotic flow has been studied as a function of the applied electrical field, ζ-potential of the sphere surface, and ionic strength. The investigation of colloidal stability and the electrokinetics in microchip devices containing fixed bed segments (particulate or monolithic structures) represents two potential applications of this work.

1. Introduction

Electrokinetic microfluidic phenomena are currently one of the fundamental and practical interest in various fields of science and engineering. This interest is stimulated, on the one hand, by rapid recent developments in microfabrication technologies and, on the other hand, by the fact that electrokinetic transport can offer distinct advantages over pressure-driven flow, e.g. reduced hydrodynamic dispersion and mass transfer resistance, as the characteristic dimensions of devices decrease to micrometer dimensions. One of the important applications of electrokinetic transport is capillary electrochromatography (CEC), where electroosmotic flow (EOF) is used as a mechanism for transporting solvent and solutes through a chromatographic column. Compared to pressure-driven liquid chromatography, CEC can provide substantially higher separation efficiency. This feature mainly results from i) the absence of pressure drop, that allows the use of very small particles packed in the chromatographic (capillary) columns and so to increase the total separation perfomance, and ii) reduced hydrodynamic dispersion, that decreases chromatographic band broadening. Figure 1 illustrates the difference between two chromatograms obtained experimentally during the separation process for the same sample and Peclet-number but with (a) pressure-driven and (b) electroosmotic flows.

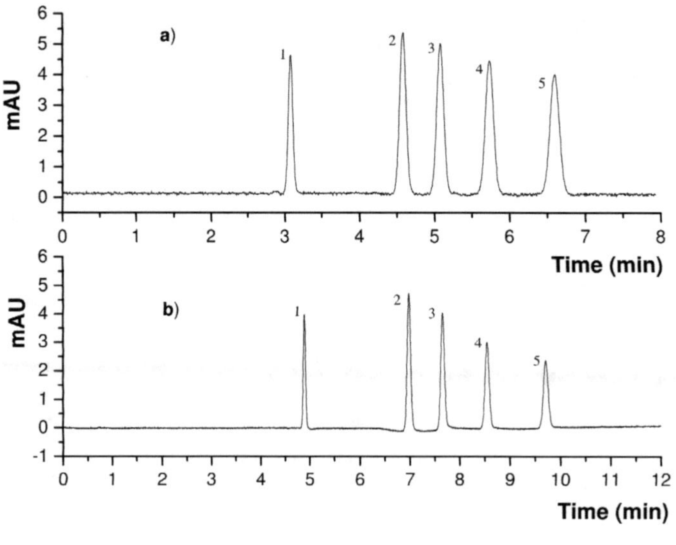

Figure 1. Two chromatograms obtained during two separation processes carried out with (a) pressure-driven and (b) electroosmotic flows. Sample: thiurea (1) and 4 alkylbenzoates (2-5).

The narrower peaks in the second chromatogram indicate the possibility to perform the analysis at higher resolving power translating it to an increased sensitivity and lower operation costs.

Since packed chromatographic columns can be represented by a capillary packed with spherical particles, the computer simulation of EOF through colloidal structures can provide complementary insight into the complex transport processes within such devices.

2. Mathematical model and implementation

The general mathematical description of the steady-state electrokinetic transport problem in an *N*-component incompressible electrolyte consists of the set of coupled Nernst-Planck, Navier-Stokes, and Poisson equations

$$\mathbf{v}\,\nabla n_i - D_i \nabla^2 n_i - \frac{e z_i D_i}{k_B T} \nabla \cdot \left(n_i \nabla \Phi \right) = 0 \quad , \tag{1}$$

$$\rho_f (\mathbf{v} \cdot \nabla) = -\nabla p + \eta_f \nabla^2 \mathbf{v} - e \nabla \Phi \sum_{i=1}^{N} z_i n_i \quad , \tag{2}$$

$$\nabla^2 \Phi = -e \sum_{i=1}^{N} \frac{z_i n_i}{\varepsilon_0 \varepsilon_r} \quad , \tag{3}$$

where $\mathbf{v}$ represents the divergence-free velocity field ($\nabla \cdot \mathbf{v} = 0$), ρ_f, η_f, p, ε_0 and ε_r denote the density of the fluid, its dynamic viscosity, hydrostatic pressure, the permittivity of vacuum and relative permeability, respectively, n_i, D_i, z_i are the number concentration, diffusion coefficient and valency of ionic species *i*, respectively, Φ is the local electrical potential, e, k_B and T represent the elementary charge, Boltzmann constant and temperature, respectively.

An iterative numerical scheme was developed for solving the aforementioned set of equations (1-3) subject to the following boundary conditions

$$\mathbf{j}_i \cdot \nu = 0, \ \ \mathbf{v} = 0, \ \ \Phi = \zeta + \varphi \quad \text{at the solid-liquid interface,} \tag{4}$$

where $\mathbf{j}_i$ and ν are the flux density and the normal to the interface, respectively. ζ and φ represent the electrical potential contributions due to the surface charge at the solid-liquid interface and applied external electric field. It should be pointed out, that the presented electric boundary condition assumes negligible ion-convective effects on the ζ-potential distribution as compared to ion-diffusion or small perturbations with respect to the equilibrium state.

At each iteration the coupled Nernst-Planck, Poisson, and Navier-Stokes equations are solved by turns, using an uniform cubic lattice. While conventional finite difference methods have been employed to resolve the first two problems, the lattice-Boltzmann algorithm [McNamara and Zanetti 1988, Higuerra et al. 1989] has been applied for solution of the latter. This allowed to treat this hydrodynamic problem with high computational efficiency for systems characterized by a complex solid-liquid interface.

3. Results of simulation

The implementation of the presented numerical scheme has been performed at a parallel computer (Hewlett-Packard Superdom). A number of EOF simulations for various capillary and colloidal systems has been carried out.

Initially this computer model was tested by calculating the electric charge distribution around a single sphere immersed in an electrolyte solution. The charge distribution determines the electromotive forces, acting on a liquid, and causing the specific fluid velocity distribution. The simulated charge density distribution in a symmetric 1:1 electrolyte has been compared (Fig. 2) with the so-called LOW-tables [Loeb et al. 1961], which contain the results of an one-dimensional numerical solution for the Poisson-Boltzmann equation describing, in particular, the equilibrium distributions of the net electrical charge for a single sphere system of radius r_S.

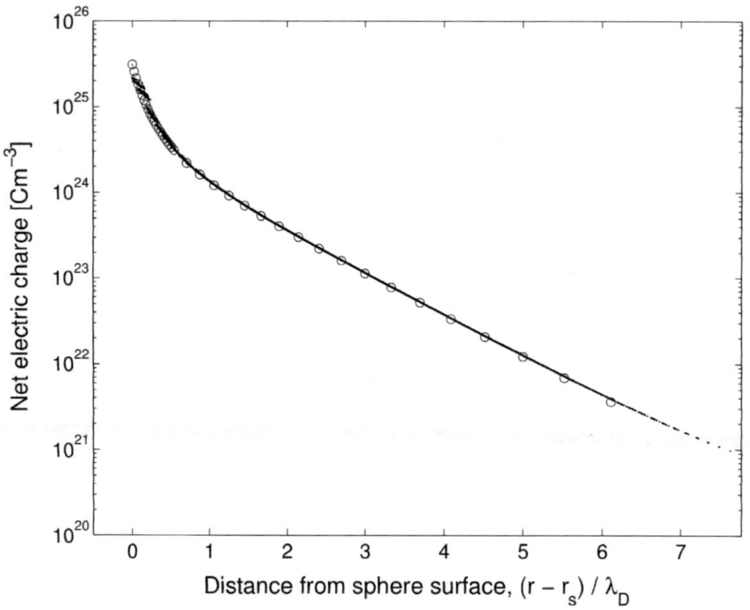

Figure 2. Point-wise comparison of the simulated electric charge distribution around a single sphere (•) with LOW's data (○) for 1:1 electrolyte. $\zeta = -102.77$ mV, the aspect ratio of sphere diameter to the electric double layer thickness $d_S/\lambda_D = 10$, bulk concentration $n_\infty = 5.678 \cdot 10^{23}$ m^{-3}.

This comparison demonstrates the good agreement between two sets of data over a wide range of the net charge magnitude.

In the next step we simulated the EOF through a simple cubic (SC) array of spheres. Although real packed chromatographic columns are represented more feasibly by random sphhere packings, we have chosen the regular structure in order to eliminate any effects related to the random nature of the medium model and to investigate clearly the relationship between simulated mean velocity and experimental conditions, for instance, ζ-potential, ionic strength, and applied electric field E_{app}. As it was mentioned in Section 2, we have assumed the ζ-potential is not affected by the applied electric field or liquid convection. A collection of EOF simulations has been carried out for various values of ζ-potential, applied electric field, computational grid resolutions and aspect ratio α of sphere diameter to the electric double layer thickness λ_D, that can be easily translated to variable ionic strength. All values of variable quantities which have been used in the aforementioned collection of simulations are given in Table 1. The complete collection consists of 512 simulations.

Table 1.

The values of variable parameters used in the simulations

ζ, mV	−25.69 −51.39, −77.07, −102.77
E_{app}, kV/m	10, 25, 50, 100
Aspect ratio, $\alpha = D_S / \lambda_D$	2.272, 22.72, 50, 100
Grid resolution (points per sphere diameter), β	25, 30, 40, 50, 75, 100, 150, 200

The lowest value of employed ζ-potential (−25.69 mV = k_BT/e) corresponds to the upper limit for the conventional use of the Debye-Hückel approximation, while the highest value (−102.77 mV = $4k_BT/e$) can be considered as the upper limit of ζ-potential for many realistic solid-liquid interfaces. The smallest value of α implies an overlap of EDL's of neighboring spheres over a quite extended area, while in the case of α = 100 this overlap is relatively small. The mean velocity of electoosmotic flow in the direction of the applied electric field was calculated for each set of variable parameters. Since there is no analytical solution and only a few number of results of other numerical simulations concerning EOF through SC array that could be used for the comparison, is published, the convergence of the mean velocity by the further refinement of the computational grid was considered as sufficient proof of the computational accuracy. A brief presentation of the information contained in the complete collection is difficult, therefore only some results will be provided in this paper.

First, we compared the simulated mean longitudinal velocity with that obtained by D. Coelho et al. [D. Coelho et al 1996]. They developed a model of electrokinetic phenomena in finely dispersed porous media under the assumptions of low ζ-potential and small perturbations from the equilibrium state and simulated EOF, in particular, through SC array of spheres. The biggest absolute values of ζ and aspect ratio in their simulations were 51.385 mV and 22.72, respectively. Figure 3 indicates the good agreement between simulated data and results presented in Ref. [4] .

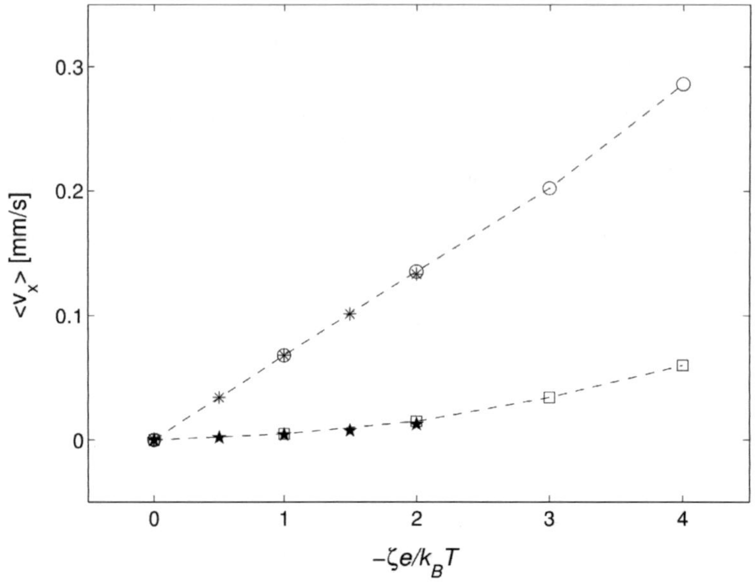

Figure 3. Point-wise comparison of simulated (□ and ○) mean EOF velocity through SC array of spheres with published one in [Coelho et al. 1996] (★and ∗). $\alpha = 2.272$ (the lower point subset) and 22.72 (the upper point subset), E_{app} = 10 kV/m, $\beta = 200$.

Then, we carried out the analysis of the discretization effects on the simulated velocity. Its dependence on the computational grid resolution for E_{app} = 10 kV/m and two limiting values of the ζ-potential, −25.69 and −102.77 mV, are shown in Figure 4. While for $\zeta = -25.69$ mV mean velocity decreases monotonically and independently of the aspect ratio α to some fixed value with the finer resolution, the velocity behavior is quite different for small and large values of α when $\zeta = -102.77$ mV. The slump at coarser resolution for the large values of α can be explained by the error in the simulated of electrical charge distribution. The net electric charge demonstrates its locally highest

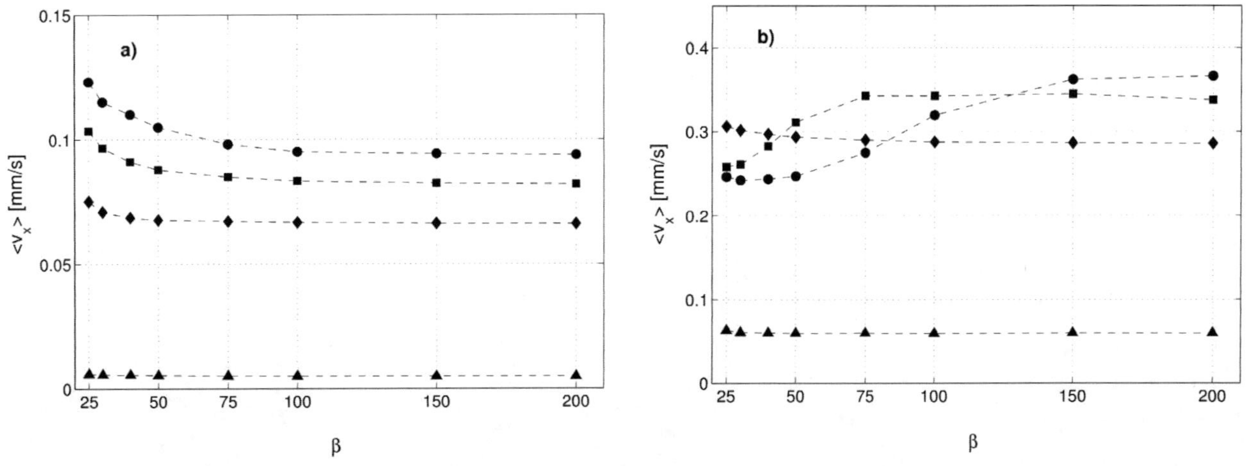

Figure 4. Mean longitudinal velocity of EOF through SC-array of spheres as a function of the computational grid resolution β (see Table 1). E_{app} = 10 kV/m, ζ = −25.69 mV (a) and −102.77 mV (b), α = 2.272 (▲), 22.72 (♦), 50 (■) and 100 (●).

density at the solid-liquid interface and then decays to zero in the bulk region. When the spatial discretization step is large with respect to the EDL thickness, even the grid point closest to the solid-liquid interface is located in a region of a relatively low electric charge density. Thus, the electromotive force is underestimated and total EOF is reduced. The effect of the ζ-potential value on the simulated velocity is illustrated in Figure 5. With increasing aspect ratio α the relation between mean velocity and applied electric field tends towards linear behaviour. This fact agrees with the "so-called" thin double layer approximation when the electroosmotic ("slip") velocity is defined by the Smoluchowski equation with a linear dependence on ζ-potential

$$V = \frac{\varepsilon_0 \varepsilon_r}{\eta_f} \zeta E_{app} \quad . \tag{5}$$

The behavior of the simulated mean velocity with changing of the aspect ratio (see Fig. 6) for different ζ-potentials can be explained along the same lines.

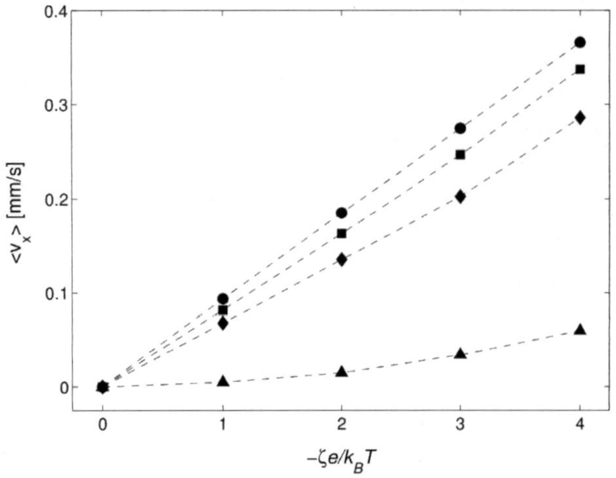

Figure 5. Simulated mean velocity as a function of ζ-potential. E_{app} = 10 kV/m, α = 2.272 (▲), 22.72 (♦), 50 (■) and 100 (●), $\beta = 200$.

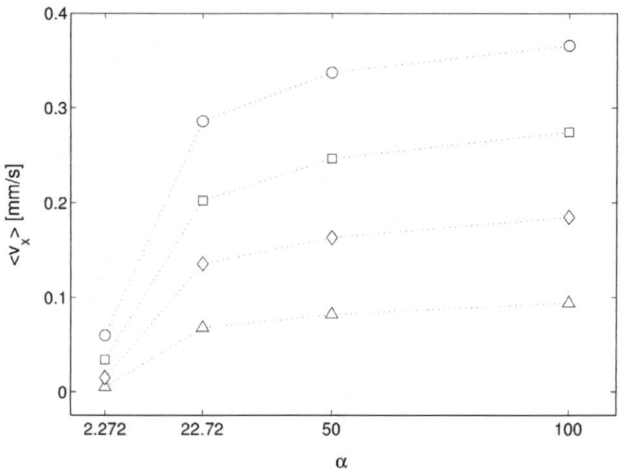

Figure 6. Simulated mean velocity as a function of α. E_{app} = 10 kV/m, ζ = −25.69 mV(△), −51.385 mV(◊), −75.07(□) and −102.77 mV (○), $\beta = 200$.

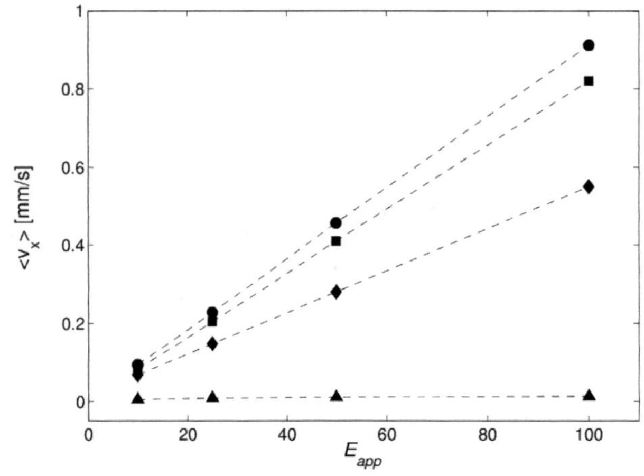

Figure 7. Simulated mean velocity as a function of E_{app}. ζ = −25.69 mV, α = 2.272 (▲), 22.72 (♦), 50 (■) and 100 (●), $\beta = 200$.

Finally, the dependence of simulated mean velocitiy on applied electric field has been examined. In Figure 7 the data for ζ = −25.69 mV are presented. The relation between electroosmotic velocity and applied electric field is almost perfectly linear for all values of α. Moreover, the effect of "saturation", when electroosmotic velocity ceases to vary with the further increase of the aspect ratio, is observed for any value of applied electric field. This corroborates the validity of the representation of a porous medium as an open channel with a certain "effective" diameter. Of course, this representation is limited to an equality of mean velocities and can become unrealistic if, for instance, hydrodynamic dispersion is evaluated.

3. Conclusion and outlook

EOF through SC sphere array has been simulated by an iterative numerical solution of the coupled Navier-Stokes, Nernst-Planck and Poisson equations assuming that the ζ-potential distribution at the solid-liquid interface is neither affected by convection, nor by the applied electric field. This novel applied combination of two different numerical techniqes for the solution of the complex problem has proven to be applicable and reliable. It can be recommended to be used in further studies.

The collection of simulations of EOF through SC sphere array has been performed for various values of applied electric field, ζ-potential and the aspect ratio of sphere diameter to double layer thickness. The dependence of mean electroosmotic velocity on those variable parameters has been examined. The simulated mean electroosmotic velocity has demonstrated good adreement with a number of frequently used approximations. At the same time the developed model can assist in resolving difficult problems requiring to hold the complete information concerning the flow velocity distribution in systems with a complex morphology, as in the evaluation of hydrodynamic dispersion in electroosmotic flow through porous media. The obtained data can be used for optimized spatial resolution in large-scale simulations, relative to computational expenses and resulting accuracy.

The possible further development of the presented model can include the dissociation of surface groups or adsorption processes which can be responsible for changes in ζ-potential or surface electric charge due to alterations in the local chemical environment caused by convection or electromigration phenomena [Pačes et al. 2003].

References

D. Coelho, M. Shapiro, J.F. Thovert, and P.M. Adler, Electroosmotic Phenomena in Porous Media, *J. Colloid Interface Sci.* **181**, 169-190 (1996)

F. Higuerra, S. Succi, and R. Benzi, Lattice Gas Dynamics with Enhanced Collisions, *Europhys. Lett.* **9**, 345-349 (1989).

A.L. Loeb, J.Th.G. Overbeek, and P.H. Wiersema, The Electrical Double Layer around a Spherical Colloid Particle, Massachusetts, MIT Press, 154 (1961).

G. McNamara and G. Zanetti, Use of the Boltzmann Equation to Simulate Lattice-Gas Automata, *Phys. Rev. Lett.* **61**, 2332-2335 (1988).

M. Pačes, J. Kosek, M. Marek, U. Tallarek, and A. Seidel_morgenstern, Mathematical Modelling of Adsorption and Transport Processes in Capillary Electrochromatography: Open-Tubular Geometry, *Electrophoresis* **24**, 380-389 (2003).

Modeling of Micro-Electric Field Flow Fractionation

Joseph J. Biernacki and Naitik Vyas
Tennessee Technological University
Department of Chemical Engineering
Box 5013, Cookeville, TN 38501

ABSTRACT

Electric field flow fractionation (EFFF) is a separation technique that utilizes an electric field perpendicular to the direction of flow to separate particulate and macromolecular materials. The electrical field is used to move particles of different size or electrical charge into different streamlines of the parabolic velocity profile of a laminar flow field, there-by causing differing particles to elute at different rates. Both constant voltage and pulsed voltage protocols have been used to generate the required cross-flow field. The advantages of each are a function of electrochemical system response and flow hydrodynamics. A computational model has been developed to explore the operating domain as a function of a wide range of system parameters.

KEY WORDS

electric field flow fractionation, EFFF, FFF, cross-flow electrophoresis, modeling, simulation

INTRODUCTION

There has been a push in the past decade to develop smaller and faster biochemical tools. These systems typically combine sample delivery and conditioning as well as detection of analytes. Extensive research has been dedicated to processes which concentrate and separate substances. Conventional electrophoresis techniques have been used for these applications but often require large field strengths. Chemically facilitated separations are also possible but may cause sample augmentations, such as denaturing of proteins. Field-flow fractionation[1] (FFF) may be a reasonable alternative to these and other technologies for many separation applications.

Electrical field flow fractionation can potentially be used to separate cells, large molecules, colloids and delicate structures such as liposome, for which a process like conventional electrophoresis cannot [1]. There are different forms of field flow fractionation. The work presented here involves one such form of FFF, electrical field flow fractionation (EFFF). EFFF was first described in 1972 [2]. One of the apparent problems, however, was polarization of the electrodes. Polarization causes the separation channel to behave like a capacitor, with most of the potential drop occurring within a small distance from the channel walls leaving a very weak field in the bulk of the channel [3]. While increasing the applied potential may help there are practical electrochemical limits. Later researchers [4] found that applying a pulsed potential enhanced separation under some conditions. EFFF is also amenability to miniaturization, upon which reduced analysis time, reduced sample volumes, and reduced system size can be realized [5]. While the theory for constant applied potential has been well developed, similar theory for pulsed potential is not available. A framework for analysis and prediction of pulsed potential protocols is presented in this study.

BACKGROUND

The concept of field flow fractionation was first proposed in 1966 [6]. Field flow fractionation is a family of techniques used to characterize macromolecules, colloids and particulate materials, primarily by size. Field flow fractionation is similar to chromatography but it has no stationary phase. In FFF, a force field is used instead of a stationary phase. FFF takes advantage of the laminar flow (parabolic profile) in narrow channels. Particles forced close to the walls by the applied cross-flow field are in the low velocity region of the flow as compared to particles further from the wall, which are in a higher velocity region. It is due to this phenomenon that particles elute at different rates.

Various types of fields can be used to push particles into the different velocity streamlines of the flow. The field is applied perpendicular to the stream-wise direction. The extent to which the particles are pushed against the walls is proportional to the magnitude of the applied field and the physical properties of the individual particles, i.e. size, surface charge, shape, etc., and the carrier fluid. The type of field applied distinguishes

[1] Also referred to as cross-flow electrophoresis.

members of the FFF family. For example, the sub technique referred to as thermal-FFF utilizes a temperature gradient and the sub-technique that utilizes a electrical potential gradient is called electrical-FFF. Thus, the different types of FFF techniques are: thermal-FFF (TFFF); sedimentation-FFF (SFFF); magnetic-FFF (MFFF); flow-FFF (FFFF); and electrical-FFF (EFFF).

EFFF separations are performed in a low viscosity liquid, typically water, which is passed through the separation channel. The relative stream-wise velocity of the particles is controlled by forcing them towards the wall with the help of an electric field.

Electric field flow fractionation was first described in 1972 [2] but the resolution was significantly poorer than the predicted theory. The field of EFFF has been revisited in the past decade. In 1993 Caldwell and Gau [7] developed a new channel design, in which current conducting walls served as electrodes for generating an electrical potential across the channel. Their channel also had problems with low effective electrical field. In 1994 Schimpf, Russell and Lewis [8] used a more conducting mobile phase (quinone-hydroquinone solution) with some success. In 1993, Giddings [5] showed that miniaturization of most classes of FFF systems had questionable performance advantages. Gale et al. more recently theorized that miniaturization of EFFF systems holds potential for significant advantages [9]. The real breakthrough came in 2001 when Gale et al. [10] fabricated and demonstrated the functioning of a miniaturized EFFF channel.

The miniaturization of the EFFF device has advantages [1], some of these advantages include the following: reduced plate heights or increased resolution, reduced analysis times, reduced sample size, reduced system size, and reduced power consumption. There are two main advantages, however, realized upon miniaturization. First, the field strength for a given system increases as the system is miniaturized, thus, a higher field is available to drive the separation over a shorter distance. Second, the shorter distances reduces analysis times.

EFFF theory is presented elsewhere in considerate detail [2, 9, 11]. Some of the basic concepts relevant to the present work, however, are developed here for clarity. Retention theory quantifies how the particles elute at different times. Immediately upon injection into the channel, the sample begins to interact with the field and migrate in the direction that allows it to minimize its potential energy. The field (electrical) induced force *F* results in a migration of the particles towards the walls and is opposed by the frictional drag of the particle, characterized by the friction coefficient *f*. Thus at steady state the particles move with a velocity [7] U=*F/f*. This so-called drift velocity U is the product of the electrophoretic mobility μ and the effective electric field ∇V_{eff}, $U=\mu\nabla V_{eff}$. The resulting potential-driven flux is countered by a dispersive flux due to *molecular* diffusion. The equilibrium concentration distribution c(y) in the direction of the applied cross flow field is determined by these two opposing fluxes, so that the field-induced flux J_f = Uc(y) at equilibrium exactly balances the diffusive flux J_D=D[dc(y)/dy], where D represents the diffusion coefficient of the analyte in the particular carrier. This condition establishes the following exponential form of the concentration distribution, valid for particles whose sizes are negligible in comparison with the dimensions of their accumulation layers [9].

$$c(y) = c(0)e^{-(yU/D)} \tag{1}$$

Therefore different analytes will populate different concentration profiles within different flow streamlines depending upon analyte molecular and electrophoretic mobility.

The degree to which an analyte's retention time is altered by this process defines the retention ratio R:

$$R = \frac{t_o}{t_r} = \frac{\left(\frac{l}{v_o}\right)}{\left(\frac{l}{v_r}\right)} = \frac{v_r}{v_0} \tag{2}$$

where, t_o = time required for an unretained sample to exit the system, t_r = time required for the particles of interest to exit the system, *l* = channel length, v_o = average velocity of the unretained particles, v_r = average velocity of the retained particles,

Since the velocity profile $v_x(y)$ is laminar:

$$v_x(y) = 1 - \frac{(y-\delta)^2}{\delta^2} \tag{3}$$

The average velocity of the unretained particles is then given by the integration of the laminar velocity profile over the channel height:

$$v_o = \frac{1}{2\delta}\int_0^{2\delta} v_x(y)dy = \frac{1}{2\delta}\int_0^{2\delta}\left[1 - \frac{(y-\delta)^2}{\delta^2}\right]dy = \frac{2}{3} \tag{4}$$

where δ = half width of the channel. Similarly, the average velocity of the retained particles can be found by integrating the concentration-weighted velocity profile:

$$v_r = \frac{1}{2\delta}\left[\int_0^{2\delta} p(y)\left[1-\frac{(y-\delta)^2}{\delta}\right]dy\right] \quad (5)$$

where *p(y)* is the following concentration-based weighting function:

$$p(y) = \frac{e^{\frac{-y}{2\delta\lambda}}}{\left[\frac{1}{2\delta}\left[\int_0^{2\delta} e^{\frac{-y}{2\delta\lambda}}dy\right]\right]} \quad (6)$$

On substitution and integration, the following analytical relationship for retention ratio is found:

$$R=\frac{t_o}{t_r} = 6\lambda\left[\coth\left(\frac{1}{2\lambda}\right) - 2\lambda\right] \quad (7)$$

where λ = D/Uw=D/μV_{eff} is a dimensionless parameter. Note that the velocity profile forms used here are dimensionless.

MODEL DEVELOPMENT

This well-established model for retention, however, is only applicable for constant applied potential. Lao et al. [4] has shown that in some cases an oscillatory potential can increase retention due to enhanced net electrical field strength. Lao et al. [4] proposed a simple RC circuit model for the observed electrical behavior of EFFF channels:

$$\text{for } t < t_+, \quad i = \frac{V_+}{R_s}e^{-t/\tau} \quad (8a)$$

$$\text{for } t_+ \le t \le t_+ + t_-, \quad i = \frac{V_-}{R_s}e^{-t/\tau} \quad (8b)$$

where $\tau = R_s C_D$ and C_D=double-layer capacitance, R_s=electrolyte resistance and R_E =charge transfer resistance of the electrolyte.

Lao et al. [4] reported the value of R_S as 423 Ω and that of C_D as 46.9 μF giving a time constant ($\tau = R_s C_D$) of 0.02 s. The frequency they used for their experimental work was 5 Hz. Figure (1) illustrates Equations 8 for the data after Lao, et al. [4].

1. Generalized Pseudo-two-dimensional Model

A generalized pseudo-two-dimensional model was based on the Nernst-Planck equation for a non-reacting system in Cartesian coordinates:

$$\frac{dC_i}{dt} + \nabla\cdot(-D_i\nabla C_i - z_i B \mathcal{F} C_i \nabla V + C_i u) = 0 \quad (9)$$

where, D_i =diffusion coefficient of species "i", C_i =concentration of species "i", $\mu_i = zB\mathcal{F}$ (electrophoretic mobility), V = applied voltage , $\mathcal{F}$ = Faradays constant (96484.1 C-mol^{-1}), u = velocity vector, and B = thermodynamic mobility. The horizontal boundary conditions are no flux:

$$\nabla C = 0 \text{ at } \pm X \quad (10)$$

The vertical boundary conditions are also no flux, however, they include the Nernstian term, which *drives* the particle transport:

$$D\nabla C = z_i \mu_i F C_i \nabla V \text{ at } \pm\delta \quad (11)$$

Equation (10) effectively makes this simulation a 1-D analysis (pseudo-2-D).

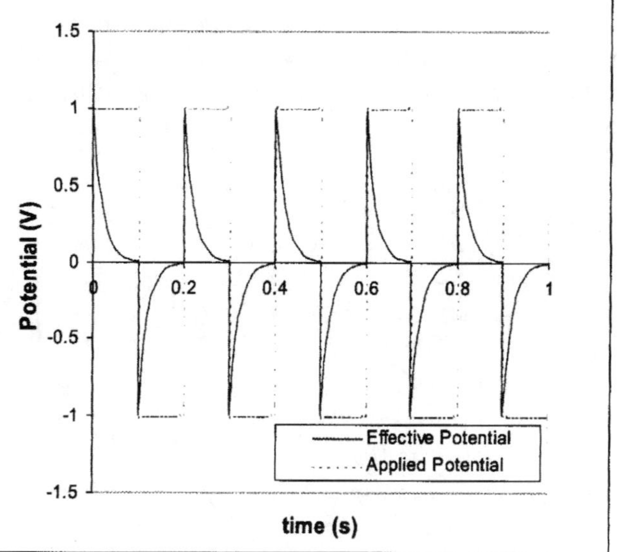

Figure 1. Pulsed voltage protocol used in this study.

The model, however, is actually a 2-D model. The no flux boundary condition applied to the left and right hand boundaries, however, effectively reduces the outcome to 1-D. The second dimension was included only so that the particle transport could be visualized in a 2-D like domain, thus making the model pseudo-2-D. Furthermore, the domain is stationary in the z coordinate. Both time dependent and steady-state simulations were performed.

Pseudo 2-D Box Model.

Equation (10) was made dimensionless, except in time, by using the following transformations:

$$X_i = \frac{C_i}{C_t} \quad (12a)$$

$$\overline{D}_i = \frac{D_i}{\delta_i^2} \quad (12b)$$

$$\overline{\nabla} = \delta\nabla \quad (12c)$$

$$\overline{v} = \frac{v}{\delta} \quad (12d)$$

where X_i = the mole fraction and the "barred" terms are the dimensionless analogues of the dimensioned quantities, and the subscript t is for total. Upon substitution the following dimensionless form of the Nernst equation was developed:

$$\frac{1}{\overline{D}}\frac{\partial X_i}{\partial t} + \overline{\nabla}(-\overline{\nabla}X_i - \frac{1}{\lambda_p}X_i + \frac{\overline{v}}{\overline{D}}X_i) = 0 \quad (13)$$

Relative orders of magnitude for different parameters we determined. Table 1 summarizes typical property and system values from which order of magnitude values for the dimensionless parameters were derived.

$$\overline{D} \cong \frac{10^{-6}cm^2/s}{\left(10^{-3}\right)^2 cm^2} = 1 \quad (14)$$

$$\frac{1}{\lambda_p} = \frac{\mu\nabla V\delta}{D} = \frac{\mu V\delta}{\delta D} = \frac{\mu V}{D} \cong \frac{(10^{-4}cm^2/V\,s)(10^{-2}\,to\,1\,V)}{10^{-6}cm^2/s} = 1\ to\ 100 \quad (15)$$

$$\frac{\overline{v}}{\overline{D}} = \frac{\delta v}{D} \cong \frac{(10^{-4}cm)(10^{-1}cm/s)}{10^{-6}cm^2/s} = 10 \quad (16)$$

Note that in the derivation of Equation (15), $\lambda_p \neq \lambda$. While both λ_p and λ have the same form, λ is defined for constant applied potential using a constant effective potential and λ_p is actually a time variant parameter since V=V(t) for pulsed voltage. A typical analyte concentration of 0.375 ppm was estimated and used as a reference value.

Table 1. Values of model parameters.

Property	Nominal Value
Electrophoretic Mobility (μ)	10^{-4} cm^2/V-s
Diffusion Coefficient (D)	10^{-6} cm^2/s
Effective Voltage (V)	1 to 10^{-2} V
Channel Half Height (δ)	10^{-3} cm

COMPUTATIONAL RESULTS

1. Steady State Concentration Profile

The pseudo-2-D steady-state model was run for a range of $1/\lambda$ value between 0.0 and 20. Figure 2 illustrates the concentration profiles for $1/\lambda$ values of 0, 1, 5, 10 and 20. A value of λ=0 implies that there in no effective voltage in the channel and so the particles are dispersed uniformly across the channel height. Figure 2 shows that as the value of $1/\lambda$ increases the particles are concentrated at one wall depending upon particle and wall polarity. As expected the resulting profiles exactly match the analytical solution, Equation (1).

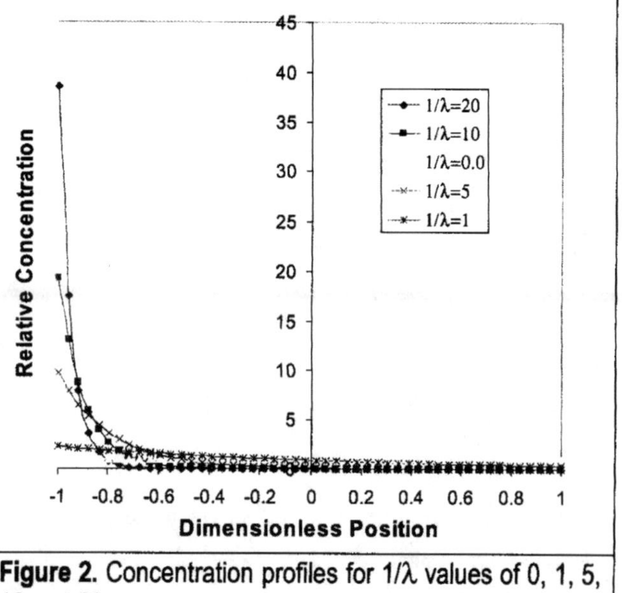

Figure 2. Concentration profiles for $1/\lambda$ values of 0, 1, 5, 10 and 20.

2. Concentration Dynamics Caused by RC Effects

When the model is run with a pulsed voltage scheme, the particles, which are charged, move from one boundary to the other depending on the pulse frequency and wave form. Figure 3 illustrates the relative concentration profile at different times during the first half cycle ($\pm$1 V square wave at 5 Hz) for 1/λ = 100. At time t=0, the concentration is uniform at 1 (the initial concentration). For this high value of 1/λ, particle transport is extremely fast. As a result, the concentration near the wall at the initial stages of the cycle (t = 0.01 s and 0.02 s) is very high. As the cycle progresses, the particles diffuse back as the effective voltage decreases and the concentration profile *relaxes*. At time t = 0.098 s, near the end of the half cycle, the particles again disperse with only a weak gradient remaining.

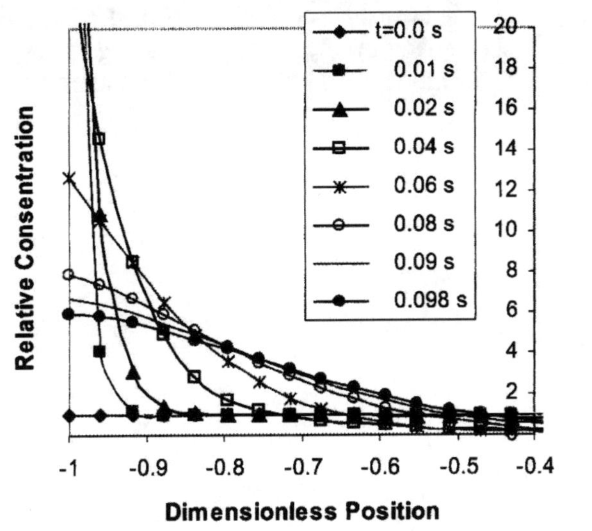

Figure 3. Concentration profile as a function of time for the first half-cycle.

3. Limit Cycle Plots

To understand the complex oscillating behavior of particle motion, limit cycles were plotted. Limit cycles are plots of response (concentration) versus the forcing function (potential) for a stationary location. The pulsed potential which follows an orderly repeating cycle, drives the concentration to oscillate. The oscillations in concentration, however, may or may not fall into a closed pattern, a closed limit cycle. Figure 4 illustrates the first five cycles for 1/λ =100 and a $\pm$1 V square wave at 5 Hz forcing function. In this case a closed loop is achieved after only one cycle. The cycle for a point on the upper electrode is shown to mirror the lower electrode.

When the pulsed voltage is applied the particles move towards the opposite wall where they concentrate until the potential falls, permitting back diffusion, (relaxation). When the polarity changes, the particles are attracted to the opposite wall where they concentrate and relax as the potential decays. When the polarity changes again the particles move towards the other wall and the cycle begins again. Figure 5 illustrates the limit cycle for points ¼ from the top and ¼ from the bottom electrodes, at y = + 0.5 and – 0.5 respectively.

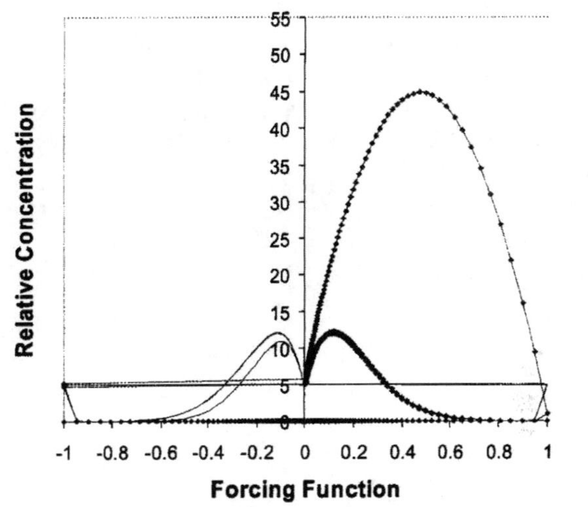

Figure 4. The limit cycle plot for $1/\lambda_p$ = 100 at y = $\pm$1.

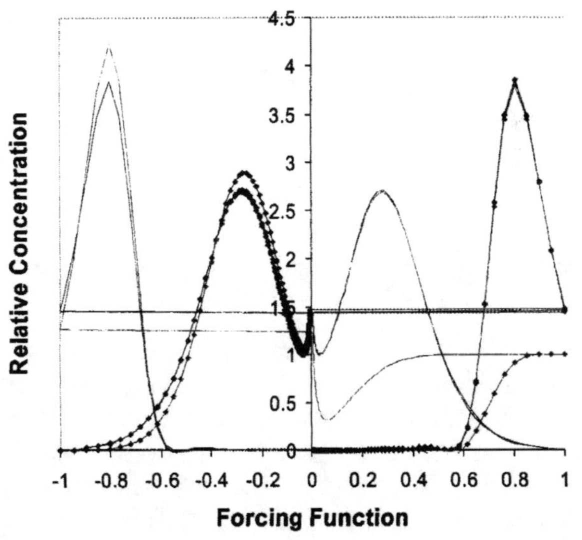

Figure 5. The limit cycle plot for $1/\lambda_p$ = 100 at y = +0.5 and -0.5.

As 1/λ decreases, increasingly more cycles are required to close the limit cycle. Figure 6 illustrate the limit cycle behavior for 1/λ=10. This type of *unstable* cycle, suggests that for some applied potential wave forms the cycles may not close, although, this may or may not affect retention (separation) negatively.

4. Retention Calculations

Calculating retention ratios for constant applied potential can be done analytically by applying equations (1) through (8). For pulsed voltage, however, C is a function of time t and position y, C(t,y) without a closed form analytical solution and so a numerical approach must be used. To obtain concentration profiles for pulsed voltage, analogous to Figure 2 for constant voltage, the dynamic (transient) concentration profiles were averaged over time:

$$\overline{C}_t = \frac{1}{t_c}\int_0^{t_c} C(y,t)\ dt \tag{17}$$

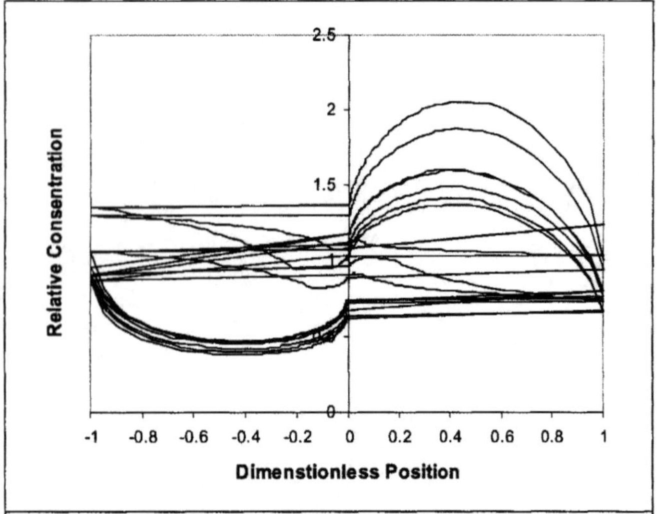

Figure 6. Illustrates the limit cycle for $1/\lambda_p = 10$.

Figure 7 illustrates the average effective concentration profile for various $1/\lambda_p$ values. λ_p is dependent on the diffusivity, mobility and the effective potential inside the channel, $\lambda_p = D/\mu V_{eff}$. Since for a given system, diffusivity and mobility are constants, the only variable is the effective potential which varies with time for pulsed potential.

As seen by Equations (3) through (7), retention ratio R is a manifestation of the interaction between the concentration and velocity profiles. Due to pulsation of the voltage, the concentration profile becomes a dynamic quantity, C(t,y). As illustrated above, an effective time-averaged profile results, $\overline{C}_t(y)$. This effective profile is analogous to the constant voltage profile in Equation (1) and subsequent equations in the derivation of R. To compare pulsed voltage performance to constant voltage performance, the pulsed voltage time averaged profiles were numerically integrated:

$$R_p = \frac{v_r}{v_o} = \frac{\displaystyle\int_0^{2\delta} \frac{\overline{C}_t(y)\left(1-\frac{(y-\delta)^2}{\delta^2}\right)}{\int_0^{2\delta}\overline{C}_t(y)\,dy}\,dy}{\frac{2}{3}} \tag{18}$$

The time dependent model was run for a particular λ_p value and the results integrated in the time domain per Equation (17), the results of which are plotted in Figure 7. The resulting time average curves were integrated per Equation (18) to find the pulsed potential retention ratio R_p.

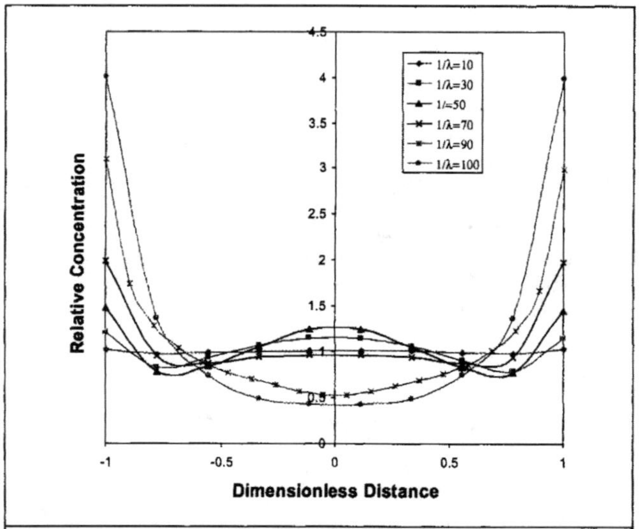

Figure 7. Average concentration profiles for pulsed potential for various $1/\lambda_p$ values.

A comparison of retention ratio for constant voltage and for one pulsed voltage protocol (± 1 V square wave at 5 Hz) for different $1/\lambda$ values is shown in Figure 8. For values of $1/\lambda = .3$ ($1/\lambda_p < 60$) the pulsed voltage protocol yields improved retention. Since λ and λ_p are not equivalent, equivalent λ values were approximated assuming that V_{eff} is proportional to $V_{applied}$. Gail [1] reports that V_{eff} is typically .25% to 1% of the applied field and so $V_{applied}/V_{eff}$ is between 100 and 400. A nominal value of 200 was chosen here for comparative purposes. Since in actuality, λ_p varies with time, no single value for $1/\lambda_p$ exists. A time averaged value, however, would most likely provide the most quantitatively correct strategy.

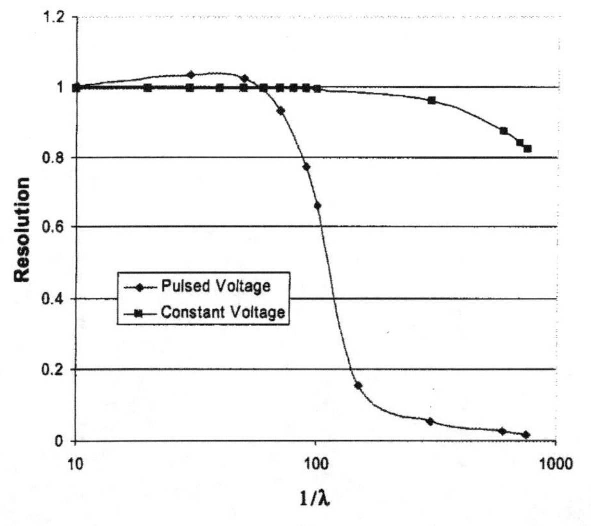

Figure 8. Comparison of retention ratios for constant voltage and one pulsed voltage protocol (± 1 V square wave at 5 Hz).

SUMMARY

A first principles mathematical model based on the Nernst-Planck equation was developed to study particle retention in electrical field flow fractionation. The model was used to simulate the behavior of constant and pulsed voltage protocols. The exercise demonstrates that complex particle trajectories that arise due to pulsed voltage lead to either stable closed loop or open loop cycles. A computational strategy was developed to approximate the retention ratio for pulsed voltage. The strategy is based on the classic theory for constant voltage, however, requires numerical integration of the time and spatially variant concentration profile. The results may provide a predictive strategy for optimization of device design and operations and shows regions of operability wherein pulsation is preferred over constant applied potential.

ACKNOWLEDGMENTS

The authors would like to acknowledge the financial support of the Tennessee Technological University (TTU) Center for Electric Power and the TTU Faculty Research Grant program.

REFERENCES

1) B. K. Gale, K. D. Caldwell and A. B. Frazier, IEEE Transactions on Biomedical Engineering, 45(12) 1459-1469 (1998).
2) K. D. Caldwell, L. F. Kesner, M. N. Myers, and J. C. Giddings, Science, 176, 296- 298 (1972).
3) S. A. Palkar and M. R. Schure, Anal. Chem. 69(16) 3223-3229 (9997).
4) A. Lao, D. Trau and I-M. Hsing, Analytical Chemistry, 74(20) 5364-5369 (2002).
5) J. C. Giddings, Microcol. Sep. 5, 497-503 (1993).
6) J. C. Giddings, Sep. Sci, 1, 123 (1966).
7) Caldwell K.D and Yu Shu Gao, Analytical Chemistry, Vol. 65(13) 1764-1772 (1993).
8) M. E. Schimpf, D. D. Russel, J. K. Lewis, J. Liq. Chrom., 17(14&15) 3221-3238 (1994).
9) B. K Gale, K. D. Caldwell, and A. B. Frazier, Analytical Chemistry, 73(10) 2345-2352 (2001).
10) B. K. Gale, K. D. Caldwell and A. B. Frazier, Anal. Chem. A (2001).
11) J. C. Giddings, Sep. Science and Tech. 13(3) 241-254 (1978).

CONTACT

J. J. Biernacki is Professor of Chemical Engineering at Tennessee Technological University. Send inquiries concerning this manuscript to Box 5013, Cookeville, TN 38501 or to jbiernacki@tntech.edu.

Paper No. 276c

An Integrated Framework for Impacts Assessment and Trade-Offs of Fuel Cell Systems

Francesco Baratto
University of Illinois at Chicago – Department of Chemical Engineering
Center for the Uncertain Systems: Tools for Optimization and Management (CUSTOM)
Room 1039, Science and Engineering Offices (SEO), 851 S. Morgan st. 60607 Chicago, IL
fbarat1@uic.edu

Urmila Diwekar
University of Illinois at Chicago – Department of Chemical Engineering
Institute for Environmental Science & Policy
Center for the Uncertain Systems: Tools for Optimization and Management (CUSTOM)
Room 1039, Science and Engineering Offices (SEO), 851 S. Morgan st. 60607 Chicago, IL
urmila@uic.edu

Prepared for presentation at the 2003 Annual Meeting, San Francisco, CA, Nov 16-21

An Integrated Framework for Impacts Assessment and Trade-Offs of Fuel Cell Systems

Francesco Baratto
University of Illinois at Chicago – Department of Chemical Engineering
Center for the Uncertain Systems: Tools for Optimization and Management (CUSTOM)
Room 1039, Science and Engineering Offices (SEO), 851 S. Morgan st. 60607 Chicago, IL
Urmila Diwekar
University of Illinois at Chicago – Department of Chemical Engineering
Institute for Environmental Science & Policy
Center for the Uncertain Systems: Tools for Optimization and Management (CUSTOM)
Room 1039, Science and Engineering Offices (SEO), 851 S. Morgan st. 60607 Chicago, IL

ABSTRACT

Fuel cells are an important technology for a potentially wide variety of applications; however, the benefits of fuel cell technology in terms of health and environmental impacts as compared to the cost of manufacturing have not been systematically identified and quantified. This work presents an integrated framework to evaluate the trade-offs between cost effectiveness and environmental & health impacts of fuel cell power systems considering various stages of the life cycle of the device. The power of this integrated framework is illustrated by a case study of introducing SOFC fuel cells as APUs in heavy-duty trucks and luxury vehicles (RVs and limos).

KEY WORDS

Fuel cell system model, Integrated framework, Environmental and health impacts, Life Cycle Assessment, Fuel cell cost model

INTRODUCTION

Since the late 1980s, there has been a strong push to develop fuel cells for use in light-duty and heavy-duty vehicle applications. A major drive for this development is the need for clean, efficient cars, trucks, and buses that can operate on conventional fuels (gasoline, diesel), as well as renewable and alternative fuels (hydrogen, methanol, ethanol, natural gas, and other hydrocarbons).

The importance of fuel cells in automotive industries is in the sheer market size. However, the benefits of fuel cell technology in terms of health and environmental impacts as compared to the cost of manufacturing have not been systematically identified and quantified.

Fuel cells are electrochemical devices that convert the chemical energy of a reaction directly into electrical energy. The basic physical structure of a fuel cell consists of an electrolyte layer in contact with a porous anode and cathode on either side. In a typical fuel cell, gaseous fuels are fed continuously to the anode compartment and an oxidant (i.e., oxygen from air) is fed continuously to the cathode compartment; the electrochemical reactions take place at the electrodes to produce an electric current. A variety of fuel cells is in different stages of development. They can be classified in different categories, depending on the combination of fuel and oxidant, whether the fuel is processed outside (external reforming) or inside (internal reforming) the fuel cell, the type of electrolyte, the temperature of operation, etc. The most common classification of fuel cells is by the type of electrolyte used in the cells and includes 1) polymer electrolyte fuel cell (PEFC), 2) alkaline fuel cell (AFC), 3) phosphoric acid fuel cell (PAFC), 4) molten carbonate fuel cell (MCFC), 5) solid oxide fuel cell (SOFC). These fuel cells are listed in the order of approximate operating temperature, ranging from ~80 °C for PEFC, ~100 °C for AFC, ~200 °C for PAFC, ~650 °C for MCFC, 800-100 °C for SOFC [1].

In this work we are presenting an integrated framework that can systematically identify and quantify trade-offs between cost effectiveness, and environmental & health impacts of fuel cell power systems. The framework has 6 components: system level modeling, environmental impacts assessment, cost modeling, health impacts assessment, life cycle assessment, and multi-objective optimization. The last component is useful for identifying trade-offs between the different objectives: low cost, high efficiency and low health and environmental impacts.

In addition to high-profile fuel cell applications such as automotive propulsion and distributed power generation, the use of fuel cells as auxiliary power units (APUs) for vehicles has received considerable attention. Auxiliary power units are devices that can provide all or part of the non-propulsion power for vehicles. They do not replace the main internal combustion engine, but they complement it. To show the usefulness of the framework, the detection of the health and environmental impacts of using SOFC fuel cells as APUs in heavy-duty trucks and luxury vehicles (RVs and limos) is presented as a case study.

In the next section, the structure of the new integrated framework with a description of all the components is presented. The third part describes the case study with an overview of the subject and the assumptions taken in modeling the system. Summary and future work is the last section of the paper.

THE INTEGRATED FRAMEWORK

The integrated framework has five main components, which will be analyzed in the following paragraphs.

SYSTEM LEVEL MODELING COMPONENT - The entire fuel cell system, comprehensive of fuel processing and fuel cell device, is simulated in Aspen Plus [2] and this constitutes the base of the integrated framework. Process simulation with Aspen Plus allows to predict the behavior of a process using basic engineering relationships such as mass and energy balances, phase and chemical equilibrium, and reaction kinetics. Aspen Plus contains thermodynamic data, components properties and unit operation models in order to simulate the actual plant behavior. With Aspen Plus it is possible to interactively change specifications, such as flowsheet configuration, operating conditions, and feed compositions and analyze the results.

The fuel cell system is simulated using mainly standard equipments such as pumps, compressors, heat exchangers, separators, equilibrium and kinetic reactors etc. Since Aspen Plus does not include any fuel cell model, two approaches are possible to overcome this problem. The first way is to use a standard reactor model, like a stoichiometric and/or equilibrium reactor, to perform energy and mass balances around the fuel cell. This unit is then coupled with a polarization model for voltage and current computations. Alternatively, only the balance of plant is simulated through standard units and a new User Model based on a FORTRAN subroutine is used to reproduce the fuel cell stack. Both the models have been developed and, depending on the need, either of the two can be used. The steady-state simulation of the system is useful to evaluate regular running emissions and to size the equipments for cost considerations.

ENVIRONMENTAL IMPACTS ASSESSMENT COMPONENT - Ecological and environmental risk impacts are quantified using the generalized WAste Reduction algorithm (WAR) by EPA (Cabezas et al. [3]). The Waste Reduction Algorithm Graphical User Interface (WAR GUI) was developed to compare the environmental friendliness of chemical process designs (Hilaly and Sikdar [4]). The WAR algorithm involves the concept of a Potential Environmental Impact (PEI) balance (analogous to a mass or energy balance):

$$I_{in} - I_{out} + I_{gen} = 0 \tag{1}$$

where I = Potential Environmental Impact

The balance involves the flow of environmental impact across system boundaries. The flow of impact can be due to mass or energy crossing the system boundaries.

The potential environmental impact of a given quantity of material and energy can be generally defined as the effect that this material and energy would have on the environment if they were to be emitted into the environment. To provide an indication of the potential environmental impact of a chemical process eight different categories of impact potential are considered: Human Toxicity Potential by Ingestion (HTPI), Human Toxicity Potential by Inhalation or Dermal Exposure (HTPE), Ozone Depletion Potential (ODP), Global Warming Potential (GWP), Photochemical Oxidation Potential (PCOP), Acidification Potential (AP), Aquatic Toxicity Potential (ATP) and Terrestrial Toxicity Potential (TTP).

The general expression of the Potential Environmental Impact is therefore:

$$I = \sum_{i=1}^{EnvCat} \alpha_i \sum_{j=1}^{Streams} M_j \sum_{k=1}^{Comps} x_{jk} \psi_{ki} \tag{2}$$

where: α_i = Weighting factor associated with the category i

M_j = Mass flow of the stream j

x_{jk} = Mass fraction of the component k in the stream j

ψ_{ki} = Impact score for the component k in the category i normalized with respect to the average impact in the category i.

WAR algorithm has been fully integrated in the ASPEN framework. This way it is no more necessary to go through the graphical interface of the EPA software (Cabezas et al. [3]) but the environmental impact is evaluated automatically running the flowsheet simulation.

COST MODELING COMPONENT – Estimating the cost of the fuel cell based system is an important task of the framework. Costs are dependent on technologies, design parameters and volume of production. Prices of standard equipments such as pumps, compressors, catalytic burners, splitters and heat exchangers are already available from vendor quotes. Commonly used chemical engineering cost models can be applied to these kinds of units.

One of the simplest possible methodologies to estimate the cost of the solid oxide fuel cell is to consider a single parameter to characterize the stack. The area of the cell appears to be a suitable measure to base the cost model on. Arthur D. Little [5] derived area based cost estimates of SOFC devices analyzing different possible production pathways.

Other units that are not commercially available, such as the fuel reformer, need to be assessed using the common rules of cost estimation. This means to identify the variable, fixed and capital costs of producing the equipment and then estimate the cost of that particular device. Earlier work on a particular size of reformer but with different production rates can be found in James et al. [6]. The data is used to link the cost of the equipment with a characteristic measure, like the reactor volume.

HEALTH IMPACTS ASSESSMENT COMPONENT - It is the characterization of the potential adverse health effects of human exposures to environmental hazards. This component of the framework quantifies the effects on population of the chemicals released in the production, use and disposal of a fuel cell. Health impacts models usually consider inhalation, ingestion and dermal contact as exposures routes and air, soil and surface water as transmission media. The effects are divided into acute/chronic and carcinogenic/non-carcinogenic. Quantifiable health effects usually include asthma, reduced respiratory functions, lung damage, blood oxygen carrying capacity reduction, cardiovascular and nervous system impairments, respiratory track impairment, retardation and brain damage carcinogenicity, toxicity, morbidity, and mortality (Garza [7]).

In general, the health risk assessment process can be distilled into the following simple model:

$$\text{Risk} = \text{Dose X Toxicity} \tag{3}$$

It involves the identification of potential health effects associated with exposure of individual or populations to hazardous materials and situations. In the above model, if either exposure or toxicity is lacking, there is no risk. The average daily dose rate can be evaluated in different ways. Generally a dispersion model is included to compute a dilution factor. This quantity represents the ratio between the ground level concentration and the emission rate. A complete database of the toxicity values of different chemicals was built by California Air Resource Board [8]. This forms the basis for the health impacts assessment component.

LIFE CYCLE ASSESSMENT COMPONENT - LCA provides a tool for identifying and evaluating environmental burdens associated with the life cycles of materials and services in a "cradle-to-grave" approach. "Cradle-to-grave" begins with the gathering of raw materials from the earth to create the product and ends at the point when all materials are returned to the earth. LCA evaluates all the stages of the life of a product from the perspective that they are interdependent, meaning that one operation leads to the next. LCA enables the estimation of the cumulative environmental impacts resulting from all stages in the product life cycle, often including impacts not considered in more traditional analyses (e.g., raw material extraction, material transportation, ultimate product disposal, etc.). By including the impacts throughout the product life cycle, LCA provides a comprehensive view of the environmental aspects of the product or process and a more accurate picture of the true environmental trade-offs in product selection [9].

For the Life Cycle Assessment of fuel cell based cycles it is possible to identify two main components: life cycle study of the fuel and life cycle study of the system itself.

Fuel cycle includes recovery or production of the feedstock, transportation of the feedstock, conversion of the feedstock to the final fuel (e.g. diesel, gasoline etc.) and subsequent storage, distribution and delivery of the fuel to the refueling station and finally the use in the vehicles. Among the different possibilities, the GREET model (Wang [10,11]) has been chosen for this part of the analysis. The impact of the final part of the cycle (the use) is not derived from GREET but evaluated by the ASPEN system model. The second component of the Life Cycle Assessment includes raw material extraction, manufacture and assembly of the stack and balance of plant (all the components other than the unit cell), their installation and eventual decommissioning. While the fuel life cycle is well documented in literature, there is a great deal of uncertainty concerning the manufacturing and end-of-life stages. Earlier studies can be found in Karakoussis et al. [12] and Zapp [13]. The collection of life cycle data is an ongoing work.

MULTI-OBJECTIVE OPTIMIZATION - The modeling structure is augmented with a new multi-objective optimization framework for identifying trade-offs between the different objectives: low cost, high efficiency and low health and environmental impacts. It is not possible to find a single solution that is optimal for all the objectives simultaneously because of the contradiction and possible incommensurability of the objective functions. Therefore, the solution of a multi-objective optimization problem is a set of solution alternatives called the Pareto set. For each of these solution alternatives, it is impossible to improve one objective without sacrificing the value of another one. This makes a range of choices available for the decision makers and provides them with the trade-off information among the multiple objectives effectively.

Multi-objective optimization problems are usually computationally very expensive and require the solution of a high number of non-linear optimization sub problems. Among the many different solution techniques, the constraint method was chosen (Cohon [14] and Zeleny [15]). The basic strategy is to transform the multi-objective optimization problem into a series of single objective optimization problems. The idea is to pick one of the objectives to optimize while each of the others is turned into an inequality constraint with parametric right-hand side. Solving for different values of these parametric right-hand sides leads to the Pareto set. In order to reduce the number of optimization problems to solve, a new efficient algorithm called MINSOOP (Fu and Diwekar [16]) (Minimizing Number of Single Objective Optimization Problems) was used. This algorithm applies the Hammersley Sequence Sampling (Kalamagnanam and Diwekar [17]) to generate the right-hand sides of the traditional constraint method. HSS is a sampling technique that guarantees good uniformity in multiple dimensions with a number of points to converge to the mean and variance of the derived distributions on average 3 to 100 times less than the Monte Carlo Sampling and Latin Hyercube methods (Kalamagnanam and Diwekar [17]).

The steps for a multi-objective problem with k objectives (to be minimized) are listed as follows:

Step 1. Solve k single objective optimization problems individually with the original constraints of the multi-objective problem to find the optimal solution for each of the individual k objectives.

Step 2. Compute the value of each of the k objectives at each of the k individual optimal solutions. In this way, an approximation of the potential range of values for each of the k objectives is determined and saved in a table (called payoff table).

Step 3. Select a single objective to be minimized. Transform the remaining k-1 objectives into inequality constraints with parametric right-hand side.

Step 4. Select a desired number of single objective optimization problems to be solved to represent the Pareto set and use the HSS technique to generate the combinations of the inequality constraint values within the range determined in Step 2.

Step 5. Solve the constrained problems set up in Step 4 for every combination of the right-hand-side values. The feasible solutions form an approximation for the Pareto set.

CASE STUDY: SOFC BASED APUS

Auxiliary power units are already in widespread use in a range of vehicle types and for a variety of applications, such as space conditioning/heating, refrigeration, lighting and other cabin amenities, communication and information equipment, entertainment (TV, radio, etc.). The main advantages of this kind of devices are to improve the power generation efficiency and to reduce emissions and noise when the vehicle is parked and to extend the life of the main engine. Now APUs are usually small internal combustion engines equipped with a generator and heat recovery to provide electricity and heat (Stodolsky et al. [18]). However, there is a good fit between APU requirements and fuel cell system characteristics. So in addition to high-profile fuel cell applications such as automotive propulsion and distributed power generation, the use of small fuel cell stacks (up to 5 kW) as auxiliary power units for vehicles has received considerable attention. APU applications may be an attractive market because they offer a true mass-market opportunity that does not require the challenging performance and low cost required for propulsion systems for vehicles.

Among the different fuel cell types, the Solid Oxide FC technology is considered the most favorable due to several characteristics:

- SOFC can use CO as a fuel and is relatively more tolerant of impurities like sulfur than other conventional fuel cells. This leads to the ability to use a variety of hydrocarbon fuels with simpler reforming processes.
- There is no need for noble or precious metals as catalysts with consequent greater cost-effectiveness (even if the manufacturing process is quite complex).
- The electrolyte is dry and so the humidification of the reactants is not necessary and no water management system is required.

APUs FOR TRUCKS AND RVs MARKET - Mainly because of cost limitations, fuel cell based APUs are predicted to penetrate the market of heavy-duty trucks and luxury vehicles (recreational vehicles and limos) first. The study conducted in this paper is therefore focused on these types of vehicles.

Concerning truck applications, fuel cell APUs can mainly reduce the common (Stodolsky et al. [18]) practice of idling heavy-duty truck engines. Drivers idle truck engines to power climate control devices (e.g., heaters and air conditioners) and sleeper compartment accessories (e.g., refrigerators, microwave ovens, and televisions) and to avoid start-up problems in cold weather. Idling of the truck large-displacement diesel engine is an extremely inefficient way to generate heat and electricity. Fuel cells, instead, offer a high-efficiency (equivalent to low consumption), low emission, and low-noise alternative that would supplant the need for truck engine idle.

Trucks traveling more than 500 miles from their home base each day are likely to be idling overnight during stopovers on long trips (Stodolsky et al. [18]). Among these, trucks with gross vehicle weight (GVW) rating of 26,000 lb or greater (classified as class 8 trucks) are the candidates for fuel cell based APUs. According to [19] in the US in 1997 there were 401,900 heavy trucks (above 26,000 lb) with range of operation greater than 500 miles. Although detailed data on idling are not available, industry sources have provided rough estimates. One source reports that a long-distance, freight-hauling, heavy-duty truck idles about 6 h/day on the average [20]. The actual extent of idling varies with the season and with the type of operation. Baseline entry could be a 6-h/d average, with a seasonal peak in the winter (eighty-five winter days at 10 h/d and 218 non-winter days at 4.5 h/d) (Stodolsky et al. [18]).

Most of the Class 8 trucks works with diesel as fuel [19]. The small amount of fuel involved in fueling APUs would likely not justify the establishment of a specialized infrastructure (e.g. a hydrogen infrastructure) for APUs alone. This means that fuel cell based APUs have to work with diesel too.

The market of recreational vehicles is in continuous expansion. A University of Michigan study (Curtin [21]) shows that the number of households that own at least one RV was projected to rise from 6.9 million in 2001 to 7.9 million in 2010 (+ 15%). This number includes any kind of recreational vehicles, of every range of price. The largest categories, commonly referred to as the conventional motorhomes, because of cost limitations are the most likely candidates for fuel cell based APUs. These categories absorb about the 20 % of the shipment market ([22],[23]) and represent about the 25 % of the RVs owned by households (Curtin [21]). For this kind of vehicles a fuel cell APU is an efficient way to produce the electricity needed for the large number of on-board accessories, like climate control (heat, A/C) and electric appliances (TV/VCR, refrigerators, lights, microwave, stereo system).

These statistics are useful to identify the extension of the possible fuel cell based APUs market and, together with the idling emission data [22,24], will be used to evaluate the health and environmental impact of installing this device on all the candidate vehicles.

The following sections describe the use of the integrated framework for evaluation of fuel cell technology as APUs. In this document we present the first two stages of analysis, namely system modeling with experimental validation and environmental impacts assessment component.

SYSTEM MODELING – This section describes the assumptions taken in building the ASPEN model of an SOFC based APU.

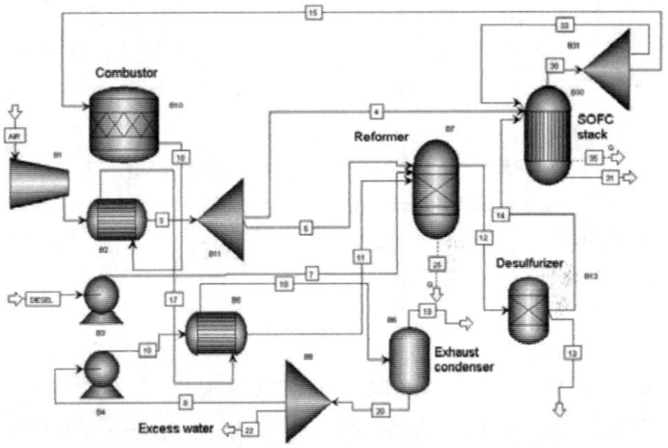

Figure 1. ASPEN model of a SOFC based APU

1. FUEL PROCESSING - Since diesel is very common on Class 8 trucks, this fuel has been chosen for the simulations, even though diesel processing is complex and the technology is still under development.

The fuel processor is a critical component of a diesel-fueled auxiliary power unit and must be able to provide a clean, tailored synthesis gas to the fuel cell stack. Key characteristics desired for the processor (and the system) include low cost, high efficiency and maximum thermal integration (Berry [25]).

As in Berry et al. [25],[26], the diesel processing system includes an autothermal reformer, a desulfurizer and a combustor that acts as a polishing bed to fully oxidize any remaining unburned hydrocarbon from the fuel cell. Authotermal reforming, where the hydrocarbon fuel is reacted with both air and water steam, is preferred because with an appropriate feed composition external heat is not required. The study from NETL (Berry [25]) shows that higher efficiency can be achieved with higher fuel/air ratio and allowing heat exchange between the reactor and the combustor. The exhausts from the combustor are then used to preheat the air for the reformer and the fuel cell and to generate the steam necessary for the reforming. All the water needed by the autothermal reactor is provided by condensation of the exhaust from the combustor.

NETL is developing a kinetic model for the reforming of diesel (Berry [25]). Since this scheme is still in the earliest stages, in this work the authotermal reformer is simulated with a Gibbs free energy equilibrium reactor.

Basic assumption in the model is that the reforming catalyst can handle any sulfur content of the fuel, while sulfur must be removed prior to the fuel gas entering the fuel cell (Berry [25]). Sulfur removal is achieved with a bed of zinc oxide, but this is just one of the several configurations that can be evaluated.

2. DIESEL SURROGATE - The fact that diesel is a complex, multi-component (>100 components) fuel that exhibits varying reaction paths is one of the aspects that makes its reforming difficult. There are mainly two qualities of diesel fuel: grade No. 1-D and grade No. 2-D. The principal differences between these two fuels are the aromatic and sulfur contents (Pereira et al. [27]). For the earlier simulations, a mixture of normal paraffins, alkylated benzenes and alkylated naphtalenes was used. This mixture, according to Amphlett et al. [28], has similar heat of formation, Gibbs free energy, and distillation curve to type 2-D diesel. Amphlett et al. [28] assumed that no sulfur was present in the mixture to avoid secondary reactions in the reforming. We included this element in the mixture and scaled the percentage of the other components.

Pereira et al. [27,29] provide experimental data on the composition of H_2, CO, CO_2 and CH_4 from the reforming of different diesel qualities. In order to fine tune the equilibrium model to the experimental results we used two approaches: (1) modifying the composition of the mixture used as surrogate of diesel (while maintaining the original physical properties) and (2) using a general temperature approach. As it is shown in Fig. 2, combining these two different approaches we were able to find a mixture that gives final concentrations of chemicals inside the experimental bounds of grade No. 2-D diesel and just slightly underestimate CH_4 concentration at 810 and 850 °C. Density, heat capacity, molecular weight and distillation curve (D86) at 10, 50, 90 °C of this new mixture are similar to original mixture within reasonable tolerance. For grade No.1-D diesel a simple temperature approach of -100 °C predicts correctly all the composition except CO_2 at 750 °C that is about 20% overestimated (Fig. 3).

We also developed an optimization approach based on the theory developed in Kalagnanam and Diwekar [30] to automate this process. Further, this method along with experimental data is being used to define uncertainties and variability in fuel as well as model predictions.

As stated in Unnasch et al. [31] for autothermal processing of gasoline, the formation of ammonia, formaldehyde, NOx, benzene and 1,3-butadiene is possible. The equilibrium composition of these components is in the order of ppm or even ppb, but there are no experimental data available to validate the results.

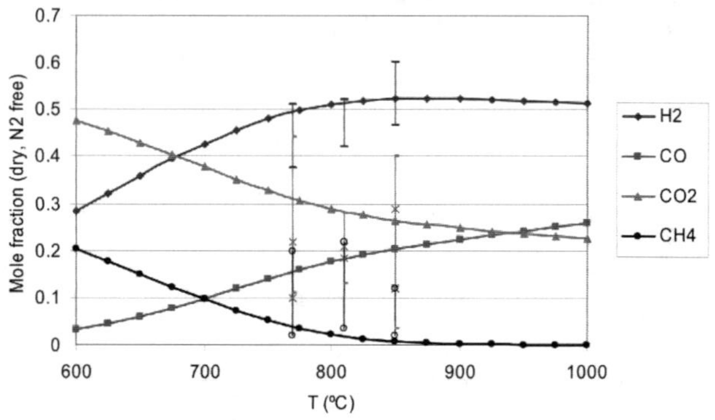

Figure 2. Reforming output composition using grade 2-D surrogate mixture and a global temperature approach of -75 °C. The vertical bars are the experimental bounds.

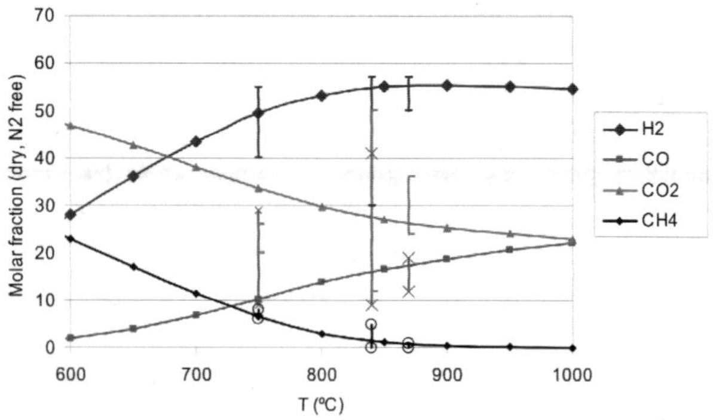

Figure 3. Reforming output composition using grade 1-D surrogate mixture and a global temperature approach of -100 °C. The vertical bars are the experimental bounds.

3. SOFC MODEL - An SOFC stack model was developed to complete the APU system. The approach used to model the stack is similar to the one utilized by Geisbrecht [32]. An equilibrium reactor at fixed temperature performs heat and material balances on the cell and then, after flowsheet convergence, an ASPEN calculator block computes voltage, current density and total cell area. Internal reforming is a common feature in solid oxide fuel cells because it provides additional cooling to the stack ([1] and Chan et al. [33]). Therefore, the reactions are: methane and ethane steam reforming, carbon monoxide water shift and hydrogen electrochemical oxidation. The first three reaction are at equilibrium (Chan et al. [33]) while hydrogen oxidation has fixed extent in order to match the given fuel utilization. Methane and ethane were chosen because they are the most prevalent hydrocarbons in diesel reforming outlet. The electrochemical oxidation of CO was neglected because in presence of water the favorable path for the oxidation of carbon monoxide is generating hydrogen by the water shift reaction ([1] and Petruzzi et al. [34]). At fixed temperature a heat balance on the reactor gives the power output of the cell (heat losses may be taken into consideration). The power output divided by the current (known once the fuel utilization is fixed) gives the voltage of the cell. At this point, an SOFC polarization model is used to find the current density of the cell at that given voltage. A 1-dimensional, steady state, algebraic polarization model derived from literature (Chan et al. [35] and Virkar et al. [36]) was used for this scope. These particular models were chosen because of their simplicity but at the same time exhaustiveness (applicable to every operating condition and sensitive to the component design of the cell). Moreover, the model has to describe an SOFC with flat-plate design that, according to Petruzzi et al. [34], will be used for APU applications. Current divided by current density gives the total cell area, important for cost estimations.

PRELIMINARY RESULTS – The simulation of the system was run with two different diesel qualities, grade No. 1-D and grade No. 2-D, using the surrogate mixtures discussed above. The common system conditions that were used are given in Table 1.

Table 1. System conditions for the simulation

Diesel intake	0.0045 kmol/hr
System pressure	2 bar
ATR temperature	800 °C
SOFC temperature	800 °C
Fuel utilization	75 %
Cathode stoichiometric ratio	5.2

The air intake is varied so that the reforming reactor is adiabatic. The oxidant intake is almost the same for the two cases.

Table 2. Results with 2 different diesel qualities

	Grade 1-D	Grade 2-D
Air intake	1.274 kmol/hr	1.255 kmol/hr
Combustor temperature	896 °C	909 °C
SOFC operating conditions:		
Power	4939.4 W	4749.8 W
Voltage	0.9787 V	0.9498 V
Current density	0.1995 A/cm^2	0.314 A/cm^2
Area	25299 cm^2	15942 cm^2
Overall efficiency	41.3 %	40.5 %

The power rating using low sulfur diesel (1-D) is a bit higher than using grade No. 2-D diesel. This is due to a better reforming of the first diesel quality, which leads to a higher hydrogen concentration. In the first case the fuel cell works at higher voltage and so higher efficiency [1], however this has the consequence of lower current density, high cell area and so high cost. The overall efficiency has been computed as the net power extracted from the system (power from the cell minus the power used by pumps and compressors) divided by the lower heating value of the fuel. The only loss considered is in the fuel cell. The value of the total efficiency that was obtained is in agreement with Berry et al. [25] with non-shared heat. The results are summarized in Table 2.

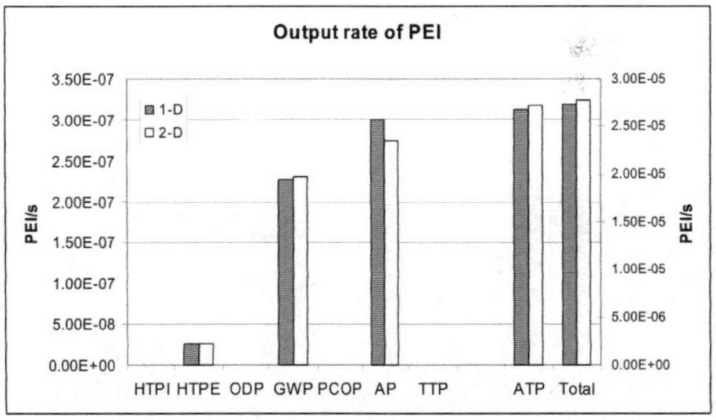

Figure 2. Output rate of PEI for the two diesel qualities. ATP and Total PEI refer to the right axis

As it is shown in Fig. 4, the greater contribution to the total Potential Environmental Impact comes from the Aquatic Toxicity Potential due to the emissions of carbon dioxide and nitric oxide. The overall output rate of PEI has been computed with all the category weights equal to 1. There is not much difference between the two diesel qualities, even if 2-D diesel total PEI is a bit higher. The sulfur content has no influence on the results because it has been assumed that all the sulfur is retained in the zinc oxide bed.

SUMMARY AND FUTURE WORK

The work presented an integrated framework to quantify the different impacts of the introduction of fuel cell power systems. The integrated framework has six main components: process modeling, environmental impact assessment, cost modeling, health impact assessment, life cycle assessment, and multi-objective optimization. This framework allows the consideration of the impacts of the regular running of the device and its production and dismissal. Moreover, it can identify the trade-offs between the different and conflicting objectives like: low cost, high efficiency, and low health and environmental impacts.

A case study of the introduction of fuel cell based APUs in US market of heavy-duty trucks and recreational vehicles illustrates the usefulness of the framework. We presented preliminary results of system modeling with experimental validation, and environmental and ecological impacts assessment of SOFC powered APUs. Future work will present results of health impacts assessment, LCA and trade-offs obtained by multi-objective optimization.

ACKNOWLEDGMENTS

This work is funded by National Energy Technology Laboratory (NETL), DOE, Morgantown

REFERENCES

1. "Fuel cell handbook", fifth edition, US Department of Energy, 2000
2. Aspen Plus version 11.1, AspenTech, October 2001
3. Cabezas, H., Bare, J.C., and Mallick, S.K., "Pollution Prevention with Chemical Process Simulators: The Generalized Waste (WAR) Algorithm - Full Version", Comp. Che. Eng. 23: 623-634 (1999)
4. Hilaly, A.K. and Sikdar, S.K., "Pollution Balance: A New Method for Minimizing Waste Production in Manufacturing Processes", J. Air & Waste Manage. Assoc., 44:1303-1308 (1994)
5. Arthur D. Little, "Assessment of Planar Solid Oxide Fuel Cell Technology", report to DOE FETC, 1999
6. B.D.James, G.D. Ariff, R.C.Kuhn, "DFMA Cost Estimates of Fuel-Cell/Reformer Systems at Low/Medium/High Production Rates", Future Car Congress 2002
7. Garza V. J. (2002), The Health Risk Assessment (HRA) Computer Program, version 2.0e. Civil & Environmental Engineering Department, University of California at Davis, California Environmental Protection Agency, Air Resources Board, Stationary Source Division
8. CEIDARS (California Emissions Inventory Data Analysis and Reporting System) LITE DATABASE, California Air Resources Board P.O. Box 2815 2020 L Street Sacramento, CA 95812-2815
9. U.S. Environmental Protection Agency and Science Applications International Corporation. LCAccess - LCA 101. 2001. Retrieved from http://www.epa.gov/ORD/NRMRL/lcaccess/lca101.htm
10. M. Wang, "Fuel choices for fuel-cell vehicles: well-to-wheels energy and emission impacts", Journal of Power Sources, 112(1):307-321 (2002)
11. GREET version 1.6, developed by Michael Wang, October 2001, Center for Transportation Research Argonne National Laboratory
12. Karakoussis V. et al., "The environmental impact of manufacturing planar and tubular solid oxide fuel cells", Journal of Power Sources, 101:10-26 (2001)
13. Zapp P., "Environmental analysis of solid oxide fuel cells", Journal of Power Sources, 61:259-262 (1996)
14. Cohon, J., "Multi-objective Programming and Planning", (1978), Academic Press, San Diego, CA.
15. Zeleny, M., "Multi criteria decision making" (1982), McGraw-Hill, Inc., New York.
16. Yan Fu and Urmila M. Diwekar, "An Efficient Sampling Approach to Multi-objective Optimization", Annals of Operations Research (accepted), 2003
17. Kalagnanam, J. R. and U. M. Diwekar, "An Efficient Sampling Technique for Off-line Quality Control", Technometrics 3:308-319 (1997).
18. Stodolsky, F., Gaines, L., Vyas, A. 2000. Analysis of Technology Options to Reduce the Fuel Consumption of Idling Trucks, ANL/ESD-43. Argonne National Laboratory, Argonne, IL.
19. U.S. Census Bureau, Economic Census: Vehicle Inventory and Use Survey: United States, EC97TV-US (Washington, DC: 1999). 1992 and 1997 editions
20. TMC, 1995, Analysis of Costs from Idling and Parasitic Devices for Heavy Duty Trucks, Truck Maintenance Council Recommended Procedure 1108, American Trucking Association, Alexandria, Va., March.
21. Richard T. Curtin, "THE RV CONSUMER", University of Michigan Survey Research Center, based on a 2001 survey
22. http://www.rvia.org/index.htm
23. http://www.fleetwood.com/corporate/99_annual/pg23.html
24. ATA, 1998, Green Truck website, American Trucking Association, http://www.truckline.com/air_emissions/1420.html

25. D. Berry et al., "Fuel Processing of Diesel Fuel for Auxiliary Power Units", Hydrogen, Fuel Cells, and Infrastructure Technologies Program, FY 2002 Progress Report, U.S. Department of Energy, pag. 337-341
26. A. Berry et al., "Fuel Processing of Diesel for Fuel Cells", 2002 SECA Core Review
27. C. Pereira et al., "Liquid Fuel Reformer Development", Proceedings of the 1999 U.S. DOE Hydrogen Program Review
28. J.C. Amphlett et al., "Simulation of a 250 kW diesel fuel processor/PEM fuel cell system", Journal of Power Sources, 71:179-184 (1998)
29. C. Pereira et al., "Liquid Fuel Reformer Development: Autothermal Reforming of Diesel Fuel", 2000 Hydrogen Program Technical Review
30. Kalagnanam J. and U. Diwekar, "An optimization approach to order of magnitude reasoning",AI for Engineering Design, 8:207-219 (1994)
31. S. Unnasch et al., "Evaluation of Partial Oxidation Fuel Cell Reformer Emissions", Hydrogen, Fuel Cells, and Infrastructure Technologies Program, FY 2002 Progress Report, U.S. Department of Energy, pag. 328-331
32. R. A. Geisbrecht, "Theory based process modeling for evaluation of fuel cells in advanced energy systems", AIChE National meeting 2002, New Orleans, LA
33. S. H. Chan et al., "Modelling of simple hybrid solid oxide fuel cell and gas turbine power plant", Journal of Power Sources 109: 111-120 (2002)
34. L. Petruzzi et al., "A global thermo-electrochemical model for SOFC systems design and engineering", Journal of Power Sources 118: 96-107 (2003)
35. S. H. Chan et al., "A complete polarization model of a solid oxide fuel cell and its sensitivity to the change of cell component thickness", Journal of Power Sources 93: 130-140 (2001)
36. A.V. Virkar et al., "The role of electrode microstructure on activation and concentration polarizations in solid oxide fuel cells", Solid State Ionics 131:189-198 (2000)

CONTACT

Francesco Baratto: fbarat1@uic.edu

Urmila M. Diwekar: urmila@uic.edu

http://www.uic.edu/labs/custom/

Simulation and Optimization of PEM Fuel Cells

Cong Xu
Chemical Engineering Department, Carnegie Mellon University
5000 Forbes Ave., Pittsburgh, PA 15213

Peter M. Follmann
Chemical Engineering Department, Carnegie Mellon University
5000 Forbes Ave., Pittsburgh, PA 15213

Lorenz T. Biegler
Chemical Engineering Department, Carnegie Mellon University
5000 Forbes Ave., Pittsburgh, PA 15213

Myung S. Jhon
Chemical Engineering Department, Carnegie Mellon University
5000 Forbes Ave., Pittsburgh, PA 15213

ABSTRACT

In this paper, we examine a 0D Direct Methanol Fuel Cell (DMFC) model, which captures the essence of electrode kinetics and methanol crossover through the polymer electrolyte membrane (PEM). We estimate certain parameters to obtain the polarization curve and these parameters were validated against published data. We also identify the key factors within the DMFC model affecting methanol crossover, e.g., PEM thickness and methanol feed concentration, which significantly impact its performance and commercialization. Since power density is a good indicator of the fuel cell performance, we attempt to establish a relationship between the methanol feed concentration and the power density under a fixed current density operational condition. To gain insight into the effect of the methanol feed concentration, we applied DASPK 3.0 to the differential algebraic equation (DAE) system resulting from the DMFC model and obtain sensitivity information between the cell voltage and the methanol feed concentration. From the stationary points revealed in the sensitivity curve, we are able to identify the optimal feed concentration, which provides the highest power density output for a set value of the current density. The dynamic response of the cell voltage to the change of the feed concentration is studied by applying a dynamic optimization tool OCC to the DAE system, where the dynamic simulation and optimization is carried out at the same time under fixed current density. Our dynamic analysis results show another promising way of improving the DMFC performance by adopting a dynamic feeding strategy.

KEY WORDS

DMFC, Simulation, DAE, Sensitivity, Methanol Crossover, Power Density, Dynamic Optimization.

INTRODUCTION

Fuel cells have attracted more and more attention recently as a possible replacement for current petroleum power generation system. For optimal performance and design, accurate system performance modeling is needed. A fundamental understanding of the principles in electrochemistry, materials & manufacturing, and heat & mass transfer is crucial for developing models, which can satisfy the goals of lower cost, high performance and high reliability for the fuel cells.

There are two types of Polymer Electrolyte Membrane Fuel Cells (PEMFCs), i.e., H_2 PEM Fuel Cell and Direct Methanol Fuel Cell (DMFC), both of which utilize PEM to transfer protons. The DMFC feeds a solution of methanol and water (typically 0.5–2 molar) to the anode, where it is internally reformed by the catalyst. Compared to H_2 PEMFC, DMFC is advantageous for its ease of fuel delivery and storage, lack of a cooling or humidification requirement, and reduced design complexity. Due to the absence of ancillary equipment, the DMFC design is ideally suited for portable electronics. Despite these compelling advantages, the DMFC's low electro-activity of methanol oxidation on the anode and large methanol crossover through the PEM from the anode to cathode still remain important technological concerns. Here some of the methanol in the anode compartment is lost through the membrane and reacted in the cathode compartment.

BACKGROUND

Methanol crossover in DMFC has been extensively studied both experimentally and theoretically. Narayanan et al. [1995] and Ren et al. [1995] measured the methanol crossover flux with different membrane thicknesses and showed that the methanol crossover rate is inversely proportional to the membrane thickness at a given cell current density, which suggests that the diffusion process dominates the methanol transport through the membrane. In addition, Ren et al. [2000] compared the diffusion with methanol electro-osmotic drag processes and demonstrated the importance of the electro-osmotic drag in the methanol transport through the membrane. In their analysis, the electro-osmotic drag is considered a convection effect and the diluted methanol moves with electro-osmotically dragged water molecules. Additionally the cathode potential is influenced by the mixed potential phenomenon due to simultaneous methanol oxidation and oxygen reduction as well as poisoning of Pt catalysts by methanol oxidation products.

Baxter et al. [2000] developed a one-dimensional model for a liquid-feed DMFC, focused primarily on the anode catalyst layer. A major assumption of their study is that the carbon dioxide is only dissolved in the liquid and hence their model of transport and electrochemical processes in the anode catalyst layer is based on single-phase. Dohle et al. [2000] presented a one-dimensional model for the vapor-feed DMFC and the crossover phenomenon. The effects of methanol concentration on the cell performance were also studied. Scott et al. [1999, 2000] developed several simplified single-phase models to study transport and electrochemical processes in liquid feed DMFC and showed that the cell performance is limited by the slow methanol diffusion in liquid.

THE DIFFERENTIAL ALGEBRAIC EQUATION (DAE) MODEL FOR DMFC

We summarize the DMFC model below. See reference [3] for details concerning the derivation of the model equations. The mathematical model considers the electrodes kinetics and the mass and charge balances. The model also accounts for the undesired methanol crossover.

REACTION KINETICS EQUATIONS AT THE ELECTRODES

Table 1. Reaction steps and kinetic equations in both electrodes.

Reaction Steps	Reaction steps and kinetic equations
(R1)	3Pt+CH3OH= Pt3-COH+3H++3e-
(1)	$r_1 = k_1 \exp\left(\frac{\alpha_1 F}{RT}\eta_a\right)\left\{\Theta_{P_t}^3 c_{CH_3OH}^{CL} - \frac{1}{K_1}\exp\left(-\frac{F}{RT}\eta_a\right)\Theta_{P_{t3}-COH}\right\}$
(R2)	3Ru+3H2O= 3Ru-OH+3H++3e-
(2)	$r_2 = k_2 \exp\left(\frac{\alpha_2 F}{RT}\eta_a\right)\left\{\Theta_{Ru} - \frac{1}{K_2}\exp\left(-\frac{F}{TR}\eta_a\right)\Theta_{Ru-OH}\right\}$
(R3)	Pt3-COH+2Ru-OH= Pt-COOH+H2O+2Pt+2Ru
(3)	$r_3 = k_3\left\{\Theta_{Pt_3-COH}\Theta_{Ru-OH}{}^2 - \frac{1}{K_3}\Theta_{Pt-COOH}\Theta_{Pt}^2\Theta_{Ru}^2\right\}$
(R4)	Pt-COOH+Ru-OH= CO2+H2O+Pt+Ru
(4)	$r_4 = k_4\left\{\Theta_{Pt-COH}\Theta_{Ru-OH} - \frac{1}{K_4}C_{CO_2}^{CL}\Theta_{Pt}\Theta_{Ru}\right\}$
(R5)	(R5) 3/2O2+6H++6e-= 3H2O
(5)	$r_5 = k_5 \exp\left(\frac{\alpha_5 F}{RT}\eta_c\right)\left\{1 - \exp\left(-\frac{F}{RT}\eta_c\right)\left(\frac{p_{O_2}}{p^\theta}\right)^{3/2}\right\}$

In equations (1)–(5) at Table 1, r_j (j=1…5) is the rate of reaction j at the electrode area, α_j is charge transfer coefficient, k_j is the rate constant of reaction j, K_j is equilibrium constant of reaction j, Θ_i is the surface fraction covered by component i, c_i^{CL} is the concentration of component i (i= CH_3OH and CO_2) in anode catalyst layer, p^θ is the standard pressure, p_{O2} is the partial pressure of O2, R is the gas constant, T is the temperature in Kelvin, F is the Faraday constant, and η_a and η_c are the anode and the cathode overpotential, respectively.

As in Sundmacher et al., we use the quasi-equilibrium approximation for reaction steps (R2)-(R4) and specie balances on the surface of the two catalyst components, Pt and Ru, to obtain the following relations:

$$\Theta_{Pt_3-COH} = \frac{c_{CO_2}^{CL}}{\widetilde{K}_2^3 K_3 K_4}\Theta_{Pt}^3, \qquad (6)$$

and

$$\Theta_{Pt}^3 + \frac{(\widetilde{K}_2 K_4 + c_{CO_2}^{CL})\widetilde{K}_2^2 K_3}{c_{CO_2}^{CL}} \Theta_{Pt} - \frac{\widetilde{K}_2^3 K_3 K_4}{c_{CO_2}^{CL}} = 0. \quad (7)$$

with the overpotential-dependent equilibrium constant defined as:

$$\widetilde{K}_2 \equiv K_2 \exp(\frac{F}{RT}\eta_a) \quad (8)$$

MASS AND CHARGE BALANCES

The mass balance equations at the anode compartment are described as:

$$\frac{dc_{CH_3OH}}{dt} = \frac{1}{\tau}(c_{CH_3OH}^F - c_{CH_3OH}) - \frac{k^{LS}A^S}{V_a}(c_{CH_3OH} - c_{CH_3OH}^{CL}) \quad (9)$$

$$\frac{dc_{CO_2}}{dt} = \frac{1}{\tau}\left(c_{CO_2}^F - c_{CO_2}\right) - \frac{\kappa^{LS}A^S}{V_a}\left(c_{CO_2} - c_{CO_2}^{CL}\right) \quad (10)$$

where A^S is cross-sectional electrode area, c_i is component i concentration in the anode compartment, c_i^F is the feed concentration of component i, k^{LS} is mass transfer coefficient, V_a and τ are the volume and mean residence time in the anode compartment, respectively.

The mass balance equations at the anode catalyst layer are described as:

$$\frac{dc_{CO_2}^{CL}}{dt} = \frac{\kappa^{LS}A^S}{V_a^{CL}}\left(c_{CO_2} - c_{CO_2}^{CL}\right) + \frac{A^S}{V_a^{CL}} r_1 \quad (11)$$

$$\frac{dc_{CH_3OH}^{CL}}{dt} = \frac{\kappa^{LS}A^S}{V_a^{CL}}\left(c_{CH_3OH} - c_{CH_3OH}^{CL}\right) - \frac{A^S}{V_a^{CL}} n_{CH_3OH}^M - \frac{A^S}{V_a^{CL}} r_1 \quad (12)$$

where V_a^{CL} is the volume of the anode catalyst layer, and n_{CH3OH}^M is the methanol mass flux in the membrane.

The charge balance equations at the anode and cathode catalyst layers are described as:

$$\frac{d\eta_a}{dt} = \frac{1}{C_a}\left(-i_{cell} - 6Fr_1\right) \quad (13)$$

$$\frac{d\eta_c}{dt} = \frac{1}{C_c}\left(-i_{cell} - 6F\left(r_5 + n_{CH_3OH}^M\right)\right) \quad (14)$$

where C_a and C_c is the double layer capacity in anode and cathode areas, respectively, and i_{cell} is the current density.

BASIC CONCEPTS ABOUT THE PARAMETER ESTIMATION

We can describe parameter estimation problems in the following manner:

$$\begin{aligned} \min_{\{\theta, y\}} \quad & f(\theta, y) \\ s.t. \quad & c(\theta, y) = 0 \\ & \theta^L \le \theta \le \theta^U, \end{aligned} \quad (15)$$

where $f(\theta, y)$ is an objective function often derived from a statistical basis, $x^T = [\theta^T, y^T]$ is the total set of variables, θ is the set of parameters, and θ^L and θ^U are the lower and upper bounds, respectively, on the parameters. Here we assume that: $f: \Re^n \to \Re, c: \Re^n \to \Re^m, \theta \in \Re^{(n-m)}, and$ n > m. The functions f and c are assumed to be at least twice differentiable. We only have the parameter bounds. Problem (15) can be solved by successive quadratic programming (SQP), repeatedly solving the following quadratic program to the kth iteration:

$$\begin{aligned} \min_d \quad & \nabla_x f(x_k)^T d + \frac{1}{2} d^T W_k d \\ s.t. \quad & c(x_k) + A(x_k)d = 0 \\ & \theta^L \le \theta_k + d_\theta \le \theta^U. \end{aligned} \quad (16)$$

Here, $d \in \Re^n, d^T = [d_\theta^T, d_y^T], d_\theta \in \Re^{(n-m)}, and \quad d_y \in \Re^m$. $W_k \in \Re^{n \times n}$ is the Hessian of the Lagrangian, $L_k = f_k + c_k^T \lambda_k \in \Re^n$.

In our parameter estimation problem, the degrees of freedom, (n-m), are few in comparison to the total number of variables, though function evaluations may be extensive. Refer to [1] for details concerning the trust region SQP algorithm applied during our parameter estimation.

PARAMETER ESTIMATION USING A TRUST REGION SQP ALGORITHM

We adopted a square sum of the differences between our calculated voltage values ($U_{cell}^{Cal.}$) and the experimental ones ($U_{cell}^{exp.}$) at ten different current densities as our objective function f.

$$f = \sum_k \{U_{cell}^{Cal.}(i_{cell}^k) - U_{cell}^{exp.}(i_{cell}^k)\}^2. \quad (17)$$

By analyzing the model equations and parameters, we found four parameters notably sensitive within the calculation; the reaction rate constants k_1 and k_5, and the equilibrium constants K_2 and K_4. We performed parameter estimation for these parameters in steady state. By applying our in-house code [1], we obtained the

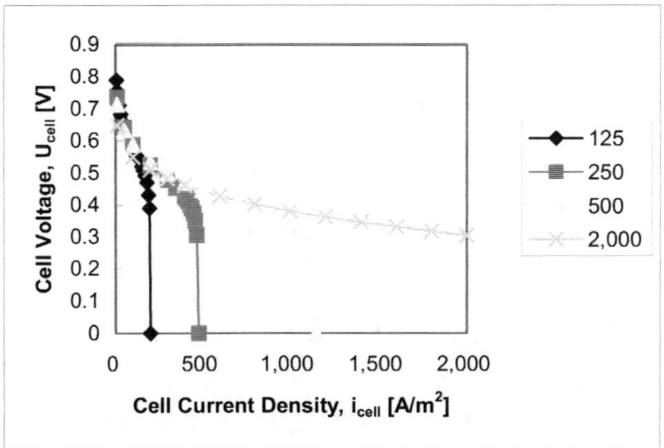

Figure 1. Steady state polarization curve for different methanol feed concentrations.

parameter values upon the convergence of the program. The parameters we used in our calculations are listed in Table 2. Further, we identified the steady state parameters via the parameter estimation program.

Table 2. Identified parameters in our simulation.

Reaction Parameters	Steady state	Dynamic
K_1	4.9×10-7	5.0×10-5
K_2	1	1
K_3	1×10-2	1×10-2
K_4	1	1
K_5	1.8×10-7	1×10-8
K_1	1	1
K_2	0.5	0.5
K_3	1	1
K_4	0.7	0.7

STEADY STATE MODELING

DAE SOLVER AND SENSITIVITY EVALUATIONS

DASPK 3.0 [2] was used as the DAE solver and sensitivity package. DASPK can solve initial value problems of the form F(t,y,y′,p)=0 via a combination of backward differentiation formulae (BDF) methods and different linear system solution methods, where F, y, and y′ are N-dimensional vectors, and p is the parameter vector. It can also produce a sensitivity analysis of the parameters dy/dpi concurrently within the DAE system.

Since all time derivative terms are set to zero in steady state, we only consider the resulting algebraic equation systems. We used Newton's method to solve for the unknowns and the steady state polarization curve (see Fig. 2) was obtained for different feeding concentrations. Our simulated results are in quantitative agreement with published data [3].

SENSITIVITY OF UCELL VS. METHANOL FEED CONCENTRATION AT VARIOUS CURRENT DENSITIES

We performed a sensitivity analysis for the DAE model using DASPK 3.0. The sensitivity calculation was carried out under different current densities. Representative results are shown in Fig. 2. From these curves, we can see that the sensitivity curve of Ucell vs. feed concentration crosses the abscissa, implying that Ucell decreases as $c^F_{CH_3OH}$ increases, which is in agreement with the experimental data. The optimal feed concentrations for cell voltage can be gleaned from the methanol feed concentrations, which correspond to the stationary points derived from the curves. If i_{cell} is assumed to be constant, power density of the fuel cell is directly proportional to U_{cell}.

Figure 2 depicts the sensitivity information of cell voltage vs. methanol feed concentration for different current densities. These curves reveal how the change in methanol feed concentration affect the cell voltage and power density under a given current density. Using these curves we are able to obtain the stationary points, which are the optimal methanol feed concentrations for the maximum power density output under constant current densities.

OPTIMAL METHANOL FEED CONCENTRATIONS AT DIFFERENT CURRENT DENSITIES

We obtained the stationary points for the methanol feed concentrations from the sensitivity curves in Fig. 2 and plotted the relationship between the optimal methanol feed concentration and current density in Fig. 3.

From Fig. 3, we can see that as the current density increases, the corresponding optimal methanol feed concentration will also increase. As the current density increases, however, the change in the optimal methanol feed concentration decreases.

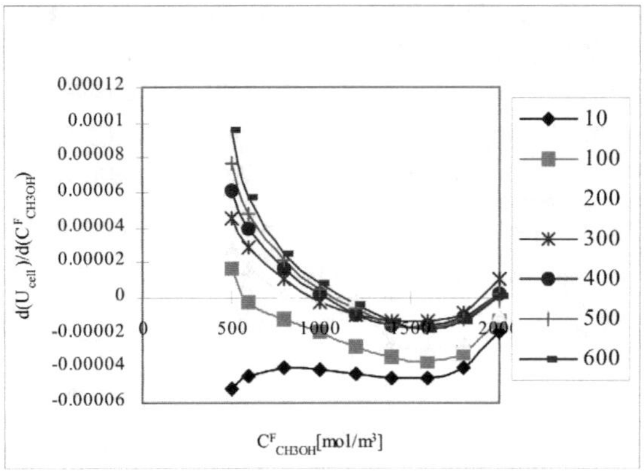

Figure 2. Sensitivities of cell voltage vs. methanol feed concentration under fixed current densities.

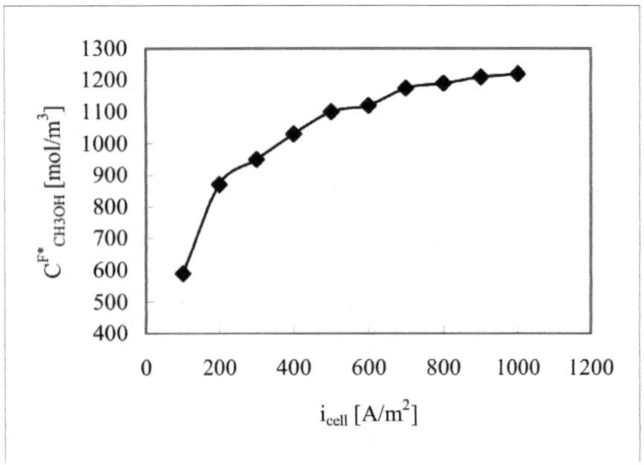

Figure 3. Optimal methanol feed concentrations at different current densities.

DYNAMIC SIMULATION AND OPTIMIZATION

METHODS AND TOOL

In general, the solution of large-scale non-linear optimal control problems can be obtained by discretizing either the control variables (Barton et al. [12], Vassiliadis et al. [13]) or both the control and state variables (Cervantes and Biegler [10]). Both approaches transform the continuous optimal control problem into a nonlinear programming (NLP) problem that can be solved with appropriate large scale NLP methods. In our study, the latter approach is adopted. The simulation and optimization is conducted simultaneously.

In order to formulate the continuous dynamic optimization problem as a NLP problem, the state variables are approximated by polynomials on finite elements. The type of polynomials and the distribution of the discrete time steps is strongly related to the well established methods for solving ordinary differential equations (ODEs). In our study the three point collocation at Radau points with piecewise constant control variables for each finite time element has been applied. A dynamic optimization tool called OCC (Optimal Control Center) was used for the analysis. In the software, to apply the discretization to general optimal control problems in a user friendly way, the software package OCOMA has been developed. With OCOMA we only define the continuous dynamic optimization problem. This can be easily done by coding the system consisting of DAEs, the inequality constraints, the bounds for variables, and the objective function in a simple syntax in MAPLE.

The main tasks are conducted by the MAPLE program in order to formulate the collocation equations and to derive the first and second derivative information. In the second step, the continuous system is discretized using the three-point Radau implicit Runge-Kutta scheme. In the third step, a FORTRAN code consisting of the objective function, the constraints and te analytical gradients is generated. The resulting files are complied and linked with the optimization code SNOPT or IPOPT to form an executable file. After the optimization, the results are printed out in a format that can be read and visualized with MATLAB.

RESULTS

We need a little modification and simplification of the anode model equation to do the dynamic simulation. From Eq.s (1) and (6), we can derive that

$$r_1 = k_1 \exp\left(\frac{\alpha_1 F}{RT}\eta_a\right)\left\{c^{CL}_{CH_3OH} - \frac{1}{K_1}\exp\left(-\frac{F}{RT}\eta_a\right)\frac{c^{CL}_{CO_2}}{\widetilde{K}_2^3 K_3 K_4}\right\}\Theta^3_{Pt} \quad (18)$$

Note that $\widetilde{K}_2 \equiv K_2 \exp(\frac{F}{RT}\eta_a)$, then we will get

$$r_1 = k_1 \exp\left(\frac{\alpha_1 F}{RT}\eta_a\right)\left\{c^{CL}_{CH_3OH} - \frac{c^{CL}_{CO_2}}{K_1 K_2^3 K_3 K_4}\exp\left(-\frac{4F}{RT}\eta_a\right)\right\}\Theta^3_{Pt} \quad (19)$$

Our dynamic simulation and optimization is carried out for a set value of current density, e.g., 300 A/m². We choose the control variable to be the methanol feeding concentration and the cell voltage to be the objective function, which is similar as what we did in the steady state case. Under above mentioned conditions, the variation of Θ_{Pt} is negligible. Thus, we treated Θ_{Pt} as a constant in the dynamic simulation and optimization, which greatly reduced the nonlinearity of the DAE system and ended up with the following form for the anode reaction kinetics, similar to the general form of Butler-Volmer equation.

$$r_1 = \widetilde{k}_1 \exp\left(\frac{\alpha_1 F}{RT}\eta_a\right)\left\{c^{CL}_{CH_3OH} - \frac{c^{CL}_{CO_2}}{K_1 K_2^3 K_3 K_4}\exp\left(-\frac{4F}{RT}\eta_a\right)\right\} \qquad (20)$$

where $\widetilde{k}_1 \equiv k_1 \Theta^3_{Pt}$, assumed to be a constant (0.385) under our dynamic simulation conditions. Here we refer to Table 2 for the relevant parameters used for the dynamic simulation and optimization of the DMFC behavior.

In our preliminary simulation, we found a jump in the cell voltage and subsequent higher performance when the methanol solution is fed dynamically (see Fig. 4). This is consistent with the observations made by other groups [3].

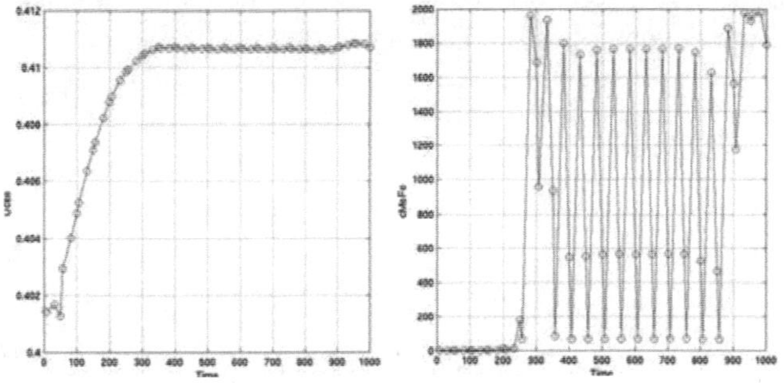

Figure 4. Dynamic simulation and optimization results for i_{cell}=300A/m^2 and methanol feed concentration limit of 3000 M.

SUMMARY

In this paper, we examined the DMFC model both in steady state and dynamically with a consideration of electrode kinetics and transport phenomena. As seen, certain important parameters are very difficult, or even impossible, to characterize via experimentation. Typically, they are fitted or estimated in the model to allow satisfactory simulation results. A trust region SQP algorithm was applied to the DMFC model for parameter estimation, and simulation results are in excellent agreement with published data.

We identified the key factors affecting the DMFC performance and energy utilization efficiency. From our analysis, we found that PEM thickness and methanol feed concentration are closely related to the methanol crossover. We also applied a sensitivity analysis to the DMFC. Our preliminary sensitivity results for cell voltage vs. methanol feed concentration at given current density are encouraging. From these curves, we were able to obtain the optimal methanol feed concentrations. Further, using these optimal feed concentrations, we were able to obtain the maximal cell voltages and power densities for a wide range of current densities, providing insight into the improvement of performance and energy efficiency by treating the methanol feed concentration as the decision variable.

A dynamic simulation and optimization of the DAE system has also been carried out, which yields promising results favoring a dynamic feeding strategy of the methanol solution in order to improve DMFC performance.

ACKNOWLEDGMENTS

PITA (ICES) is gratefully mentioned for the financial support of this project.

REFERENCES

1. N. Arora and L. T. Biegler, "A Trust Region SQP Algorithm for Equality Constrained Parameter Estimation with Simple Parameter Bounds", December, 2002, submitted for publication.
2. S. Li and L. Petzold, "Software and Algorithms for Sensitivity Analysis of Large-Scale Differential Algebraic Systems", J. Comput. and Appl. Math. 125 (2000), 131-145.
3. K. Sundmacher, T. Schultz, S. Zhou, K. Scott, M. Ginkel, and E.D. Gilles. Dynamics of the Direct Methanol Fuel Cell (DMFC): Experiments and Model-based Analysis. Chemical Engineering Science, 56, 333-341 (2001).
4. M. M. Mench, C.Y. Wang, and S. Thynell, "An Introduction to Fuel Cells and Related Transport Phenomena", The International Journal of Transport Phenomena, Volume 3, Number 3, 2001 2001
5. Larminie, J., and Dicks, A. (2000) Fuel Cell Systems Explained, John Wiley and Sons, NewYork, NY.
6. Cruickshank, J. and Scott, K. (1998) The Degree and Effect of Methanol Crossover in the Direct Methanol Fuel Cell, Journal of Power Sources, Vol. 70, pp. 40-47.
7. D. M. Bernardi and M. W. Verbrugge, "A Mathematical Model of the Solid-polymer-electrolyte Fuel Cell", J. Electrochem. Soc., 139, 2477 (1992).
8. Newman, J. (1991) Electrochemical Systems, Prentice-Hall, New York, NY.
9. J. Bausa, G. Tsatsaronis and T. Jockenhovel, "Reducing the Energy Demand of Continuous Distillation Processes by Optimal Controlled Forced Periodic Operation".
10. A. M. Cervantes, A. Wachter, R. Tutuncu and L. Biegler, "A Reduced Space Interior Point Strategy for Optimization of Differential Algebraic Systems".
11. S. Zhou, T. Schultz, M. Peglow and K. Sundmacher, "Analysis of the nonlinear dynamics of a direct methanol fuel cell", Phys. Chem. Chem. Phys., 2001, 3, 347-355.

12. Barton, P.I., Allgor R.J., Feehery, W.F., and Galan, S., "Dynamic optimization in a discontinuous world", Ind. Eng. Chem. Res., 37, 966-981, 1998.
13. Vassiliadis, V.S., Sargent, R.W.H., and Pantelides, C.C., "Solution of a class of multistage dynamic optimization problems, part II-problems with path constraints", Ind. Eng. Chem. Res., 33, 2123-2133, 1994.
14. H. A. Gasteiger, N. Markovic, P. N. Ross Jr. And E. J. Cairns, "Electro-oxidatin of small organic molecules on well-characterized Pt-Pu alloys", Electrochimica Acta, Vol. 39, No. 11/12, 1825-1832, 1994.
15. S. R. Narayanan, H. Frank, B. Jeffries-Nakamura, M. Smart, W. Chun, G. Halpert, J. Kosek, and C. Cropley, in Proton Conducting Membrane Fuel Cells I, S. Gottesfeld, G. Halpert, A. Landgrebe, Editors, PV 95-23, p. 278, The Electrochemical Society Proceedings Series, Pennington, NJ (1995).
16. X. Ren, T. A. Zawodzinski Jr., F. Uribe, H. Dai, and S. Gottesfeld, in Proton Conducting Membrane Fuel Cells I, S. Gottesfeld, G. Halpert, A. Landgrebe, Editors, PV 95-23, p. 284, The Electrochemical Society Proceedings Series, Pennington, NJ (1995).
17. X. Ren, T. E. Springer, T. A. Zawodzinski, and S. Gottesfeld, Journal of Electrochemical Society, 147, 466 (2000).
18. S. F. Baxter, V. S. Battaglia, and R. E. White, Journal of Electrochemical Society,146, 437 (2000).
19. H. Dohle, J. Divisek, and R. Jung, Journal of Power Sources, 86, 469 (2000).

CONTACT

Cong Xu

cxu@andrew.cmu.edu

412-2682238(O)

Aug. 2000-present, Ph.D. Student, Department of Chemical Engineering, Carnegie Mellon University.

Sep. 1996-Jul. 2000, Department of Chemical Engineering, Tsinghua University, Beijing, China.

Multiobjective Optimization under Uncertainty of Hybrid Fuel Cell Power Plants

Karthik Subramanyan
University of Illinois at Chicago
Center for Uncertain Systems : Tools for Optimization and Management
Department of Chemical Engineering
851 S Morgan street Rm.1036, Chicago, IL 60607

Urmila M. Diwekar
University of Illinois at Chicago
Center for Uncertain Systems : Tools for Optimization and Management
Department of Chemical Engineering
Institute for Environmental Science & Policy

Prepared for Presentation at :

AIChe Annual Meeting 2003
Monday, November 17, 2003
Topical TI
Session : Modeling Fuel Cell Behaviour
Paper No. : 276f

Unpublished

Multi-objective optimization under uncertainty of hybrid fuel cell power plants

Karthik Subramanyan and Urmila M. Diwekar
University of Illinois at Chicago
851, S Morgan street, Chicago, IL 60607

ABSTRACT

One of the major applications of fuel cells is as onsite stationary electric power plant. Several types of configurations have been hypothesized and tested for these kinds of applications at the conceptual level but hybrid power plants are one of the most efficient. These are designs that combine fuel cell cycle with other thermodynamic cycles to provide higher efficiency. Generally, the heat rejected by the fuel cell at higher temperature is used in a bottoming cycle to generate steam. In this work we are considering a conceptual design of a SOFC-PEM hybrid power plant in which the high temperature fuel cell acts both as electricity producer and fuel reformer for the low temperature fuel cell. The exhaust from the PEM cell goes to a waste hydrogen burner and heat recovery steam generator that produces steam for further utilizations. Optimizing this conceptual design involves consideration of a number of objectives. The process should have lower pollutant emissions as well as cost or profits competitive with the existing technology. Moreover some specialized objectives specific to the problem also need to be considered. For example the fuel cell current density and system efficiency are very important and also need to be optimized along with cost and emissions. These objectives results in a multi-objective optimization problem where the optimum goals are simultaneously achieved. Since this technology is new, uncertainties are inherent in it. We have obtained the optimal trade-off design solutions or the Pareto set for this plant through a multi-objective optimization framework. Uncertainty in the current density characteristics have been characterized, quantified and propagated through this framework to analyze the effect on the various objectives like cost , profit, power output and emission.

KEY WORDS

Fuel cells, hybrid power plants, multi-objective optimization, uncertainty, pareto surface

INTRODUCTION

Fuel cells are an important technology for a potentially wide variety of applications including micropower, auxiliary power, transportation power, stationary power for buildings and other distributed generation applications, and central power. These applications can be used in a large number of industries worldwide. Research efforts are being made to design fuel cell plants with low emissions, high efficiencies, high performance and low costs. High efficiency hybrid fuel cell power plants are being conceptualized in order to meet the great expectations from this technology. Also, a large number of materials need to be considered in the fuel cells for obtaining the desired properties for electrolyte issues, electrode performance issues, and for different configurations. Moreover, the development in the area of new materials and other technologies where the performance and economic data is scarce and/or incomplete, calls for consideration of uncertainties in the design and optimization. Furthermore, the problem involves multiple objectives such as cost, process efficiency and pollutant emissions. The conflicting nature of these objectives in the face of uncertainties makes the handling of this problem an even more formidable task. The focus of this paper is effect of uncertainties on the multi-objective designs of a new SOFC-PEM hybrid power plant. In section 2, we discuss the salient features of a hybrid SOFC-PEM hybrid power plant. The concept of multi-objective optimization is explained in section 3. This section also describes the deterministic and stochastic modeling framework and discusses the results obtained through these frameworks and finally section 4 presents a summary of the paper.

THE HYBRID FUEL CELL POWER PLANT

Distributed power generation is one of the most attractive applications of the fuel cell technology. A major bottleneck in the design of fuel cell power plants for this application is to package them in a system balance of plant (BOP) that allows them to function effectively. All fuel cells, especially those operating at high temperatures like Molten Carbonate Fuel Cell (MCFC) and Solid Oxide Fuel Cell (SOFC) need spent fuel utilization or waste heat recovery subsystems to increase process efficiency. Low temperature fuel cells (e.g. Polymer Electrolyte Membrane, PEM) also require fuel-reforming subsystems. A common approach for providing the BOP has been to integrate the fuel cell with a heat engine which gives rise to the fuel cell/heat engine hybrid systems. This is also a very effective means to utilize the waste heat of the fuel cell but compromises on several features of the fuel cell and distributed power generation in the first place. The plant is no longer a compact module and it is now difficult to scale it up. The addition of the heat engine bottoming cycle with moving parts lead to increase in noise and calls for more maintenance. Moreover, the dynamics and turndown characteristics of the fuel cell plant no longer remain simple. These considerations have fueled the development of fuel-cell/fuel-cell hybrid power plants as a very effective way to achieve the BOP. Generally, a high temperature internally reforming fuel cell (SOFC) needs to be combined with a low temperature fuel cell (PEM) having a complementary electrolyte (anion-conducting versus cation-conducting electrolytes) to get such a synergistic effect and Figure 1 shows the schematic of such a hybrid power plant. This SOFC-PEM hybrid plant [1][2] uses the following process integration steps to enhance the efficiency further.

Process Integration - The plant contains two fuel cells (one each of SOFC and PEM) combined with a heat recovery steam generation cycle. The use of two fuel cells makes this cycle up to 37.8 % more efficient than the case where only a SOFC is used (maximum efficiency of 52.4%). Natural gas is used as the fuel. It is processed in a fuel pre reformer to scrub H2S and other sulphur compounds and fed to the SOFC. The solid oxide fuel cell acts both as an electricity producer as well as a fuel reformer for PEM. The exhaust fuel from the SOFC is cooled and shifted in a low temperature shifter that also functions as a low-pressure steam boiler.

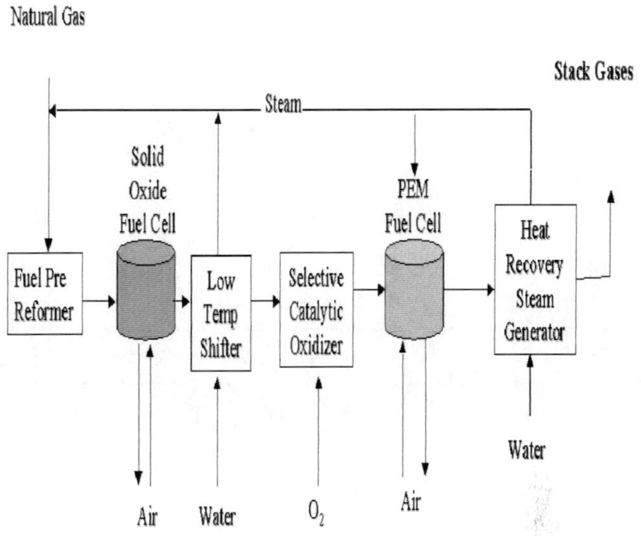

Figure 1: SOFC-PEM Hybrid Power Plant

Shifted fuel gas is then treated with pure oxygen in a selective catalytic oxidizer to reduce CO from several hundred parts per million (ppm) to below 10 ppm. The utilization of this reformed fuel is completed in the PEM. The exhaust from the PEM cell goes to waste hydrogen burner and heat recovery steam generator to utilize the waste heat of the exhaust stream to make steam from water. This steam produced in the low temperature shifter and the heat recovery steam generator is used in both the SOFC and PEM. In the SOFC, steam is used as a reactant for the reforming and downstream shift reactions and to control against the carbon. In the PEM, steam is used to humidify fuel and air streams to maintain water balance in the electrolyte and electrodes. The fuel utilization in SOFC can be limited to a range necessary, only to reform the natural gas and make the exhaust gas suitable for use by PEM after some treatment, but not completely oxidize it. This way, the reformed fuel (SOFC exhaust) can be completely oxidized in the PEM where more favorable thermodynamics apply rather than forcing a higher fuel utilization in the SOFC which can cause a drop in current density and cell voltage. Another important SOFC design aspect is internal recuperation, whereby efficient heat management is used maximize system compactness and power density by minimizing the required air-fuel equivalence ratio. The base case designs for this hybrid power plant is shown in table 1.

Table 1. : Base Case Design and Performance of the SOFC-PEM Hybrid Plant

Power Rating	1472 kW	Overall Efficiency	72.6 %
Capital Cost	1773 \$/kW	CO_2 Emissions	5.61 kg/kWh
Cost of Electricity	6.35 c/kWh		
SOFC		**PEM**	
Temperature	1750 °F	Temperature	176 °F
Pressure	20 psi	Pressure	25 psi
Current Density	75 mA/cm^2	Current Density	190 mA/cm^2
Fuel Utilization	70%		
Equivalence Ratio	1.25		

MULTI-OBJECTIVE OPTIMIZATION FRAMEWORK

In the case of the hybrid power plant, there are seven objectives to be optimized simultaneously which involves highly complex non-linear models. Moreover some objectives are conflicting and incommensurable i.e. it is not possible to improve them simultaneously and there is always a trade-off. Hence a multi-objective optimization framework is required to obtain the Pareto set. The objectives to be optimized, along with the constraints and decision variables are shown in table 2 [2][3].

The concept of multi-objective optimization - A generalized Multi-objective optimization (or Multi-Objective Programming, MOP) problem can be formulated as follows:

$$\min Z = Z_i, i = 1,\ldots\ldots, p, p \geq 2; \quad (1)$$

$$\text{s.t. } h(x, y) = 0,$$

$$g(x, y) \leq 0,$$

where x and y are continuous and discrete decision variables, and p is the number of objective functions. The functions h(x, y) and g(x, y) represent equality and inequality constraints, respectively. There are a large array of analytical techniques to solve this MOP problem; however, the MOP methods are generally divided into two basic types: preference-based methods and generating methods. Preference-based methods like goal programming attempt to quantify the decision maker's preference, and with this information, the solution that best satisfies the decision maker's preference is then identified[7]. Generating methods, such as the weighting method and the constraint method, have been developed to find the exact Pareto set or an approximation of it. As is well known, mathematics cannot isolate a unique optimum when there are multiple competing objectives. Mathematics

Table 2. : The objectives, constraints and decision variables

Objectives :

- ✓ Min Capital cost (CAP)
- ✓ Min Cost of electricity (COE)
- ✓ Min CO2 emissions (CO2EM)
- ✓ Max Current density SOFC (CDSOFC)
- ✓ Max Current density PEM (CDPEM)
- ✓ Max Overall efficiency (ACEFF)

Subject to

- ✓ *Mass & energy balance constraints*
- ✓ *Power rating of 1472 kW (Base case)*

Decision variables

- ✓ *Fuel utilization (UTIL)*
- ✓ *Equivalence ratio (ERAT)*
- ✓ *Pressure of the PEM (PPEM)*

Uncertain variables

- ✓ *PEM and SOFC current densities*

can at most aid designers to eliminate design alternatives dominated by others, leaving a number of alternatives in what is called the Pareto set [8]. For each of these designs, it is impossible to improve one objective without sacrificing the value of another relative to some other design alternative in the set. From among the dominating solutions, it is then a value judgment by the customer to select which design is the most appropriate. At issue is an effective means to find the members of the Pareto set for a design problem, especially when there are more than two or three objectives; the analysis per design requires significant computations to complete, and there are an almost uncountable number of design alternatives. A pure algorithmic approach to solving is to pick one of the objectives to minimize while the remaining others are turned into an inequality constraint with a parametric right-hand-side, L_k. The problem takes on the following form:

$$\min\ Z_j, \tag{2}$$

$$\text{s.t. } h(x, y) = 0,$$

$$g(x, y) \leq 0,$$

$$Z_k \leq L_k,\ \ k = 1,\ldots,j-1, j+1,\ldots,p \quad p \geq 2$$

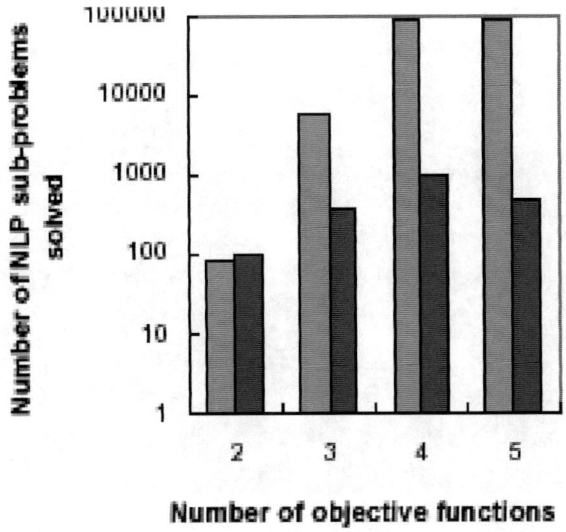

Figure 2. : Computational speed-up through MINSOOP

where Zj is the chosen j-th objective that wished to be optimized. Solving repeatedly for different values of L_k

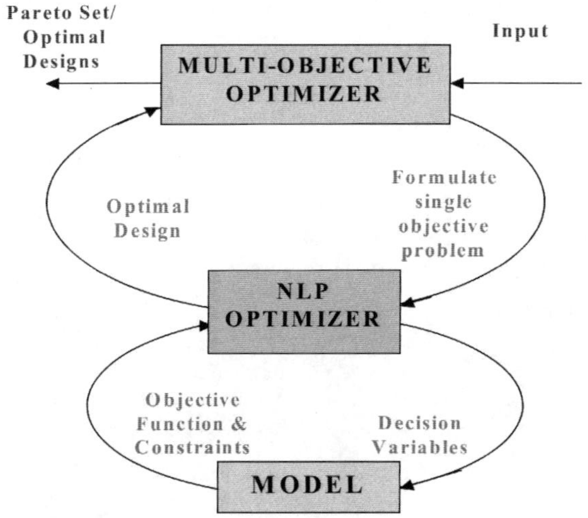

Figure 3. : Multi-objective optimization framework

chosen between the upper, $Z_L(j)$ and lower bounds $Z_U(j)$

leads to the Pareto set. The multi-objective optimization algorithm, MINSOOP[9] used in this work uses the Hammersly sequence sampling to generate combinations of the right-hand-side. MINSOOP exploits the n-dimensional uniformity of the HSS technique to obtain greater efficiency. Figure 2 shows how this MINSOOP algorithm improves efficiency for a simple, nonlinear, convex optimization problem, as the number of objectives increases

Deterministic Framework - A deterministic multi-objective optimization formulation was used to obtain the Pareto surface without accounting for uncertainty. Figure 3 shows a brief schematic of the deterministic multi-objective optimization framework to address this problem with 7 objective functions. The inputs are given to the multi-objective optimizer, which formulates single objective NLP optimization problems based on the new efficient low discrepancy Hammersley sequence sampling technique. These problems are passed to the NLP optimizer (SQP optimizer in ASPEN Plus), which goes through the model repeatedly until an optimum solution is reached and then passes back the optimum value of the objective functions and the decision variables back to the multi-objective optimizer. The output of this framework is the Pareto set of solutions

Table 3. : Bounds for all objectives : a deterministic pay-off table

OBJECTIVES	MAX ACEFF	MIN ACEFF	MIN CAP	MIN COE	MAX CO2EM	MIN CO2EM	MAX CDPEM	MIN CDPEM	MIN CDSOFC
PWRTG	**1475.1856**	**1465.51**	**1568.32587**	**1469.0955**	**1634.293405**	**1470.4051**	**1500.24142**	**1496.3793**	**1471.963**
ACEFF	**0.7232446**	**0.52**	0.6008321	0.58273	0.5426174	0.7066749	0.57530437	0.7077066	0.726258777
CAP	1456.4933	1599.80938	994.783579	**993.38655**	1444.397	1281.92286	739.9451	1289.6887	**1664.6514**
COE	5.67E-02	5.63E-02	4.16E-02	4.28E-02	5.05E-02	**4.15E-02**	3.68E-02	5.15E-02	**6.28E-02**
CO2EM	0.2728659	**0.37885684**	0.3284591	0.3386584	0.36369784	0.2793638	0.343033	0.27885679	**0.2717334**
CDSOFC	101.86	672.009	531.131	616.0684	**737.209689**	157.6117844	678.179	149.0192	**76.34382**
CDPEM	**290.3345**	304.847	**308.62476**	307.62895	294.63637	294.38629	319.5917	287.22287	293.602876
UTIL	0.69993	0.55147	0.43617	0.40266	0.41114	0.7	0.4	0.7	0.7
PPEM	23.3115	36.8556	39.7391	36.189	20.1208	26.3631	75	20	25.3265
ERAT	1.37434	5.69168	1.25	1.25	3.01218	1.79611	1.88651	1.70141	1.25
FUEL	20.1683	27.8188	25.8102	24.9279	29.7814	20.5743	25.7853	20.9073	20.0408
AIR	189.276	851.883	137.29	122.408	359.824	252.367	189.831	242.929	171.08

which gives the trade-offs between objectives. As a first step towards obtaining the whole Pareto set, the bounds for all the objective functions are generated, this is the payoff table. A payoff table is widely used in decision analysis, where it specifies the alternatives, acts, or events. Especially in MOP, a payoff table shows a potential range of values of each objective. In more detail, a payoff table contains individual objective values (Zk*) for single optimization problems (k), and also provides potential ranges of the objectives on the Pareto surface (i.e., ZL to ZU). In this way, an approximated range of the right-hand-side Lk in the Pareto surface is determined. These are presented in Table 3. It has been found that the Capital cost (CAP) and Cost of electricity (COE) follows the same trend so these objectives can be combined in one. Again, the SOFC current density and CO2 emissions are seen to have the same trend and can be combined as one while the PEM current density can be dropped because of not much change. The reduced objective set is then used to obtain the complete Pareto Set of a power plant of fixed capacity, i.e. 1472 kW. The algorithmic framework shown in Figure 3 is used to obtain the complete Pareto set using only 150 single optimization problem instead of 10,000 optimization problems required by other algorithms. The results are plotted in Figures 4.1- 4.2. These plots provide great insights into the design especially in the face of such diverse objectives. Figure 4.1 gives the trade-offs between Overall efficiency, CO2 emissions and Capital cost which is in the vertical axis. Interpretation of the graph is thus : if the decision maker wants a high overall efficiency and low CO2 emissions, he will have to pay a high capital cost which corresponds to the

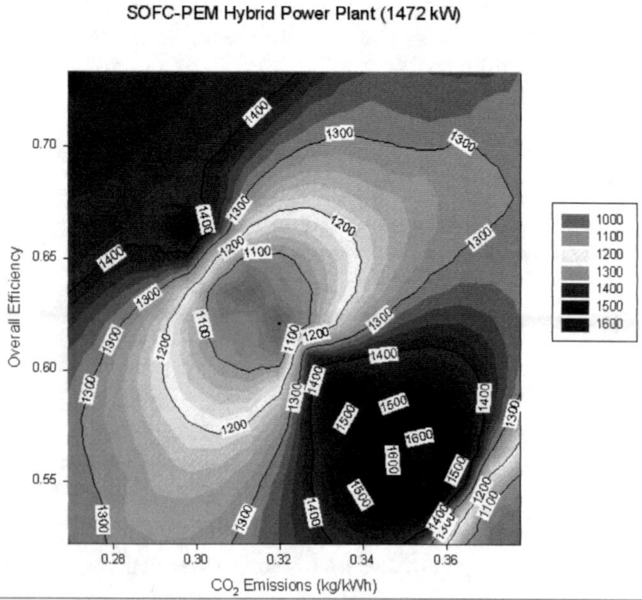

Figure 4.1 : Pareto Set showing trade-offs between CO2 emissions, overall efficiency and capital cost

region at the top-left part of the graph. He can operate at low cost, while having moderate efficiency and CO2 emissions . Hence, these are good operating regions. Figure 4.2 shows the trade-offs between overall efficiency, SOFC current density and capital cost. There are low cost regions with

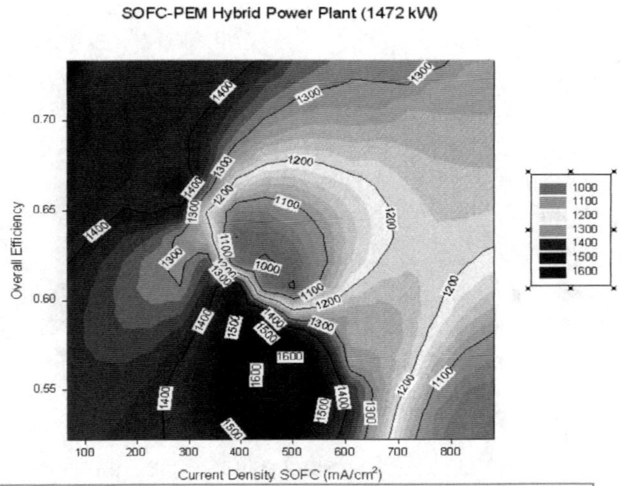

Figure 4.2 : Pareto set showing trade-offs between SOFC current density, overall efficiency and capital cost

moderate efficiency and SOFC current density and moderate cost regions with high efficiency and current density. What operating point to work in is the decision maker's choice.. There are some low cost regions at high SOFC current density but involve low efficiency and high emissions. Another major low cost region is with SOFC current density between 350-500 mA/cm2, overall efficiency between 60-65% and CO2 emissions between 0.30 and 0.32 kg/kWh. The decision maker can also identify operating regions which have to be avoided, for eg. in Figure 4.1 there are regions of high capital cost but low efficiency and high CO2 emissions. These plots provide a flexibility to the decision maker as to where he/she wants to operate and also identifies quantitatively similar design groups. These contour graphs have been plotted using the results from deterministic MOP framework.

Stochastic Framework – This framework is used to account for uncertainty in the data. The first step in uncertainty analysis is characterization. We have characterized uncertainty as the ratio between the experimental current density data and the current density calculated by the model. The next step is quantification. Uncertainties can be quantified in terms of probability distributions. We quantified the uncertainties in SOFC current density as triangular distributions and that in PEMFC current density as log-normal distribution. Once probability distributions are assigned to the uncertain parameters, the next step is to perform a sampling operation from the multi-variable uncertain parameter domain. For this we used an efficient sampling technique – Hammersely sequence sampling (HSS) developed by our

Table 4: Bounds for different objectives : a stochastic Payoff table

OBJECTIVES	MAX ACEFF	MIN ACEFF	MIN CAP	MIN COE	MAX CO2EM	MIN CO2EM	MAX CDP	MIN CDP	MIN CDS
PWRTG	1467.3817	1495.6275	1471.4098	1471.624	1467.3298	1472.9992	1475.6765	1469.188	1468.2194
ACEFF	0.7284426	0.5106597	0.5921345	0.5824547	0.4957409	0.7286441	0.5831194	0.7246442	0.7317909
CAP	1716.7441	1479.569	990.11317	994.29749	1611.0741	1645.4066	740.39783	1692.2209	2407.4501
COE	6.33E-02	5.30E-02	4.28E-02	4.28E-02	5.62E-02	6.21E-02	3.72E-02	6.35E-02	7.91E-02
CO2EM	0.2709188	0.3864585	0.3666121	0.3388226	0.3980886	0.2708439	0.3384363	0.2723389	0.2696792
CDSOFC	100.4153	970.2553	660.79064	720.6015	1006.6466	92.311412	754.07279	87.39595	86.283898
CDPEM	254.44981	255.9232	253.00722	253.85462	255.99008	247.36991	263.64304	236.90655	255.62836
UTIL	0.69	0.4	0.48	0.4	0.43	0.7	0.4	0.7	0.7
PPEM	29.02	20	36.78	36.23	20	33.03	74	20	64.85
ERAT	1.25	2.8	1.25	1.25	4.96	1.37	2.01	1.3	1.25
FUEL	20.2	27.7	23.6	27.04	29.4	20.12	25.69	20.33	20.03
AIR	172.16	296.37	146.6	132.38	611.27	189.02	202.25	179.92	171.06

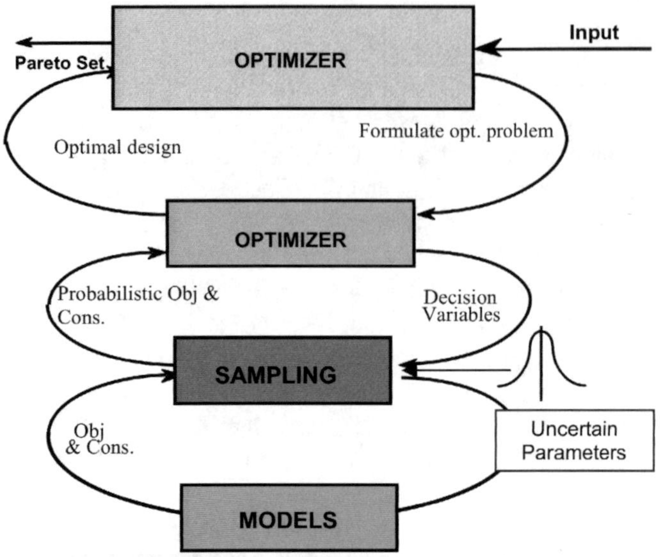

Figure 5 : Stochastic optimization framework

group[5][6]. To study how uncertainties change the multi-objective designs, we have used the expected value of each function calculated by sampling the uncertain variables. The conceptual framework for this stochastic optimization is shown in Figure 5 where the deterministic model in Figure 3 is replaced by the stochastic modeling framework which has an additional sampling loop. As was done before, the bounds for all the objective functions are generated, this is the payoff table shown in table 4. The HSS scheme ensures that the sample set is more representative of the population, showing uniformity properties of random variables in multi-dimensions, unlike Monte Carlo, Latin hypercube[7] (LHS), and its variant, the Median Latin hypercube sampling technique. The stochastic framework shown in Figure 5 is used to obtain the Pareto surface giving the trade-offs between the objectives. The surfaces are shown in Figure 6.1-6.2.There is a considerable difference in the contour shape and levels between the stochastic and deterministic case. This can be attributed to the fact that the optimum decision variables are different for the two cases as is evident from tables 3 and 4. For example, for the maximum overall efficiency case, airflow is higher for stochastic case than the deterministic case. Minimum efficiency designs have higher current density for stochastic case and lower airflow. This decreased the capital cost in the stochastic case more in the stochastic case than in the deterministic case which changed the trade-off surface. Similarly in case of maximum CO2 emission designs, the airflow in the stochastic case is almost double of the deterministic case, hence this has increased the capital cost thereby changing the pareto surface.In case of

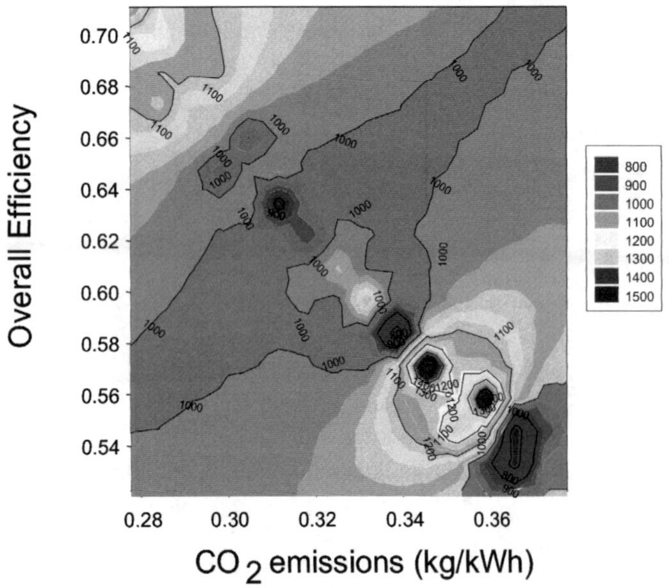

Figure 6.1 : Stochastic Pareto surface for Overall efficiency, CO2 emissions and capital cost

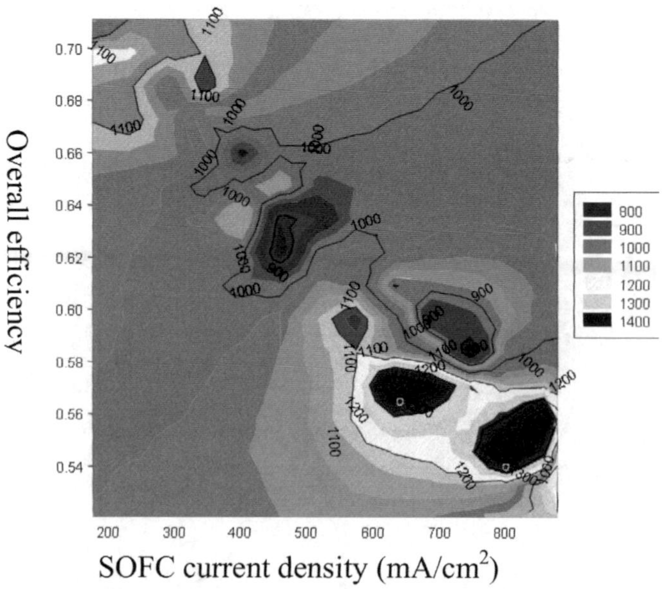

Figure 6.2 : Stochastic Pareto surface for SOFC current density, Overall efficiency and capital cost

Figure 6.1, the costs overall have shifted towards lower levels. The high cost regions have shrunk at some regions and at other regions have disappeared all together. The low cost regions are more distributed over the surface than being concentrated at the centre as was in the deterministic case. In case of Figure 6.2, though the costs have generally decreased, the lowest cost regions have shrunk in size. Low costs regions in the bottom right region of the deterministic surface have been converted to high cost regions. Good operating regions can be identified at the centre of the plot with moderate efficiency and current density. High cost regions with high efficiency but low current density are also seen.

SUMMARY

This paper presented a hybrid fuel cell fuel cell power plant with multiple objectives to be optimized. A deterministic multi-objective optimization framework was presented and the resulting trade-off surface was analyzed. The characteristics of the surface were explained and their use in predicting good operating regions were discussed. A stochastic multi-objective optimization framework to incorporate uncertainties into the model was presented. Finally, the effect of uncertainty on these surfaces was shown through comparison of surfaces obtained through deterministic and stochastic multi objective optimization.

ACKNOWLEDGMENTS

We would like to thank Amit Goyal for previous work Thanks also to the National Energy Technology Laboratory (NETL) for their funding of this project and for the base case model.

REFERENCES

1. Geisbrecht, R.; Theory based process modeling for evaluation of fuel cells in advanced energy systems. *In Fuel Cells Topical Conference, AIChE, Spring 2001 National Meeting;* New Orleans, LA, 2001.
2. Goyal, A.; Diwekar, U.; Geisbrecht, R. Multi-objective for hybrid fuel cells power systems design. *In Fuel Cells Topical Conference, AIChE, 2001 Fall Meeting*; Reno, NV, 2001.
3. Goyal, A;. *Multi-objective optimization for effcient environmental friendly design of benchmark and new processes"*, Master's thesis, Carnegie Mellon University, 2002.
4. Diwekar, U. M.; *Process design for environment problems. In Tools and Methods for Pollution Prevention*; Sikdar, S. K.; Diwekar, U. M., Eds.; NATO Science Series Environmental Security Kluwer Academic Publishers: Dordrecht, Netherlands, 1999.
5. Kalagnanam, J. R.; Diwekar, U. M. *Technometrics* 1997, 39, 308-319.
6. Diwekar, U. M.; Kalagnanam, J. R. *AIChE J.* 1997, 43, 440-447.
7. Diwekar, U. M. *Introduction to Applied Optimization*; Kluwer Academic Publishers: Dordrecht, 2003.
8. Hwang C. L., S. R. P.; Masud, A. S. M. *Computers and Operations Research* 1980, 7, 5-31.
9. Fu, Y.; Diwekar, U. M., *An Efficient Sampling Approach to Multi-objective Optimization,* Annals of Operations Research (accepted) 2003,
10 AspenTech, *Aspen Plus Documentation Version* 11.1-0; AspenTech: Cambridge, MA, 2002.

Meniscus Shape, Position and Stability in Straight Pores

Carlos Amador, Panagiota Angeli and Asterios Gavriilidis
University College London (UCL)
Torrington Place, London WC1E 7JE, UK

John Shaw and David Wenn.
Central Research Laboratories (CRL)
Dawley Road, Hayes, Middlesex, UB3 1HH, UK

ABSTRACT

In this paper a model is presented for determining the position and stability of interfacial menisci within cylindrical and polygonal pores in thin solid layers that separate two fluids. Stability is defined as the range of pressure differences across the interfacial meniscus that the interface can cope with without one phase breaking through into the other (convective mixing). The model is based on the variation of the system contact angle due to hysteresis and on meniscus rotation at the round pore ends. The meniscus area is also estimated as a function of contact angle, pore size and geometry. Experimental work carried out in single cylindrical and square capillaries confirmed the assumption of a 90-degree rotation of the meniscus at the round capillary ends while the measured range of pressure difference the meniscus could cope with without breakthrough agreed with the predicted one.

KEY WORDS

Meniscus; stability; mesh; pore; polygonal; microreactor.

INTRODUCTION

Multiphase operations in microscale systems may require different approaches than in conventional macrosystems to ensure high interfacial area, ease of phase separation and flexibility. Since separation by gravity in these scales is not very effective, alternative ways have been used to bring in contact the phases in multiphase systems. One such way is to have the phases flowing in different channels separated by a thin solid layer with holes through which the two phases can come in contact [1,2]. The fluid layers must be thin for efficient mass transfer. The solid layer should be thin enough to reduce species diffusion time in it and have large open area to increase interfacial mass transfer. At the same time it should have structural stability. The geometry of the openings in the solid layer (pores), pressure difference between the phases, interfacial tension and contact angle determines the shape of interfacial area available for mass transfer and the position of the interfacial meniscus within the pore. To avoid mixing of the two phases (breakthrough), the meniscus should remain within the solid pores for the range of pressure differences between the two phases established during operation (stability).

BACKGROUND

A straight pore can be considered as a capillary and the interface separating both phases inside each pore can be modelled by using the concept of capillary rise [3]. Figure 1 represents the classical capillary effect in the left side and the real situation found in straight pores in the right side. The system consists of two phases, a wetting phase, *W*, which has the smaller contact angle with the solid material and is considered to be below the mesh, and a non-wetting one that is above the mesh, *NW*.

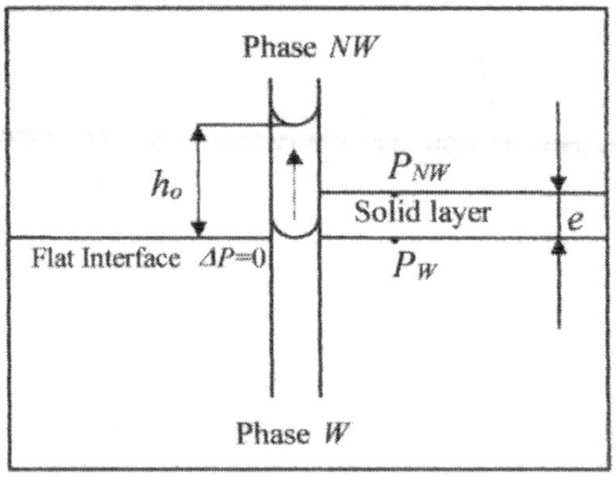

Figure 1. Capillary rise model inside the pores.

If a capillary is introduced in a two-phase system the interface inside the capillary curves as a function of the wettability properties of the system. The shape of the interface depends on the contact angle of the phases with the solid wall, θ, the capillary geometry and gravity, whose relative effect is a function of the capillary radius and phase properties. Initially, the interface is not in equilibrium since the pressure in each side of the interface is the same and it is known that the pressure difference across a curved interface in equilibrium is given by the Young-Laplace equation, which applies for *each point at the interface* [3,4],

$$\Delta P = \gamma \cdot \left(\frac{1}{R_1} + \frac{1}{R_2} \right) = \gamma\kappa \qquad (1)$$

where γ is the interfacial tension, R_1 and R_2 are the radii of curvature for each point at the interface and κ is the surface curvature at the specific point. As a consequence, the meniscus rises inside the capillary until this pressure difference between the phases, is reached. The meniscus apex reaches the equilibrium height h_0 with respect to the flat interface and the pressure difference across the apex, ΔP_0, which is found from Equation (1), is also given by Equation (2).

$$\Delta P_0 = \left(P_{NW,I} - P_{W,I}\right)_0 = \left(\rho_W - \rho_{NW}\right) \cdot g \cdot h_0 \qquad (2)$$

where ρ is the density, g is the acceleration of gravity, the subscript I refers to the meniscus interface and h_0 is the capillary rise. The pressure difference across the rest of the surface, ΔP, will vary as a function of the height to the apex. If a solid layer with openings separates both phases (right side in Figure 1), the pressure in each phase can be independently modified. The two phases are only in contact via the interfacial menisci inside the pores, which are assumed to be identical to each other. Pressure changes in one of the phases (P_{NW} or P_W) will move the meniscus in the direction that makes the pressure difference across the meniscus interface equal to that given by the Young-Laplace equation, Equation (1). The pressure difference across the meniscus apex can be expressed as a function of the phase pressures,

$$\Delta P_0 = \left(P_{NW,I} - P_{W,I}\right)_0 = \left(P_{NW} - P_W\right) + \left(\rho_W - \rho_{NW}\right) \cdot g \cdot h_I + \rho_{NW} \cdot g \cdot e \qquad (3)$$

where P_{NW} and P_W are the pressures in each phase at the solid layer surface (see Figure 1), h_I is the height of the meniscus apex from the lower pore end and e is the pore length. Therefore, in order to keep the meniscus inside the pore avoiding the breakthrough of one phase into the other, it is necessary that the pressure difference between the two phases, $(P_W - P_{NW})$, remains within the values given by Equation (3) where h_I can only varies approximately from 0 to e. Assuming a constant value of ΔP_0 independent of the meniscus position, the available range of pressure differences through the mesh (stability), that is the pressure difference when the meniscus is at the non-wetting phase pore end minus the pressure difference when the meniscus is at the wetting phase pore end, is given by Equation (4).

$$\left(P_{NW} - P_W\right)_{NW\,end} - \left(P_{NW} - P_W\right)_{W\,end} = \left(\rho_{NW} - \rho_W\right) \cdot g \cdot e \qquad (4)$$

Considering a typical pore length e = 5 - 200 μm, a very precise control of pressures would be necessary to keep the menisci inside the pores. However, ΔP_0 can vary inside the pore due to the contact angle hysteresis and the round pore ends, allowing a larger range of pressure differences across the meniscus interface. The meniscus shape and area as well as the range of pressure difference it can cope with without breaking through to one or the other phase (stability) are analysed below.

STABILITY MODEL IN CYLINDRICAL PORES

MENISCUS SHAPE AND AREA IN CYLINDRICAL PORES – The pressure difference across the interfacial meniscus, together with the pore geometry defines the meniscus shape and therefore the contact area between phases. The pressure difference at any point of the meniscus surface, $\Delta P = P_{NW,I} - P_{W,I}$ can be expressed as a function of the pressure difference across the meniscus apex plus a hydrostatic term, assuming a concave meniscus since the wetting phase is below the non-wetting phase,

$$\Delta P = \Delta P_0 + \left(\rho_W - \rho_{NW}\right) \cdot g \cdot y \qquad (5)$$

where y is the height from the meniscus apex to the point considered. To find ΔP from Equation (1), it is necessary to know the two radii of curvature R_1 and R_2 that describe the curved surface at any point. In cylindrical pores, the meniscus is axisymmetric, which means that the 3D shape is generated by rotation of a 2D $y(x)$ profile around the y-axis. According to Figure 2, R_1 is the radius of curvature of the profile at the point in question, P, on the paper plane. R_2 must be chosen on the plane perpendicular to that of the paper and, for axisymmetric figures, it is given by extending the normal to the profile until it cuts the axis of revolution (y-axis).

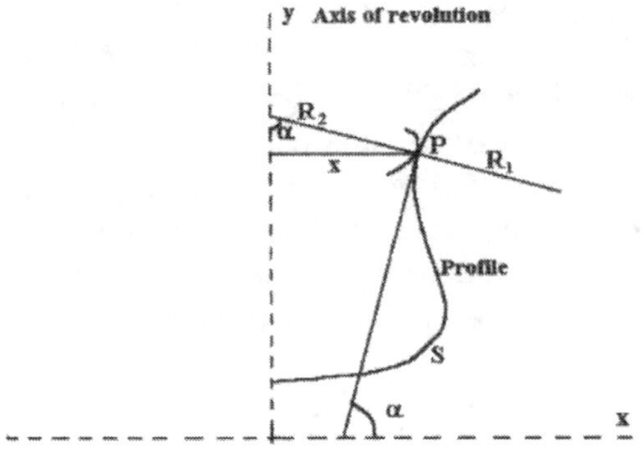

Figure 2. Radii of curvature for a 2D axisymmetrical surface.

The definition of radius of curvature R_1, at the point *P*, is given by Equation (6), where α is the angle of the tangent to the profile in the point *P* with the *x*-axis and *S* is the arc length of a α range, as given by [5].

$$\frac{1}{R_1} = \frac{\partial \alpha}{\partial S} = \frac{\partial x}{\partial S}\frac{\partial \alpha}{\partial x} \tag{6}$$

Considering that $\partial x/\partial S = \cos\alpha$, Equation (6) becomes Equation (7).

$$\frac{1}{R_1} = \frac{\partial \sin\alpha}{\partial x} \tag{7}$$

For an axisymmetric shape (3D), the second radius of curvature, R_2, can be simply written as a function of the co-ordinates *x* and *y* according to Figure 2, obtaining Equation (8).

$$\frac{1}{R_2} = \frac{\sin\alpha}{x} \tag{8}$$

Substituting Equations (5), (7) and (8) into Equation (1), substituting ΔP_0 by Equation (2) and applying a dimensionless analysis with the pore radius as the characteristic length, Equation (9) is obtained. Equations (10) and (11) complete the set of three ordinary differential equations giving the relationship between *x*, *y* and α with *S*, which is solved by Matlab. The Bond number, *Bo*, which represents the ratio of gravitational to surface tension forces, is given by Equation (12).

$$\frac{d\alpha}{dS^*} = Bo(h_0^* + y^*) - \frac{\sin\alpha}{x^*} \tag{9}$$

$$\frac{dx^*}{dS^*} = \cos\alpha \tag{10}$$

$$\frac{dy^*}{dS^*} = \sin\alpha \tag{11}$$

$$Bo = \frac{(\rho_W - \rho_{NW})gr^2}{\gamma} \tag{12}$$

This system of ODEs is subjected to the initial conditions ($S = 0$, apex of meniscus): $x = 0$, $y = 0$ and $\alpha = 0$. As a boundary condition, the angle α at the pore wall ($x = r$) is given by $\alpha = 90°-\theta$, where θ is the contact angle of the wetting phase with the solid wall. Figure 3 shows profiles at different *Bo* numbers in a system with $\theta = 0°$.

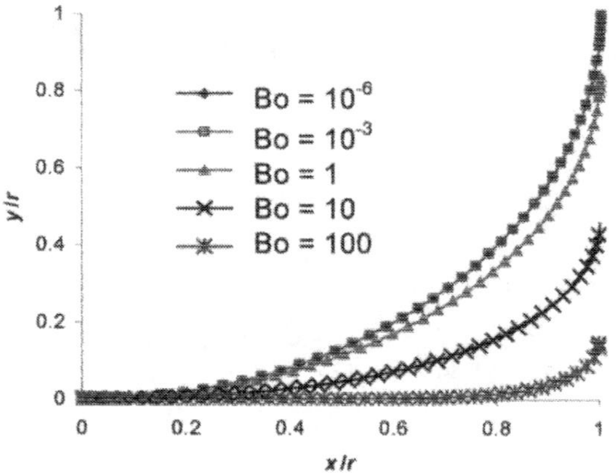

Figure 3. Meniscus profiles in a cylindrical pore for different *Bo* number at zero contact angle.

As Figure 3 shows, spherical meniscus profiles are obtained for $Bo \ll 1$. In an extreme system with very low interfacial tension $\gamma = 0.001$ N/m and large density difference, ($\rho_{NW} - \rho_W$)= 5000 kg/m³, pore diameters smaller than $d = 90$ μm give $Bo < 0.1$.

Therefore, generally the system can be assumed not to be influenced by gravitational effects and a constant curvature applies throughout the meniscus as $\Delta P = \Delta P_0$, which in cylindrical pores implies a spherical meniscus ($R_1 = R_2 = r/\cos\theta$) with Equation (1) becoming Equation (13) according to Figure 4.

$$\Delta P = \gamma \cdot \left(\frac{1}{R_1} + \frac{1}{R_2}\right) = \frac{2\gamma\cos\theta}{r} \tag{13}$$

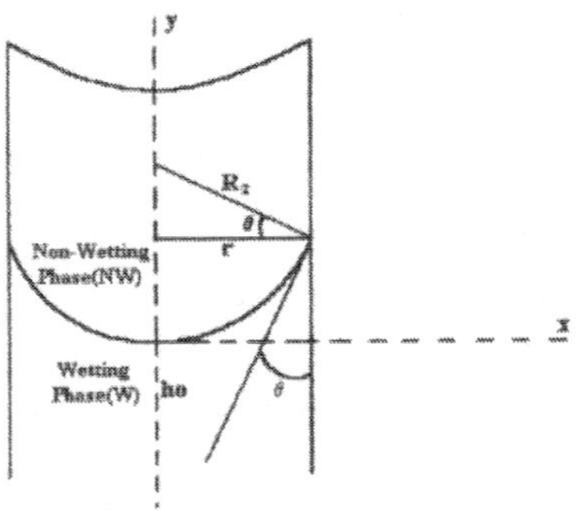

Figure 4. Relation between radius of curvature $R_1 = R_2$ for a gravity-free meniscus, pore radius r and contact angle θ in a cylindrical pore.

The interfacial contact area, A, defined by the meniscus is found from Equation (14) for the area of a surface generated by a 2D profile revolution [6]. This equation can also be used when the meniscus is not spherical. For a spherical meniscus, $y = R-(R^2-x^2)^{1/2}$, Equation (14) results in Equation (15).

$$A = 2 \cdot \pi \cdot \int_{y_0}^{y_F} x \cdot \sqrt{1+(dx/dy)^2}\, dy \tag{14}$$

$$A = 2\pi \frac{r^2}{\cos^2\theta}(1-\sin\theta) \tag{15}$$

MENISCUS POSITION AND STABILITY IN CYLINDRICAL PORES - Interface stability is defined as the available range of pressure differences that the interface can cope with, without breakthrough of one phase into the other. The next considerations are taken into account:

- The wetting phase is below the mesh, although the opposite case, i.e. non-wetting phase below the mesh, would be similar.
- The pores are supposed to have smooth surfaces so that the contact angle is uniform around the pore wall. For non-planar surfaces, wetting properties are poorly understood [7].
- Negligible static pressure variation due to gravity is considered inside the pore, although Equation (3) can be applied to take into account this term. The meniscus is also assumed to be spherical although the same concepts discussed below can be applied for a non-spherical meniscus, determining the ΔP across the meniscus via Equations (9-11).
- Contact angle hysteresis: Almost all systems present this phenomenon that is generally attributed to surface roughness, surface heterogeneity and solution impurities on the surface or swelling, rearrangement or alteration of the surface by the solvent [3]. The contact angle measured for a liquid advancing across a surface, θ_A, exceeds that of the liquid receding from the surface, θ_R and all intermediate contact angles correspond to metastable equilibrium states [8]. Therefore, an interface remains static by changing its curvature when contact angles vary from θ_A to θ_R. This change in curvature allows the interface to support larger pressure changes avoiding breakthrough (increase of stability). The difference between receding and advancing contact angles can be quite large in some systems [9].
- The corners at the end of a pore can be either *sharp* or *round*; the latter should be closer to reality (see Figure 5). The concept of *virtual contact angle* θ_V is introduced when the interface is at a round pore end, to define the angle between the interface and the vertical line that prolongs the inner pore wall. For theoretically sharp corners the virtual contact angle can only change by 90° in one single step, which takes the meniscus out of the pore resulting in breakthrough. For round corners the virtual contact angle θ_V can smoothly change up to 90° before the meniscus enters the other phase (as it is shown in Figure 5), which allows the meniscus to change its curvature while keeping the spherical shape. When referring to virtual contact angles, a subscript that denotes the angle of the tangent of the interface in the corner with respect to the inner pore wall is used. For example, $\theta_{A,15}$ stands for a virtual contact angle with respect to the inner pore wall of θ_A + 15°. It should be stressed here that the real contact angle of the system only changes between θ_A and θ_R.

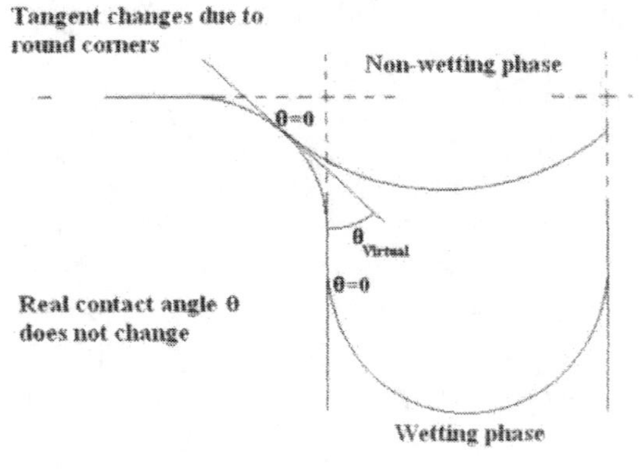

Figure 5. Virtual contact angle change at the pore end. The round corner is exaggerated for illustration.

- The limiting meniscus curvature at which one phase breaks into the other is assumed to occur at θ_V = 180° (non-wetting end) or θ_V = 0° (wetting end). These values correspond to a hemispherical meniscus where the largest absolute values of the curvature κ are obtained.

- Sharp corners without contact angle hysteresis:

In this case, the meniscus cannot rotate at the pore ends and the virtual contact angle can only take the exact values of θ and θ_{90} or θ_{-90} and no intermediate ones. The available range of pressure differences the interface can cope with is given by Equation (4) due to hydrostatic changes and because of the small pore height, e, this range is very small. Since the interface cannot rotate at the pore ends it will breakthrough to one or the other phase as soon as it reaches the respective pore end.

- Sharp corners with contact angle hysteresis:

The contact angle must reach either the receding value θ_R (movement towards the wetting phase) or the advancing one θ_A (movement towards the non-wetting phase) before the interface starts to move at any part of the pore, that is $\theta_R < \theta < \theta_A$. The interface can cope with pressure difference changes by changing its curvature from $2\cdot\cos\theta_A/r$ to $2\cdot\cos\theta_R/r$. Neglecting static pressure inside the pore, the stability range is given by Equation (16), where the subscripts *A* and *R* stands for advancing and receding respectively.

$$(P_{NW}-P_W)_A-(P_{NW}-P_W)_R=\frac{2\cdot\gamma\cdot(\cos\theta_A-\cos\theta_R)}{r} \qquad (16)$$

- Round corners without contact angle hysteresis:

Round corners allow the virtual contact angle θ_V to change gradually within 90° at the pore ends. In the non-wetting phase end, θ_V can change up to θ_{90}, which can give a negative curvature before breakthrough of the wetting phase into the non-wetting phase occurs. For example, with a contact angle of θ = 45°, the virtual contact angle can change up to θ_{90} = 135° at the non-wetting phase end. At this pore end, the maximum virtual contact angle is 180°, for the case of θ = 90° (maximum). In the wetting phase end, θ_V can change up to $\theta_{-\theta}$ = 0 (hemisphere-shaped meniscus). The maximum change of θ_V is given by $(\Delta\theta_V)_{Max}$ = 90° + θ. The pressure difference variation that the interface can cope with can be calculated by Equation (16) using θ_{90} and $\theta_{-\theta}$ (0°) rather than θ_A and θ_R.

- Round corners with contact angle hysteresis:

This case is similar to the previous one but considering that there is no movement of the interface until either θ_A or θ_R is reached. If θ_A < 90°, in the non-wetting phase pore end θ_V can change from $\theta_{A,0}$ to $\theta_{A,90}$. In the wetting phase end θ_V can change from $\theta_{R,0}$ to 0°. It must be taken into account that it is possible to find systems where θ_A > 90° when θ_R < 90°. Although these liquids would not normally be considered as wetting, because they do not advance spontaneously on the pore surface, the same concepts can be applied to them. In such a case the phase with the smallest θ_A value is called the wetting phase and in the non-wetting phase end θ_V can change from $\theta_{A,0}$ to 180°. The maximum change of θ_V when θ_A < 90° is given by $(\Delta\theta_V)_{Max}$ = 90° + θ_A and for θ_A > 90°, $(\Delta\theta_V)_{Max}$ = 180°. Equation (16) still can be used to determine the stability of the system using $\theta_{A,90}$ (or 180° if θ_A > 90°) and 0°.

Real systems are closer to the last case above (round corners and contact angle hysteresis), which fortuitously give the largest available range of pressure differences between the phases. Equation (15) that gives the meniscus area can be written as a function of the pressure difference across the interface, ΔP, obtaining Equation (17), where the values of ΔP are restricted to the virtual contact angles explained above.

$$A=2\cdot\pi\cdot\left(\frac{2\cdot\gamma}{\Delta P}\right)^2\cdot\left(1-\sqrt{1-\left(\frac{\Delta P\cdot r}{2\cdot\gamma}\right)^2}\right) \qquad (17)$$

As an example, the following system is considered: γ = 0.03 N/m; θ_A = 45°, θ_R = 15° and r = 5 μm. Figure 6 represents meniscus area and position versus pressure difference considering that the virtual contact angle can change from 0° to $\theta_{A,90}$ = 135°. At contact angles $\theta_R < \theta < \theta_A$, the meniscus is pinned at any part of the pore, at $\theta < \theta_R$ the meniscus is at the wetting end and at $\theta_V > \theta_A$ the meniscus is at the non-wetting end. As the contact angle gets close to 0° or 180°, small changes in the pressure difference modify the meniscus area and curvature significantly. Working under these conditions can result in meniscus instabilities and the full meniscus rotation at the pore end may be difficult to achieve.

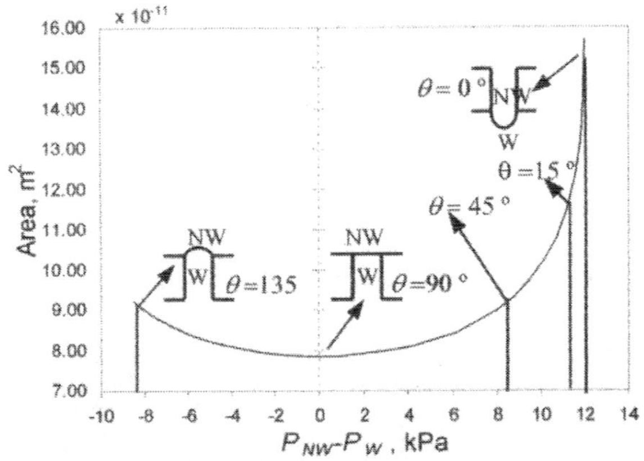

Figure 6. Meniscus area and position as a function of the pressure difference between phases (P_{NW}-P_W).

STABILITY MODEL IN POLYGONAL PORES

The analysis of polygonal pores has two objectives. Firstly, to give a more realistic view of the interface stability model as typical microstructures do not necessarily have cylindrical pores [1]. Secondly, to identify the most suitable pore geometry in relation to meniscus stability and contact area.

MENISCUS SHAPE AND AREA IN POLYGONAL PORES – In polygonal pores two different types of meniscus can appear: the main terminal meniscus (MTM) that is always present and the arc (or wedge) menisci (AMs) at the corners [9], which appear at certain contact angles (see Figure 7).

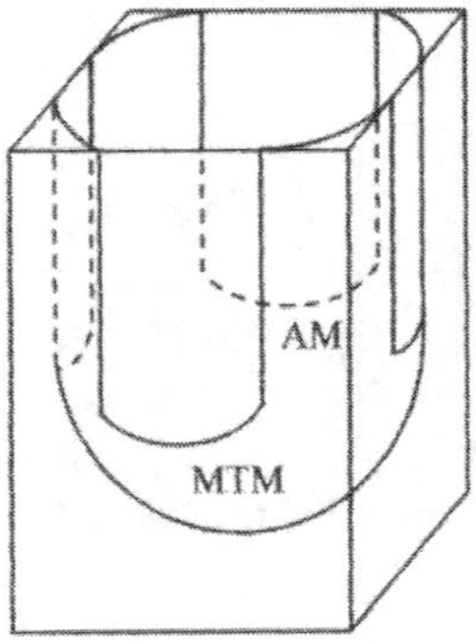

Figure 7. Types of menisci inside a square pore.

MTM is the region that spans and fills the tubular space of the pore. Under gravity absence and certain contact angles, a cross section of the meniscus interface within the pore will show an arc of a circle in each corner, R_{Arc}, which represent the arc menisci (AMs). These have an infinite radius of curvature in the axial direction and therefore their curvature is expressed as $\kappa_{AM} = 1/R_{Arc}$. Negligible gravity effect is the basis for the Mayer-Stowe and Princen method [10] used to solve the mean curvature κ of complex menisci in which the mean curvature of the main terminal meniscus is equated to that of the arc menisci, that have only one radius of curvature, $\kappa = \kappa_{AMs}$. Carrying out a work balance to displace the meniscus by dx, Equation (18) is obtained, where A' is the main terminal meniscus area projected in a perpendicular plane, $p_{NW\text{-}S}$ and $p_{W\text{-}S}$ are the contact perimeters of the non-wetting and wetting phases respectively with the pore solid walls (identical in this case) and p_L is the contact perimeter of the interface with the arc menisci.

$$\Delta P \cdot A' \cdot dx = \left(\gamma_{NW-S} \cdot p_{NW-S} - \gamma_{W-S} \cdot p_{W-S} + \gamma \cdot p_L\right) \cdot dx \qquad (18)$$

Although the balance in Equation (18) applies to the main terminal meniscus, the pressure difference across the interface ΔP is solved using the curvature of the AMs. The surface tensions of the wetting and non-wetting phases, $\gamma_{W\text{-}S}$ and $\gamma_{NW\text{-}S}$ respectively, are related to the interfacial tension of the system by Young's equation,

$$\gamma \cdot \cos\theta = \gamma_{NW-S} - \gamma_{W-S} \qquad (19)$$

Equation (18) is reduced to Equation (20) where p' is given by $p' = p_S \cos\theta + p_L$ and r_h is the distance from the centre of the polygonal tube to the middle of any side and is used to non-dimensionalise the mean curvature, $\kappa^* = \kappa r_h = r_h/R_{Arc}$.

$$\kappa^* = \frac{r_h p'}{A'} \qquad (20)$$

Considering the geometrical relationships in an n-side polygonal tube, Equation (20) results in Equation (21) that is used to solve the mean meniscus curvature at $\theta \leq (90° - \beta)$ [11] where AMs are present. β is half the angle of the polygon corner given by Equation (22). Therefore, the arc menisci exist as a function of the wettability properties and number of sides in the pore.

$$\kappa^* = \cos\theta + \left[\frac{\pi/2 - \beta - \theta + \sin\theta \cdot \cos\theta}{\tan(\pi/2 - \beta)}\right]^{1/2} \qquad (21)$$

$$\beta = 90 \cdot \left(\frac{n-2}{n}\right) \qquad (22)$$

At $\theta > (90° - \beta)$, arc menisci do not exist and the curvature of the main terminal meniscus, κ^*, is solved by Equation (20), where $A' = n \cdot r_h^2 \cdot \tan\beta$ (polygon area), $p_S = 2n \cdot r_h \cdot \tan\beta$ (n-side polygon perimeter) and $p_L = 0$, both expressed as a function of the in-circle radius, r_h. The dimensionless mean curvature expressed by Equation (23) corresponds to that of a spherical meniscus of in-circle radius $R = r_h/\cos\theta$, where the liquid-liquid interface enters each tube corner.

$$\kappa^* = \frac{2nr_h^2 \tan\beta\cos\theta}{nr_h^2 \tan\beta} = 2 \cdot \cos\theta = \frac{2r_h}{R} \qquad (23)$$

The meniscus in a polygonal pore at contact angles $\theta > (90° - \beta)$ can be drawn by cutting a sphere of radius $R = r_h/\cos\theta$ with a polygonal domain as Figure 8 shows for a hexagonal pore with $\theta = 45°$ and $r_h = 5.20$ μm. The contact angle along the contact line of the meniscus with the pore wall, θ, keeps constant but 6 points of this meniscus reach the non-wetting pore end first than the rest, which can imply a different stability range due to the meniscus rotation when compared to cylindrical pores.

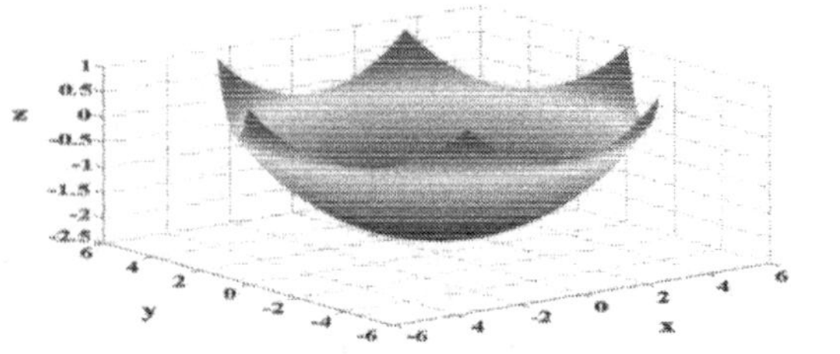

Figure 8. Meniscus in a hexagonal pore with $r_h = 5.20$ μm and $\theta = 45°$.

The interfacial area of menisci in polygonal pores at $\theta > (90° - \beta)$ is solved by a surface integral given by Equation (24) where dS is a scalar differential of the surface, **n** is a normal unit vector to the surface dS, dA is the projected area differential of dS and **k** is a normal unit vector to dA. **F** is the implicit function of a sphere $\mathbf{F} = x^2 + y^2 + z^2 = 0$.

$$\mathbf{Area} = \int_S d\mathbf{S} = \int_S \mathbf{n}\, dS = \int_S \mathbf{n} \frac{|\nabla\mathbf{F}(x,y,z)| \cdot dA}{\partial F(x,y,z)/\partial z} \qquad (24)$$

The polygonal domain is divided into simple subdomains (triangles and squares) where the surface integral is solved. Applying this method, the interfacial area of a meniscus inside a hexagonal pore is shown in Figure 9 for different contact angles $\theta > (90-\beta) > 30°$ as a function of the meniscus area in a cylindrical pore of radius r_h which is found by Equation (15).

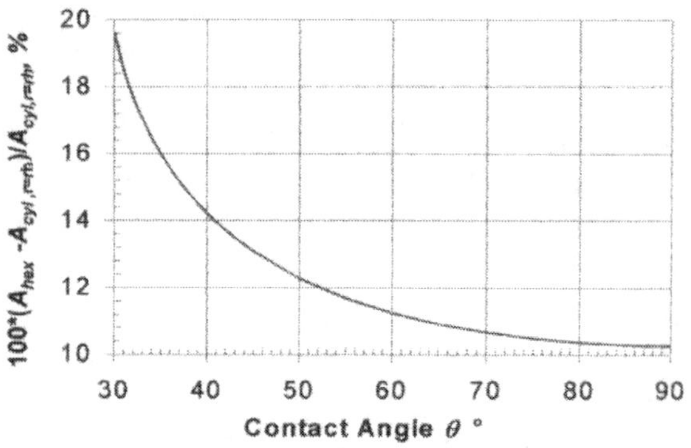

Figure 9. Interfacial area for a meniscus inside a hexagonal pore at contact angles $\theta > (90°-\beta) > 30°$.

At contact angles $\theta \leq (90°- \beta)$ AMs are formed, which decrease the mean curvature of the MTM although the contact area between phases can increase due to the length of the arc menisci in long pores. The Mayer-Stowe and Princen method allows the mean curvature at $\theta \leq (90°- \beta)$ to be calculated. However, the computation of the meniscus shape at $\theta \leq (90°- \beta)$ requires a three-dimensional analysis, as the geometry is not axisymmetric. The common boundary condition of a constant contact angle along the contact line between the meniscus and the capillary wall can be avoided using the augmented Young-Laplace Equation (25), which incorporates thin-film forces (i.e. conjoining/ disjoining pressure) that repel the meniscus at small distances from the capillary wall [12],

$$\nabla \cdot \mathbf{n} + \Pi^*(h^*) = \kappa^* \qquad (25)$$

where $\Pi^* = r_h\Pi/\gamma$ is the dimensionless conjoining/disjoining pressure isotherm and h^* is the dimensionless location of the meniscus measured normal to the wall. Therefore, a thin film is assumed to separate the meniscus from the capillary walls and the solution with a contact line along the capillary wall is retrieved when the film thickness tends to zero [12].

The area of a meniscus inside a polygonal pore at contact angles $\theta \leq (90°- \beta)$ consists of two contributions and can be approximated as follows. First, the area of the arc menisci, which is the product of the meniscus length arc by the length of the arc menisci from the lower point of the meniscus contact line with the pore wall (at $r = r_h$) to the top, l, is given by Equation (26),

$$A_{AMs} = 2n\frac{l}{r_h}\cdot(\pi/2-\beta-\theta)\cdot\frac{r_h^2}{(R_{Arc}\cdot r_h)} \quad (26)$$

where $R_{Arc} = 1/\kappa$ is found from Equation (21). The second contribution is the MTM area, which is solved assuming a spherical main terminal meniscus applying Equation (15) when $r = r_h$ and $\theta = \text{acos}(\kappa^*/2)$, whose contact angle is not the real one but the one that a spherical meniscus would have in order to produce the same curvature that the main terminal meniscus in the polygonal pore. However, in order to obtain the value κ^* via Equation (21) the real contact angle θ is used.

MENISCUS POSITION AND STABILITY IN POLYGONAL PORES – As for cylindrical pores, Equation (1) defines the pressure difference across the meniscus in equilibrium, when gravity is negligible. The normalised meniscus mean curvature or dimensionless pressure difference ($\kappa^* = \Delta P^* = \Delta P r_h/\gamma$) is constant throughout the interface and is found from Equation (21) or (23) depending on whether or not the contact angle θ is smaller or larger than (90°-β) respectively. Figure 10 shows the dimensionless pressure difference across the meniscus as a function of the contact angle and number of sides of the polygonal pore. Contact angle hysteresis would also apply to polygonal pores in the same way. For example, in a system with θ_A = 75° and θ_R = 25°, the available range of pressure differences without considering round pore ends, ΔP^*, is contained by the solid lines in Figure 10 for cylindrical and triangular pores, the latter results in a stability reduction of 11.48 %. In this case the meniscus can be at any part of the pore.

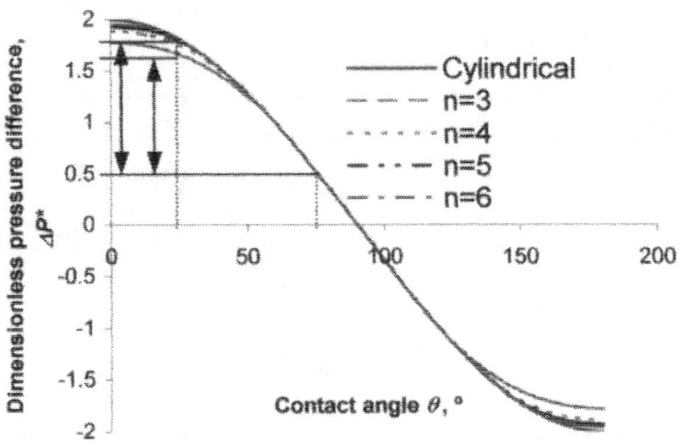

Figure 10. Pressure difference as a function of contact angle in cylindrical and polygonal (n=3, 4, 5 and 6) pores.

Considering round pore ends however, the stability of the meniscus may be different to that of cylindrical pores, since its contact line is not of uniform height in the whole pore. In a hexagonal pore as that shown in Figure 8, 6 points (equal to number of sides) reach first the round pore end and start rotating first than the rest of points at the meniscus contact line. Therefore, at these points there will be a virtual contact angle different from the contact angle at the rest of the contact line as the pressure of the wetting phase increases, which also makes other points of the contact line start rotating too. As the rotation of meniscus contact line occurs at different degrees the meniscus stability could decrease, and breakthrough might first take place at the highest points of the meniscus contact line. If the meniscus recedes towards the wetting phase a similar situation will take place but now the lower points of the meniscus contact line reach the round pore end first. When θ < (90° - β), AMs appear and they extend along the whole pore length reaching the non-wetting pore end where they are retained by the round corners. In polygonal pores, it also applies that at a virtual contact angle $\theta_R < \theta < \theta_A$, the meniscus is pinned at any part of the pore, at $\theta < \theta_R$ the meniscus is at the wetting end and at $\theta_V > \theta_A$ the meniscus is at the non-wetting end.

EXPERIMENTAL WORK

Experimental work is undertaken to investigate whether the assumption of a 90-degree rotation at the pore ends is valid in cylindrical and polygonal capillaries.

Air (ρ = 1.19 kg/m³) is used as non-wetting phase while two different liquids are tested as wetting phase: octane (γ = 21.49 mN/m, ρ = 703 kg/m³) and ethanol GPA (γ = 22.15 mN/m, ρ = 789 kg/m³). The surface tension was measured at T = 23.3 °C via the plate method. Two cylindrical glass capillary sizes are used with diameters d = 556 μm and d = 282 μm. The square glass capillaries used have an inner side length of $2r_h$ = 500 μm.

Figure 11 shows the set up for measuring the range of pressure differences at the non-wetting capillary end. The capillary, attached to a micrometer that allows the microscope to focus, is at the end of a plastic tube connected to the reservoir that contains the liquid phase. A needle fixed to the reservoir indicates the height, which can be modified by means of a precision stage. Images of the meniscus are taken via a video camera Kodak ES 1.0 connected to the microscope and are downloaded into computer through a DT3157 framegrabber. The height increments, Δh, are converted into a range of pressure difference across the meniscus by Equation (27). The precision of the ruler is ± 0.25 mm, which for the experiments with the 556 μm and 500 μm capillaries results in errors smaller than 2 % while for the 282 μm capillary the errors are smaller than 1 %.

$$\Delta P_{Rot} = \left(P_{NW} - P_W\right)_{Advancing} - \left(P_{NW} - P_W\right)_{Breakthrogh} = (\rho_l - \rho_{Air}) g \Delta h \quad (27)$$

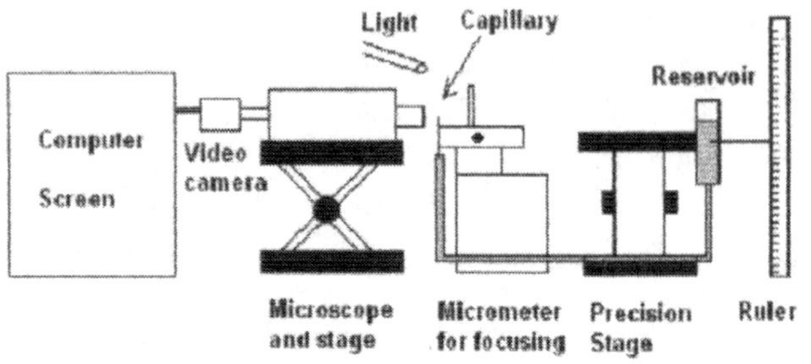

Figure 11. Experimental set up for measurement.

The advancing and receding contact angles are measured inside the 556 μm diameter round capillaries by determining the liquid rise in the capillary with a micrometer of precision ± 10 μm. For this capillary size the capillary rise is h_0 = 2.190 cm and h_0 = 1.980 cm for octane-air and ethanol-air systems respectively for receding contact angles. The Bond numbers of both systems are $Bo_{O\text{-}A}$ = 0.02476 and $Bo_{E\text{-}A}$ = 0.02696. Although $Bo \ll 1$, the exact solution given by Equations (9-11) is used to find an accurate value of the contact angles, which are shown in table 1. Figure 12 shows the meniscus advancing and breakthrough positions inside a cylindrical capillary, whose intermediate situations are due to the meniscus rotation at the capillary end.

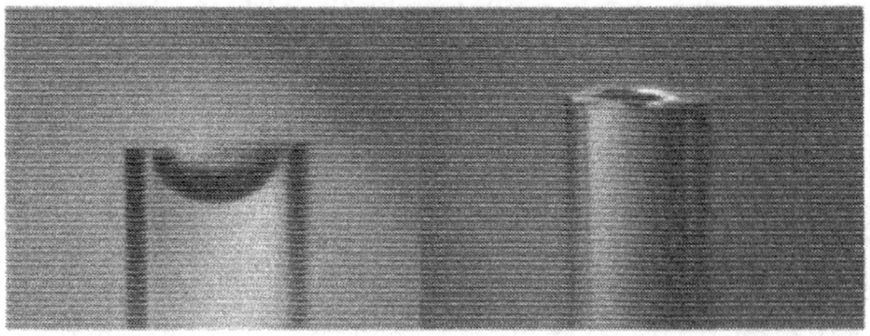

a) b)

Figure 12. a) Meniscus advancing and b) meniscus position just before breaking through. Ethanol-air system in a cylindrical capillary of diameter d = 556 μm.

Figure 13 shows the same positions in a square capillary. As it can be seen, arc menisci are present as the contact angle is smaller than (90°-β) = 45°. While the main terminal meniscus is within the capillary then, irrespective of its position, the arc menisci do not go beyond the capillary end (see Figure 13a). Breakthrough only happens when the main terminal meniscus reaches the pore end and the whole contact line rotates at it.

a) b)

Figure 13. Ethanol-air meniscus in a square capillary of inner side 500 μm, a) advancing, b) imminent breakthrough.

Table 1 shows the results obtained for the pressure difference range that the meniscus is able to cope with at the capillary end due to the virtual contact angle change, ΔP_{Rot}, and it is compared to the theoretical value for a 90-degree meniscus rotation. Three different capillaries of each kind and 3 different runs for each condition were used. The standard deviation of the measured pressure differences in each capillary was smaller than 2 % in all cases. Only average values are here presented.

Table 1. Comparison of experimental results for the range of pressure difference due to the meniscus rotation and the capillary end and theory (90° rotation).

System	θ_R, °	θ_A, °	$\Delta P_{Hys.}$ Pa	ΔP_{Rot} Pa	$\Delta P_{Rot,T}$ Pa	Rot ±2 °
O-A 556C	12	12	0	179	183	88
O-A 282C	12	12	0	342	362	86
O-A 500S	12	12	0	183	195	85
E-A 556C	15	21	5.2	214	206	93
E-A 282C	15	21	10.2	407	406	90
E-A 500S	15	21	3.9	204	223	83

O = Octane, A = Air, E = Ethanol, C = Cylindrical capillary, S = Square capillary, *Rot.* = Actual meniscus rotation, *T* = ideal (90 ° rotation)

The last column shows the actual rotation of the meniscus, which in all cases is in good agreement with the theoretical value of 90 degrees for a round pore end. Interestingly the square capillaries also seem to give similar results, showing a slightly smaller rotation than the cylindrical ones, although their finishing was not so smooth as in the cylindrical capillaries.

The small variations, generally resulting in a smaller rotation that the theoretical one can be attributed to perturbations in the system like vibrations and pressure air fluctuations as well as irregularities in the capillaries.

SUMMARY

The shape, position and stability of interfacial menisci inside cylindrical and polygonal capillaries or straight pores have been analysed. The work is applicable when operating with microcontactors where two phases are brought into contact via a solid layer with openings [1,2]. The meniscus shape, which is a function of the wettability of the system, capillary geometry and pressure difference between the phases, defines the available contact area for mass transfer. The model for determining the stability or *available range of pressure differences* across the phases that the meniscus can cope with without breakthrough of one phase into the other, is based on the contact angle hysteresis and the ability of the meniscus to rotate 90 degrees at the capillary end. The larger the extent of these two phenomena is, the better the meniscus stability.

Polygonal capillaries present an increasing stability reduction when compared to cylindrical ones as the number of sides decreases when arc menisci are present (AMs), $\theta < (90^{o}-\beta)$, where β is half the angle of the polygon corner. This is as a result of the meniscus mean curvature reduction that takes place because of the AMs [11]. Equation (15) for cylindrical pores and Equations (24) and (26) for polygonal ones are used to determine the interfacial contact area as a function of the wettability properties of the system and pressure difference across the phases. The interfacial meniscus area in polygonal pores can increase substantially at contact angles $\theta < (90^{o}-\beta)$ because of the AMs, which extend along the pore.

Experimental work has confirmed that real cylindrical and square capillaries can be considered to have round ends where the meniscus is allowed to rotate 90 degrees. Square capillaries showed a slightly smaller meniscus rotation at the capillary end although the finishing was not so smooth as in the cylindrical ones. Therefore, a proper design of the pore ends that bring the fluid phases into contact can be useful to increase system stability.

ACKNOWLEDGMENTS

The project was funded by the European Commission (Contract No. G1RD-CT2000-00469).

REFERENCES

1. Turner C., Shaw J., Miller B. and Bains V., *Solvent Extraction using Micro-Mesh Reactors*, IMRET 4, Poster Session 1, Atlanta (2000).
2. Cypes S.H. and Engstrom J.R., *Construction, Analysis and Evaluation of a Microfabricated Stripping Column*, IMRET 7, Lausanne, Switzerland (2003).
3. Adamson A.W. and Gast A.P., *Physical Chemistry of Surfaces*, Chapter II – Capillarity, Sixth Edition, John Wiley & Sons, Inc. (1997).
4. Matijevic E., *Surface and Colloid Science*, **Volume II**, pp. 2-63, Wiley-Interscience (1969).
5. Bronshtein I.N. and Semendyayev B., *Handbook of Mathematics*, pp. 551-556, 3rd Edition, Springer (1997).
6. Ardanuy A., Del Rio J., Menendez M.D., Perez C., Suarez L. and Verde J.M., *Análisis y Probabilidad C.O.U*, ICE, University of Salamanca (1986).
7. Wapner P.G. and Hoffman W.P., *Partial Wetting Phenomena on Nonplanar Surfaces and in Shaped Microchannels*, Langmuir, **18(4)**, 1225-1230 (2002).
8. Volpe C.D., Maniglio D., Morra M. and Siboni S., *The Determination of a 'Stable Equilibrium' Contact Angle on Heterogeneous and Rough Surfaces*, Colloids and Surfaces A, **206**, 47-67 (2002).
9. Shouxiang M., Geoffrey M. and Norman R.M., *Effect of Contact Angle on Drainage and Imbibition in Regular Polygonal Tubes*, Colloids and Surfaces, A **117**, 273-291 (1996) .
10. Princen H.M., *Capillary Rise in Systems with more than Two Cylinders*, J. of Colloid and Int. Sci., **30(3)**, 359-371 (1969).
11. Mason G. and Morrow N.R., *Meniscus Curvatures in Capillaries of Uniform Cross-Section*, J. Chem. Soc., Faraday Trans., **80**, 2375-2393 (1984).
12. Wong H., Morris S., and Radke J., *Three Dimensional Menisci in Polygonal Capillaries*, J. of Colloid and Int. Sci., **148(2)**, 317-336 (1992b).

CONTACT

Corresponding author: Panagiota Angeli.
Email address: p.angeli@ucl.ac.uk
Tel: 0044 (0)20 7679 3832
Fax: 0044 (0)20 7383 2348

Development of a Proton Exchange Membrane Fuel Cell

I. Eroglu
Middle East Technical University
Department of Chemical Engineering, Middle East Technical University, Ankara, 06531, Turkey

O. Yazaydin
Middle East Technical University
Department of Chemical Engineering, Middle East Technical University, Ankara, 06531, Turkey

L. Türker
Middle East Technical University
Department of Chemistry, Middle East Technical University, Ankara, 06531, Turkey

ABSTRACT

The proton exchange membrane fuel cell (PEMFC) is one of the strongest contenders as a power source for powering future cars, residential areas and small combined power units due to their modular construction and the high energy densities that can be attained. In this study, a H_2-O_2 PEMFC was constructed and effects of cell temperature, humidification temperature of gases, flow rates of gases and flow rate ratio of O_2 to H_2 on the performance were investigated. It was found that performance increases significantly with increasing temperature of cell and humidification temperature. High O_2 to H_2 flow rate ratio had also a positive effect on the performance up to an extent. On the other hand, excess flow rate of gases decreased the performance.

KEY WORDS

Fuel Cell, Hydrogen, Proton Exchange Membrane

INTRODUCTION

Fuel cells offer the potential of revolutionizing electrical energy production by affording highly efficient and largely pollution free power generation systems for both transportation and stationary applications [1]. Proton exchange membrane fuel cell (PEMFC), operating on H_2 and O_2, is particularly attractive because of mild operating conditions required (50-80 °C temperature, 1-3 atm pressure) low Pt loadings, relative robustness, long life [2]. It comprises two gas-diffusion layers that allow simultaneous transport of gases and water while collecting current, and a membrane electrode assembly (MEA) that has two supported-catalyst (typically Pt/C) layers on both sides of a perfluorosulfonic acid (PFSA) polymer membrane such as Nafion®. The MEA and gas diffusion layers are sandwiched between two graphite plate current collectors, with machined channels for gas distribution [3]. In the present study, a H_2-O_2 PEMFC was constructed using commercially available gas diffusion layers and membrane electrode assembly. We investigated the effect of operating parameters such as the effects of cell temperature, humidification temperature of gases, flow rates of gases and flow rate ratio of O_2 to H_2 on the performance.

EXPERIMENTAL

1. CONSTRUCTION OF THE PEMFC

A membrane electrode assembly (MEA) with a Nafion® membrane thickness of 25 µm coated with a platinum catalyst loading of 0.3 mg/cm^2 was purchased from Ion-Power, USA. The MEA was 9 x 9 cm^2 where the catalyst coated active area is 50 cm^2. Gas diffusion layers (GDL 10-BB type) were obtained from SGL Carbon, Germany. Gas distribution channels with serpentine flow geometry were machined on the 2mm thickness graphite plates (SGL Carbon BPP4 type) with CNC technology. An illustrative drawing of the graphite plate is shown in Figure 1. Channels were in 1mm depth and width and located within a square area of 50 cm^2 at the center.

Figure 2 illustrates the structure of the PEMFC constructed in the present work. Two cupper plates with a thickness of 3 mm and two cupper films with a thickness of 0.5 mm were used as current collector. Cupper films were placed on the cupper plates to compensate nonuniformity of the surface that may cause ohmic loss. MEA and gas diffusion layers were placed between two graphite plates. Then graphite plate - gas diffusion layer -

MEA structure was placed between the cupper plates. Finally steel plates (12 x 12 cm^2) were used to hold the whole structure.

Figure 1. Illustrative drawing of graphite plates with machined channels.

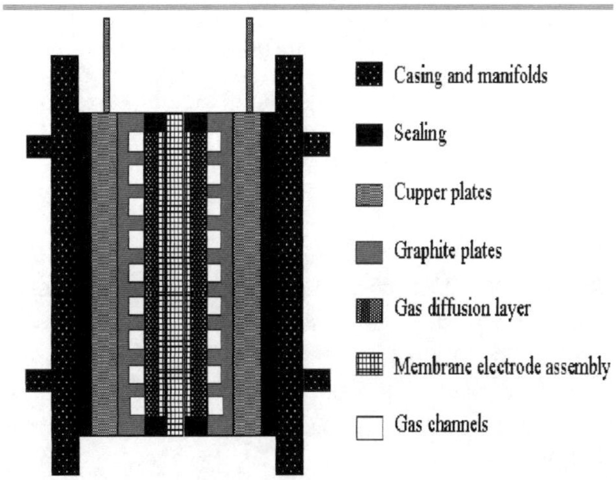

Figure 2. The structure of the PEMFC constructed in the present work.

2. CONSTRUCTION OF THE TEST STATION

Two Gas Mass Flow Controllers (Aalborg) were used to control the flow rates of the gases. Humidification of the gases was performed by bubbling them through bottles containing distilled water. Bottles were placed in a temperature controlled hot water bath. A temperature controlled heating band was wrapped around the PEMFC to control its temperature. A very thin thermocouple had been placed between the cupper plate and graphite plate before it was assembled which was linked to the controller unit of the heating band. A voltmeter, an ampermeter and a variable resistor were properly connected to the PEMFC to complete the circuit. A representation of the test station built is given in Figure 3.

3. PROCEDURE

A standard procedure was applied to obtain the V-i diagrams. Flow rate of the gases were set by gas mass flow controllers (Aolborg). Water bath was heated to the desired humidification temperature. Fuel cell was heated to its operating temperature by setting the temperature controller.

After humidification and fuel cell temperatures reached the set points, PEMFC was operated between 0.6 V (for 30 minutes) and 0.4 V (for 30 minutes) and then for 1 minute at the open current voltage (OCV) conditions. To reach the pseudo steady state condition this cycle was repeated at least four times until no further significant increase was observed on the performance. Then voltage-current data were collected by varying the resistor to obtain V-i diagrams.

RESULTS

Effects of humidification temperature were investigated at two different flow rates. The results are shown in Figure 4 and Figure 5 that were obtained keeping the cell temperature (50°C) and gas flow rates constant and changing the humidification temperature between 35 and 60 °C. Flow rates of H_2 and O_2 gases were 0.4 and 0.2 sLm (standard liters per minute), respectively for the excess amount condition, whereas 0.1 and 0.05 sLm, respectively for the other one.

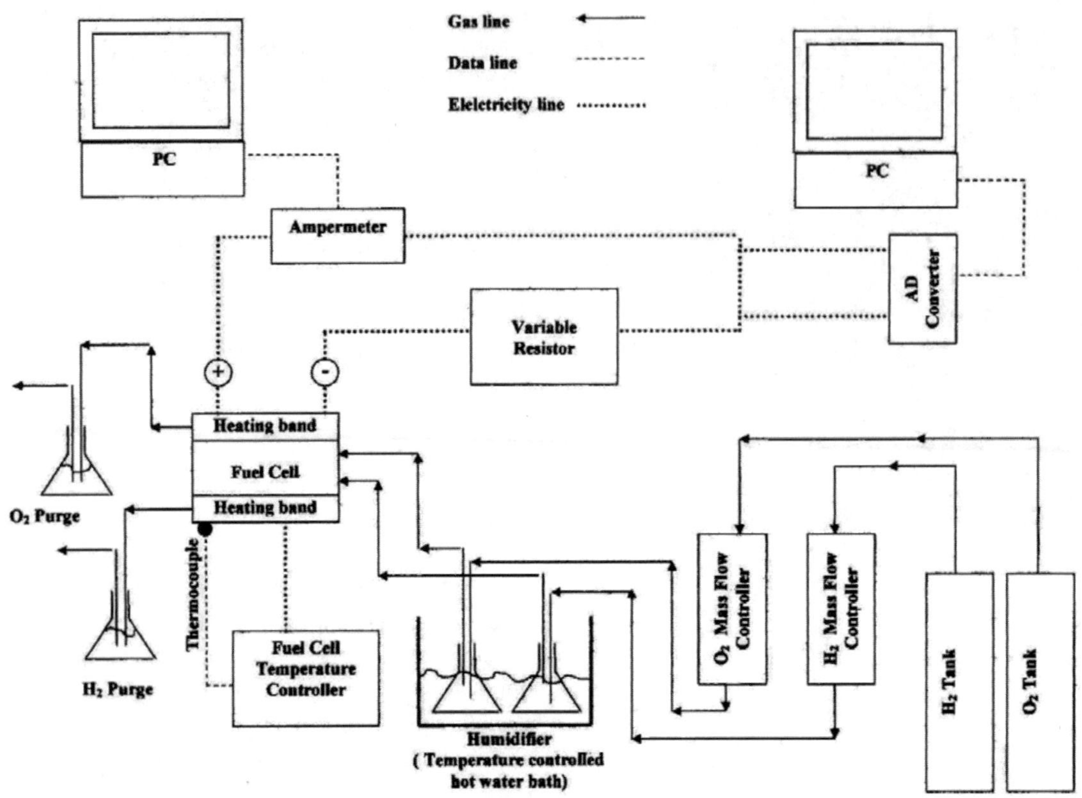

Figure 3. Schematic diagram of the test station.

The effect of cell temperature was investigated at two different humidification conditions. Cell temperature were changed between 30 and 60 °C while keeping the humidification temperature 10 °C more than the cell temperature in first experiment and 5 °C less in the second case. Results are shown in Figures 6 and 7.

The effect of O_2 to H_2 ratio was investigated by keeping the H_2 flow rate constant and changing the O_2 flow rate. Results are shown in Figure 8.

Ion-Power [4] reported a V-i curve obtained at 60 °C, operated with hydrogen and air with a relative humidity of 100%. Figure 9 illustrates the performance reported by Ion-Power with the one obtained in the present study at 60 °C fuel cell temperature and 70 °C gas humidification temperatures. Ion-power presents significantly a better performance although the same MEA was used in both performance tests. This shows the critical importance of other parts of the PEMFC on the performance. The linear part of the V-i curve is attributed to the ohmic losses. From the slope of the lines the internal resistances of the two PEMFCs are estimated. Internal specific resistance of the upper curve is 0.23 Ω/cm^2 and the lower curve is 0.63 Ω/cm^2.

DISCUSSION

The positive effect of increasing humidification temperature can clearly be seen at Figures 4 and 5. For the transportation of protons through the membrane water is acting as the carrier molecule. Therefore the membrane should be wetted sufficiently. As the temperature of the water bath increased, humidity of gases is expected to increase carrying more water vapor

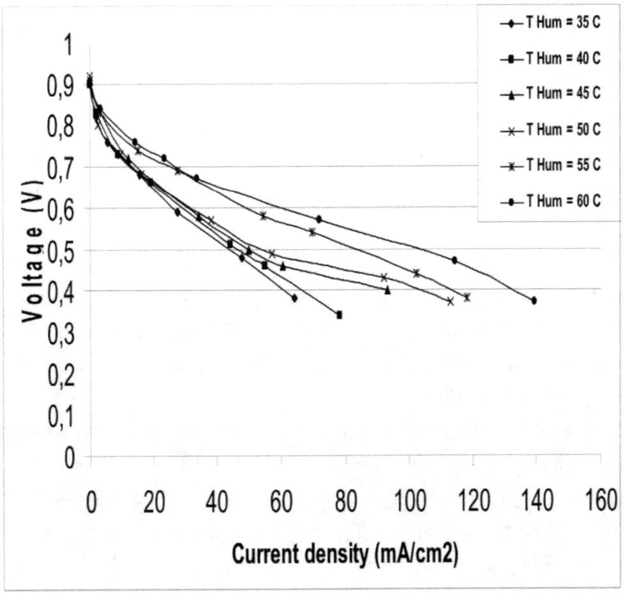

Figure 4. Effect of gas humidification temperature with excess flow rate of reactants.
(T Cell = 50 °C; H_2: 0.4 sLm; O_2: 0.2 sLm).

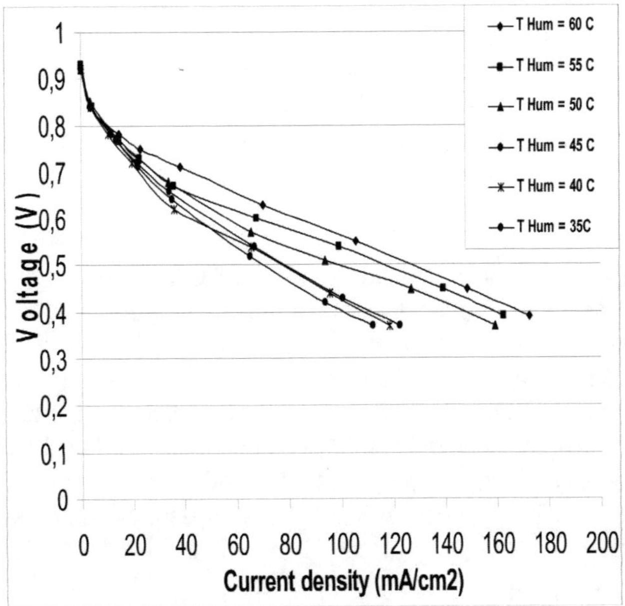

Figure 5. Effect of gas humidification temperature.
(T Cell = 50 °C; H_2: 0.1 sLm; O_2: 0.05 sLm)

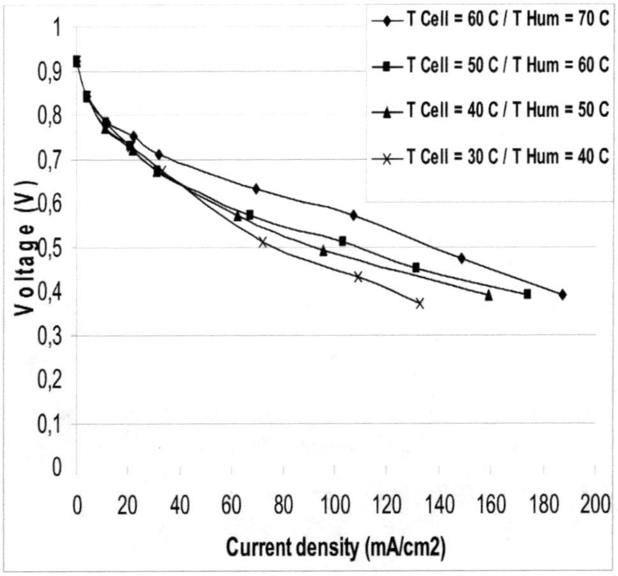

Figure 6. Effect of cell temperature at high humidification level.
(T Hum = T Cell + 10 °C;H_2: 0.1 sLm;O_2:0.05 sLm)

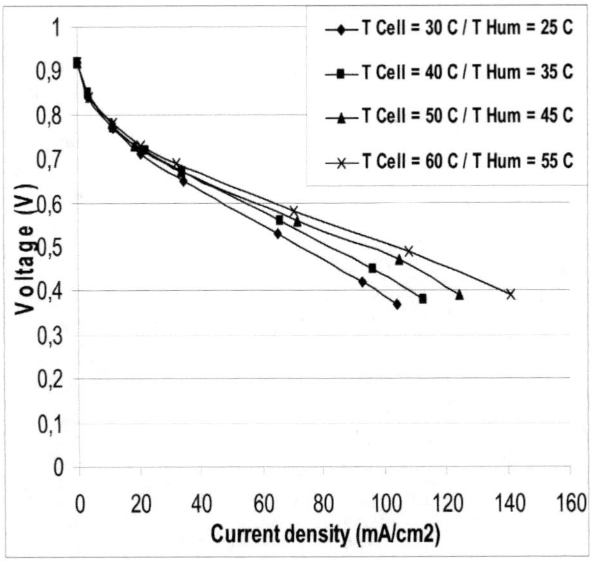

Figure 7. Effect of cell temperature at low humidification level.
(T Hum = T Cell - 5 °C; H_2: 0.1 sLm; O_2: 0.05 sLm)

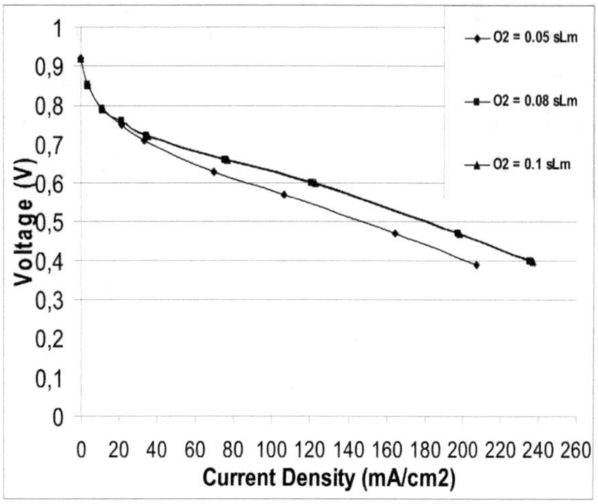

Figure 8. Effect of O_2 to H_2 ratio.
(T Cell = 60 °C; T Hum = 70 °C; H_2: 0.1 sLm)

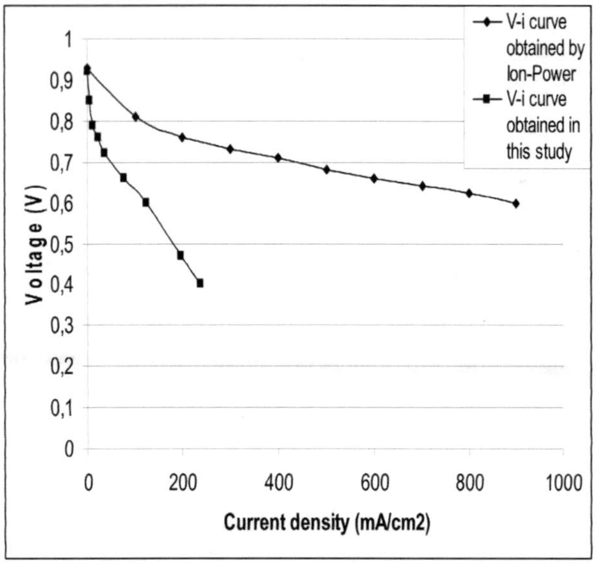

Figure 9. Comparison of the V-i curves provided by Ion-Power and obtained in this study with the same MEA.

to the cell. When two figures are compared it can be clearly realized that cell performance is low with the excess flow rate of gases. When the gas flow rate is fast, the contact time is not enough for sufficient humidification. Therefore, the gases are not humidified as much as they are humidified at low flow rates. When the gases are not sufficiently humidified, they may cause the membrane to dry up too much which result in low performance.

Figures 6 and 7 show that increase in the cell temperature had a positive effect on the performance at both high and low humidification conditions. Since, the reaction taking place in the PEMFC is basically a redox reaction, the kinetics is positively affected in the case of a temperature increase as long as there is sufficient humidification. If sufficient humidification is not provided then increasing temperature will have an adverse effect since the membrane is dried. The overall phenomena taking place within the MEA is strictly limited by the diffusion of protons through the membrane.

The increase in the flow rate ratio of O_2 to H_2 had also a positive effect on the performance. It is well known that the cathode reaction, which is the reduction of oxygen, has a low rate of kinetics when compared with the anode reaction, hydrogen oxidation [5]. A higher O_2 flow rate might have increased the mass transfer rate of O_2 resulting a higher performance. On the other hand, this positive effect of the increasing flow rate of O_2 has a limit as can be seen from Figure 8. An increase of the flow rate from 0.08 sLm to 0.1 sLm did not make any significant improvement on the performance.

The significant difference between the performances of the same MEA tested in different fuel cells (as shown in Figure 9) indicate that other parts of the fuel cell together with the testing conditions also have critical roles on the performance. The ohmic loss experienced in the PEMFC used in this study is almost three times larger than the reported value. Therefore, it is recommended that the test cell should be improved for further studies. All V-i curves obtained in the present study declined rapidly from OCV when high external resistance was applied; on the other hand, this fall in the V-i curve provided by the manufacturer company is relatively small. Although the same OCV was measured a rapid decline in the voltage was observed in the present study. High voltage drop at low current densities might be due to activation losses and fuel crossover. The cause of activation losses are catalyst effectiveness, electrode structure, pressure and concentration of reactants. However, these factors were quite similar. Fuel crossover could be the main reason of this high voltage drop. Therefore, sealing and the design of the graphite plate should be improved.

CONCLUSION

In the present study the effects of the cell and humidification temperatures, flow rates of the reactant gases and flow rate ratio of O_2 to H_2 on the performance of the PEMFC were investigated. It was found that the higher the cell and humidification temperatures, the increased performance is. High O_2 to H_2 ratios favored the reaction positively up to an extent. Excess flow rate of reactant gases did not make any positive effect on the performance since the membrane is dried due to the fact that the gases were not sufficiently humidified.

ACKNOWLEDGMENTS

This study is supported by EAE Elektrik A.S. Istanbul/Turkey and Turkish Scientific and Research Council (TUBITAK-MISAG 230). Gas diffusion layers and graphite plates were supplied by SGL Carbon, Germany.

REFERENCES

1. Jacoby M., C&EN, June 14, 31, 1999.
2. Kordesh K.V.,. Simader G.R, Fuel Cells and Their Applications, VCH, Weinheim, 1996.
3. Tamphan T., Malhotra S., Zhang J., Datta R., Catalysis Today, 67, 15, 2001.
4. www.ion-power.com
5. Larminie, J., Dicks, A., "Fuel Cell Systems Explained", 2nd ed., John Wiley and Sons, London, 2003.

CONTACT

Inci Eroglu (Prof.)

e-mail: ieroglu@metu.edu.tr

Web address: http://www.metu.edu.tr/home/wwwbiohg/

CHITOSAN BASED FUEL CELL MEMBRANE

Khantong Soontarapa and Unchalee Intra

Fuel Research Center, Department of Chemical Technology,
Faculty of Science, Chulalongkorn University
Phayathai Rd., Phathumwan, Bangkok, 10330, Thailand
khantong@sc.chula.ac.th

Key words: Chitosan, composite membrane, ion exchange capacity, ionic conductivity

Prepared for presentation at the AIChE 2003 Annual Meeting, November 16 – 21, [278] – Experimental Analysis of Fuel Cell Behavior

October 1, 2003

Abstract

Cast films of uncrosslinked chitosan, crosslinked chitosan with sulfuric acid, chitosan composite containing lithium nitrate as doping salt, and plasticized-salt complexes composing of chitosan as the host polymer, lithium nitrate as the doping salt and ethylene carbonate as the plasticizer were prepared and studied as candidates for fuel cell membranes. The shift of NH_2 band at 1590 cm^{-1} to lower wave number proved the complexation between chitosan and the lithium salt. The increase of T_g from 49.7 °C for uncrosslinked membrane to 50.0 °C for crosslinked membrane and to 51.9 °C for crosslinked salt-complexed membrane was the consequence of the interactions of components in composites. EC did not interact with chitosan but reduced the T_g. The ion exchange capacity of uncrosslinked chitosan was 2.5±0.03 meq g^{-1}. It was enhanced to 5.22±0.04 meq g^{-1} when crosslinking with 4 %w/w sulfuric acid and to 5.85±0.11 meq g^{-1} when further doping with 50 %w/w of $LiNO_3$. It was found that the ionic conductivity of 4%-crosslinked 50%-salt-complexed chitosan membrane was as high as 6.3×10^{-2} S cm^{-1} at 100% relative humidity and room temperature. The ethylene carbonate offered no significant advantages on ion exchange capacity and ionic conductivity.

Key words: Chitosan, composite membrane, ion exchange capacity, ionic conductivity

1. INTRODUCTION

The fuel cell is an electrochemical device that converts the chemical energy of its reactant directly to electricity and heat without combustion. Fuel cells are commonly classified according to the electrolyte used. This work focuses on the proton exchange membrane (PEM) fuel cells, which is a low temperature fuel cell employing a solid ion conducting polymer membrane as the electrolyte. Perfluorosulfonic acid (PFSA) is the most commonly used membrane material for PEM fuel cells. Currently, the development of new membrane materials is extensively done. The foremost challenge suggested by Rikukawa and Sanui[1] is to produce materials that are cheaper than PFSA. They note that some sacrifice in material lifetime and mechanical properties may be acceptable, providing the cost factors are commercially realistic. Hydrocarbon membranes provide five advantages over PFSA: (1) they are less expensive, (2) many types are commercially available, (3) polar groups can be formed to have high water uptakes over a wide temperature range with the absorbed water restricted to the polar groups of the polymer chain, (4) decomposition of hydrocarbon polymers can be depressed to some extent by proper molecular design, and (5) they are recyclable by conventional methods.

Zaidi et al.[2] reported that the undoped POD membrane possessed a rather low conductivity of the order of 10^{-7} S cm^{-1} and the conductivity was almost independent on the temperature in the studied range of 25-90 °C. This signified that the observed conduction occurs via liquid-like mechanism and completely resulted from water, adsorbed in the membrane pores and on surface. In the case of doped POD membranes, the conductivity of uncompressed samples was always higher than that of the compressed ones reaching 1.1×10^{-1} S cm^{-1}. This could be attributed to the fact that concentrated H_3PO_4 fills the pores of these materials, thus providing a conduction pathway through liquid phase. The higher void fraction of uncompressed materials led to a higher acid content and therefore to a higher conductivity.

Kerres group[3-5] has developed acid-base blend membranes for fuel cell application. These membranes exist specific interactions between the blend components: hydrogen bridges and ionic crosslinking by proton transfer from the acid to the basic groups as following:

$$\text{Polymer-}SO_3H + \text{Polymer-}NH_2 \rightarrow \text{Polymer-}SO_3^- \, {}^+H_3N\text{-Polymer}$$

Arof group[6-8] presented the plasticized chitosan-salt complexes as an ionic conductor. The studies have proven that chitosan can be used as polymer matrix for ionic conduction.

Chitosan is the N-deacetylated derivative of chitin. The structures of chitin and chitosan are shown in Figure 1. Chitosan has a high potential for development into sophisticated functional polymers since it has both free amino groups and hydroxyl groups on its backbone. Each of the nitrogen and oxygen atoms in chitosan has a lone pair electron where complexation can occur.

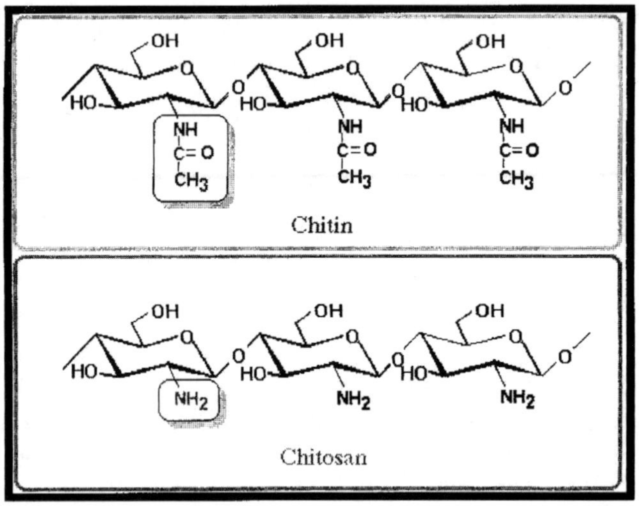

Figure 1. Structures of chitin and chitosan[9]

In this paper, a study was carried out on a chitosan (CS) biopolymer synthesized from shrimp shells in our laboratory. Crosslinked chitosan membranes with different amounts of sulfuric acid were investigated. The interaction between a lithium-cation and a nitrogen donor in chitosan expected to enhance the ionic conductivity were investigated with different amounts of lithium nitrate ($LiNO_3$). A fixed amount of ethylene carbonate (EC) plasticizer expected to increase segmental motion of the polymer resulting in ionic conductivity enhancement was also performed.

2. EXPERIMENTAL

2.1 Membrane Preparation

Chitosan was done through the deproteinization of ground shrimp shells with 3 %w/v of NaOH at 100°C for 1 hour. Demineralization was executed with 1.25 N HCl at ambient temperature for 1 hour, drying at 65°C. The solid materials obtained were chitin. Deacetylation was performed with 50 %w/v of NaOH at 100°C under vacuum for 30 minutes. The resulting chitosan flakes were washed with tap water until neutral and dried at 65°C. Molecular weight, viscosity, and degree of deacetylation were the normal parameters to characterize chitosan[10-12]. Molecular weight was measured by gel permeation chromatography. Several methods were proposed for N-deacetylation of chitin and chitosan[10-17] of which NMR measurement is as a choice.

The chitosan flake was dissolved in aqueous acetic acid solution at a concentration of 1.5 %w/w. The concentration of acetic acid solution was 1 %w/w. It was then cast onto a glass plate and dried in air oven at 40 °C. For uncrosslinked membranes, the dry membrane was immersed in NaOH solution of 4 %w/w, washed with water, dried in air oven at 40 °C and stored in desiccator until use. The membranes were immersed in different concentration of H_2SO_4 solution before drying at 40 °C in case of preparing crosslinked membranes. The chemical structure of which was shown in Figure 2. For the chitosan-salt membranes, $LiNO_3$ was added to the chitosan solution in different quantities and stirred until complete dissolution was obtained. The solutions were then formed as membrane. The plasticized chitosan-salt membranes were prepared by adding a fixed amount of EC of 40 %w/w to the chitosan solution. Table 1 listed the membranes prepared in this study.

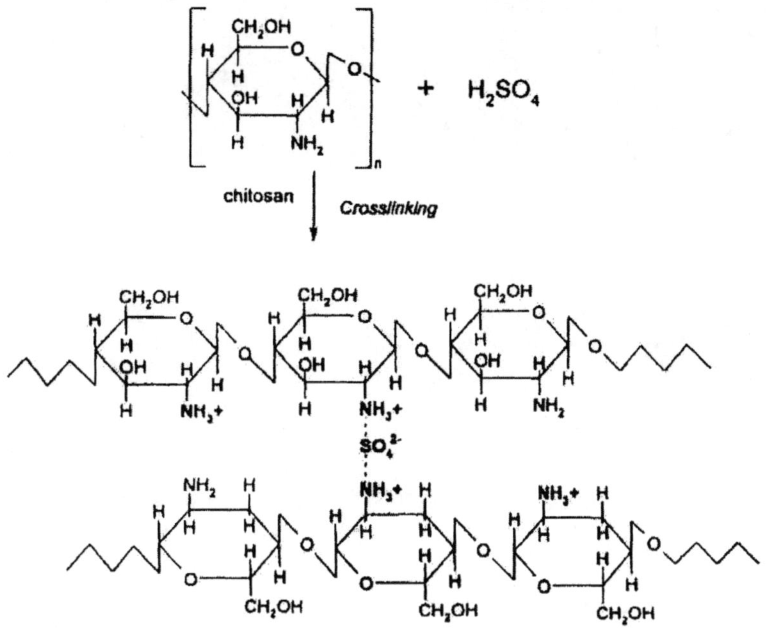

Figure 2. Crosslinking of chitosan with sulfuric acid[18]

Table 1. Composition of the complexes

Types of Membranes	$LiNO_3$ (%w/w of CS)						H_2SO_4 (% w/w of solution)			EC (40 %w/w of CS)
	0	5	10	20	50	100	2	4	6	
Crosslinked	+						+			
	+							+		
	+								+	
Salt-complexed		+	+	+	+	+				
Crosslinked salt-complexed		+	+	+	+	+	+			
		+	+	+	+	+		+		
		+	+	+	+	+			+	
Plasticized salt-complexed			+		+	+				+
Plasticized crosslinked salt-complexed			+		+	+	+			+
			+		+	+		+		+
			+		+	+			+	+

2.2 Membrane Characterization

2.2.1 Physical Properties

The tensile strength and elongation were measured according to ASTM D638 by Universal testing machine. The water content as the weight fraction of membrane was calculated from the following equation by using the weight of membrane swollen and equilibrated in water at ambient temperature (W_s) and the dry membrane weight (W_d):

$$\text{water content} = \frac{W_s - W_d}{W_d}$$

Similarly, the change of membrane thickness was calculated from that of swollen membrane (T_s) and of dry membrane (T_d).

$$\text{thickness change} = \frac{T_s - T_d}{T_d}$$

2.2.2 Thermal properties

The glass transition temperature (T_g) of the membranes was determined by differential scanning calorimeter (DSC). The heating rate was 20 °C min^{-1} under N_2 atmosphere.

2.2.3 Fourier Transform Infrared Spectroscopy (FTIR)

FTIR was carried out to verify the occurrence of complexation in the wave number region between 4000 and 400 cm^{-1}.

2.3 Ion Exchange Capacity Measurement

The ion exchange capacity was measured by immersing 20 mg of the membrane in a 25-ml solution of 0.005 N NaOH overnight, with continuous stirring. A 10-ml aliquot was back-titrated with 0.005 N HCl to a pH of 7.0. Ion exchange capacity was reported as milli-equivalent per gram of dry membrane.

2.4 Ionic Conductivity Measurement

The ionic conductivity was measured by direct current, four-point-probe method under 100% humidity hydrogen atmosphere at room temperature. A four-point-probe cell as shown in Figure 3 consisted of two platinum foil outer current-carrying electrodes and two platinum wire inner potential-sensing electrodes. Membrane samples were cut into strips that were approximately 1.0 cm wide and 4 cm long prior to mounting in the cell. The cell was placed in a thermo-controlled humidic chamber to measure the temperature and humidity dependence of ionic conductivity.

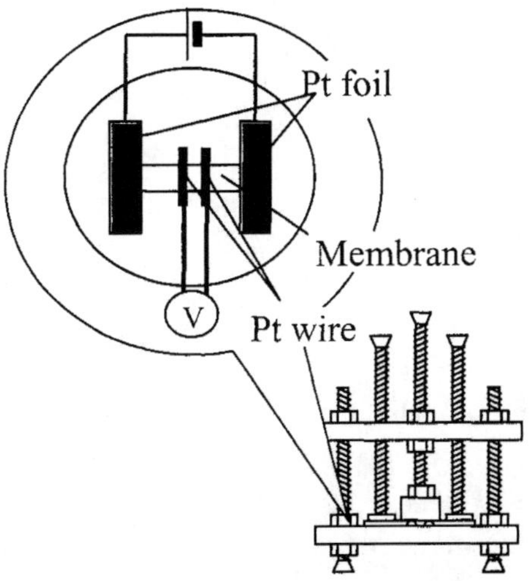

Figure 3. Schematic diagram of 4-point-probe cell

3. RESULTS AND DISCUSSION

3.1 Polymer Property

Molecular weight and degree of deacetylation of chitosan prepared from this study were 1.7×10^{6} Da and 94.2±1.1%, respectively. The viscosity value of solutions of 1.5% chitosan by weight in 1% by weight aqueous acetic acid was about 4000 cP.

3.2 Membrane Characteristics

3.2.1 Physical Characteristics

All membranes had a thickness of 15.3±3.2 μm. The tensile strength, swelling degree, and thickness change of various types of chitosan membranes were shown in Figure 4-6, respectively. In Figure 4(a), the crosslinked membranes lost their tensile strength with the increase in crosslinking degree. This is due to the stress induced by crosslinking. However, the crosslinked salt-complexed membranes of 50 and 100% showed the recovered tensile strength and increased with crosslinking degree up to 4% crosslinking. The tensile strength of plasticized uncrosslinked membranes showed improved tensile strength compared with their unplasticized membranes but dropped off with increasing crosslinking degree as shown in Figure 4(b). This behavior may suggest that the suitable membranes were crosslinked salt complexes with crosslinking degree of either 2% or 4% and containing $LiNO_3$ of 50%.

Due to an increase in tightness of the network, the swelling degree decreased with increasing crosslinking degree as shown in Figure 5(a). The compact structure and chain stiffness of crosslinked membrane led to decrease of thickness change as shown in Figure 6(a). The swelling property of the membrane will affect the ion permeability through the membrane by allowing ions to go through the membrane more easily in the swollen state. On the other hand, a relatively high swelling degree of the membrane increased the membrane thickness and resulted in increased cell resistance. The complexation with 50% salt showed the lowest swelling degree and thickness change. EC seemed to hinder the swelling of the membrane and thickness change as shown in Figure 5(b) and 6(b), respectively.

3.2.2 Glass Transition Temperature (T_g)

The T_g of uncrosslinked membrane was 49.7°C and the T_g of 4% crosslinked membrane was 50.0 °C, as shown in Figure 7. A little change in T_g was a consequence of the acid-base interaction between H_2SO_4 and NH_2 group. The T_g increase from 50.0 °C of crosslinked membrane in Figure 7(b) to 51.9 °C of complexed membrane in 7(c) indicated the interaction between chitosan and salt of the complexes. The EC plasticizer softened the polymer backbone resulting in reduction in T_g to 45.5 °C, as shown in Figure 7(d). In Figure 8 the T_g in dependence of the $LiNO_3$ content of the membranes was shown. It was obvious from Figure 8 that the T_g markedly increased with $LiNO_3$ more than 40 %w/w expecting in enhancement of IEC and ionic conductivity in these membranes.

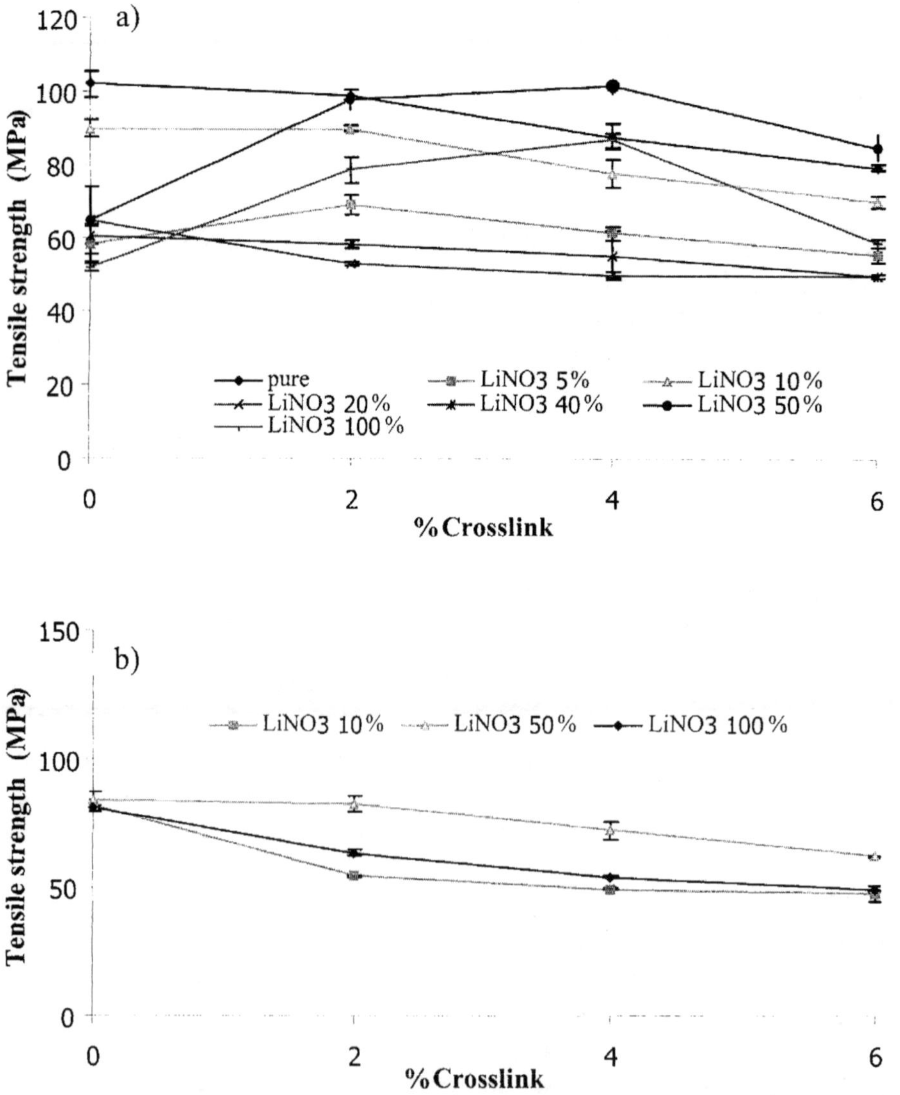

Figure 4. Variation of tensile strength with %crosslink in:
(a) membrane without EC; (b) membrane with EC

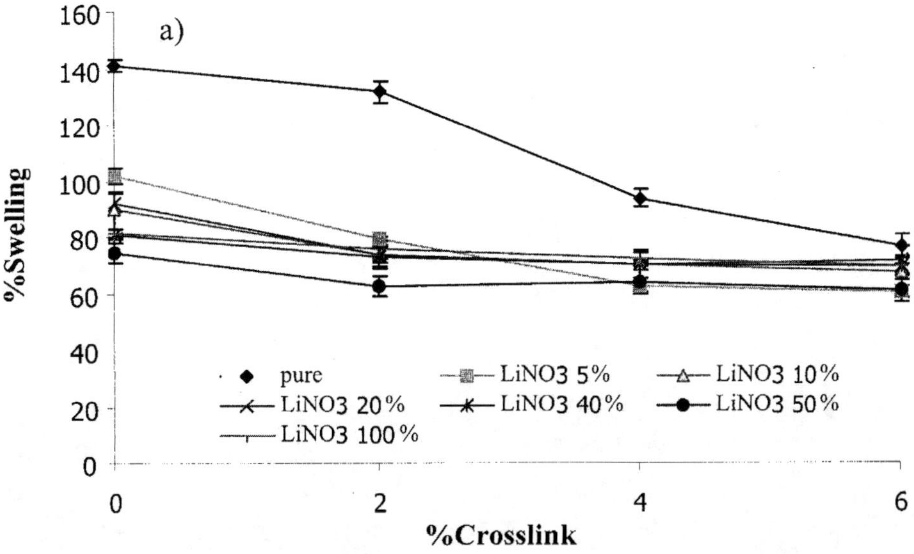

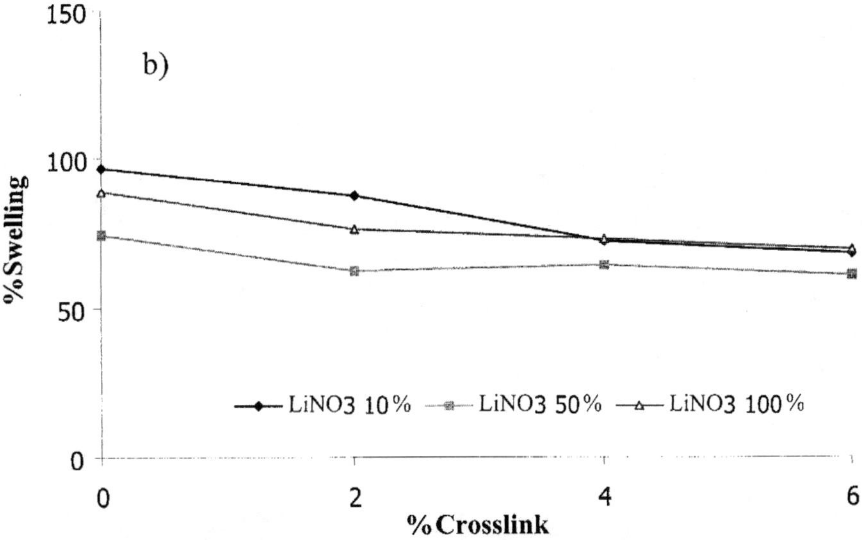

Figure 5. Variation of %swelling with %crosslink in:
(a) membrane without EC; (b) membrane with EC

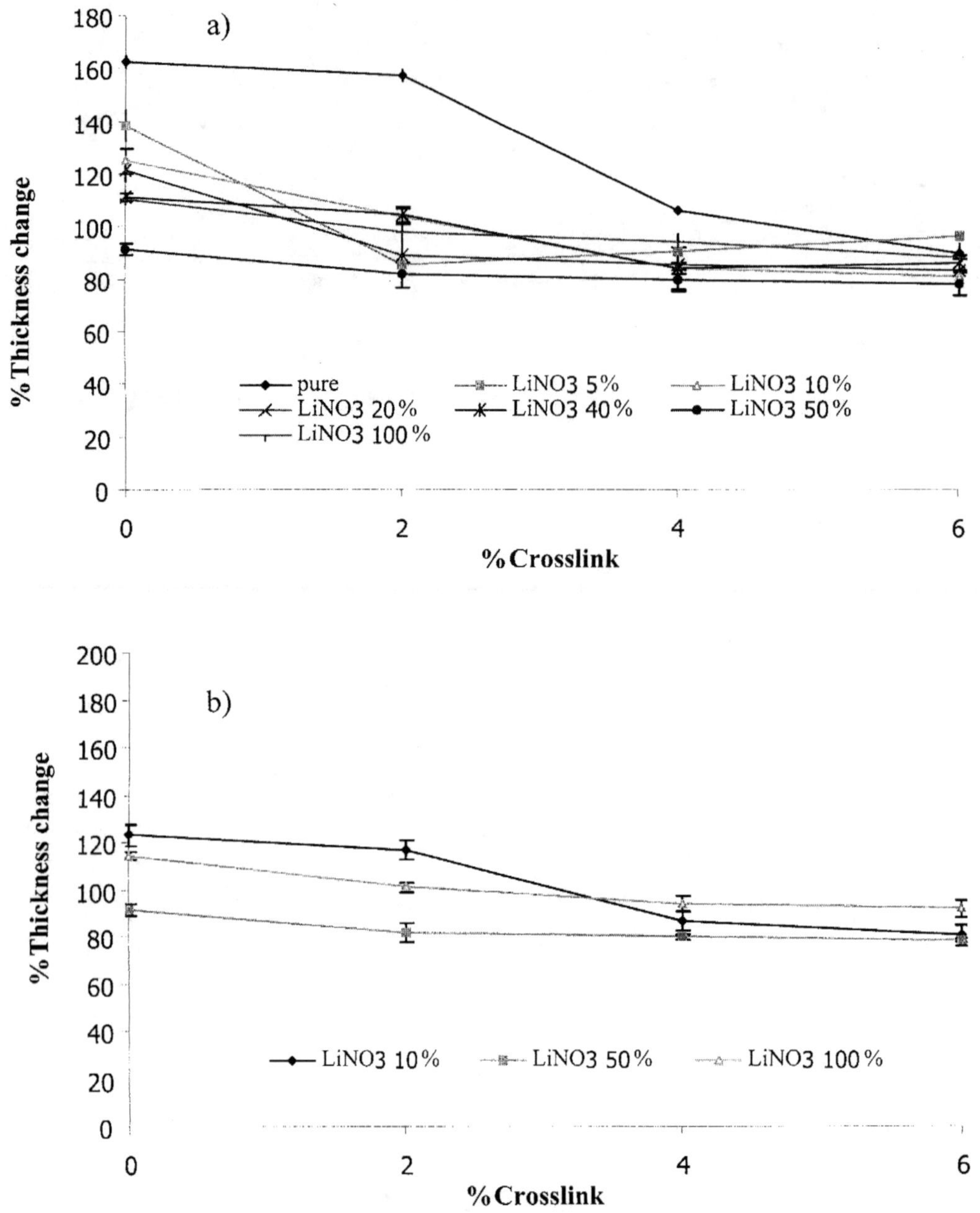

Figure 6. Variation of %thickness change with %crosslink in:
(a) membrane without EC; (b) membrane with EC

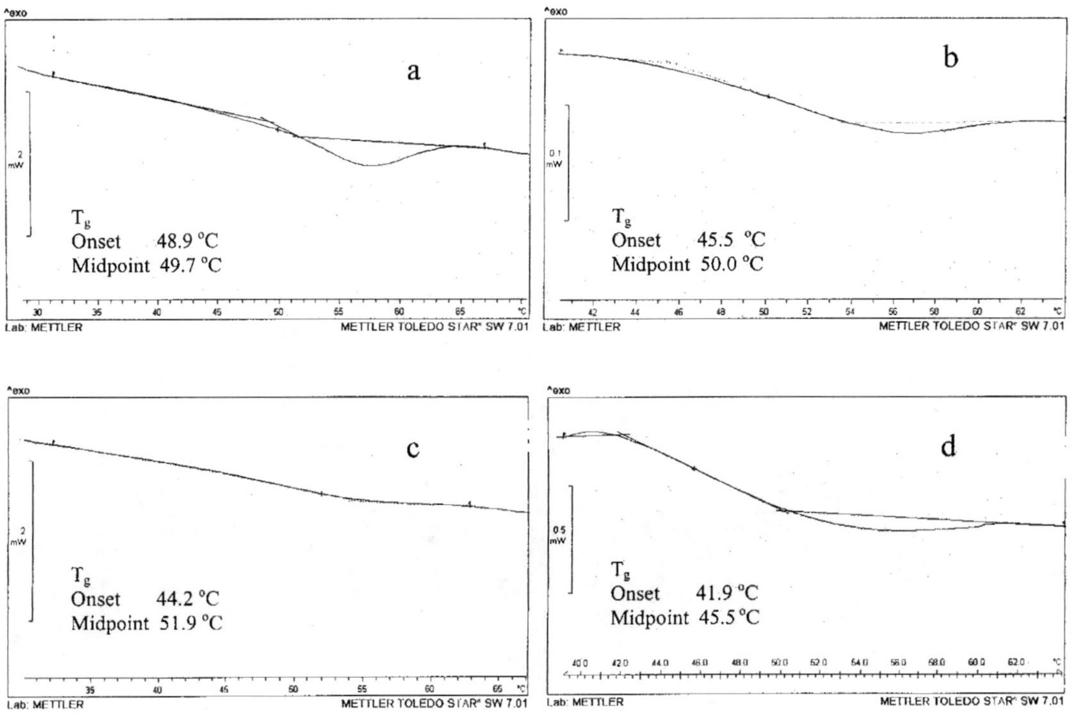

Figure 7. DSC traces of membranes a) uncrosslinked; b) 4% crosslinked; c) 4% crosslinked + 50%$LiNO_3$; d) 4% crosslinked + 50%$LiNO_3$ + EC

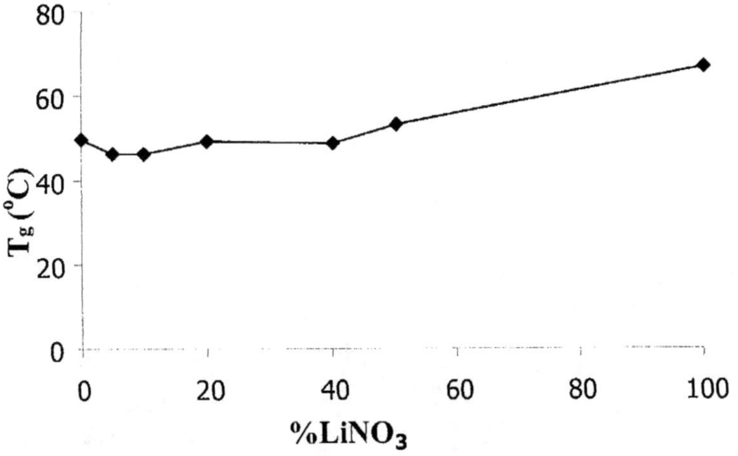

Figure 8. Variation of Tg with %$LiNO_3$ in membrane without EC

3.3 Identification of Prepared membranes through FTIR

Fig. 8 showed the FTIR spectra for chitosan-salt complexed membranes compared to the uncrosslinked chitosan membrane. It could be observed that the NH_2 band at 1590 cm^{-1} of uncrosslinked membrane in 8(a) was shifted to lower wave numbers. As the salts increased the shift increased further. This indicated that some complexation occurred between nitrogen atom of the NH_2 group and Li^+. Fig. 9 showed the FTIR spectra for crosslinked chitosan membranes compared to the uncrosslinked membrane. The shift in NH_2 band indicated that some interaction occurred between chitosan and sulfuric acid. The S-O band at about 618 cm^{-1} appeared in crosslinked membranes.

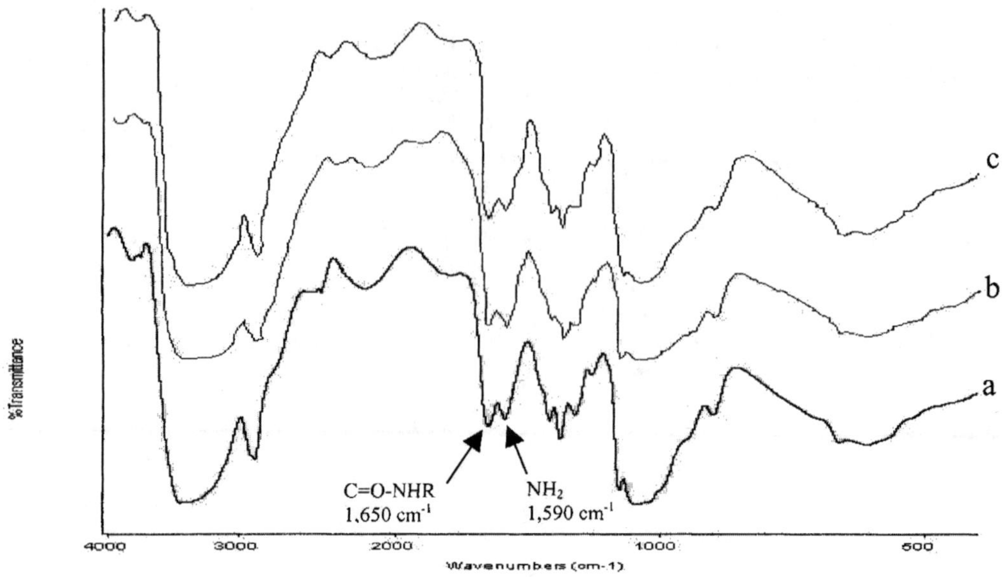

Figure 8. FTIR spectra of membranes a) CS; b) CS + 50%wt. $LiNO_3$; c) CS + 100%wt. $LiNO_3$

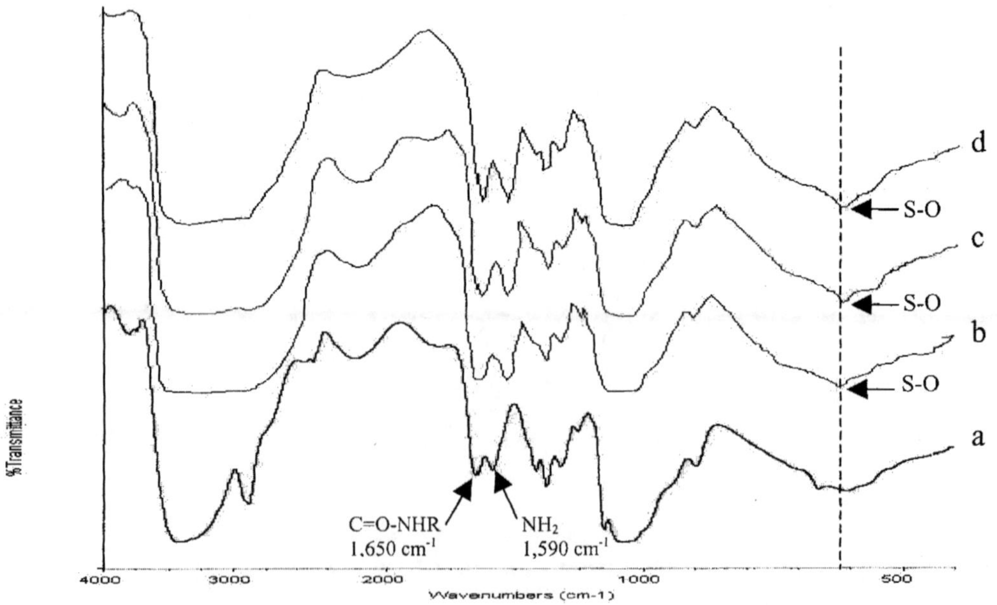

Figure 9 FTIR spectra of membranes a) CS; b) 2%crosslinked; c) 4% crosslinked; d) 6% crosslinked

3.4 Ion Exchange Capacity

The effect of complexation on the ion exchange capacity was shown in Figure 10. It showed that the higher crosslinking degree, the higher ion exchange capacity. It could be explained that the H_2SO_4 introduced protonic sites into the membranes. The ion exchange capacity of chitosan membrane increased from 2.5±0.03 for uncrosslinked membrane to 5.15±0.24, 5.22±0.04 and 5.66±0.19 meq g^{-1} for 2%, 4% and 6% crosslinking, respectively. Even uncrosslinked membrane possessed higher ion exchange capacity than Nafion® 117 of 0.91 meq g^{-1}. The chitosan membranes showed the potential as candidates for fuel cell electrolyte membranes.

Complexation with $LiNO_3$ as shown in Figure 10(a) showed the optimum content of 50 %w/w for 2% and 4% crosslinked membranes. Plasticizer EC did not contribute to the IEC determination, as shown in Figure 10(b), which was performed via acid-base titration.

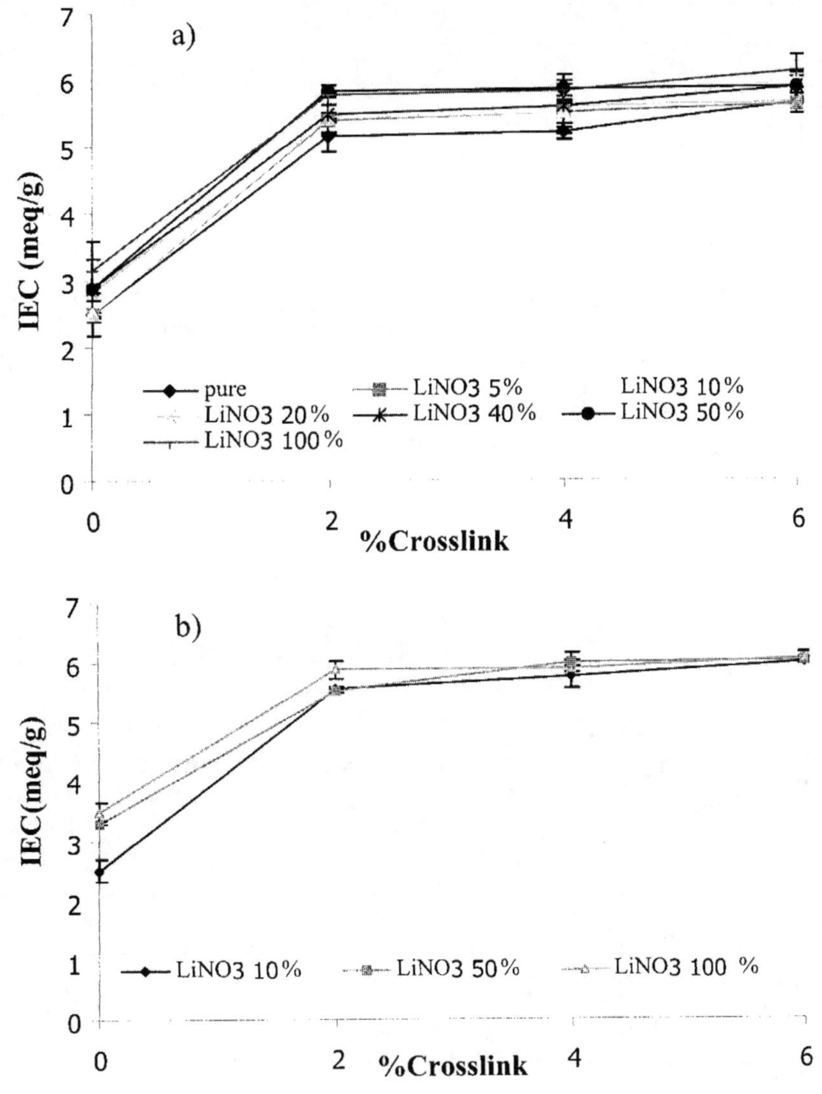

Figure 10. Variation of IEC with %crosslink in:
(a) membrane without EC; (b) membrane with EC

3.5 Ionic Conductivity of Membrane

The ionic conductivity needed to be measured because the high ion exchange capacity of chitosan membrane may be due to various ionic groups in molecular structure. The results were shown in Table 2. This may be explained based on the following tentative mechanism. The conductivity in uncrosslinked membranes was attributed to a liquid-like proton conductivity mechanism where protons were transported as free hydronium ions through swollen structure of the membranes. The higher conductivity in crosslinked membranes was attributable to the higher IEC due to the introduction of sulfuric acid onto the backbone of chitosan, resulting in higher conductivity. In case of salt-complexed membranes, the Li^+ may complex with the OH^- in water and leaving free H^+ to move and give an additional ionic current under the action of DC signal.

Table 2. Ionic conductivity of some selected membranes

Type of Membranes	Ionic conductivity ($S\ cm^{-1}$)
Uncrosslinked chitosan	8.7×10^{-4}
2% crosslinked chitosan	1.2×10^{-2}
4% crosslinked chitosan	2.8×10^{-2}
4%-crosslinked 50%-$LiNO_3$ chitosan	6.3×10^{-2}

4. CONCLUSION

Various chitosan membranes were prepared. It was observed that the physical properties of membranes prepared in this study were mainly affected by the chemical crosslinking of chitosan matrix with sulfuric acid and by the introduction of lithium salts. T_g measurement and FTIR results indicated the chitosan-salt interaction. The ionic conductivity of chitosan membranes was enhanced by crosslinking with sulfuric acid and addition of lithium salt. The suitable membranes were the crosslinked salt complexes containing 50 %w/w $LiNO_3$ and crosslinking with 4 %w/w sulfuric acid. Further work is being carried out to improve the ionic conductivity of the membranes.

ACKNOWLEDGEMENTS

The authors would like to gratefully acknowledge The Petroleum and Petrochemical Technology Consortium and The National Metal and Materials Technology Center (MTEC) for financial support of this research. K. Soontarapa thanks K.-I. Ota, and members in Division of Materials Science and Chemical Engineering, Yokohama National University, Japan for providing 4-point-probe cell and corporations.

REFERENCES

1. M. Rikukawa, K. Sanui, Proton-conducting polymer electrolyte membranes based on hydrocarbon polymers, Prog. Polym. Sci. 25 (2000) 1463-1502.
2. S. M. J. Zaidi, S. F. Chen, S. D. Mikhailenko and S. Kaliaguine, Proton conducting membranes based on polyaxadiazoles, J. New Mat. Electrochem. Systems 3 (2000) 27-32.
3. J. Kerres, Development of ionomer membranes for fuel cells, J. Membr. Sci. 185 (2001) 3-27.
4. J. Kerres, A. Ulrich, Th. Häring, M. Baldauf, U. Gebhardt and W. Preidel, Preparation, characterization and fuel cell application of new acid-base blend membranes, J. New Mat. Electrochem. Systems 3 (2000) 229-239.

5. W. Walker, K.-M. Baumgärtner, M. Kaiser, J. Kerres, A. Ullrich, E. Räuchle, Proton-conducting polymers with reduced methanol permeation, J. Appl. Polym. Sci. 74 (1999) 67-73.
6. M. Z. A Yahya, A. K. Arof, Effect of oleic acid plasticizer on chitosan-lithium acetate solid polymer electrolytes, Eur. Polym. J. 39 (2003) 897-902.
7. M. Z. A Yahya, A. K. Arof, Studies on lithium acetate doped chitosan conducting polymer system, Eur. Polym. J. 38 (2002) 1191-1197
8. N. M. Morni, A. K. Arof, Chitosan-lithium triflate electrolyte in secondary lithium cells, J. Power Sources 77 (1999) 42-48.
9. http://user.chollian.net/~chitin/prechitosan.gif, "What is Chitin/Chitosan?"
10. A. Domard, M. Rinaudo, Preparation and characterization of fully deacetylated chitosan, Int. J. Biol. Macromol. 5 (1983) 49-52.
11. H. Miyoshi, K. Shimura, K. Watanabe, K. Onodera, Characterization of some fungal chitosans, Biotech. Biochem. 56 (19922) 1901-1905.
12. M. T. Qurashi, H. S. Blair, S. J. Allen, Studies on modified chitosan membranes. I. Preparation and characterization, J. Appl. Poly. Sci. 46 (1992) 255-261.
13. F. Nanjo, R. Katsumi, and K. Sakai, Enzymatic method for determination of the degree of deacetylation of chitosan, Anal. Biochem. 193 (1991) 164-167.
14. H. Sato, S. Mizutani, and S. Tsuge, S., Determination of the degree of acetylation of chitin/chitosan by pyrolysis-gas chromatography in the presence of oxalic acid, Anal. Chem. 70 (1998) 7-12.
15. A. Baxter, M. Dillon, K. D. A. Taylor, Improved method for I.R. determination of the degree of N-acetylation of chitosan, Int. J. Biol. Macromol. 14 (1992) 166-169.
16. T. Sannan, T., K. Kurita, and Y. Iwakura, Studies on chitin, 2, Makromol. Chem. 177 (1976) 3589-3600.
17. J. G. Domszy, G. A. F. Roberts, Evaluation of infrared spectroscopic techniques for analysing chitosan, Makromol. Chem. 186 (1985) 1672-1677.
18. Y. M. Lee, H. B. Park, S. Y. Nam, J. M. Won, H. Kim, Effect of deacetylation degree in chitosan composite membranes on pervaporation performance, Separ. Sci. and Technol. 33 (1998) 1255-1269.

Microfibrous Entrapped Supported-ZnO Sorbents with High Contacting Efficiency for Trace H_2S Removal in PEMFC Applications

Yong Lu and Bruce J. Tatarchuk
Center for Microfibrous Materials Manufacturing
Department of Chemical Engineering, Auburn University, Auburn, AL 36849

ABSTRACT

H_2S removal is a key step in ensuring the activity of various fuel processing catalysts and high value membrane electrode assemblies. To accomplish this goal a sintered microfibrous carrier consisting of 2.0 - 3.0 vol% of 4 and 8 μm diameter Ni fibers is used to entrap from 20 to 30 vol% of 150-250 μm diameter carbon and SiO_2 support particulates. ZnO is then placed onto these supports by incipient wetness impregnation at loadings from 15 to 20 wt% (including the mass of the Ni fibers). At equivalent bed volumes, ZnO/Carbon for low temperature applications (R.T. to 100°C) and ZnO/SiO_2 for regenerable use all provide 2-3 times longer breakthrough time for H_2S with a 67% reduction in sorbent loading, compared to packed beds of commercially available 1-3 mm extrudates. In both cases the highly dispersed nature of the ZnO and the small support particulates promote both high ZnO utilization and high contacting efficiency, and also facilitates regeneration in air. Microfibrous sorbents (ca. 1 mm in thickness) can provide greater than 5-log reduction in H_2S concentrations (at 1.2-1.7cm/s face velocities) until ≥67% ZnO utilization has been achieved. Furthermore, microfibrous entrapped sup-ported ZnO sorbent also envisioned to use as a polisher behind packed beds of large extrudates (ca. 1-3 mm dia.). This approach synergistically combines the high volume loading of packed beds and the overall contacting efficiency of small particulates.

KEY WORDS Microfibrous materials carrier, H_2S removal, ZnO sorbent, Fuel Cell

INTRODUCTION

1-100 kW PEMFC power systems have become the focus of intense interest because of expending markets including stationary, transportation, military, residential, APU, telecommunication, and UPS/battery replacement. A significant challenge confronting these efforts is the development of on-board reformation technologies utilizing high energy density commercial-grade hydrocarbon fuels. Recently, Tatarchuk's group has invented a new type of microfibrous composite materials that allow entrapment of micron sized sorbent and catalyst particulates[1-8]. Sintered microfibrous carrier materials have substantial potential for enhanced heat/mass transfer, improved contacting efficiency and regenerability, and avoid high pressure drop and high intraparticle mass transport resistance that typically limit packed-bed performance. Use of these materials allows reduction of both reactor weight and volume while permitting development of small and robust fuel processors. Indeed, weight and volume as well as efficiency are very important fuel cell design considerations. Microfibrous entrapped 16% Ni/Al_2O_3 composite catalysts for toluene hydrogenation in a trickle bed reactor have demonstrated 2-6 times higher specific activities than conventional packed bed catalysts on a gravimetric basis, while volumetric activities of 40 vol% composite catalysts were 80% higher than conventional extrudates[1-8]. ABB has advanced catalysis with microfibrous composite materials and developed novel catalytic structured packings for catalytic distillation applications. Two-phase mass-transfer testing indicated that microfibrous composite catalysts take advantage of both high gas-liquid contacting and bulk mixing at low pressure drop with the potential to provide enhanced catalyst utilization[9].

H_2S removal is a key step in ensuring activity maintenance of various fuel processing catalysts and high value membrane electrode assemblies. For PEMFC application, H_2S content less than 0.1ppmv is crucially required[10,11]. Zinc oxide is well known as a highly efficient desulfurizer due to extremely low equilibrium H_2S concentration and is widely used for sulfur removal from various gas sources. Normally, packed bed of large ZnO sorbent extrudates with high sorbent inventory, are employed to remove trace H_2S from the reformate gas for fuel cell application[12]. Nevertheless, large particles result in low ZnO utilization and poor regenerability because of low contacting efficiency, pore and lattice diffusion limitation[13-15]. Requisite high sorbent inventory requires larger reactor weight and volume than necessary.

In this paper, sintered microfibrous carrier was used to entrap 150-250 μm supported ZnO sorbent particulates for high contacting efficiency, high ZnO utilization and low diffusion limitation. Two different ZnO/support recipes were developed. ZnO/Carbon for low temperature applications is envisioned to operate as a last line of defense at stack temperatures (i.e., R.T. to 100°C), and ZnO/SiO_2 for regenerable use in a continuous batch mode at ca. 400°C. Furthermore, microfibrous entrapped supported ZnO sorbent is also envisioned to be used as a polishing sorbent behind packed beds of large extrudates (ca. 1-3 mm dia.).

EXPERIMENTAL

PREPARATION OF MICROFIBROUS ENTRAPPED SUPPORTED ZnO SORBENT PARTICULATES - Sintered microfibrous carrier was used to entrap 150-250 μm diameter support particulates firstly by regular wet layer paper-making/sintering procedure[1-8], where SiO_2 (300m²/g, Grace Davison), BPL activated carbon (560m²/g, Calgon), and γ-Al_2O_3 (220m²/g, Alfa) were chosen. ZnO was then placed onto this support by incipient wetness impregnation. Detailed procedure was described as following sample preparation: 7.5g of 8μm Ni fibers, 2.5g 4μm Ni fibers, and 2.5g of cellulose were added into water and stirred vigorously to uniform dispersion. The produced slurry and 18g 150-250 μm SiO_2 were added into the headbox of the 1.0 ft² M/K sheet former under aeration. 1.0 ft² preform (1mm thick) was then formed by vacuum filtration followed by drying on a heated drum. Such preform was pre-oxidized in air flow for 2h at 500 °C followed by sintering in H_2 flow for 1h at 900 °C. The resulted product is comprised of 2.0vol% Ni fiber and 29vol% SiO_2 (including its own pore volume) support particulates. In order to couple the zinc with the prepared microfibrous entrapped SiO_2 support particulates, a piece of microfibrous entrapped SiO_2 support particulates was immersed into a ZnO sol-gel for 10min. The ZnO sol-gel was prepared by adding 70ml $NH_3{\cdot}H_2O$, 42g $(NH_4)_2CO_3$, and 66g $Zn(Ac)_2{\cdot}6H_2O$ in series into 56ml water under vigorously stirring. After subsequent vacuum filtration, the product was calcined for 0.5h at each temperature of 80°C, 100°C, 120°C, 140°C, 160°C and 180°C to form a microfibrous entrapped ZnO/SiO_2 sorbent.

The composition, physico-chemical properties of as-prepared microfibrous entrapped supported ZnO sorbents are summarized in Table 1.

Table 1. Composition, physico-chemical properties of microfibrous entrapped supported ZnO sorbents

Microfibrous Sorbent	Composition, wt%			Voidage %
	ZnO	Support	Ni fiber	
ZnO/SiO_2	18[a]/17.3[b]	43	39	82
ZnO/Carbon	19[a]/18.6[b]	44	37	80
ZnO/γ-Al_2O_3	18[a]	43	39	83

[a] by wieghing; [b] by ICP-AES

A commercial ZnO sorbent (3/16" pellet, 25m²/g, 90wt%ZnO) from Sud Chemie, and several carbon-based sorbents taken from state-of-the-art respiratory cartridges were crushed, sieved into desired particle size for comparative study use.

MICRO-REACTOR EVALUATION

1 H_2S PULSE REACTION - H_2S pulse reaction was carried out on a microreactor system consisting of a quartz tube reactor (i.d. 5mm) followed by a moisture trap packed with P_2O_5, two six-port valves, a tube furnace, a temperature controller and an online GC equipped with a TCD. For microfibrous entrapped supported ZnO sorbent, 0.11g sample (containing 0.02g ZnO) was packed. For Sud-Chemie extrudates (1-2mm), 0.022g sample (containing 0.02g ZnO) diluted with α-Al_2O_3 particulates to the same volume as the above sample was used. The sorbent bed was heated to appointed temperature in GC carrier gas (He, 30ml/min) flow. After obtaining a straight GC baseline, 50μL H_2S was pulsed and carried to sorbent bed by He carrier gas every 2min.

2 H_2S REMOVAL FROM MODEL REFORMATE IN PRESENCE OF H_2O - Adsorption reaction was carried out on the same microreactor system as above just using a quartz tube reactor with 11 mm i.d. instead of the smaller one. An H_2S-containing model reformate (50ppmv $H_2S/40\%H_2/10\%CO/20\%CO_2/30\%N_2$, BOC GASES) was used as challenge. Before contacting sorbent bed at face velocity of 1.7cm/s, challenge gas was saturated with 5-30vol% H_2O by passing a water vapor saturator. For microfibrous entrapped supported sorbent, 11mm (dia.)×3mm(thick) disc sample was used. For particle sorbents, equivalent volume (0.29mL) of particulates was packed. It is defined that breakthrough point for H_2S appeared when H_2S concentration in outlet reached 1 ppmv. The breakthrough point was checked using lead acetate strip, of which the detection limitation for H_2S is 1 ppm.

3 MULTICYCLE TEST AND COMPOSITE BED TEST - High-temperature adsorption/regeneration reaction was carried out on the same microreactor system with a SS tube reactor with 7mm i.d., using 2vol%H_2S/98vol%H_2 as challenge. Sorbent loading was always 0.4g for either microfibrous entrapped supported ZnO (including the mass of Ni fiber) or Sud-Chemie ZnO sorbent extrudates (1-2mm). Sud-Chemie sorbent extrudates were diluted with α-Al_2O_3 to the equivalent volume as microfibrous sorbents. H_2S adsorption was performed at 400 °C and a face velocity of 1.2cm/s. Regeneration was performed at 500-600 °C in a pure air flow (50ml/min) for indicated time length. The breakthrough point was also checked using lead acetate strip. Online GC with TCD (H_2 as carrier gas) was used to determine the H_2S concentration in outlet when H_2S concentration was higher.

For composite bed, the tests for H_2S adsorption were carried out on the following four cases on the same reaction system under same reaction conditions as above.

A: Microfibrous entrapped supported ZnO, 0.1g

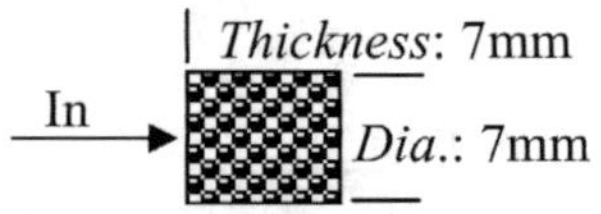

B: Sud-Chemie sorbent particulate (1-2mm), 0.27g

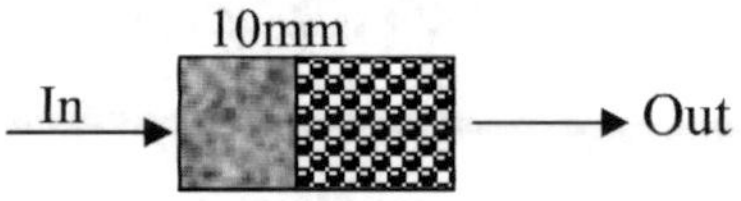

A+B

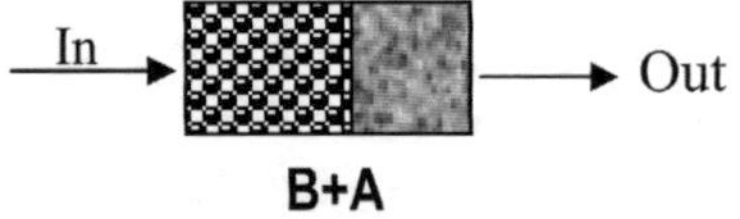

B+A

2.3 XRD ANALYSES - Power XRD patterns were recorded on Rigaku instrument using a scanning speed of 4°/min and an accelerating voltage of 40kV.

RESULTS AND DISCUSSION

H_2S PULSE REACTION OVER MICROFIBROUS ENTRAPPED ZNO/SIO_2 AND SUD-CHEMIE SORBENT - Fig. 1 shows the H_2S pulse reaction results over microfibrous entrapped ZnO/SiO_2 and Sud-Chemie ZnO sorbent extrudates at 400°C. ZnO/SiO_2 entrapped materials and 100-150μm Sud-Chemie ZnO sorbent particulates provided higher ZnO utilization and sharp breakthrough curves, compared to ~1mm Sud-Chemie ZnO sorbent extrudates. It is well known that H_2S reaction with ZnO sorbent is controlled by both pore and lattice diffusion limitation[13-15]. The results described above suggested that diffusion was eliminated with reduction of the size of sorbent particulates. On the other hand, the reduction of the size of sorbent particulates allowed enhanced contacting efficiency. This is also a reason for high efficiency H_2S adsorption over micro-sorbent particulates.

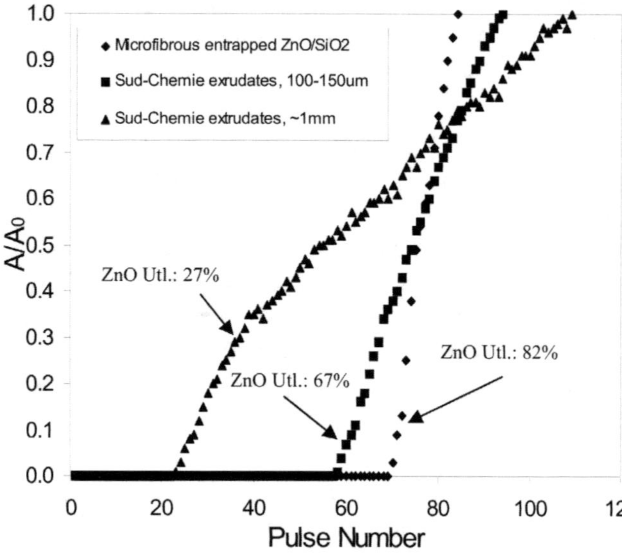

Figure 1. H_2S pulse reaction results of microfibrous entrapped ZnO/SiO_2 with Sud-Chemie ZnO sorbent extrudates at 400°C.

HIGH EFFICIENCY H_2S REMOVAL FROM MODEL REFORMATE IN THE PRESENCE OF H_2O

1 H_2S ADSORPTION OVER MICROFIBROUS ENTRAPPED ZnO/CARBON AT STACK TEMPERATURES - Table 2 shows the performance of microfibrous entrapped ZnO/Carbon for trace H_2S adsorption from model reformate at stack temperatures (R.T. to 100°C) and 1.7cm/s face velocity. Over 11mm(dia.)×3mm(thick) disc sorbent with loading of 0.02g ZnO, in the presence of 5%H_2O breakthrough time was 8.7h at 70°C and 9.5h at 100°C. ZnO utilization was 36% and 39%, respectively. However, for PEMFC application, up to 30vol% H_2O in reformate gas is required to keep membrane electrode assemblies from dry. In the case of the presence of 30% H_2O in reformate, this disc sorbent provided breakthrough time of 2.2h at 25°C and 8h at 70°C. ZnO utilization was 3.3% and 32% respectively. The above results suggested that the presence of 30% H_2O in reformate has a little negative effect on the H_2S adsorption efficiency. Blank disc sample without loading ZnO provided quite shorter breakthrough time for H_2S, compared to the disc sorbent sample. This suggested that ZnO captured most part of H_2S at 25°C to 100°C.

Table 2. Performance of microfibrous entrapped ZnO/Carbon for H_2S adsorption from model reformate at R.T. to 100°C in the presence of H_2O*

Sorbent	Breakthrough Time, hr.	ZnO Utl., %
Microfibrous entrapped ZnO/Carbon	9.5@100°C in 5%H_2O	39
	8.7@70°C in 5%H_2O	36
(11mm dia. × 3mm thick)	8.0@70°C in 30%H_2O	32
	2.2@25°C in 30%H_2O	3.3
Blank**	1.0@70°C in 5%H_2O	—
	1.4@25°C in 5%H_2O	—

* Face velocity: 1.7cm/s@50ppmv H_2S in model reformate (40%H_2, 10%CO, 20%CO_2, 30%N_2)
** Microfibrous entrapped carbon Impregnated with a solution consisting of 70ml $NH_3{\cdot}H_2O$, 42g $(NH_4)_2CO_3$, and 56ml H_2O

The accurate H_2S concentration in outlet was not known before breakthrough point for H_2S (1ppmv). Nevertheless, it has been noticed that the wet lead acetate strip did not change its color after in the effluent for >10min. This suggested that the real H_2S concentration at outlet was far from 1ppmv. The thermodynamic equilibrium data for reaction of H_2S with ZnO indicate that equilibrium concentration for H_2S in the presence of 30% becomes lower than 0.1ppmv at <200°C[13].

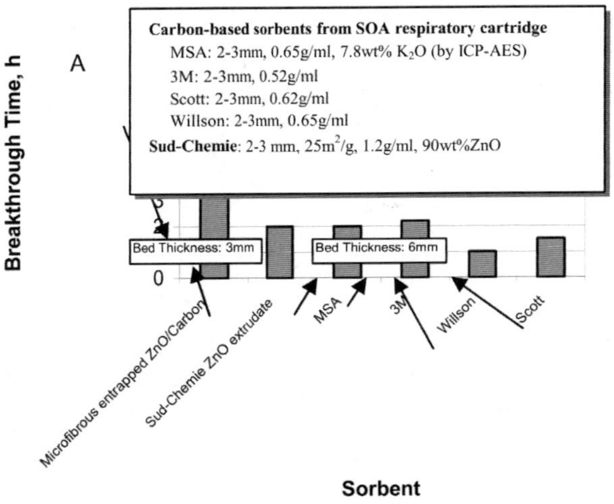

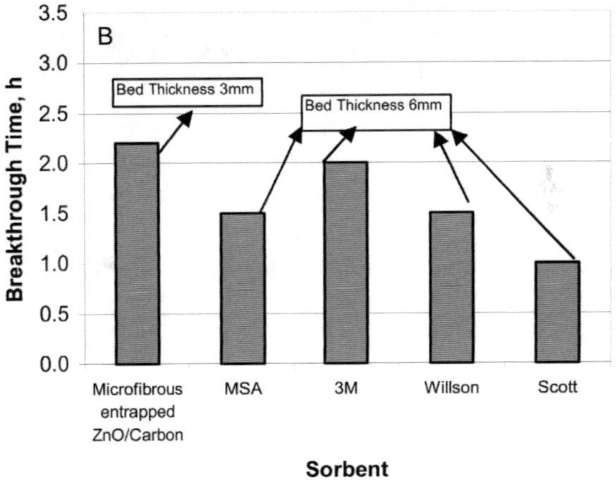

Figure 2. Comparison of microfibrous entrapped ZnO/Carbon with several commercially available sorbent particulates for H_2S adsorption from model reformate in 30% at (A) 70°C and (B) 25°C under a face velocity of 1.7cm/s.

Fig.2 compares microfibrous entrapped ZnO/Carbon with several commercially available sorbents for trace H_2S adsorption from model reformate at 25°C and 70°C respectively. Since there were no real low-temperature sorbents, several carbon-based sorbents taken from SOA respiratory cartridge and Sud-Chemie ZnO sorbent for high-temperature application were used for this comparison. As shown in Fig.2, the microfibrous entrapped ZnO/Carbon always provided longer breakthrough time for H_2S than those commercial sorbents. At equivalent bed volume ZnO/Carbon sorbent provided 3 times longer breakthrough time than 1-2mm Sud-Chemie ZnO sorbent extrudates at 70°C. For 2-3mm carbon-based sorbent particulates, bed volume was doubled but breakthrough time was still shorter than that for ZnO/Carbon sorbent at 25°C and was just a quarter of that for ZnO/Carbon sorbent at 70°C.

2 H_2S ADSORPTION OVER MICROFIBROUS ENTRAPPED ZnO/SiO_2 AT 400°C - Table 3 compares microfibrous entrapped ZnO/SiO_2 with 1-2mm Sud-Chemie ZnO extrudates for H_2S adsorption removal from model reformate at 400°C in the presence of 30% H_2O. At equivalent bed volume microfibrous entrapped ZnO/SiO_2 provided about 2 times longer breakthrough time with about 67% reduction of sorbent loading. ZnO utilization was 57% for ZnO/SiO_2 and 4% for Sud-Chemie ZnO extrudates. The above results further confirmed that small support particulates promote both high ZnO utilization and high contacting efficiency. The equilibrium concentration for H_2S in the presence of 30% is ~0.6ppmv at 400°C[13].

Table 3. Comparison of microfibrous entrapped ZnO/SiO_2 with Sud-Chemie ZnO sorbent extrudates for H_2S removal from model reformate at 400°C in the presence of 30% H_2O*

Sorbents	Microfibrous entrapped ZnO/SiO_2	Sud-Chemie extrudates (1-2mm)
Weight, g	0.12	0.36
Volume, ml	0.29	0.29
Bed thickness, mm	3	3
Breakthrough time, hr.	12	4.5
ZnO Utilization, %	57	3

* Face velocity:1.7cm/s@50ppmv H_2S in model reformate (40%H_2, 10%CO, 20%CO_2, 30%N_2)

3 XRD ANALYSES - Fig. 3 shows the XRD patterns of ZnO/Carbon and ZnO/SiO_2 and corresponding supports. Supported ZnO sorbents or supports were collected by unlocking the network of sintered microfibrous carriers followed by grinding to powder for XRD analyses. Unlike the mixture of ZnO and the support with the same composition as the corresponding supported-ZnO sorbents, no sharp peaks assigned to ZnO crystals were observed on both ZnO/Carbon and ZnO/SiO_2, indicating that ZnO was highly dispersed onto the surface of the supports. Like the small support particulates, the highly dispersed nature of the ZnO further promotes both high ZnO utilization and high contacting efficiency, compared to packed beds of commercially available 1-3 mm extrudates.

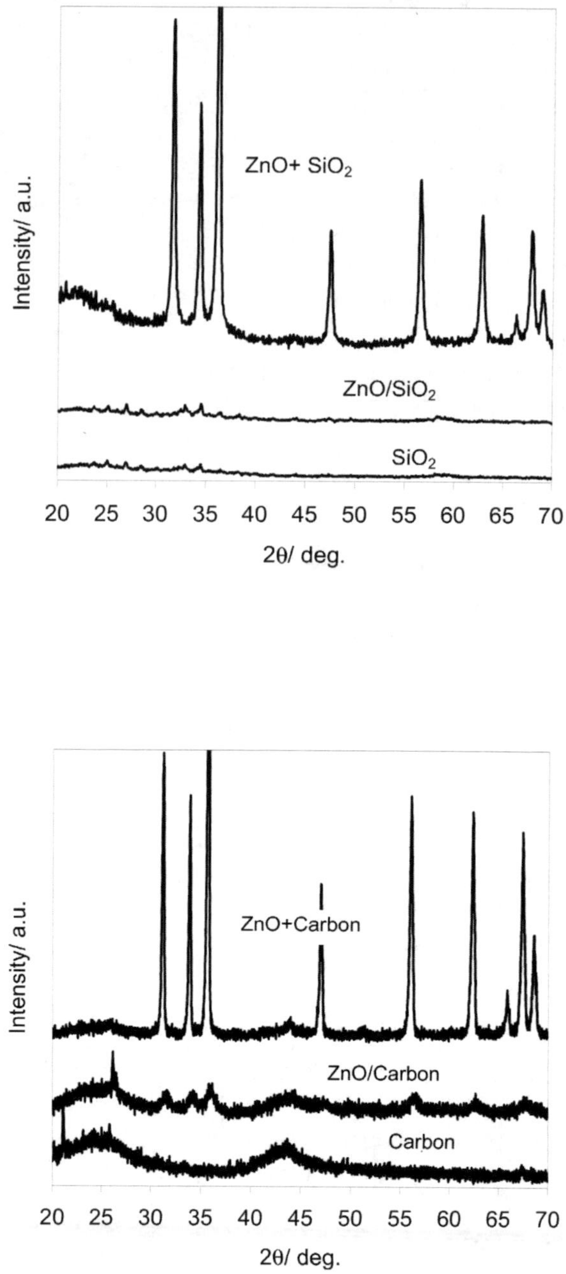

Figure 3. XRD patterns of microfibrous entrapped ZnO/SiO_2 and microfibrous entrapped ZnO/Carbon.

REGNERABILITY OF MICROFIBROUS ENTRAPPED ZnO/SiO_2 AND MULTICYCLE TEST RESULTS

1 COMPARISON OF REGENERABILITY - Table 4 compares the regenerability of microfibrous entrapped supported ZnO sorbents and 1-2mm Sud-Chemie ZnO sorbent extrudates. ZnO utilization for both fresh microfibrous entrapped supported ZnO was 67% and 65%, twice times as high as that of fresh Sud-Chemie extrudates. After regeneration at 600 °C for 1-3h, ZnO utilization decreased from 67% to around 60% for microfibrous entrapped ZnO/SiO_2, and from 33% to 12-20% for Sud-Chemie extrudates. After regeneration at 550 °C for 3h, ZnO utilization was 32% for microfibrous entrapped ZnO/SiO_2, and just 4% for Sud-Chemie extrudates. The above results suggested that the highly dispersed nature of the ZnO and the small SiO_2 particulates also facilitate regenerability in air.

Table 4. Comparison of regenerability of microfibrous entrapped supported ZnO and Sud-Chemie ZnO sorbent extrudates*

Sorbent	ZnO Utilization, %		
	Microfibrous entrapped supported ZnO sorbent		Sud-Chemie, 1-2mm
	ZnO/SiO_2	ZnO/Al_2O_3	
Fresh	67	65	31
3h@600°C in air**	60	8	21
1h@600°C in air**	57	12	13
3h@550°C in air**	44	15	4

* H_2S adsorption was carried out at face velocity of 1.2cm/s@2%H_2S/98%H_2 at 400°C

** Regeneration conditions

Table 4 also shows that microfibrous entrapped $ZnO/\gamma\text{-}Al_2O_3$ exhibited poor regenerability compared to microfibrous entrapped ZnO/SiO_2. XRD analyses show that very strong peaks assigned to ZnO crystals were observed on $ZnO/\gamma\text{-}Al_2O_3$ after regeneration at 550°C for 3h. New peaks assigned to $ZnAl_2O_4$ spinel crystals were clearly observed on $ZnO/\gamma\text{-}Al_2O_3$ after regeneration at 600°C for 3h (see Fig. 4). For ZnO/SiO_2, however, just weak peaks assigned to ZnO crystals were also observed after regeneration and peak intensity increased slightly with regeneration temperatures but without any detectable new peaks (see Fig. 4). To our knowledge, the surface is quite reactive for $\gamma\text{-}Al_2O_3$ but very inert for SiO_2. As a result, strong interaction even solid reaction of ZnO with $\gamma\text{-}Al_2O_3$ occurred, but hardly with SiO_2 during regeneration in air. This is the real reason for the poor regenerability of $ZnO/\gamma\text{-}Al_2O_3$.

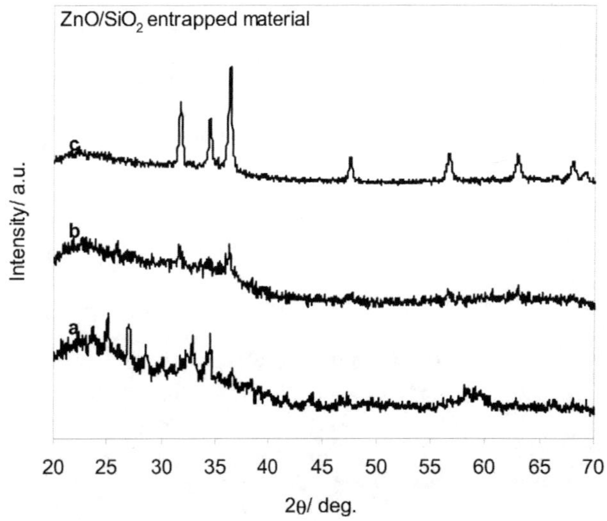

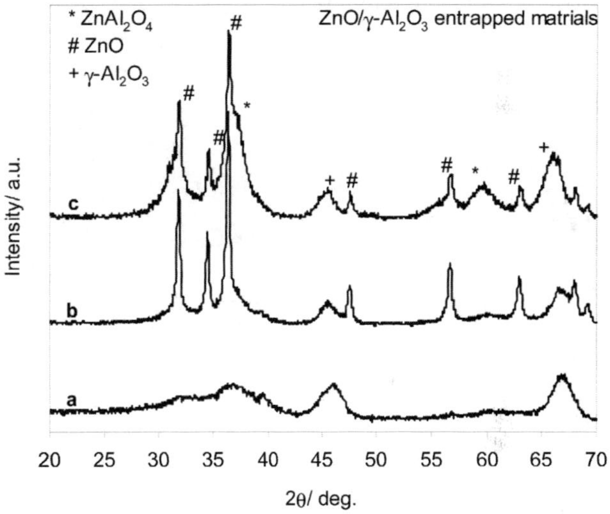

Figure 4. XRD patterns of microfibrous entrapped supported ZnO sorbents (a) fresh, (b) after regeneration at 550°C in air for 3h, and (c) after regeneration at 600°C in air for 3h.

2 MULTI-CYCLE TEST RESULTS - Fig. 5 shows the multicycle test results over microfibrous entrapped ZnO/SiO_2. The breakthrough times almost remained unchanged in 10 cycles at regeneration temperature of 600 °C, even in 20 cycles at regeneration temperature of 500°C. The results suggested that microfibrous entrapped ZnO/SiO_2 sorbent has good stability and can be operated in duty cycles.

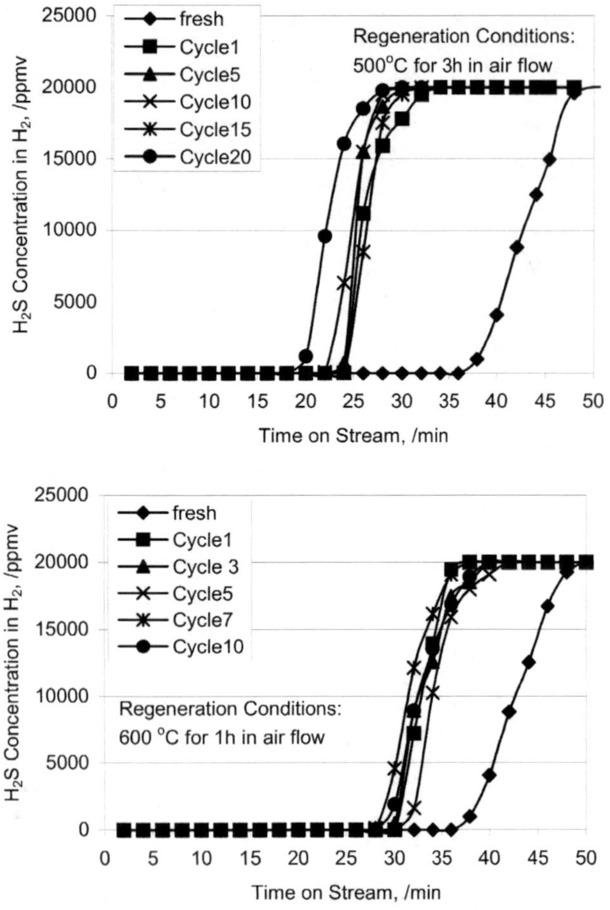

Figure 5. Adsorption/regeneration multi-cycle test results of microfibrous entrapped ZnO/SiO_2 sorbent. H_2S adsorption reactions were carried out at 400°C and a face velocity of 1.2cm/s@2%H_2S/98%H_2.

OPTIMIZATION OF MICROFIBROUS ENTRAPPED PARTICLE SOLID

PARTICLE SOLID - Fig. 6 compares the performance of single and composite bed for H_2S adsorption from 2%H_2S/98%H_2 at 400°C. Two single beds, 7mm(dia.)×3mm(thick) microfibrous entrapped ZnO/SiO_2 (**A**) and 7mm(dia.)×7mm(thick) packed bed of 1-2mm Sud-Chemie ZnO sorbent extrudates (**B**), provided breakthrough times of 5min and 28 min, respectively; the breakthrough curve was very sharp for bed A but quite mild for bed **B**. The mild breakthrough curve of bed **B** suggested that H_2S started out at 28min time point and its concentration increased slowly with time. On the bed of **A+B**, breakthrough time was 32 min, very close to the value of 5 plus 28; breakthrough curve was still mild like single bed B. When the polishing sorbent A was placed behind the bed **B** (bed **B+A**), low concentration H_2S out from sorbent **B** was encountered immediately with high contacting efficiency sorbent **A**. As a result, breakthrough time length was extended to 52min, but the adsorption capacity of the polishing sorbent A remained unchanged because area 1 and area 2 were found almost equal. Clearly, the composite bed consisting of the packed bed of large extrudates followed by microfibrous entrapped polishing sorbents, synergistically combine the high volume loading of packed bed and the overall contacting efficiency of small particulates.

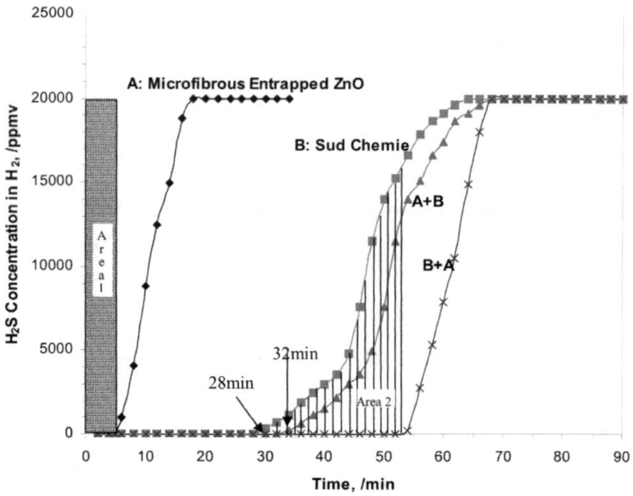

Figure 6. Performance of single and composite beds for H_2S adsorption from 2%H_2S/98%H_2 at 400°C and a face velocity of 1.2cm/s.

SUMMARY

Sintered microfibrous carrier consisting of 2-3vol% of 4 and 8 μm diameter Ni fibers is used to entrap from 20-30vol% of 150-250 μm supported ZnO sorbent particulates for enhanced contacting efficiency and improved regenerability. Two different ZnO/support recipes have been developed. ZnO/Carbon for low temperature applications envisioned to operate as a last line of defense at stack temperatures (i.e., R.T. to 100 °C), and ZnO/SiO_2 for regenerable use in a continuous batch mode at ca. 400°C

At equivalent bed volumes, microfibrous entrapped supported ZnO sorbents provide 2-3 times longer breakthrough time for H_2S with a 67% reduction in sorbent loading, compared to packed beds of commercially available 1-3 mm extrudates

For ZnO/Carbon entrapped materials, H_2S concentrations of 50 ppmv can be reduced to as little as 0.1 ppmv at 25°C in 30% H_2O at face velocity of 1.7cm/s for layers as thin as 1.0 mm in thickness. ZnO utilization ranges from 3.3 to 39% at R.T. to 100°C, respectively

For ZnO/SiO_2 entrapped materials, H_2S concentrations of up to 20,000 ppmv can be reduced to as little as 0.6 ppmv at 400°C in 30% H_2O at face velocities of 1.2 - 1.7cm/s for layers as thin as 1.0 mm in thickness. ZnO utilization ranges from 57 to 67% at 400°C, respectively

The highly dispersed nature of the ZnO and the small support particulates promote both high ZnO utilization and high contacting efficiency, and also facilitate regeneration in air at 500 to 600°C and therefore, can be used to even further extend the magnitude of this volumetric and mass difference.

Composite bed consisting of packed bed of large extrudates (ca. 1-3 mm dia.) followed by microfibrous entrapped polishing sorbents, synergistically combine the high volume loading of packed bed with the overall contacting efficiency of small particulates.

ACKNOWLEDGMENTS

This work was performed under a U.S. Army contract at Auburn University (DASG 60-00-C-0070) administered through the U.S. Army Space & Missile Defense Command.

REFERENCES

1. Bruce J. Tatarchuk, M.F. Rose, A. Krishnagopalan, J.N. Zabasajja, and D. Kohler, US Patent 5,304,330, 04/19/94; US Patent 5,080,963, 01/14/92.
2. Rudolf A. Overbeek, Ali M. Khonsari, Yung-Feng Chang, Lawrence L. Murrell, Bruce J. Tatarchuk, Michael W. Meffert, US Patent 6,231,792, 05/15/01.
3. Bruce J. Tatarchuk. US Patent 5,096,663, 03/17/92; US Patent 5,102,745, 04/07/92
4. Donald R. Cahela and Bruce J. Tatarchuk. Catal. Today, 69 (2001) 33.
5. Christopher J. Marrion, Donald R. Cahela, Soonho Ahn, and Bruce J. Tatarchuk, J. Power Sources 47 (1994) 297.
6. D. Kohler, J.N. Zabasajja, A. Krishnagopalan, and Bruce J. Tatarchuk, J. Electrochem. Soc. 137(1) (1990) 136.
7. S. Ahn and Bruce J. Tatarchuk, J. Appl. Electrochem. 27 (1997) 9.
8. M.W. Meffert. Preparation and characterization of sintered metal microfiber-based composite materials for heterogeneous catalyst applications, PhD Dissertation, College of Engineering, Auburn University, Auburn, AL, 1998.
9. Robert E. Trubac, Frits M. Dautzenberg, Timothy A. Griffin, Bettina Paikert, Verena R. Schmidt, Rudolf A. Overbeek, Catal. Today 69 (2001) 17.
10. T. Zawodzinski, F. Uribe, W. Smith, T. Springer, J. Valerio, D. Vernon, T. Rockward, D. McMurry, G. Bender, M. Hickner, E. Brosha, B. Orler, C. Adams. "R&D on optimized cell performance for operation on refomate and air". In: Proceedings of the Annual National Laboratory R&D Meeting, DOE Fuel Cells for Transportation Program, Oak Ridge, Tennessee, 6-8 June 2001.
11. Chunshan Song, Catal. Today 77(1-2) (2002) 17.
12. M. Cervi and D. Hoffman, 2nd DOD Diesel Fuel Reforming Coordination Meeting, Fort Belvoir, Virginia, (Dec. 2001).
13. Martyn V. Twigg, in: Catalyst Handbook (2nd ed.), eds. Martyn V. Twigg, (Wolfe, London, 1989) p.209.
14. P.J.H. Carnell and D.P. Denny, AICE Ammonia Safety Symposium, San Francisco, Paper 94c (Nov. 1984).
15. P.J.H. Carnell and P.E. Starkey, Chem. Eng. 408 (1984) 30.

Fuel Processor Utilizing PSA for Generating Hydrogen for Fuel Cell Applications

Craig S. Gittleman, John B. Ruhl, Yan Zhang, and Kiran Mallavarapu
General Motors
Fuel Cell Activities
10 Carriage Street
Honeoye Falls, NY 14472

ABSTRACT

This paper discloses a system in which pressure swing adsorption (PSA) is used to purify the H_2 generated in an autothermal reformer (ATR) based fuel processing system for fuel cell applications. While PSA systems have been used for purifying H_2 generated by the reforming of hydrocarbon fuel for decades, the PSA equipment and cycles of industrial H_2 PSAs are not ideal for purifying H_2 for fuel cell applications where ATR is used to generate H_2. Industrial PSAs operate at relatively high pressures, often greater than 20 bara. Processes that require air compression for ATR would be inefficient at these elevated pressures, and would preferably operate at pressures of 5 bara or less. Typical industrial H_2 PSAs also operate at lower temperatures than those suited for integration with fuel cells systems, where it is not desired to cool the H_2 stream to ambient temperature only to subsequently reheat the H_2 to the fuel cell stack operating temperature of 60-100°C. Additionally, industrial H_2 PSAs in steam reforming plants are designed to purify a stream which contains greater than 75% H_2, whereas, the product from an ATR typically contains no more than 50% H_2.

This paper describes a fuel processor system that utilizes rapid-cycle PSA to purify H_2 from a gas mixture generated by autothermal reforming of a hydrocarbon fuel. A PSA device was developed for selectively adsorbing CO, CO_2, N_2, H_2O, and light hydrocarbons at pressures below 5 bara and temperatures between 20 and 100 °C. The PSA purifies a reformate stream which consists of less than 50% H_2. The PSA H_2 product contains less than 50 ppm CO, at least 95% H_2, and 70% of the H_2 in the feed is recovered in the product stream.

KEY WORDS

Fuel cell, fuel processing, reforming, H_2 PSA, adsorption, hydrogen

INTRODUCTION

In proton exchange membrane (PEM) fuel cells, hydrogen is the anode reactant (i.e. fuel) and oxygen is the cathode reactant (either a pure form or air). The Pt catalysts typically used in the fuel cell electrodes demand control of catalyst fouling constituents in the fuel feed stream, such as carbon monoxide (CO), which block the catalyst sites required for H_2 oxidation in the anode.

For vehicular applications, it is desirable to use a liquid fuel such gasoline or methanol as the source of hydrogen for the fuel cell. Such liquid fuels for the vehicle are easy to store onboard, and there is a nationwide infrastructure for supplying liquid fuels. Gaseous hydrocarbons such as methane, propane, natural gas, or LPG are also suitable fuels for both vehicular and non-vehicular fuel cell applications. However, hydrocarbon-based fuels must be dissociated to release the hydrogen content thereof for fueling the H_2 fuel cell. The dissociation reaction is accomplished within a chemical fuel processor or reformer. The fuel processor contains one or more reactors wherein the fuel reacts with steam (and sometimes air) to yield a reformate gas comprising primarily hydrogen and carbon dioxide. Industrial scale H_2 is typically produced by steam reforming. However, steam reforming is not suitable for on-board vehicle applications due to the large thermal mass and subsequent slow startup times and transient response of such systems. A preferred alternative utilizes autothermal reformation, where steam, air and hydrocarbon fuel are reacted in a fuel processor to generate a reformate comprising hydrogen, carbon dioxide, carbon monoxide, nitrogen and water. Typically, the H_2-containing reformate is subsequently fed to a water gas shift (WGS) reactor to further reduce the CO level and to increase the H_2 content of the reformate.

However, the reformate still requires further purification to remove the CO to a level tolerable in a PEM fuel cell.

This is commonly achieved by Preferential Oxidation (PrOx), where air is fed into the reformate over a noble metal catalyst. However, the PrOx produces a dilute H_2 stream (< 50%), which is not suitable for compression and storage. Other disadvantages of PrOx systems include the use of noble metal catalysts and the relatively long startup times required before stack grade H_2 becomes available.

BACKGROUND

The objective of this project was to develop a fuel processor system that could generate near pure hydrogen from natural gas for a 10 kW electric fuel cell system. The requirement for near pure H_2 was at least 95% H_2 purity. This near pure H_2 could then be either sent directly to a fuel cell stack to generate electricity or stored for later use. One possible application for such a system would be a fuel processor for a residential fuel cell system that could also be used as a H_2 generator for providing H_2 for a fuel cell vehicle.

The principal technology selected for hydrogen generation was autothermal reforming, as this was the primary technology GM had developed for its automotive fuel cell systems. However, the PrOx based systems of GM's automotive fuel processor could not meet the H_2 purity requirements. Additionally, compressor limitations constrained the overall system pressure to about 3.5 bara. H_2 purification systems, such as PSAs and membranes, that could purify H_2 at such low pressures were not available. Therefore, a novel method for purifying a H_2 stream from an ATR-based fuel processor at relatively low pressures needed to be developed.

FUEL PROCESSOR SYSTEM

A schematic of the fuel processor system of the invention is shown in Figure 1. Desulfurized natural gas is fed into the autothermal reformer (ATR) along with air and steam, where it is reformed into H_2 (~30%). The ATR effluent also contains CO, CO_2, N_2, H_2O, and trace CH_4, and leaves the ATR at about 750°C. The ATR effluent passes through a post-ATR heat exchanger where the reformate is cooled to about 320°C. The post-ATR heat exchanger also serves to pre-heat the steam-air mixture that is fed to the ATR. Additionally, there is a water spray available to further cool the reformate leaving the post-ATR heat exchanger if necessary.

The reformate then flows into a series of medium temperature water-gas shift reactors, WGS1 and WGS2, where CO and H_2O in the reformate are converted to H_2 and CO_2. A heat exchanger after WGS1 cools the reformate before it enters WGS2 and preheats steam before it is mixed with the feed air to the ATR. The reformate leaves WGS2 at about 300°C, depending on the requested power level, and contains ~30%H_2, ~1% CO, CO_2, N_2, H_2O, and trace CH_4. A heat exchanger optionally cools the reformate leaving WGS2 by exchanging heat with water which is vaporized to create a secondary steam supply for the ATR. As described below, the primary steam supply is vaporized in the post-combustor heat exchanger.

The reformate then flows through an air-cooled condenser, which cools the reformate to between 60-80°C. The cooled reformate flows to a water separator to remove liquid water, which can be collected and refed into the fuel processor for steam generation. A three-way valve then directs the reformate either to the PSA for standard operation or to the combustor for system start-up.

The reformate is fed to the PSA at about 35 psig and between 60-80°C. The PSA raffinate product contains at least 95% H_2 with the balance primarily N_2 and no more than 50 ppm CO, and leaves the PSA at essentially the feed pressure and temperature. The H_2 product may be either fed directly to a fuel cell stack or may be compressed and stored.

The extract from the PSA contains nearly all of the CO, CO_2, N_2, H_2O, and CH_4 from the reformate and a fraction of the H_2. The extract, which is at near ambient pressure, is mixed with air from the combustor blower, and fed to the combustor where the remaining H_2, CO, and CH_4 are burned. The hot combustor exhaust is fed to a heat exchanger where water is vaporized as the primary steam source for the ATR.

The volume of the fuel processor system excluding the condenser, water separator and PSA is about 7 L. This includes all reactors, heat exchangers and tubing. The system was designed and demonstrated to operate with a 5:1 turndown ratio. The optimum flows of fuel, air and water were determined upon system operation. At full power the optimum O/C was 0.9 and the optimum S/C was 3.2.

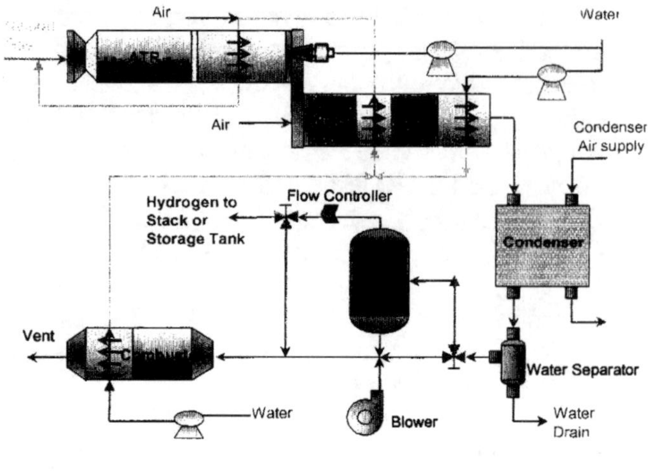

Figure 1. Schematic of Fuel Processor System for generating Near Pure H_2.

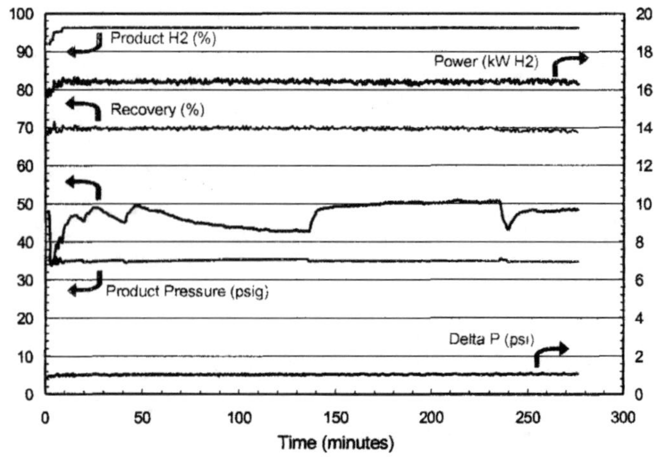

Figure 2. Steady Performance of HX-9100g H_2 PSA: 70°C Feed: 47.4 % H_2, 1 % CO, 15.2 % CO_2, 5.9 % H_2O, balance N_2

Table 1. Performance of HX-9100g H_2 PSA dependence on operating pressure

Operating Pressure	3.5 bara	5 bara
Product H_2 (%)	96	98
Product CO (ppm)	50	5
Max Power (kW H_2)	17	35
H_2 Recovery (%)	70	79

PSA OPERATION

The HX-900g H_2 PSA is a rotary valve device that was co-developed and built by SeQual Technologies Inc[1-3]. Its rapid cycle design enables the device to be very compact. It contains a series of columns packed with an adsorbent that selectively adsorbs CO, CO_2, N_2, H_2O, and CH_4. Rotary valves in direct contact with manifolds at the feed and product ends of the columns direct the flow of gases into and out of the columns for the various stages of the PSA cycle. The total volume of the PSA system is 33L including the adsorbent columns, feed and product manifolds, motor, and a product reservoir.

The speed of the rotary valves is controlled by a stepper motor. A mass flow controller in the product line from the PSA controls the amount of product drawn from the PSA. By controlling the setting on the product mass flow controller and the speed of the rotary valves, the purity of the H_2 product can be maintained over a range of system power levels.

Figure 2 shows stable performance of the PSA over 280 minutes. Reformate containing 47.4% H_2, 1% CO, 15.2% CO_2, 5.9% H_2O and the balance N_2 was fed into the PSA at 70°C and about 36 psig. The PSA product contained about 96% H_2 with the balance N_2. The product CO level fluctuated between 40 and 50 ppm and the H_2 recovery was steady at 70%. The PSA produces about 17 kW (lower heating value) of H_2. The pressure drop from the reformate to product was steady at about 1 psi.

The PSA also demonstrated the ability to purify an inlet reformate with the composition of the effluent from the ATR. The PSA was run at 25°C with an inlet stream containing 9.5% CO and produced a H_2 product which met specifications (<95% H_2, <50 ppm CO). Thus the system can operate as soon as the ATR is up to operating temperature, even if the WGS reactors and the PSA are still cold. This enables rapid startup of the fuel processor system to generate H_2.

Tests to evaluate the impact of operating pressure on PSA performance were performed at up to 5 bara. The results are shown in Table 1. At 5 bara, recoveries of nearly 80% can be reached with H_2 purities of 98% and CO levels below 5 ppm.

SYSTEM MODEL

A detailed model of the fuel processor subsystem was developed using ASPEN Plus. Laboratory constraints limited testing of the fuel processor system to about 3.5 bara, but the model shows that there are system efficiency benefits to operating at higher pressures based on PSA performance data. Table 2 shows ASPEN Plus model results at 3.5 and 5 bara PSA operating pressure.

The results show that, at 5 bara PSA operating pressure, there is an efficiency gain achieved due to higher H_2 recovery. The efficiencies are 63 and 71% respectively for the 3.5 and 5 bara cases, based on the lower heating values of natural gas and H_2. The efficiency gain at 5 bara more than compensates for the higher parasitic losses in the 5 bara case due to greater air compression requirements. Thus the net FP system efficiencies for the 3.5 and 5 bara cases are 59 and 64%, respectively.

Table 2. ASPEN Plus Model Results of Fuel Processor

Operating Pressure	3.5 bara	5 bara
O/C	1.0	1.0
S/C	4.7	3.6
PSA Recovery	70	79
Efficiency - H_2 out/NG in (%)	63	71
Net FP System Efficiency	59	64

SUMMARY

A fuel processor system was designed and built which generates near pure H_2 from natural gas using autothermal reforming for generating H_2 and PSA for purifying H_2. The systems operates under 3.5 bara and produces 95% H_2 with less than 50 ppm CO. The fuel processor subsystem including the PSA is 40L, and it is capable of producing 17 kW of H_2 with 5:1 turndown.

The PSA purifies the H_2 from a reformate containing less than 50% H_2. It can operate effectively at temperatures between 20 and 80°C and at pressures as low as 3 bara. The full power H_2 recovery at 3.5 bara is 70%. At 5 bara recoveries approaching 80% have been measured with product CO levels below 5 ppm. A detailed model indicates that the improved PSA recovery at 5 bara more than compensates for the higher parasitic loads compared to the 3.5 bara case.

These results demonstrate the capability to provide a compact fuel processing system for generating near pure H_2 for fuel cell applications. This is accomplished by using autothermal reforming of natural gas to generate hydrogen and using low pressure, rapid-cycle PSA for H_2 purification.

REFERENCES

1. Hill et al., U.S. Patent No. 5,112,367
2. Hill et al., U.S. Patent No. 5,268,021
3. Hill et al., U.S. Patent No. 5,366,541

ACKNOWLEDGMENTS

The authors gratefully acknowledge Scot Appel, Brian Sward, and David Winter of SeQual Technologies Inc. for their role in the design and build of the HX-9100g PSA device.

CONTACT

Craig S Gittleman, Ph.D.
Senior Research Engineer
GM Fuel Cell Activities
10 Carriage Street
Honeoye Falls, NY 14472
Phone: (585) 624-6738
FAX: (585) 624-6680
craig.gittleman@gm.com
Honeoye Falls, NY 14472

DEFINITIONS

ATR: Autothermal Reformer

WGS: Water Gas Shift

O/C: Elemental ratio of oxygen to carbon in ATR feed

PSA: Pressure Swing Adsorber

S/C: Molar ratio of steam to carbon in FP feed

Hydrogen Recycle for High Performance SOFC Systems

Bowie Keefer, Mark Kirby, and Edward Rode
QuestAir Technologies Inc., 6961 Russell Avenue, Burnaby, B.C., Canada V5J 4R8

Alberto LaCava, Felician College, Department of Computer Science, 262 South main St., Lodi, NJ 07644, USA

ABSTRACT

Fuel utilization, power density and overall efficiency of an internal reforming SOFC power plant canl be enhanced by carbon dioxide separation and enriched hydrogen recycle from the anode exhaust. These gas separation duties are efficiently and compactly achieved by advanced pressure swing adsorption (PSA) equipment using a rotary adsorber. As electrochemical power output is boosted, SOFC heat output and heat exchanger loads are reduced. High grade SOFC heat can be recovered for driving mechanical compression loads of the PSA process. This approach is a variation of hybrid fuel cell power plants, here maximizing electrical output and efficiency of the SOFC system, while optionally facilitating capture of carbon dioxide or coproduction of hydrogen. In another configuration, the rotary adsorber can be used without mechanical compression to transfer CO_2 from the anode recycle stream to the cathode exhaust stream, with the main benefit of increasing fuel utilization from the typical 85% to about 95%.

KEY WORDS

SOFC, PSA, hydrogen recycle, rotary adsorber

1. INTRODUCTION

All types of fuel cells are subject to performance degradation by depletion of fuel reactants and the accumulation of reaction products in the anode and cathode channels, as well as by the presence of inert gas components. The performance of solid oxide fuel cells (SOFC) is especially sensitive to reactant partial pressures because of their typical high operating temperature. This degradation is manifested by a loss of voltage efficiency, which is most severe for high temperature fuel cells such as SOFC's. With lower voltage at given current density, the fuel cell stack delivers less electrical power and more heat. To avoid excessive decline of voltage because of fuel starvation and steam buildup toward the end of the anode channel, the fuel utilization within the stack must be limited. Hence, a small conventional SOFC will be severely compromised in electrical efficiency, while delivering a large amount of relatively high grade waste heat. Contributions to this waste heat arise from the reduction of electrochemical free energy by reactant depletion and steam accumulation, polarization and internal resistance losses, and the combustion of unreacted fuel.

While the overall electrical generation efficiency of larger SOFC power plants can be enhanced by hybridization with a gas turbine for thermal bottoming, the efficiency of the gas turbine is much lower than the efficiency of direct energy conversion within the fuel cell stack.

Another area of difficulty with conventional SOFC plants (typically fueled by natural gas) is the limited degree to which satisfactory internal reforming may be performed within the anode channel, without problems in carbon deposition and steep thermal gradients. While steam accumulation over the anode reduces voltage efficiency, conventional SOFC plants require elevated steam concentrations at the front end to achieve satisfactory fuel processing. Anode exhaust gas recirculation without hydrogen enrichment is standardly practiced to supply steam for prereforming and internal reforming; but does not alleviate fuel starvation at the anode outlet, unless supplementary recirculation of enriched hydrogen is also provided.

2. HYDROGEN RECYCLE STRATEGY

The above challenges are being addressed with advanced gas separation equipment based on QuestAir rotary adsorbers to improve management of hydrogen, steam and carbon dioxide. Some previous studies [1, 2, 3, 4] have considered anode exhaust gas separation by chemical or physical absorption, or by conventional presure swing absorption, in SOFC power plants particularly for CO_2 capture. The QuestAir rotary adsorber in this application contains carbon dioxide selective adsorbent in a proprietary high surface area and high density laminated format, with rotary valve logic for sequencing cyclic adsorption and desorption steps at high frequency up to 150 cycles per minute.

Hydrogen from the anode outlet is recycled as carbon dioxide is stripped to a discharge stream and excess water is condensed. A portion of the anode gas may be recycled directly without removal of H_2O or CO2, in order to maintain a desired steam concentration at the anode inlet or in a prereformer.

High hydrogen concentration in the anode channel boosts voltage efficiency directly, while reducing the need to add steam for suppression of carbon deposition. Internal reforming conditions are improved. With excess hydrogen circulating in the anode channels, fuel starvation at the outlet is averted so that much higher fuel utilization than the typical 85% can be achieved.

Theoretical Background - The ideally available electrochemical energy is given by the Gibbs free energy,

$$\Delta G = \Delta H - T\Delta S = -nFE_{oc} \qquad (1)$$

where E_{oc} is the open circuit cell potential and n = 2 for hydrogen as the dominant reactive fuel component after internal reforming. The standard open circuit cell potential at any temperature T is the free energy of the oxidation reaction referred to atmospheric partial pressure of each component.

$$E_{oc}^{\ o} = \frac{\Delta G^o}{2F} \qquad (2)$$

$$E_{oc} = E_{oc}^o + \frac{RT}{2F}\ln\left[\frac{H_2}{H_2O}\right] + \frac{RT}{4F}\ln[O_2] \qquad (3)$$

The operating cell voltage E is then determined by subtracting the polarization terms η_i corresponding to electrode activation, ohmic resistance, and concentration polarization.

$$E = E_{oc} - \eta_{activation} - \eta_{ohmic} - \eta_{concentration} \qquad (4)$$

It is to be noted that E_{oc}, as well as activation polarization and concentration polarization, may be manipulated favourably by gas separation to boost the local ratio of hydrogen to steam over the anode and the partial pressure of oxygen over the cathode.

The present strategy is to "pump" the free energy by enriched hydrogen recycle to raise the hydrogen/steam ratio and thus boost cell voltage. By pumping the ΔG term in equation (1) through the partial pressure dependencies of equation (3), the heat generation term $T\Delta S$ is correspondingly reduced. This overcomes the common perception that high temperature fuel cells such as SOFC cannot attain high cell voltage, because higher temperature expands $T\Delta S^o$ to reduce E_{oc}^o in proportion to T. Actually, higher temperature directly enhances the ability of the present strategy to boost E_{oc}, as may be seen by inspection of equation (3).

Figures 1 and 2 show illustrative calculations for the example of a pressurized high temperature internal reforming SOFC power plant fueled by methane. The theoretical open circuit voltage E_{oc} is plotted versus the local fuel utilization along the anode channel. The fuel utilization is the extent of oxidation of the fuel, so that 100% fuel utilization corresponds to reaction of two moles of oxygen to oxidize each mole of original fuel methane completely. These calculations are idealized to assume local equilibrium of the steam reforming and water gas shift reactions as electrochemical oxidation proceeds. Water gas shift is assumed to be equilibriated to 300 C.

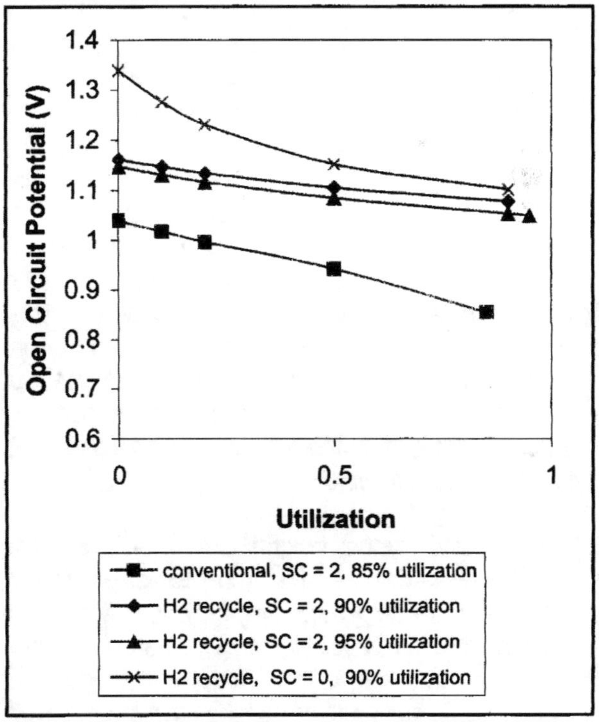

Figure 1. Open circuit potential versus fuel utilization for SOFC at 1000°C and 10 bara.

In Figure 1, the open circuit voltage along the anode channel of a conventional SOFC, operating with a steam/carbon ratio of SC = 2 and a fuel utilization of 85%, is compared with three examples of the same SOFC with enriched hydrogen anode recirculation.

The enriched hydrogen examples in Figures 1 – 3 are all based on the same high temperature pressurized tubular SOFC, with 98% hydrogen recovery by the rotary adsorber gas separation system. This high recovery level can readily be achieved by advanced pressure swing adsorption (PSA) processes with recompression of a fraction of the CO2 enriched exhaust as a supplementary "heavy reflux" stream. With 98% hydrogen recovery from the anode outlet gas, the recycle flow will be rather large, particularly with lower values of fuel utilization at the anode outlet. With lower levels of hydrogen recovery from the feed to the rotary adsorber, the recycle flow will be smaller. With hydrogen recycle and the same steam carbon ratio of SC = 2, fuel utilization can readily be increased to 90% or 95%, while open circuit voltage along the anode channel is increased by about 160 mV on average. The gradient of open circuit voltage decline along the anode channel is much reduced, because the excess recycled hydrogen keeps the ratio of hydrogen to steam relatively high.

An example is also shown of the same SOFC operating with enriched hydrogen recycle at 90% fuel utilization, but with no steam addition or recycle so that SC = 0. With conventional nickel anodes, an extremely low steam carbon ratio would not be practicable with natural gas fuel even with hydrogen recycle, because the catalytic activity of nickel for steam reforming also provokes carbon laydown in the absence of excess steam. However, this example illustrates the severe thermodynamic penalty of excess steam addition in reducing the attainable open circuit voltage. It also provides an upper bound on the enhancement of open circuit voltage which may be achieved by reducing the steam carbon ratio through use of alternative anode compositions in combination with enriched hydrogen recycle.

Figure 2 provides further insights. An SOFC fueled by pure hydrogen is contrasted with the earlier example of an internal reforming SOFC fueled by methane with a steam/carbon ratio of 2 for the conventional case without enriched hydrogen recycle. For the hydrogen and methane fueled cases without enriched hydrogen recycle, theoretical E_{oc} is plotted up to 100% fuel utilization in order to show the steep decline of voltage above about 85% fuel utilization.

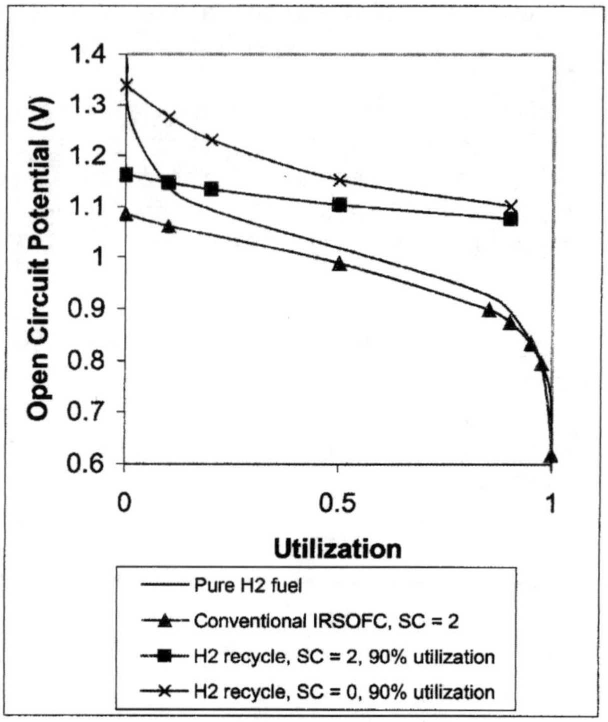

Figure 2. Open circuit potential versus fuel utilization for SOFC at 1000°C and 10 bara, comparing hydrogen and methane fuel options.

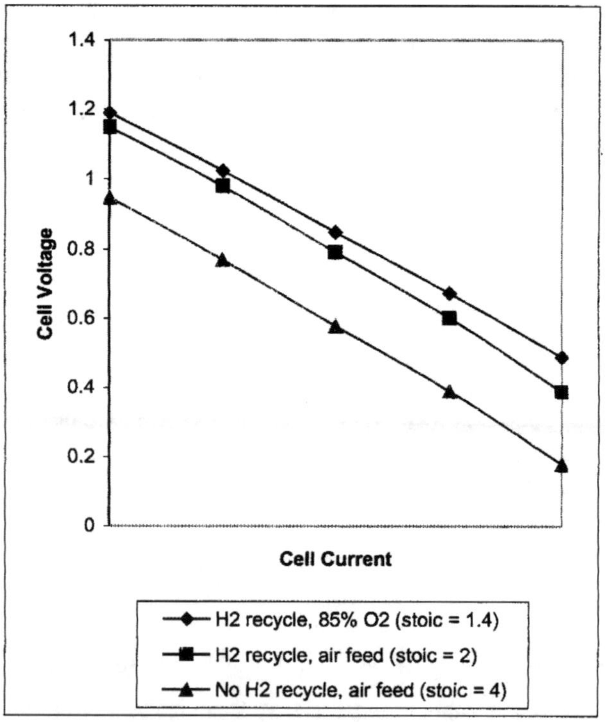

Figure 3. Cell potential versus current for SOFC at 1000°C and 10 bara. Effect of enriched hydrogen anode recycle with and without oxygen enrichment to cathode.

Contrary to the popular view that pure hydrogen is the ideal fuel for fuel cells, Figure 2 indicates that a fuel mixture of methane with recycled hydrogen may achieve much higher voltage efficiency (and therefore power density) compared to either hydrogen or methane separately.

Figure 3 shows expected cell voltage/current characteristics for the same pressurized high temperature SOFC example. These characteristics are based on averaging equation (4) over the anode channel, with activation and concentration polarization terms typical for a tubular design with cathode-supported electrolyte. Cases are shown for enriched hydrogen recycle to the anode with air feed or enriched oxygen feed to the cathode, contrasted to conventional operation without hydrogen or oxygen enrichment.

As can be seen from Figure 3, gas management by enriching recycle hydrogen to the anode can raise the cell voltage at any current, so that the SOFC can be operated at given cell voltage with much higher current density or at a given current density with significantly boosted voltage. The SOFC can thus be operated with much improved power density and voltage efficiency, and with reduced excess heat generation. These benefits could be augmented by the additional option of oxygen enrichment to the cathode, as discussed below.

Oxygen Enrichment - While anode enriched hydrogen recycle provides a major boost of cell voltage (for the assumed case of a heavy reflux PSA achieving 98% hydrogen recovery), oxygen enrichment provides a less dramatic voltage boost. It may be noted that the voltage boost from oxygen enrichment increases at high currents, owing to the reduction in concentration polarization within the thick porous cathode layer. This benefit applies to cathode-supported rather than anode-supported configurations.

In the conventional SOFC, stack heat output is relatively large, while the anode flows are relatively small so that a large cathode air flow (typically about 4 times the stoichiometric air flow) is required for stack cooling.

With enriched hydrogen recycle, larger anode flows will contribute to stack convective cooling, while less stack heat is generated owing to higher voltage efficiency. Hence the cathode air flow may desirably be about twice stoichiometic, without severely reducing the average partial pressure of oxygen over the cathode.

Oxygen enrichment allows the oxidant flow to be not much greater than stoichiometric, depending on the feed oxygen purity. Convective cooling of the SOFC stack will then be achieved predominantly on the anode side. This approach reduces the recuperative heat exchange duty. While oxygen enrichment VPSA (vacuum pressure swing adsorption) equipment based on QuestAir rotary adsorbers is both compact and efficient, the mechanical compression power demand will still largely cancel the net benefit of enriched oxygen in overall energy efficiency. Oxygen enrichment may be considered for the benefits of enhanced stack power density, consequent reduced SOFC stack capital cost, and smaller thermal recuperators. Enriched oxygen may also be applied to partial oxidation fuel processors, or to tail gas burners particularly where CO_2 capture from the rotary adsorber exhaust stream is desired.

Synergetic Anode Compositions- Special opportunities for the hydrogen recycle technology arise with so-called "direct oxidation" anodes, which are being developed for lower temperature SOFC systems capable of operation with dry hydrocarbon fuel albeit at seriously impaired low power density. Much enhanced SOFC performance may be achieved by recycling hydrogen with the hydrocarbon fuel feed, with the entrance section of the anode operating relatively dry and using copper ceria cermet or other "direct oxidation" anode material.

In this entrance anode section, the most reactive fuel component will be the recycle hydrogen while hydrocarbon reaction rates to either steam reforming or direct oxidation are relatively slow. Product steam will build up further downstream along the anode channel, where the anode material may be based on a more conventional material such as nickel zirconia cermet with good steam reforming activity in the presence of steam.

The downstream anode operates in an internal reforming mode to consume a portion of product steam and stack heat to generate hydrogen. A high hydrogen/steam ratio is mantained in both entrance and downstream zones. The anode composition may be varied continuously or discretely from relatively dry entrance conditions to humid downstream conditions, with hydrogen being the active fuel consumed in all anode zones for highest efficiency and safety from carbon deposition.

3. HYDROGEN RECYCLE TECHNOLOGY

The hydrogen recycle process includes the steps of recuperative heat exchange, water gas shift reaction, CO_2 removal by adsorption, excess water removal and low grade heat rejection by condensation. For superior efficiency and compactness, the adsorption process for CO_2 removal and H_2 enrichment is preferably conducted with a rotary adsorber apparatus as illustrated in Figure 4.

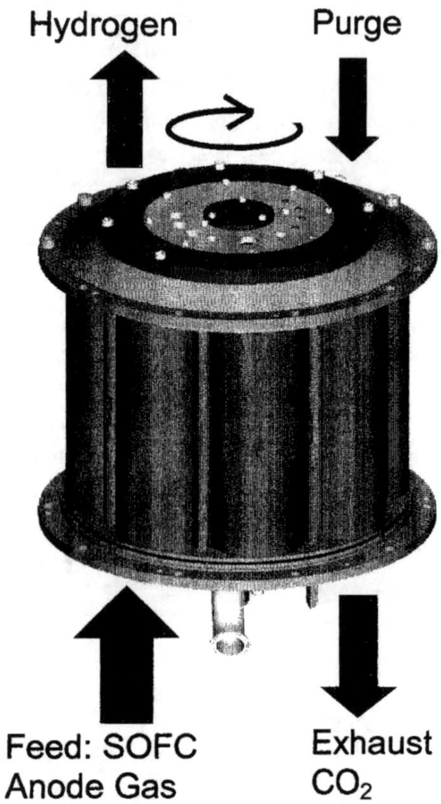

Figure 4. QuestAir rotary adsorber module.

The rotary adsorber may be operated in a pressure swing (PSA) or vacuum swing (VSA) cycle with purging by a portion of the enriched hydrogen product gas at reduced pressure, to release CO_2 as a concentrated exhaust stream. Since CO_2 and H_2 partial pressures are both increased by the separation process, mechanical compression work must be exerted against the free energy of mixing. The PSA or VSA will achieve higher recovery and also consume more compression power, with a higher pressure ratio between the pressure of feed admitted to the rotary adsorber and the pressure of exhaust withdrawn from the rotary adsorber.

A portion of the exhaust stream from the PSA or VSA rotary adsorber may be recompressed as a "heavy reflux" stream returned to a rotary adsorber feed port, in order to increase hydrogen recovery for recycle, and also to increase the CO2 concentration in the final exhaust stream. A simplified schematic of a SOFC with enriched hydrogen recycle by "heavy reflux" PSA is shown in Figure 5.

Alternatively, the rotary adsorber may be regenerated by temperature swing (TSA) with purging with the cathode exhaust gas, so that the CO_2 is discharged after dilution into oxygen-depleted air. A fraction of the anode exhaust gas is hydrogen enriched by water gas shift followed by CO_2 transfer to the cathode exhaust stream. A simplified schematic of the corresponding SOFC system is shown in Figure 6. This approach avoids mechanical compression, except for circulation blowers or ejectors. The separation process is assisted by the exergy of mixing CO_2 into air, an efficiency credit which has seldom if ever been exploited for power generation from fossil fuels.

Waste Heat Recovery - The first use of heat generated within an internal reforming SOFC stack is of course for the endothermic stream reforming reaction within the anode channels. Inlet gases to the SOFC stack and prereformer may be heated by heat exchange from cathode and anode exhaust streams, or in part by an adsorber tail gas catalytic burner so that the internal reforming SOFC system is thermally self-sustaining.

Excess heat may be exported or else recovered, for example by a thermal bottoming engine such as a gas turbine. As discussed above, conventional SOFC plants are penalized by lower open circuit potential and constrained fuel utilization, so that a large fraction of the original fuel heating value must be delivered as heat. Higher electrical generation efficiencies can be achieved by hybridization of the SOFC with a heat engine.

The enriched hydrogen recycle strategy augments the direct electrochemical conversion of fuel heating value, while reducing heat generation. The main sources of heat in SOFC power plants with hydrogen recycle include the SOFC stack excess heat after internal reforming load, heat recoverable from adsorber tail gas combustion, and steam that may be generated with the exothermic water gas shift reactor.

As hydrogen recycle by PSA or VSA intensifies the electrical output of the stack through (1) elevated open circuit voltage, (2) reduced activation and concentration polarizations, and (3) enhanced fuel utilization in the fuel cell stack, it is desirable to generate all delivered export power through the fuel cell, and thus use recovered high grade heat only to drive internal loads. Internally generated heat must firstly sustain all heat exchange and fuel reforming loads, and then may be applied to power PSA gas compression loads thermally (e.g. by turbocharger equipment). At least a portion of the gas compression loads may be powered electrically, as would be necessary with a "heavy reflux" PSA or VSA if the SOFC is operated at cell voltage much higher than about 0.7 volts.

Alternative thermally powered drivers for compression loads of the SOFC with enriched hydrogen recycle include gas turbines, Stirling engines, steam expanders or ejectors, and internal combustion engines burning PSA tail gas.

Gas turbines may be directly heated on the cathode side of pressurized SOFC's using air as the oxidant, as in the example of Figure 5. Rather than driving a generator, the gas turbine may drive a heavy reflux centrifugal compressor directly. A twin spool gas turbine would resemble a pair of turbochargers, with one expander driving the heavy reflux compressor, and the other expander driving the air compressor to pressurize the SOFC and sustain the gas turbine cycle.

Anode side direct gas turbines may also be considered for pressurized SOFC's with hydrogen recycle, in view of the large anode loop circulation flow cooling the SOFC stack. Gas turbines using anode loop gas as working fluid may be useful for very large SOFC power plants, or when enriched oxygen is supplied to the cathode so that the SOFC is cooled from the anode side. Low molecular weight of anode loop gas (especially with high hydrogen recycle) as gas turbine working fluid would require special expanders that may be adapted from steam turbines.

For atmospheric SOFC's, a gas turbine using feed air as working fluid may be indirectly heated by heat exchange to both cathode and anode sides of the SOFC with enriched hydrogen recycle. Indirect gas turbines are a well known option for unpressurized high temperature fuel cells.

Stirling engines are attractively efficient while accepting heat in the temperature range of medium to high temperature SOFC's. They have the important advantage of more favourable part load efficiency than gas turbines, complementary to the good performance of rotary adsorber PSA under turndown conditions. Hydrogen is an ideal Stirling engine working fluid, and may conveniently be replenished from the PSA unit.

A heat recovery steam generator can provide steam to a prereformer via gas circulation ejectors, or to a steam expander driving compression loads, while recovering heat from water gas shift reactor and hot exhaust streams such as a gas turbine exhaust.

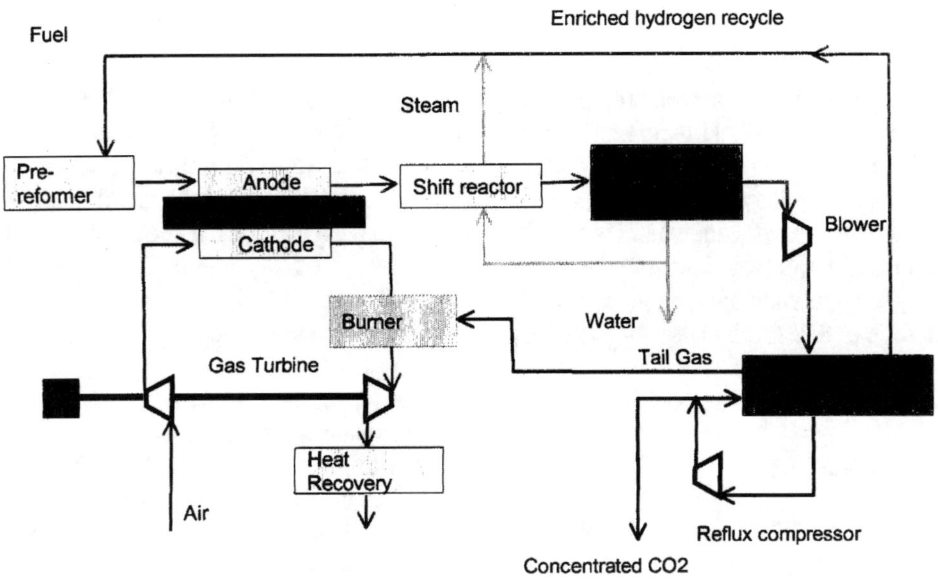

Figure 5. Simplified schematic of SOFC with anode hdyrogen recycle and rotary adsorber with "heavy reflux" PSA cycle.

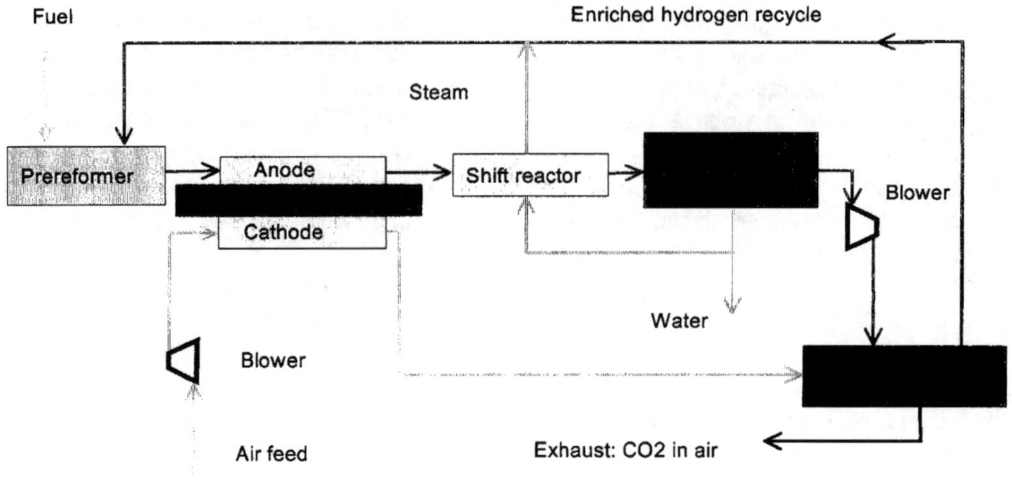

Figure 6. Simplified schematic of SOFC with anode hdyrogen recycle and rotary adsorber purged by cathode exhaust.

There are some challenges in heat recovery with hydrogen recycle. The temperature rise across the SOFC may be relatively is small, while heat contribution from the tail gas burner is also small with high fuel utilization in the range of 90% to 95%. This implies a low pressure ratio for direct gas turbines.

As the recuperative heat exchange load is smaller than with conventional hybrid SOFC systems, it may be economic to design the recuperator for relatively high effectiveness to widen the ΔT available for the gas turbine.

Heat recovery from water gas shift is available at only relatively low temperature.

Benefits - With higher voltage and more uniform conditions along the anode channel, higher fuel utilization can be achieved within the fuel cell stack. Less waste heat and waste fuel gas are produced, so that the size of any thermal bottoming heat recovery equipment is greatly reduced. The thermal bottoming machinery may be used to drive the auxiliary gas separation equipment, so that all power exported from the SOFC plant is generated at highest efficiency by the fuel cell stack.

PERFORMANCE ESTIMATES

Maximum power density of SOFC stacks is achieved when operating voltage is about half of the open circuit voltage. The maximum power density is proportional to the square of open circuit voltage, divided by the stack electrical resistance. Hydrogen recycling and optional oxygen enrichment can increase open circuit voltage of natural gas fueled SOFC systems by about 15 %, while also reducing activation polarization and concentration polarization contributions to electrical resistance. Consequently, the maximum power density of the SOFC stack may be increased by about 33 %.

Exploratory perfomance calculations have been performed for diverse examples of SOFC operation. The following cases illustrate the main alternatives of heavy reflux PSA as in the schematic of Figure 5, or purging the rotary adsorber with cathode exhaust gas as in the schematic of Figure 6.

(1) Rotary adsorber with "heavy reflux" PSA. In typical cases, a PSA hydrogen recovery of 98% was assumed, with relatively low H_2 purity ~99% as 5% slip of CO_2 to the hydrogen recycle stream was tolerated in order to minimize compression. If the ability to export cogenerated hydrogen is desired, 99.99% hydrogen purity can be achieved with a modest penalty of about 2.5% on overall efficiency.

For net power rating of 1 MW on natural gas fuel, it is estimated as a baseline case that a conventional direct gas turbine hybrid internal reforming SOFC (operating at 0.7 volts cell voltage and 85% utilization) would achieve about 65% efficiency based on lower heating value. With QuestAir PSA rotary adsorber recycle hydrogen enrichment, the same SOFC stack could deliver 1 MW net power output at estimated 70% efficiency, or alternatively could deliver 1.33 MW net power output at estimated 65% efficiency. Higher efficiency to about 75% could be achieved by larger plants.

(2) Rotary adsorber with cathode exhaust purge. Table 1 shows performance parameters calculated for an atmospheric tubular SOFC at 900°C, showing performance enhancement with supplementary hydrogen recycle with CO_2 transfer to cathode exhaust. Steam generated by the exothermic water gas shift reaction is used as motive fluid for a steam ejector to recycle a portion of the anode outlet gas directly to the prereformer, while the balance of the anode outlet gas is fed to the rotary adsorber. The enhancement of cell voltage is much smaller than could be achieved by a heavy reflux PSA or VSA, since the hydrogen recovery achieved with cathode exhaust purge is about 90%, much less than the 98% assumed in the examples of Figures 1 – 3. However, the major parasitic power load of a heavy reflux compressor is avoided, and fuel utilization is increased from 85% to 95%.

Table 1.

Performance enhancement of high temperature SOFC with enriched hydrogen recycle by rotary adsorber transferring CO_2 to cathode exhaust.

Description	Reference Case	Rotary Adsorber Case
Recycle/feed ratio (H_2/CH_4 mole ratio)	0.0	1.68
Fuel utilization	85%	95%
Cell potential (V)	0.690	0.711
Net Power (kWe)	120.2	137.7
Electrical generation efficiency (LHV)	56.3%	64.6%

CONCLUSIONS

The QuestAir hydrogen separation technology on the anode side may be provided in a simplified format to achieve lowest capital cost and highest overall efficiency, with dilute carbon dioxide discharged to the atmosphere. In alternative PSA or VSA formats, carbon dioxide may be captured as a concentrated product for external delivery or sequestration, while the plant may also deliver purified hydrogen with load-switching capability between energy conversion from fossil fuel to electricity or hydrogen or both.

The enhanced performance of the fuel stack may be exploited to achieve higher efficiency at the same or modestly increased power output, or much greater power density with modestly improved efficiency. For large SOFC plant high energy efficiency competitive with the best hybrid power plants can be achieved, while substantially reducing the power rating and complexity of heat recovery equipment. For small SOFC fuel cells without thermal bottoming or CO_2 sequestration, electrical generation efficiency can be greatly enhanced.

As manufacturing costs of SOFC stacks are at present a severe barrier to market entry, a breakthrough to reduce stack cost per kilowatt by increasing stack power density may have greatest near term impact, while enhanced efficiency will improve the already strong performance leverage of SOFC technology.

ACKNOWLEDGMENTS

Process modeling for the results of Table 1 by Jan Wilco Dijkstra of ECN (Energy research Centre of the Netherlands) is gratefully acknowledged.

REFERENCES

1. J.L. Carson, Proc. 30th Intersociety Energy Conversion Engineeering Conference (1995) 229.
2. S. Campanari and P. Chiesa, Proc. 5th International Conference on Greenhouse Gas Control Technologies, GHGT-5, Australia (2000).
3. S. Campanari, J. Power Sources 112 (2002) 273.
4. J.W. Dijkstra and D. Jansen, Proc. 6th International Conference on Greenhouse Gas Control Technologies, GHGT-6, vol. I (2003) 161.

CONTACT

keefer@questairinc.com

Frequency and Wavelet Analysis of Electrochemical Noise from Aluminum Alloy 1100 in Chloride Environment

Idelfonso J. Arrieta
Department of Chemical Engineering, University of South Florida, 4202 E. Fowler Ave, ENB 118, Tampa, FL 33620.

Babu Joseph, Deepak Srinivasagupta
Department of Chemical Engineering, University of South Florida, 4202 E. Fowler Ave, ENB 118, Tampa, FL 33620.

Alberto A. Sagüés
Department of Civil Engineering, University of South Florida, , 4202 E. Fowler Ave, ENB 118, Tampa, FL 33620.

ABSTRACT

Electrochemical Noise (EN) includes the spontaneous fluctuations of the potential of a freely corroding metal with respect to its surrounding electrolyte. The fluctuations reflect random spatial and temporal separation between the anodic and cathodic processes inherent to the electrochemical character of corrosion. In recent years EN has been increasingly used as an indicator of the presence and rate of corrosion. However, EN is a novel monitoring technique still under development and reliable EN signal analysis methods to identify specific corrosion processes and rates are still being sought. Wavelet analysis of EN is a promising approach to that end. The phase plane plot obtained from the coefficients of the wavelet transforms has revealed promising results for identifying noise transients. The corrosion events coincide with high intensity regions in the phase plane plot. We also investigated the ability of quantitative measures based on the phase plane plots to correlate with the observed corrosion behavior.

KEY WORDS

Aluminum alloy 1100; electrochemical noise; corrosion monitoring; wavelet transforms

1 INTRODUCTION

The Electrochemical Noise (EN) used in this study consists of recording the spontaneous fluctuations of the Open Circuit Potential (OCP) of corroding metals versus a Reference Electrode (RE) [1]. EN often reflects transients where a predominantly anodic dissolution stage is followed by a predominantly cathodic recovery period. The transients may be seen taking place individually, or as superimposed events if they happen frequently enough [2]. In recent years, EN has been increasingly used as an indicator of the presence and rate of corrosion which is an important (~3% of the U.S. gross domestic product (GDP)) deterioration mode of engineering components. Time and frequency domain analysis of the EN records have been employed to try to characterize the corrosion behavior of different electrochemical systems [3, 4, 5]. For example, in the time domain, Cottis et al. [5] reported that the ratio of the standard deviation of the potential noise records to the standard deviation of the current noise records called "noise resistance" is equivalent to the polarization resistance and can be used to determine the corrosion rate of the material. In the frequency domain, Hladky et al. [4] has reported that roll-off slopes obtained from the Power Spectrum Density (PSD) plots can be related to the type of corrosion mechanism (e.g. ~ -20 db/decade has been correlated experimentally to pitting corrosion). However, an explicit interpretation of the EN signals has not been reached with the approaches already used [6]. Lately, wavelet analysis has been proposed as an alternative tool to obtain valuable information of the corrosion process from the EN records. This has been done using the wavelet based on the phase plane plot (PPP) [7] and/or energy distribution plot (EDP) [8]. The variance of the coefficients of each frequency level used during the decomposition has also been related by Wharton et al. [9] to the different corrosion processes present for a particular material over a time period.

2 BACKGROUND

2.1 ANALYSIS OF EN WITH WAVELET TRANSFORMS

Wavelet transforms have become a promising tool in the analysis of EN since they represent the EN signals in the combined frequency – time domain. In a wavelet transform the signals are decomposed into two sets of coefficients [7,8] by eq. 1.

$$W_{x,y} = \langle \varphi(t), \psi \rangle \equiv c_0^{-x/2} \int_{-\infty}^{+\infty} \varphi(t)\psi\left(c_0^{-x}t - yd_0\right)dt \qquad (1)$$

where $\varphi(t)$ is the signal in the time domain, c_0 and d_0 are constants, y the function of translation in time, x the degree of dilation that is applied to the mother wavelet (ψ) and $W_{x,y}$ the resulting set of coefficients which can be defined as follows:

1. The smooth coefficients set which contains information about the general trend of the signal.
2. The detail coefficients set which contains information about the local fluctuations in the signal.

These coefficients can be used in several ways to provide information about the EN signals.

The most important graphical representations of the wavelet coefficients are as follows:

1. The set of coefficients that corresponds to points on a two-dimensional grid or lattice of discrete points in the time-frequency domain. The wavelet lattice can be represented in a phase plane plot [7]. From visual examination of the phase plane plots, an indication of some specific patterns can be obtained for different types of corrosion. It was reported by Joseph et al. [7] that pitting and crevice corrosion led to distinct patterns on the time-frequency phase plane. Also, the intensity of the pattern in the phase plane was consistent with the type of corrosion observed on the electrode surface for simple systems; for example, a highly corrosive environment produces a high-intensity pattern.
2. The Energy Distribution Plot (EDP) which can be obtained to compare the amount of energy accumulated in each coefficient [8]. Aballe et al. [8] reported that the energy of each frequency scale provides a measure of the activity of the events or events that are operating at the associated scale.

The phase plane plot which is the most common representation of the wavelet coefficients was proposed by Wickerhauser [10] and each transform coefficient is associated with a rectangular patch and the phase plane is generated by using the power of the signal component present in each rectangle. The energy of the coefficients is represented in a grayscale coding where the minimum intensity is represented by the "black" end of the scale and maximum intensity by the "white" end of the scale. The "x" axis of the plot represents the time variable and the "y" axis are the frequency scales where the lowest scale represents the highest frequency (sampling frequency divided by two) and the highest scale corresponds to the lowest frequency.

The goal of this study is to explore the wavelet results obtained from EN signals that have been recorded for aluminum alloy 1100 in chloride media in order to assess the potential of this technique to discriminate between the different possible corrosion mechanisms. The main approach will be the analysis of the standard deviation and mean of the coefficients that correspond to the different frequency levels of the wavelet decomposition.

3 METHODOLOGY

Aluminum alloy 1100 was used for this study and it is representative of the commercially pure aluminum. The alloy was obtained as strip coupons from Metal Samples, Inc. [11]. Portions of the coupons were cut out, embedded in metallographic epoxide cylinders and wet-ground to a 600 grit finish to expose the flat portions with surface area of ~ 0.2 cm^2. Contact to the exterior was made by a copper wire that was attached to the surface of the aluminum alloy specimen in the embedded region. The wire was otherwise isolated. The finalized specimen is shown in Fig. 1.

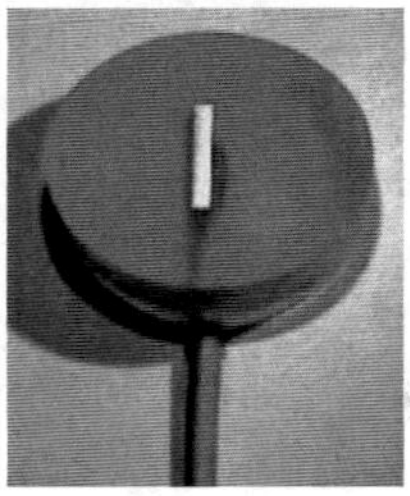

Figure 1. Finalized aluminum alloy specimens

The specimen was placed in a test cell with the exposed metal surface in a vertical plane. The test cell included typically the specimen in a beaker containing ~ 200 cm^3 of distilled water to which 3.5% NaCl was added. The test solution was naturally aerated and at ambient temperature (25 ± 5 °C). A Saturated Calomel Electrode (SCE) was placed near the specimen. The difference of potential between the specimen and the SCE was monitored with a data acquisition system at 0.333 samples per second.

3.1 EN EXPERIMENTAL SETUP

EN measurements for this investigation were based on recording the spontaneous fluctuations of OCP of aluminum alloy 1100 in an aqueous 3.5 wt% NaCl solution at ~ 25 °C. The EN data were recorded in packets of 86400 data points at 0.333 samples per second during the first three days after immersion. The experimental setup was implemented using a combination of Vernier products [12] and devices built at the University of South Florida. A schematic of the experimental setup is shown in Fig. 2.

The OCP was measured with respect to a Saturated Calomel Electrode (SCE). The SCE was connected to the Voltage Compensator (VC) which was used to remove the DC component by dialing the appropriate voltage.

The DC compensated signal was connected to the Voltage Follower (VF) which was used to effectively increase the input impedance of the Vernier Instrumentation Amplifier (IA) to > $2x10^9$ ohms. Using the polarity shown in Fig. 2, the raw potential records listed the reference electrode potential minus the metal potential (thus negative potential excursions were in the upward direction). This polarity was inverted to report the result in the manner usual in the electrochemical literature.

The output of the VF was connected to the IA which had selectable ranges of input voltage, providing gain settings to optimize signal processing. The IA was set to provide a ± 200 mV recording window for the smallest surface area specimens, and a ± 20 mV window for the other two areas. The VC was adjusted at the beginning of each 300 s EN recording interval to obtain a starting compensated potential close to the zero of the recording window. The potential value indicated by the VC was manually recorded as the nominal OCP value for that EN record.

The IA output was sent to the input of the Vernier Lab Pro interface which was used for data collection. The Vernier interface was used with a computer in combination with the Logger Pro 2.2 software to display the actual EN signals as a potential-time plot.

The test cell was open to the atmosphere and placed inside a grounded Faraday's cage in order to avoid spurious signals due to external noise.

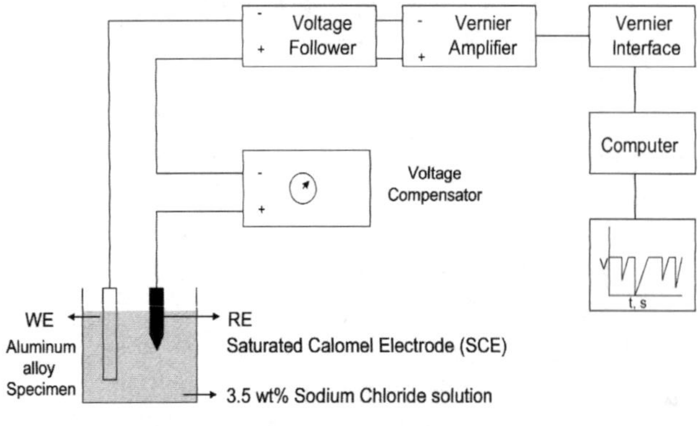

Figure 2. Schematic of the EN experimental setup

4 RESULTS AND DISCUSSION

The EN records for aluminum alloy 1100 in a 3.5 wt% NaCl solution consisted of a series of transients (see Fig. 4) that were composed of two stages [13]: a fast anodic stage where the potential suddenly dropped due to the metal dissolution into metal ions and release of electrons into the metal surface. In the case of aluminum, the anodic reaction is as follows:

$$Al \rightarrow Al^{3+} + 3e^- \quad (2)$$

The anodic stage ended with the repassivation of the active site. It was followed by an exponential-like recovery of the OCP stage which corresponded to the consumption of the electrons accumulated at the metal-electrolyte interface by the cathodic reaction (assumed to be mainly the reduction of oxygen from the aerated electrolyte):

$$O_2 + 2H_2O + 4e^- \rightarrow 4OH^- \quad (3)$$

From Fig. 3, we notice a noise transient taken from the EN records obtained during this research showing the two stages described above at ~ 1108 minutes after immersion. This type of noise transients where the two stages described above are present has been typically ascribed to pitting corrosion [13]. Arrieta [14] also reported that the noise events were very frequent shortly after immersion and the number of events per time period tended to decrease as time progressed. However, both the shape of the transients and the number of events per time period were not easily identifiable when the EN

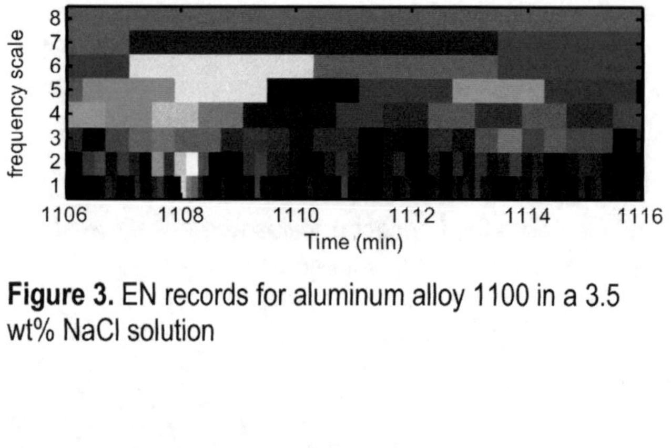

Figure 3. EN records for aluminum alloy 1100 in a 3.5 wt% NaCl solution

Table 1. Frequency scales and their frequency values in Hz.

Frequency Scale	Frequency (Hz)
1	0.167
2	0.084
3	0.042
4	0.021
5	0.011
6	0.006
7	0.003
8	0.002

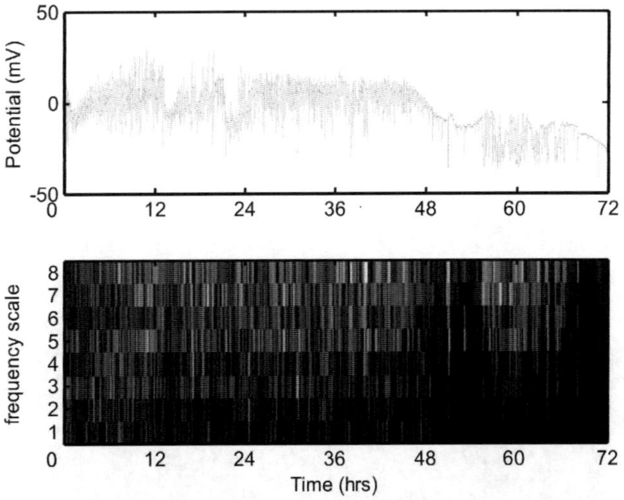

Figure 4. EN signal and its phase plane representation for aluminum alloy 1100 in a 3.5 wt% NaCl solution

signal was recorded for a long period and composed of a large number of data points. Therefore, the wavelet analysis was used to try to discriminate between the types of corrosion present during the experiments and the progression of each mechanism. This analysis was done using "Daubechies 1" as the mother wavelet and eight frequency scales which are shown in Table 1. The calculations were done using the wavelets toolbox included in Matlab.

The individual noise transients (see Fig. 3) can be identified if isolated in the phase plane representation (e.g, Fig. 3 at ~ 1108 min) by lighter patches (high intensity) of the high frequency coefficients when the anodic stage takes place followed by high intensity of the low frequency coefficients during the cathodic stage.

The EN signal for an aluminum alloy 1100 specimen and its phase plane representation is shown in Fig. 4. The first step was to visually examine the phase plane generated by the wavelet transform. Comparison of several phase plane plots indicated that the coefficients of the first three levels decreased in intensity as time progressed. The decrease in intensity as time progressed confirmed that the number of high frequency events which have been correlated to pitting corrosion decreased as time progressed. This decrease of the number of high frequency events may be caused by a gradual depletion of nucleation sites (with subsequently smaller pitting rate as time progresses) on the exposed surface of the metal [15]. To confirm the visual observations, the data was divided in 144 segments of 30 minutes and the wavelet transform was applied to such segments. Then, the standard deviation of the coefficients of each frequency scale corresponding to each segment was computed and plotted as a function of time. The standard deviation of the coefficients of the first two frequency scales plotted as a function of time is shown in Fig. 5. From these graphs, we can see a downward trend for these two coefficients which indeed confirmed the decrease of activity of the high frequency component of the EN signal as time progressed.

The standard deviation of the coefficients of the 6th and 8th scale were also plotted and shown in Fig. 6. These graphs revealed that the standard deviation reached a peak at ~ 40 hrs after immersion. This behavior may be correlated to a type of corrosion that generates low frequency EN signals like uniform corrosion, stable pitting and/or crevice corrosion.

Detailed visual observation of the specimens was not attempted since the 600-grit surface finish was very rough and identification of pits with sizes of the order of a micron which have been reported [16] for this alloy would have been very difficult to see. However, the specimens showed small signs of crevices at the end of each experiment. Therefore, we assumed from the EN signals observed during this investigation and previous works [13,15] that pitting was the main corrosion mechanism present during the experiments. Further investigation will involve optical techniques to identify the actual corrosion mechanism on the surface of the material.

The frequency analysis was also conducted using the Power Spectrum Density (PSD) plot which was obtained from the Welch's method in Matlab (see Fig. 7). The roll-off slope [4] obtained from the PSD is ~ -20 db/decade.

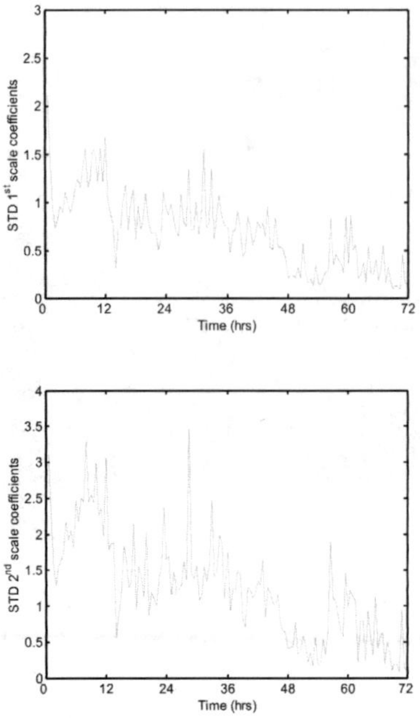

Figure 5. Standard deviation of the coefficients of the first two frequency scales as a function of time

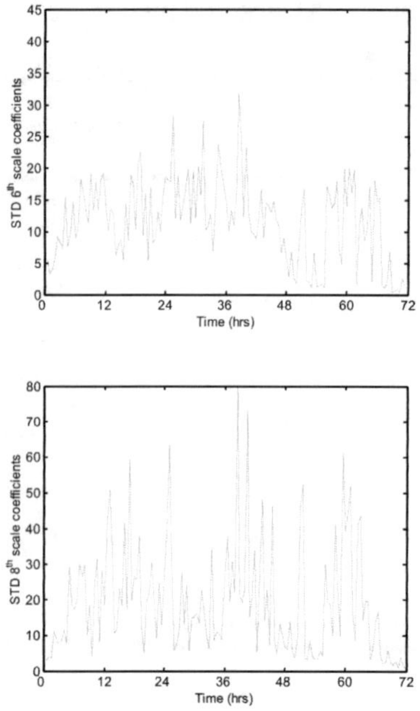

Figure 6. Standard deviation of the coefficients of the 6th and 8th frequency scale as a function of time

This type of roll-off slope can be associated with the frequency domain response of a resistor-capacitor in parallel circuit. Therefore, the modeling stage should be based on an equivalent circuit like that described above. Also, this type of roll-off slopes (~ - 20 db/decade) has been correlated experimentally to pitting corrosion by previous investigators [4].

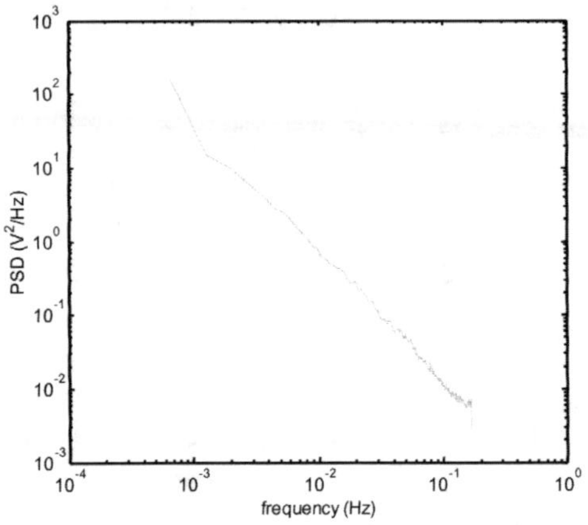

Figure 7. PSD plot of the EN signal shown in Fig.4

5 CONCLUSIONS

Based on the results found in this investigation, we can draw the following conclusions:

- Wavelet analysis provides a promising procedure to characterize EN signals.
- The wavelet analysis of long duration EN data show changes in the frequency content as time progresses indicating a change in the corrosion mechanism that take place during the experiments.
- The intensity and pattern of the signal in the phase plane show that pitting corrosion is probably the main corrosion mechanism for the aluminum alloy 1100 in a chloride media in the beginning stages of the experiment.
- The decrease of high frequency activity as time progressed indicates that the number of pitting events gradually decreases and other corrosion mechanisms are contributing to the EN.
- Further investigation is needed to correlate the corrosion mechanism to the characteristics of the EN records in the time-frequency domain.

REFERENCES

1. Iverson, W.P., “Transients Voltage Changes Produced in Corroding Metals and Alloys”, Journal of Electrochemical Society, Vol. 115, pp. 617, (1968).
2. Roberge, P.R., ”Quantifying the Stochastic Behavior of Electrochemical Noise Measurement During the Corrosion of Aluminum”, Electrochemical Noise Measurement for Corrosion Applications, ASTM STP 1277, Kearns, J.R., Scully, J.R., Roberge, P.R., Reichert, D.L. & Dawson, J.L., Eds., American Society for Testing and Materials, pp. 142-156, (1996).
3. Perdomo, J.J. & Singh, P.M., “Electrochemical Noise (ECN) Measurements as a Corrosion Monitoring Tool: A Review”, Institute of Paper Science and Technology, Vol. 20, No. 4-5, pp. 359-378, (2002).
4. Hladky, K. & Dawson J.L., “The Measurement of Corrosion Using Electrochemical 1/f Noise”, Corrosion Science, Vol. 22, No.3, pp. 231-237, (1982).
5. Cottis, R.A., Al-Awadhi M.A.A, Al-Mazeedi, H. & Turgoose, S., “Measures for the Detection of Localized Corrosion with Electrochemical Noise”, Electrochimica Acta, Vol. 46, pp. 3665-3674, (2001).
6. Bertocci, U., Gabrielli, C., Huet, F., Keddam, M. & Rousseau, P., “Noise Resistance Applied to Corrosion Measurements Part II. Experimental Tests”, Journal of Electrochemical Society, Vol. 144, No. 1, pp. 37-43, (1997).
7. Joseph, B., Dai, X.D., Motard, R.L. & Silverman, D.C., “Improved Discrimination of Electrochemical Noise Signals Using Wavelets Analysis”, Corrosion, Vol. 57, No.5, pp. 394-403, (2001).
8. Aballe, A., Bethencourt, M., Botana, F.J. & Marcos, M., “Using Wavelets Transform in the Analysis of Electrochemical Noise Data”, Electrochimica Acta, Vol. 44, pp. 4805-4816, (1999).
9. Wharton, J.A., Wood, R.J.K. & Mellor B.G., “Wavelet Analysis of Electrochemical Noise Measurements During Corrosion of Austenitic and Superduplex Stainless Steel in Chloride Media”, Corrosion Science, Vol. 45, pp. 97-122, (2002).
10. Wickerhauser, M.V., “Acoustic Signal Compression with Wavelet Packets” in Wavelets – A Tutorial in Theory and Applications, ed. C.K. Chui (Boston, MA: Academic Press, 1992), pp. 679 – 700.
11. Metal Samples Inc., 152 Metal Samples Rd., Munford, AL, USA, 36268.
12. Vernier, 13979 SW Millikan Way, Beaverton, OR, USA, 97005 – 2886.
13. Williams, D.E., “The Analysis of Current and Potential Fluctuations in Corroding Systems”, DECHEMA Monographien., Vol. 101, pp. 253-264, (1986).
14. Arrieta, I., “The Effect of Surface Area on Electrochemical Noise of Aluminum Alloys 1100, 2024 and 5052 in Chloride Media”, Thesis for M.S. in Chemical Engineering, University of South Florida, 2003.
15. Uruchurtu, J.C. & Dawson J.L., “Noise Analysis of Pure Aluminum Under Different Pitting Conditions”, Corrosion, Vol. 43, No.1, pp. 19-25, (1987).
16. Pride, S.T., Scully, J.R. & Hudson, J.L., ”Metastable Pitting of Aluminum and Criteria for the transition to Stable Pit Growth”, Journal of Electrochemical Society, Vol. 141, No.11, pp. 3028-3040, (1994).

Optimization and Control using a Proton Exchange Membrane Fuel Cell Model

Vinay Prasad
H2Gen Innovations, Inc.
4740 Eisenhower Avenue, Alexandria, VA 22304

B. Wayne Bequette
Howard P. Isermann Department of Chemical Engineering
Rensselaer Polytechnic Institute
110 8th Street, Troy, NY 12180.

ABSTRACT

Fuel cells, and proton exchange membrane (PEM) fuel cells in particular, are being seriously considered as alternative energy sources for stationary and transportation applications. In order to optimize performance of these fuel cells, it is important to study the effect of various operating parameters on the behavior of the fuel cell. We present a model for a PEM fuel cell that includes descriptions of the anode and cathode kinetics, and transport through the membrane. The electrochemical model is used to describe the dynamic behavior of the fuel cell. The model is used to study the parametric sensitivity of the performance of the fuel cell, and the response to changes in the feed characteristics (partial pressure of hydrogen fed to the anode, amount of CO contaminant). The final aim of this work is to study techniques for optimization and control of PEM fuel cells based on first-principles modeling.

KEY WORDS

Proton exchange membrane, fuel cell, electrochemistry, kinetics, control

INTRODUCTION

The need for lower polluting, more efficient sources of energy has sparked considerable interest in fuel cells in the recent past. Fuel cells convert chemical energy directly into electrical energy, and offer better conversion efficiency than most other energy conversion devices. When hydrogen feed is used as the feed in a fuel cell, the only byproduct from the fuel cell is water, and the cell is totally pollution-free. Of the various types of existing fuel cell systems, the proton exchange membrane (PEM) fuel cell is the most promising, especially for terrestrial applications such as power generation and transportation, because of the simplicity of its design and operation. Some of the characteristics of the PEM fuel cell system are tolerance to CO_2, self-starting at lower temperatures, and low-cost construction materials (Yi and Nguyen, 1999). Another significant advantage is that there is no problem of electrolyte leakage, since the electrolyte used in this system is a solid polymer proton conducting membrane.

While significant advances have been made in fuel cell systems over the past few years, especially in the reduction of catalyst loading and in the development of new membrane and electrode assembly preparation techniques, these systems are still not fully optimized with respect to performance and cost of production. Mathematical modeling is a powerful tool for analysis of the behavior of the fuel cell system, and can be used in developing design, optimization and control techniques for fuel cells.

Bernardi and Verbrugge (1992) and Springer et al. (1991) were among the first researchers to attempt to model PEM fuel cell behavior on a phenomenological basis. Their work represented a departure from earlier analysis of a more empirical nature. In these studies, the membrane electrode assembly was divided into several regions, and equations for charge and mass transport were developed for each region. The models were one-dimensional, since the essential features of the performance curve of a MEA (cell potential vs. cell current density) are well described with such models, and the computation complexity is modest (Murgia et al., 2002).

Later modeling studies have improved on these early models in different ways. Murgia et al. (2002) developed a modified version of the Bernardi-Verbrugge model, with the aim to establish a numerically stable and fast one-dimensional model which lends itself as a starting point for generalization to two and three dimensions, as well as the inclusion of energy transport, water management and transport phenomena. Pisani et al. (2002a) present a semi-empirical derivation of a performance equation with the goal of having the largest number of derived mechanistic parameters. Their aim is to use the values of the mechanistic coefficients to reproduce experimental data and to provide information on the phenomena occurring in the fuel cell as the current is changed. In another paper, Pisani et al. (2002b) present a model that includes a description of the flooding of the electrodes and of the diffusion mechanisms in the cathode reactive region. Yi and Nguyen (1998) present a model that includes a description of water transport in the anode and cathode channels and through the membrane.

Other studies on PEM fuel cells include the analysis of two and three-dimensional effects such as gas distribution at the electrodes and the resulting flow fields. Natarajan and Nguyen (2001) and He et al. (2000) study the effects of using conventional and interdigitated gas distributors on the performance of PEM fuel cells. Roos et al. (2003) use a numerical volume averaging method to perform three-dimensional simulations of fuel cell stacks and obtain effective parameters for the transport equations. Costamagna (2000) also presents a three-dimensional model of a PEM fuel cell and discusses critical conditions for flooding, membrane drying and degradation due to temperature peaks.

Another aspect of fuel cell modeling involves the study of operation using different feeds. Springer et al. (2001) study the performance of a PEM fuel cell operating on hydrocarbon reformate instead of pure hydrogen. Shukla et al. (2002) compare the performance of a PEM direct methanol fuel cell using aqueous methanol and air mixed at the anode with one using only aqueous methanol.

Most of the above-mentioned models are steady-state models that do not describe the dynamic behavior of the fuel cells. Some of the fuel cell models that do attempt to describe their dynamic behavior include the work of Sundmacher et al. (2001) and Ceraolo et al. (2003). Sundmacher et al. obtain the cell voltage response to dynamic feeding of methanol, and observed that a significant voltage increase can be obtained from dynamic changes in the methanol feed concentration. Ceraolo et al. study the cell voltage response to steps in the current, and obtain good agreement with experimental data. The model used in our work borrows heavily from these studies, and attempts to describe the dynamic transient performance of PEM fuel cells to current steps and changes in feed properties. The ultimate aim of this work is to place the fuel cell system model in an optimization and control framework to develop optimal operating profiles in a manner similar to that of Godat and Marechal (2003).

MODEL DEVELOPMENT

Our aim is to develop a model that explains the transient, dynamic behavior of PEM fuel cells. Much of our work is based on the model of Ceraolo et al. (2003), who study the dynamic behavior of the fuel cell with a model consisting of mechanistic and empirical equations.

We consider the main elements comprising the fuel cell to be the conductor plates, the electrodes and the membrane. We also consider a cell that utilizes hydrogen as fuel and air as oxidant, both humidified.

The reaction occurring at the catalytic layer of the anode involves the decomposition of hydrogen. This releases electrons to the platinum electrode and protons to the membrane.

$$H_2 \rightarrow 2H^+ + 2e^-$$

At the cathode, the oxygen reacts with the protons transported through the membrane and with electrons supplied by the catalyst to form water.

$$\frac{1}{2} O_2 + 2H^+ + 2e^- \rightarrow H_2O$$

Similar to Ceraolo et al. (2003), we consider a one-dimensional model with the quantities of interest varying in the direction orthogonal to the anode and cathode surfaces. We also assume an isothermal fuel cell, and thus do not consider the energy balance for the fuel cell. The air total pressure is assumed to be uniform, and the variation of the partial pressures of its components is calculated in the model.

The current created by the electrochemical reaction at the cathode is described using the Butler-Volmer equation:

$$j_r = j_0 A_r \left\{ \frac{p_1}{p_{10}} \frac{[H^+]}{[H^+]_0} \exp\left(\frac{\eta}{b}\right) - 1 \right\}$$

$$\eta = E_{0c} - \Delta\phi_{ce} = E_0 - V_{cell} - R_{ohm} j$$

The cell current density is the sum of the reaction current added to the contribution due to the charge storage in the electrical double layer at the interface between the catalyst and diffusion layers (Ceraolo et al., 2003).

$$j = j_r + C_{dl} \frac{\partial \eta}{\partial t}$$

The oxygen partial pressure in the gas diffusion layer at the cathode is given by the following equation

$$\frac{\partial p_1}{\partial t} = \omega \frac{\partial^2 p_1}{\partial \xi^2} - \psi \frac{j_r}{4F} \frac{\partial p_1}{\partial \xi}$$

where p_1 is the oxygen partial pressure, ξ the dimensionless spatial coordinate, and ω and ψ are functions of the cell temperature and air pressure as given by Ceraolo et al. The proton concentration in the catalyst layer is considered to be a function of the cell current, similar to the expression in Ceraolo et al. (2003). The internal resistance of the cell is calculated using the same empirical relation used by Ceraolo et al. For water transport through the membranes, we consider that the water vapor in the macropores is in equilibrium with the surrounding liquid phase, and the partial water pressure is uniform. As with Ceraolo et al., the membrane is considered as being completely saturated with water, and its conductivity is assumed to be only a function of temperature.

Another consideration is the operation of fuel cells on reformate feed, or on hydrogen feeds that are not completely CO free. This is an important consideration in PEM fuel cells where the fuel cell is directly connected to a fuel reformer such as a steam methane reformer. Another situation where CO contamination might be present in the H_2 feed is when the reformate is purified in a pressure swing adsorption system that is not optimized for purity. In such as a case, CO levels in the fuel cell feed of up to tens of ppm are not uncommon. To describe this situation, we use the processes described by Springer et al. (2001) to determine the interfacial kinetics:

$CO + M \leftrightarrow (M\text{-}CO)$

$H_2 + 2M \leftrightarrow 2\,(M\text{-}H)$

$(M\text{-}H) \rightarrow H^+ + e^- + M$

$H_2O + (M\text{-}CO) \rightarrow M + CO_2 + 2H^+ + 2e^-$

We use the rate expressions developed by Springer et al. to develop anode side material and charge balances.

RESULTS

In this section, we present dynamic results for the transient response of the PEM fuel cell to step changes in cell current and to changes in feed composition and flow rates.

In the first case, we describe the cell dynamic response to a current step in which overshoot is observed in the cell voltage in response to a sudden reduction in the cell current. The results are shown in Fig. 1, and these results are in agreement with those of Ceraolo et al. (2003). The explanation for the overshoot, as given by Ceraolo et al., is that the proton concentration decreases more slowly than the cell current as it follows the slow dynamics of excess water removal from the cathode catalyst layer.

Figures 2 and 3 describe the cell voltage behavior in response to cell current steps at different feed conditions. Figure 2 shows the cell voltage response to steps in the cell current with the cell temperature at 60 C, reactant pressure at 3 bar, and air and H_2 flows at 1000 scc/min and 500 scc/min respectively. This result is in good agreement with that of Ceraolo et al. Figure 3 shows the cell voltage response to a different current step with the cell temperature at 70 C, air and H_2 pressures at 3 and 2 bar respectively, and air and H_2 flows at 1500 scc/min and 1000 scc/min respectively.

Figure 4 presents results for changes in feed composition. In this case, the CO level at the anode changes from 0 to 25 ppm. The cell voltage response shows a decrease in the voltage with the increase in CO level and corresponding decrease in H2 partial pressure. This change shows the same characteristics as the steady-state results for change in the cell voltage-current characteristics shown by Springer et al. (2001).

The results shown in this section represent preliminary results for the dynamic response of PEM fuel cells to changes in cell current and feed characteristics. In future

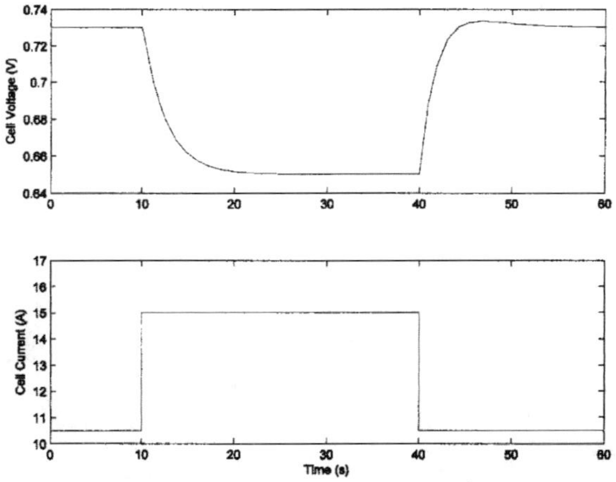

Figure 1. PEM fuel cell dynamic response to a current step, with overshoot in the voltage response to current reduction.

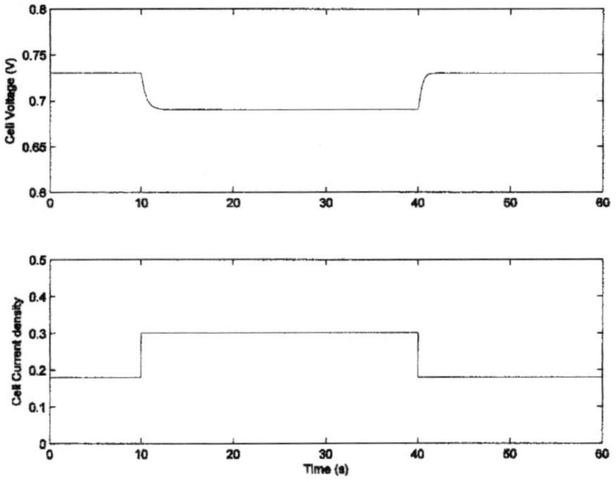

Figure 2. Cell dynamic response for a current step with cell temperature 60 C, reactant pressure 3 bar and air and H_2 flow 1000 scc/min and 500 scc/min.

work, we will address optimization and control issues related to obtaining optimal operating parameters.

SUMMARY

We have presented a model and simulation results for the behavior of proton exchange membrane fuel cells. The model includes descriptions of electrode kinetics and transport through the membrane. Water and charge transport considerations are included in the model. The model is used to study the transient behavior of the PEM fuel cell. In particular, the cell dynamic response to current steps and feed parameters is studied. The ultimate aim of this work is to use the developed model to conduct parametric sensitivity studies and develop optimization and control techniques for PEM fuel cells.

REFERENCES

1. T. Springer, T. Zawodzinski and S. Gottesfeld, "Polymer Electrolyte Fuel Cell Model", *J. Electrochem. Soc.*, **138**(8), 2334-2341 (1991).
2. D. M. Bernardi and M. Verbrugge, "A Mathematical Model of the Solid-Polymer Electrolyte Fuel Cell", *J. Electrochem. Soc.*, **139**(9), 2477-2490 (1992).
3. J.S. Yi and T.V. Nguyen, "An Along-the-Channel Model for Proton Exchange Membrane Fuel Cells", *J. Electrochem. Soc.*, **145**(4), 1149-1159 (1998).
4. J. S. Yi and T. V. Nguyen, "Multicomponent Transport in Porous Electrodes of Proton Exchange Membrane Fuel Cells Using the Interdigitated Gas Distributors", *J. Electrochem. Soc.*, **146**(1), 38-45 (1999).
5. W. He, J.S. Yi and T.V. Nguyen, "Two-Phase Flow Model of the Cathode of PEM Fuel Cells Using Interdigitated Flow Fields", *AIChE J.*, **46**(10), 2053-2064 (2000).
6. P. Costamagna, "Transport Phenomena in Polymeric Membrane Fuel Cells", *Chem. Eng. Sci.*, **56**, 323-332 (2000).
7. D. Natarajan and T.V. Nguyen, "A Two-Dimensional, Two-Phase, Multicomponent, Transient Model for the Cathode of a Proton Exchange Membrane Fuel Cell Using Conventional Gas Distributors", *J. Electrochem. Soc.*, **148**(12), A1324-A1335 (2001).
8. K. Sundmacher, T. Schultz, S. Zhou, K. Scott, M. Ginkel and E.D. Gilles, "Dynamics of the Direct Methanol Fuel Cell (DMFC): Experiments and Model-based Analysis", *Chem. Eng. Sci.*, **56**, 333-341 (2001).
9. T.E. Springer, T. Rockward, T.A. Zawodzinski and S. Gottesfeld, "Model for Polymer Electrolyte Fuel Cell Operation on Reformate Feed", *J. Electrochem. Soc.*, **148**(1), A11-A23 (2001).
10. G. Murgia, L. Pisani, M. Valentini and B. D'Aguanno, "Electrochemistry and Mass Transport in Polymer Electrolyte Membrane Fuel Cells", *J. Electrochem. Soc.*, **149**(1), A31-A38 (2002).
11. L. Pisani, G. Murgia, M. Valentini and B. D'Aguanno, "A New Semi-empirical Approach to Performance Curves of Polymer Electrolyte Fuel Cells", *J. Power Sources*, **108**, 192-203 (2002a).
12. L. Pisani, G. Murgia, M. Valentini and B. D'Aguanno, "A Working Model of Polymer Electrolyte Fuel Cells", *J. Electrochem. Soc.*, **149**(7), A898-A904 (2002b).
13. A. K. Shukla, C. L. Jackson, K. Scott and G. Murgia, "A Solid-Polymer Electrolyte Direct Methanol Fuel Cell with a Mixed Reactant and Air Anode", *J. Power Sources*, **111**, 43-51 (2002).
14. M. Roos, E. Batawi, U. Harnisch and Th. Hocker, " Efficient Simulation of Fuel Cell Stacks with the

Volume Averaging Method", *J. Power Sources*, **118**, 86-95 (2003).
15. M. Ceraolo, C. Miulli and A. Pozio, "Modelling Static and Dynamic Behaviour of Proton Exchange Membrane Fuel Cells on the Basis of Electro-chemical Description", *J. Power Sources*, **113**, 131-144 (2003).
16. J. Godat and F. Marechal, "Optimization of a Fuel Cell System Using Process Integration Techniques", *J. Power Sources*, **118**, 411-423 (2003).

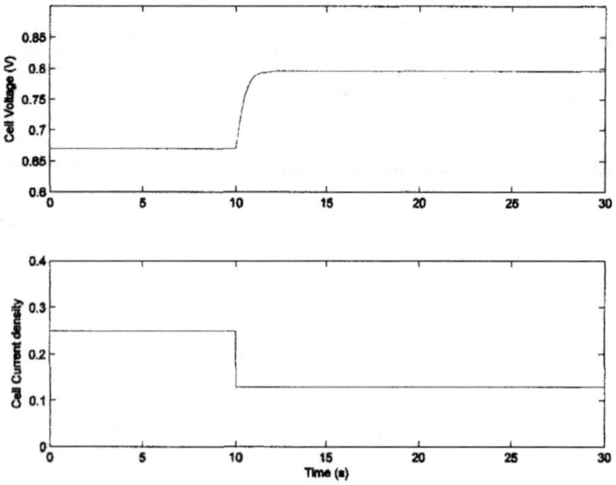

Figure 3. Cell dynamic response for a current step with cell temperature 70 C, air pressure 3 bar, H_2 pressure 2 bar and air and H2 flow 1500 scc/min and 1000 scc/min .

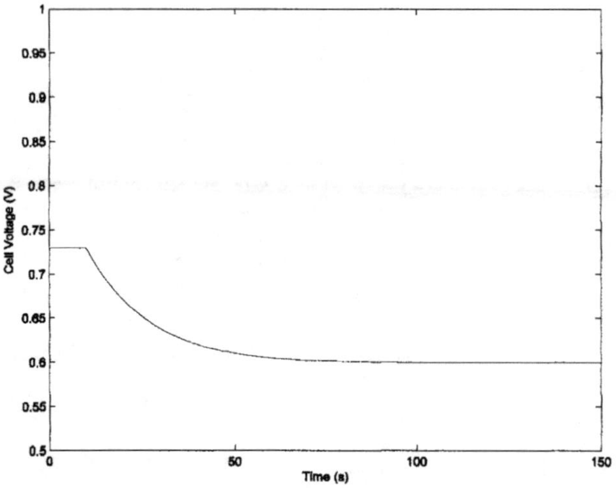

Figure 4. Cell dynamic response for a step in CO level at anode from 0 to 25 ppm.

CONTACT

Vinay Prasad: prasav@alum.rpi.edu

B. Wayne Bequette: bequette@rpi.edu

Hydrocarbon Reforming for Fuel Cell Systems: Fuel Effects on Fuel Processor and Light-off

Rod Borup, Michael A. Inbody, José I. Tafoya, Will J. Vigil and Dennis Guidry

Los Alamos National Laboratory
MST-11, MS J579
P.O. Box 1663
Los Alamos, NM 87545

ABSTRACT

Fuel cell systems must show long-term durability under challenging conditions, which includes rapid start-up to enable commercialization for transportation applications. Major gasoline and diesel fuel constituents, such as aliphatic compounds, napthanes, and aromatics have been compared for their effect on the fuel processing performance, and their propensity for carbon formation and for their light-off characteristics.

Fuel effect on catalytic light-off has been measured for a large number of different components. Straight chained hydrocarbons have a lower light-off temperature than similar carbon number branched chained hydrocarbons. Aromatics require higher light-off temperatures, while oxygenated compounds lower the light-off temperatures.

Carbon formation was measured for various hydrocarbon compositions, and operating conditions. Carbon formation was also measured during the fuel processor start-up transient. Aromatics show the highest tendency for carbon formation, while oxygenates help suppress carbon formation during light-off.

Introduction

Fuel cells have high efficiency for conversion of fuel to electricity. However, most types of fuel cells do not have high power density and efficiency operating directly from the hydrocarbon fuels currently widely available for transportation applications (gasoline and diesel), thus require fuel reforming. The choice of fuel for fuel cells is likely to be different for differing fuel cell applications, and debates exist about what the future fuels for fuel cell transportation applications will be. Potential on-board vehicle applications for fuel cells include both prime mobility power and auxiliary power production, and the fuel requirements will vary depending upon the power production application.

The goal of this research is to explore the effects of fuels, fuel constituents, additives and impurities on the performance of on-board hydrogen generation devices. Understanding the performance of fuels and their effects on the durability of fuel cell systems is key for their commercial introduction and use. Start-up of fuel processor systems is also crucial for their use in automotive fuel cell applications. Energy consumption, carbon formation, start-up time and limits on durability are all potential effects of fuel composition on the fuel processor during the start-up and light-off of fuel processors.

Experimental

To examine the effect of the fuel on hydrogen production devices, various fuel components and real fuels have been tested in fuel auto-thermal reformers (ATR) and fuel reforming systems. Fuel reforming has been conducted by catalytic partial oxidation and reforming using air as oxidant. Typical ATR catalysts which are noble metal based (such as platinum and rhodium) have been employed in this work. The catalysts are supported onto a high surface area substrate, either a monolith honeycomb, or a reticulated foam. For gasoline reforming, the gasoline, or gasoline component is vaporized, then mixed in vapor phase with steam. The fuel/water mixture is then mixed with air, preheated and introduced to the reforming catalyst. Carbon formation during light-off was measured by monitoring carbon via *in situ* laser absorption and scattering measurements. The oxidation catalyst is preheated to a temperature of about 250 – 300 °C so that fuel light-off occurs when the fuel/air mixture is introduced.

The analysis of the fuel effect on the product composition stream is conducted by various analytical techniques, specifically laser scattering and laser fluorescence, gas chromatography, FTIR (Fourier transform infrared), NDIR (non-dispersive infrared), paramagnetic oxygen and GC/MS (gas chromatography/mass spectrometry).

Various fuel compositions and components have been measured in these reactor systems. Main components include aliphatic (straight-chained and branched chained), aromatic, napthenic, olefinic. Hydrotreated naptha, reformulated gasoline, kerosene, de-odorized kerosene, low sulfur diesel, diesel fuel and bio-diesels are some of the fuels which have been used.

Results and Discussion

In autothermal reforming, proper fuel/steam/air mixture preparation is critical to avoid hot/cold spots, carbon formation, and catalyst degradation and to obtain proper conversion of hydrocarbons. Vaporization is one method to introduce fuel into a reformer, however it may lead to fuel pyrolysis, thus forming particulates or carbon during vaporization. Figure 1 compares the carbon formation measured during vaporization and refluxing of fuels and fuel components. In general, pure hydrocarbons did not show carbon formation, while real fuels did show carbon formation during the vaporization. The higher hydrocarbon streams such as kerosene and diesel fuel had more carbon formation than kerosene and diesel fuel in which sulfur has been removed and RFG, reformulated gasoline. To suppress carbon formation with these fuels that pyrolyze, direct fuel injection into the authothermal reformer may be required instead of prior fuel vaporization.

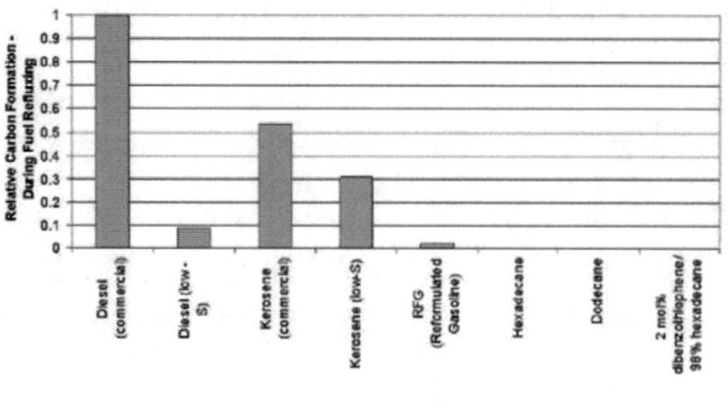

Figure 1. Carbon formation measured during fuel vaporization and refluxing of fuel components

The start-up of fuel processors for transportation applications is one challenging area requiring development for commercialization. During the start-up process, water may not be available. Without water it is impossible to light off a fuel processor at rich operating conditions and avoid regimes where carbon formation is favored by chemical equilibrium. Fuel effects on carbon formation during startup was quantified for different fuel compositions using an ATR reactor. These measurements were made at steam/carbon ratios of 0.0 and 0.5, simulating low water light-off of the reactors. The carbon formed is summarized in figure 2. Ethanol addition to iso-octane and RFG helps suppress carbon formation, while the addition of aromatic compounds or longer chained hydrocarbon (dodecane) led to increased carbon production.

In addition to the potential for carbon formation during start-up, other challenges in starting the fuel processor system include: the speed of start-up to limit on-board energy storage, start-up without stored water onboard the vehicle, and limiting the energy consumed during the start-up. These start-up challenges can be affected by the fuel composition used for the reformer and the ease at which the fuel oxidation reaction 'lights-off', i.e. at which the reaction exotherm become self-sustaining. Hydrocarbon components and fuel blends were measured for their propensity for light-off.

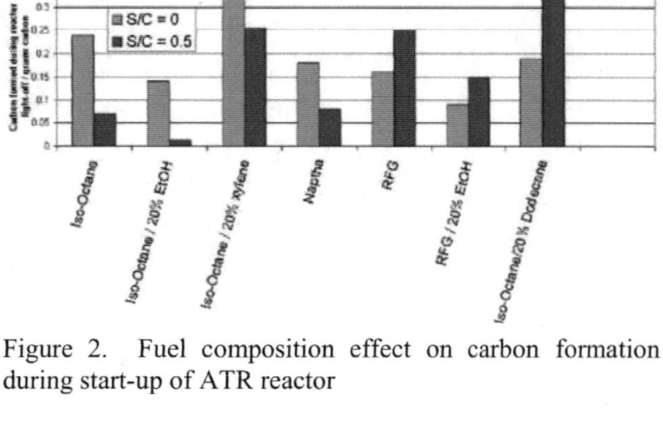

Figure 2. Fuel composition effect on carbon formation during start-up of ATR reactor

The reformer light-off was measured by introducing fuel/air/steam mixture to the reformer at a O/C = 0.9, S/C = 1.0, then ramping the temperature of the catalyst bed in the reformer. The outlet composition was measured in real time by a mass spectrometer to observed when oxygen consumption started, and hydrogen, carbon monoxide and carbon dioxide were generated. An example light-off for iso-octane with 20% pentane is shown in fig. 3.

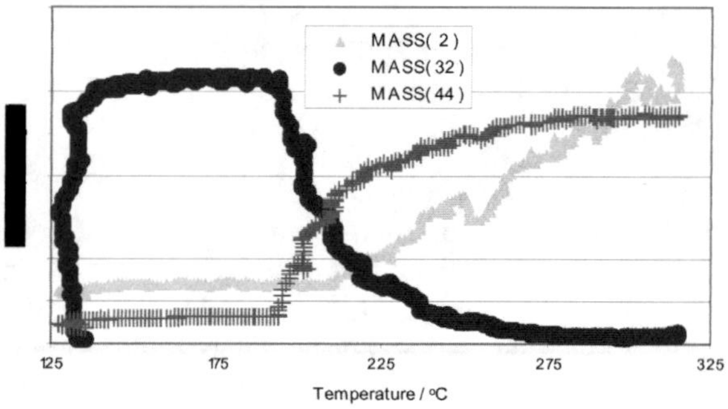

Figure 3. Reformer light-off with iso-octane and 20% pentane at O/C = 0.9 and S/C = 1.0.

Monitoring the light-off temperature of the reformer as a function of fuel helps determine the relative fuel effect on the reformer start-up. Iso-octane is typically used as a gasoline fuel simulant, and is thus used in all measurements for comparison purposes. Figure 4 shows the light-off of various straight-chained hydrocarbons, from n-pentane to n-decane, iso-octane and mixtures of iso-octane and pentane and hexane at 80% iso-octane. As the carbon length of the straight chained molecules is increased, the light-off temperature of the reformer increases. However, the light-off temperature of all of the straight-chained hydrocarbons are lower than the light-off temperature of iso-octane. As iso-octane is mixed with the straight-chained hydrocarbons, the light-off temperature is reduced.

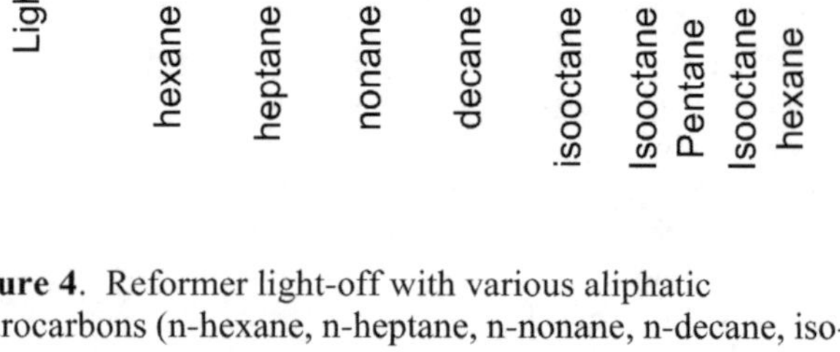

Figure 4. Reformer light-off with various aliphatic hydrocarbons (n-hexane, n-heptane, n-nonane, n-decane, iso-

octane, iso-octane + 20 % n-pentane and iso-octane + 20 % n-hexane).

A ubiquitous impurity in hydrocarbon fuels tends to be sulfur, in various compounds, and has been shown to be detrimental to the reforming of hydrocarbons.[1] To examine the effect of sulfur and sulfur compounds on the light-off of reformers, thiophene and dibenzothiophene were added to iso-octane at levels of 100 ppm and 1000 ppm. These results are shown in figure 5. The sulfur at levels of 100 ppm for both thiophene and dibenzothiophene are not detrimental to the light-off of the reformer, in fact decrease the light-off temperature of the catalyst by 10 to 15 °C. However, when the compounds are added at a concentration of 1000 ppm, favorable effect disappears with thiophene, and dibenzothiophene increases the light-off temperature by over 30 °C.

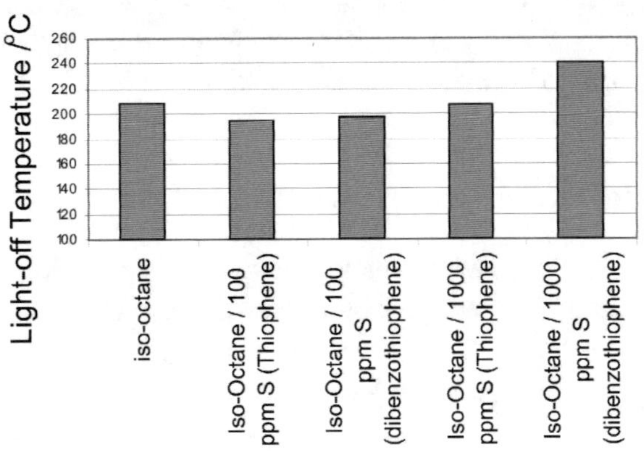

Figure 5. Reformer light-off with iso-octane and iso-octane with 100 ppm thiophene, 100 ppm dibenzothiophene, 1000 ppm thiophene and 1000 ppm dibenzothiophene (ppm by weight of S).

The current fuel infrastructure for transportation applications consists mainly of gasoline and diesel fuel. These fuels have many constituents, including generally relatively high content of aromatic components (~ 20 %). Aromatic components have been shown to slow the reforming kinetics and to increase carbon formation.[2] Figure 6 shows the light-off measurements of iso-octane, and iso-octane mixed with various aromatic compounds found in gasoline (xylene) and diesel fuel (anthracene, naphthalene). In addition, reformulated gasoline (RFG), kerosene, low sulfur diesel fuel and a commercial diesel fuel were measured for their light-off.

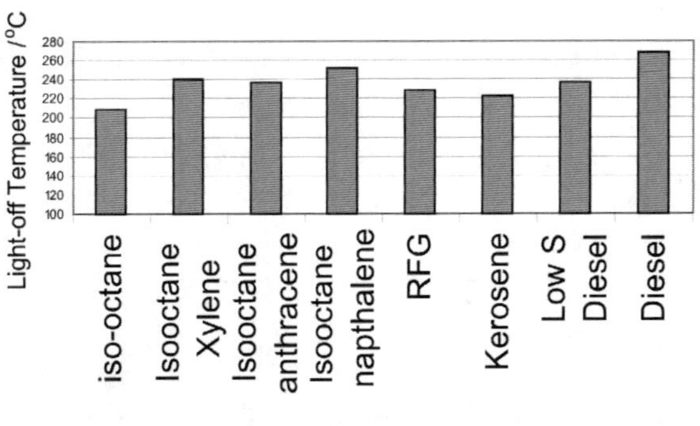

Figure 6. Reformer light-off with various aromatic hydrocarbons and fuels (iso-octane, iso-octane + 20 % xylene, iso-octane + 5 % anthracene, iso-octane + 5 % naphthalene, RFG – reformulated gasoline, kerosene, Low sulfur diesel fuel, and commercial diesel fuel).

The addition of aromatic compounds all increased the light-off temperature of iso-octane. The fuel blends all showed higher light-off temperatures than did iso-octane.

Conclusions

Fuel composition is important to reforming of fuel for transportation fuel cell systems. The fuel composition can affect carbon formation, the reformer light-off and start-up, the oxidation and reforming kinetics. Tailoring of the fuel composition can help the fuel processor start-up and long-term durability.

Aliphatic hydrocarbons tend to have favorable light-off and reforming characteristics compared with aromatic components. Addition of oxygenates lower the light-off temperature for catalytic oxidation and also tend to reduce carbon formation during the start-up sequence. The effect of sulfur containing-hydrocarbons does not make a large impact at low levels of sulfur (100 ppm) with regards to the fuel processor start-up.

Acknowledgments

This work was supported by the Department of Energy Hydrogen, Fuel Cells and Infrastructure Technologies Program, and from DOE SECA (Solid state Energy Conversion Alliance).

References

(1) Paul Berlowitz et al. 2003 Spring AIChE Meeting, New Orleans, March 30 – April 3.
(2) Borup et al. Fuel Composition Effects on Fuel Processor Dynamics, Office of Advanced Automotive Technologies Transportation Fuel Cell Power Systems FY 2000 Accomplishments Report.

Novel Circulating Fluidized Bed Membrane Reformer-Regenerator System for the Production of Hydrogen for Fuel cells from Heptane

Zhongxiang Chen *[1], **Yibin Yan** [2], **and Said S.E.H. Elnashaie** [3]
[1,2,3] Department of Chemical Engineering, Auburn University, Auburn, AL 36849-5127, USA
* E-mail: chenzho@auburn.edu, Tel: 1-334-844-2051, Fax: 1-334-844-2063.

ABSTRACT

Steam reforming of heptane for hydrogen production is investigated in a novel circulating fluidized bed membrane reformer-regenerator system (CFBMRR) utilizing a number of hydrogen and oxygen selective membranes. It is shown that although the amount of carbon deposition in the bed is significant, the effect on catalyst deactivation is negligible due to the large solid to gas mass feed ratio and the continuous catalyst regeneration in the system. The combustion of the deposited carbon in the catalyst regenerator supplies the heat needed for the endothermic steam reforming as well as the combustion of flammable gases from the riser reformer. Autothermal operation is achievable for the entire adiabatic reformer-regenerator system when the exothermic heat generated from the regenerator is sufficient to compensate the endothermic heat consumed in the reformer. Multiplicity of the steady states exists in the range of steam to carbon feed ratio of 1.444-2.251 mol/mol. The novel configuration has the potential advantages not only with respect to hydrogen production but also energy minimization.

KEY WORDS autothermal system, bifurcation, carbon formation, catalyst deactivation, catalyst regeneration, circulating fluidized bed, heptane, hydrogen, membrane reformer, multiplicity, steam reforming.

INTRODUCTION

Hydrogen is forecast to become one of two principal energy carriers with electricity in the 21st century [1], which is mainly produced by steam reforming of hydrocarbons in classical fixed-bed steam reformers [2-6]. The hydrogen produced can be used by fuel cells to produce electricity efficiently with efficiency of 45-55% and with zero emissions [7,8]. Recently, President Bush announced in his State of the Union Address $1.2 billion in research funding for developing clean, hydrogen-powered automobiles and fuel cells. However, fixed-bed steam reformers are usually inefficient, polluting and suffers from carbon formation [2,5,9]. This fixed bed configuration also limits the range of hydrocarbons to be used as feedstocks. In order to produce pure hydrogen for fuel cells more efficiently and from a wide range of higher hydrocarbons including gasoline, diesel and bio-oils, a novel process consisting of a circulating fluidized bed membrane reformer and a catalyst regenerator has been proposed [6]. In this paper, we investigate the carbon deposition and hydrogen production in the circulating fluidized reformer-regenerator system.

REACTIONS AND MODEL EQUATIONS

For the circulating reformer-regenerator system, the following reactions have been considered in the reformer:

$$C_7H_{16} + 7H_2O \rightarrow 7CO + 15H_2 \quad (1)$$

$$CO + 3H_2 \rightleftharpoons CH_4 + H_2O \quad (2)$$

$$CO + H_2O \rightleftharpoons CO_2 + H_2 \quad (3)$$

$$CH_4 + 2H_2O \rightleftharpoons CO_2 + 4H_2 \quad (4)$$

$$C_7H_{16} + 3.5O_2 \rightarrow 7CO + 8H_2 \quad (5)$$

$$CH_4 + 2O_2 \rightarrow CO_2 + 2H_2O \quad (6)$$

$$CH_4 + CO_2 \rightleftharpoons 2CO + 2H_2 \quad (7)$$

$$C_7H_{16} \rightarrow 7C + 8H_2 \quad (8)$$

$$CH_4 \rightleftharpoons C + 2H_2 \quad (9)$$

$$2CO \rightarrow C + CO_2 \quad (10)$$

$$C + H_2O \rightarrow CO + H_2 \quad (11)$$

$$C + 0.5O_2 \rightarrow CO \quad (12)$$

$$C + CO_2 \rightarrow 2CO \quad (13)$$

The reaction kinetics for these 13 reactions is obtained from the published literatures [3-6,10-15].

The riser reformer is modeled as a plug flow reactor with co-current flow in the reformer and membranes. The material and energy balance equations are as follows:

$$\frac{dF_i}{dl} = \rho_C(1-\varepsilon)A_f\sum_{j=1}^{13}\sigma_{i,j}r_j - aJ_{H_2}\pi N_{H_2}d_{H_2} + bJ_{O_2}\pi N_{O_2}d_{O_2} \quad (14)$$

$$\frac{dT}{dl} = \frac{\sum_{j=1}^{13} r_j(-\Delta H_j)\rho_C(1-\varepsilon)A_f + \dot{Q}}{\sum F_i Cp_i} \quad (15)$$

where, a and b are the control indices for the membrane permeation fluxes. For hydrogen, a=1, b=0; for oxygen, a=0, b=1; and for any other components, a=0, b=0.

Due to the carbon formation, the catalyst deactivates in the riser reformer. Accordingly, the reaction rate equations are reformulated as follows:

$$r_j = r_{j0} \cdot \phi_j \qquad (16)$$

$$\phi_j = \begin{cases} \exp(-\alpha_C C_k) \\ 1.0 \end{cases} \qquad (17)$$

The value of ϕ_j for reaction j is calculated by Eq.17 [16] or equal to 1.0 depending on whether the jth reaction is affected by the catalyst deactivation or not.

For the catalyst regenerator, we assume that the burning of carbon and the combustion of flammable gases are highly efficient or the catalyst regenerator is over designed that we can consider these reactions to be complete. We also assume that the heat generated in the regenerator is used for pre-heating the clod feed water (to steam), heptane and the recycle catalyst. Therefore the energy balance equation in the regenerator is as follows:

$$\sum F_i(-\Delta H_c) = \sum F_k [(T_{k,b} - T_{room}) Cp_{k(l)} + \Delta H_{vap} + (T_0 - T_{k,b}) Cp_{k,b(g)}] + G_{cat} Cp_{cat} (T_0 - T_{exit}) \qquad (18)$$

where, subscript k stands for feed water and heptane. The hydrogen or oxygen permeation flux in hydrogen or oxygen permselective membranes can be calculated using the equation reported by Shu et al. or Tsai et al. [17,18], respectively. Thus based on these equations, the complete reformer-regenerator system is simulated. The standard simulation conditions are similar to the data reported by Chen et al. [6].

CARBON DEPOSITION AND CATALYST DEACTIVATION IN ISOTHERMAL REFORMER

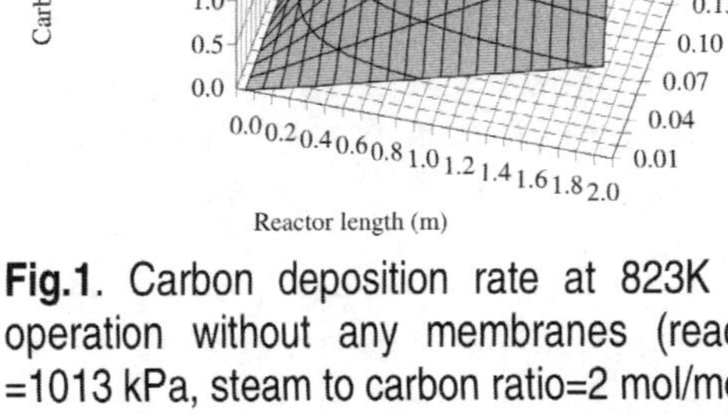

Fig.1. Carbon deposition rate at 823K for isothermal operation without any membranes (reaction pressure =1013 kPa, steam to carbon ratio=2 mol/mol).

First we investigate the carbon deposition and catalyst deactivation in the isothermal riser reformer. Fig.1 shows the carbon deposition rate at 823K for the case without any hydrogen or oxygen membranes. The carbon deposition rate can be as high as 2.56 kg/h. It increases firstly along the reactor length and also increases when the solid catalyst fraction increases in the bed. However, after a certain reactor length or above a certain solid catalyst fraction, the carbon deposition rate decreases. This phenomenon is due to the fact that carbon formation is a catalytic process related to the amount of catalyst. The less the amount of solid catalyst, the slower the carbon formation rate by heptane cracking. At the same time with the proceeding of steam reforming of heptane, the by-products carbon monoxide and methane will also contribute to the formation of carbon on the nickel catalyst. Similarly, the less the solid catalyst in the bed, the slower the steam reforming of heptane and the formation of carbon monoxide and methane. As a result the less the solid catalyst, the lower the carbon formation rate from carbon monoxide and methane. Summation of the above factors results in the fact that the carbon deposition rate decreases when the solid catalyst fraction decreases and vice versa. However, when the solid catalyst fraction is high, the steam reforming rate of heptane is high, which makes the reactor length for full heptane conversion short. After the full conversion of heptane at a certain reactor length, the major source of carbon formation from heptane stops, as a result the overall carbon deposition rate decreases along the rest of the reactor length as shown in Fig.1. For example, at solid catalyst fraction of 0.2, the maximum carbon deposition rate is at the reactor length of 0.3 m. This position is the place where heptane is fully converted to hydrogen and carbon monoxide in the steam reformer. Fig.1 shows that a maximum carbon deposition rate (2.56 kg/h) is developed at a solid catalyst fraction close to 0.05 v/v. The carbon deposition rate increases when the solid catalyst fraction increases from 0 to 0.05 v/v, then the carbon deposition rate decreases above 0.05 v/v. Fig.2 shows the catalyst activity as a function of the solid catalyst fraction in the bed. Although the carbon deposition rate can be as high as 2.56 kg/h in Fig.1, the catalyst activity is almost constant (0.984-1.0). Why is the change of the catalyst activity very small although so large amount of carbon deposit on the catalyst? This is due to the high mass flow ratio of the solid catalyst to the gas stream in the circulating fluidized bed and also the continuous catalyst regeneration in the regenerator in this novel process. Obviously this is a potential advantage of the circulating fluidized bed over conventional fixed-bed steam reformers.

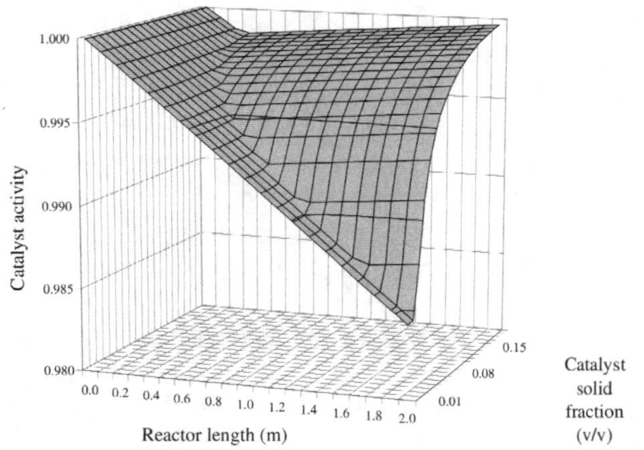

Fig.2. Catalyst activity at 823K for isothermal operation without any membranes (reaction pressure=1013 kPa, steam to carbon ratio=2 mol/mol).

AUTOTHERMAL REFORMING SYSTEM AND BIFURCATION BEHAVIOR

Because of the continuous catalyst regeneration, there is exothermic burning of carbon and flammable gases (such as unreacted heptane, by-products methane and carbon monoxide, and remaining product hydrogen) in the regenerator, the heat generated can be used for the endothermic steam reforming in the riser reformer. In order to investigate the thermal effect as well as the efficient production of hydrogen, the whole circulating fluidized bed membrane reformer-regenerator system is investigated under adiabatic conditions and we assume any remaining hydrogen in the exit gas stream is further separated before entering the catalyst regenerator. If heat balance is well controlled, then the whole reformer-regenerator system can be operated under autothermal condition. That is, at steady state, the heat generated in the catalyst regenerator is equal to the heat consumption in the riser reformer. In this section, hydrogen and oxygen permseletive membranes are used. Chen and coworkers have discussed the effects of these membranes [6]. The numbers of hydrogen and oxygen membranes are 20 and 80, respectively. The reaction pressure in the reformer is 1013 kPa. Figs. 3-6 show the performance of the reformer-regenerator system when the entire adiabatic process achieves autothermal condition. Figs. 3-6 also show that there are multiplicities (three steady states) in the range of steam to carbon feed ratio of 1.444-2.251 mol/mol. In this paper we present a preliminary static bifurcation analysis of the autothermal reformer-regenerator system. A detailed bifurcation investigation is underway and will be published shortly.

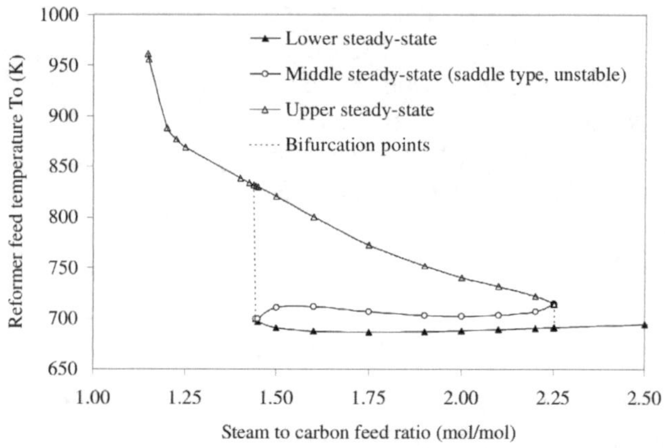

Fig.3. Circulating feed temperature to the reformer and its bifurcation behavior at different steam to carbon feed ratio when the circulating reformer-regenerator system is operated under autothermal condition.

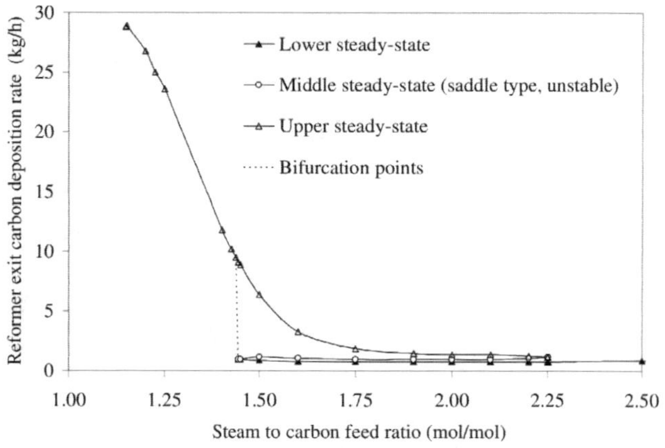

Fig.4. Carbon deposition rate at the reformer exit and its bifurcation behavior at different steam to carbon feed ratio when the circulating reformer-regenerator system is operated under autothermal condition.

Usually, high steam to carbon feed ratio increases the hydrogen yield and hydrogen production. On the other hand, the higher the steam feed, the higher the energy consumption for preheating the cold water to generate steam. Thus in order to produce hydrogen efficiently and economically, the steam to carbon feed ratio is an important parameter for energy consumption in industrial applications. Under autothermal conditions the adiabatic riser reformer and catalyst regenerator system produces hydrogen with "zero" external heat input. That is, when the circulating reformer-regenerator system is operated under autothermal conditions, the heat consumed for preheating the cold feed water (to generate steam), heptane and hot catalyst is just supplied by the heat of reactions from the catalyst regenerator, in which it supplies the necessary heat for endothermic steam reforming in the adiabatic riser reformer.

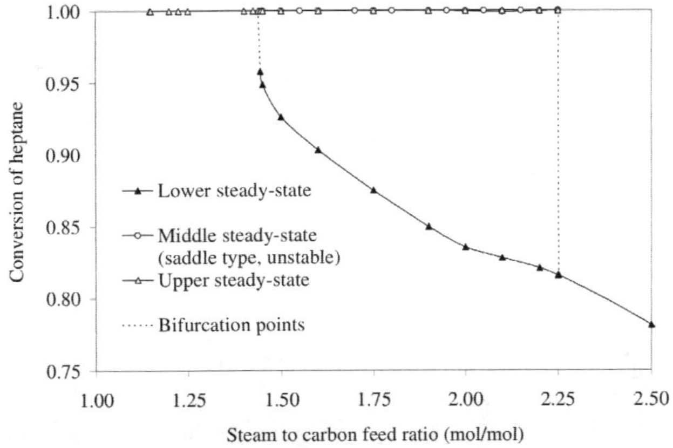

Fig.5. Conversion of heptane and its bifurcation behavior at different steam to carbon feed ratio when the circulating reformer-regenerator system is operated under autothermal condition.

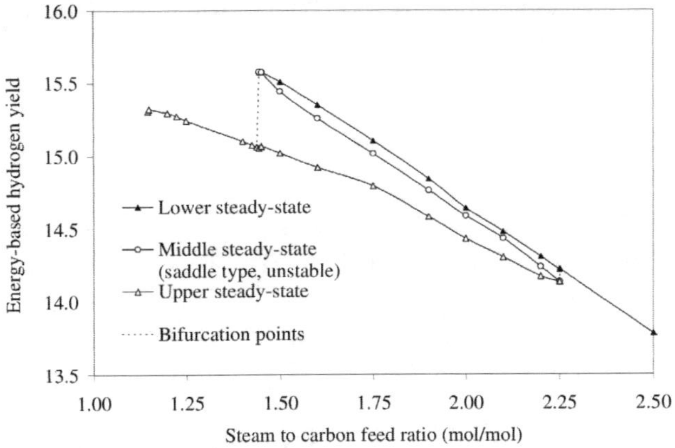

Fig.6. Energy-based hydrogen yield and its bifurcation behavior at different steam to carbon feed ratio when the circulating reformer-regenerator system is operated under autothermal condition.

Figs. 3 and 4 show the circulating feed temperature to the reformer and reformer exit carbon deposition rate at different steam to carbon feed ratio when the entire adiabatic reformer-regenerator system is operated under autothermal conditions. Note, in the autothermal process the reformer feed temperature is a result from the heat supply in the regenerator by preheating the cold water and heptane, and recycled catalyst. It is a system variable instead of feed parameter. According to the value of reformer feed temperature to the reformer, the multiple steady states are classified with regard to temperature as lower, middle and upper steady state, respectively. When steam to carbon feed ratio is higher than 2.251 mol/mol, the autothermal circulating feed temperatures are almost constant with the change of steam to carbon feed ratio. This is due to the fact that the higher the steam feed, the higher the reaction extents of endothermic steam reforming (of heptane and methane) and water gas shift reaction, and the lower the carbon formation. In order to maintain the reformer-regenerator system operate autothermally under these conditions, the conversion of heptane has to be low to provide enough "fuel" for combustion in the regenerator to supply the necessary heat for the endothermic steam reforming. Therefore under autothermal condition, the circulating feed temperature in this case is the lowest to ensure that the conversion of heptane is low. For example, at steam to carbon feed ratio of 2.5 mol/mol, the conversion of heptane is 78.11%, the net carbon deposition rate is 0.9 kg/h in the riser reformer and the autothermal circulating feed temperature is 694.31 K. If the energy-based hydrogen yield is defined as the amount of hydrogen produced per mole of heptane fed under autothermal conditions. The more the feed heptane is consumed for generating the necessary heat in the regenerator for the endothermic process, the smaller the energy-based hydrogen yield. Fig. 6 shows the energy-based hydrogen yield in the autothermal reformer-regenerator system. At higher steam to carbon feed ratio, the energy-based hydrogen yield is shown to be the lowest. For example, at 2.5 mol/mol steam to carbon feed ratio, the yield of hydrogen is 13.78 mole of hydrogen per mole of heptane feed. While when steam to carbon feed ratio is between 1.444-2.251 mol/mol, multiple steady states occur with lower, middle and upper autothermal feed temperatures. For example, at steam to carbon feed ratio of 2.0 mol/mol, three steady states exist at reformer feed temperatures of 687.99, 702.15 and 740.34 K. Fig. 5 shows that at middle and upper steady state, the conversion of heptane is 100%. While at lower steady state, the conversion of heptane is between 81.60-95.80%. The conversion of heptane decreases when steam to carbon feed ratio increases. This is the opposite of the usual operation where the performance improves as the steam to carbon feed ratio increases. However, in this autothermal operation, we have to notice that high steam to carbon feed ratio prevents carbon formation causing the heat supply by burning carbon to decrease, forcing the system to lower conversion. Fortunately, due to the large mass flow ratio of catalyst in this configuration and the continuous catalyst regeneration, the catalyst activity is always kept high (>0.985), even at low steam to carbon feed ratio of 1.1475 mol/mol. The controlling factor in this novel autothermal process is the heat supply associated with the burning of carbon in the regenerator. In the multiplicity range of steam to carbon feed ratio, the higher the carbon deposition rate, the higher the autothermal feed temperature to the reformer. But the trend of hydrogen yield is totally reversed. As shown in Fig.6, the higher the autothermal feed temperature to the reformer, the lower the energy-based hydrogen yield. That is, at the upper temperature steady state, the energy-based hydrogen yield is the lowest. At the lower temperature steady state, the energy-based hydrogen yield is the highest. For example, for steam to carbon feed ratio of 2 mol/mol, at the upper feed temperature of 740.34 K, the energy-based hydrogen yield is 14.43. At the middle feed temperature of 702.15 K, the energy-based hydrogen yield is 14.59. While at the lower feed temperature of 687.99 K, the energy-based hydrogen yield is 14.64. This reverse relationship between the hydrogen yield and the autothermal feed temperature can be explained by the carbon formation-burning process during the steam reforming of hydrocarbons and regeneration of the catalyst. Since steam reforming not only extracts hydrogen from hydrocarbons but also extracts hydrogen from the reactant steam. While carbon formation only extracts hydrogen from hydrocarbons. Therefore, the larger the part of the hydrocarbon which cracks for carbon formation, the more the carbon generation, the higher the autothermal feed temperature and the lower the hydrogen production. According to the bifurcation theory [19], only the lower and upper steady states are stable and the middle steady state is an unstable saddle type state. If the start-up is above the middle steady state, the system will finally settle at the upper steady state. If the start-up is below the middle steady state, the system will settle at the lower steady state. When steam to carbon feed ratio is below 1.444 mol/mol, only one stable steady state exists in the reformer-regenerator system. Since in this region (<1.444 mol/mol) low steam to carbon feed ratio is fed, the extent of steam reforming becomes small. On the other hand, the carbon formation and deposition competes for the reactant heptane. Since carbon is burned-off by excess air in the catalyst regenerator, the more the carbon deposition, the higher the heat supply. As a result, the autothermal feed temperature to the reformer is higher. Since the feed temperature and carbon formation affects each other, at low steam to carbon feed ratio, the autothermal feed temperature and carbon deposition rate are high as shown in Figs. 3 and 4. Fig. 3 shows that if the steam to carbon feed ratio is smaller than 1.1475 mol/mol, the autothermal circulating feed temperature is very high, leading to thermal "run away". Figs. 3-6 show that if the steam to carbon feed ratio is higher than 2.251 mol/mol or lower than 1.444 mol/mol, the reformer-regenerator system will have only one steady state. In the region of 1.444-2.251 mol/mol, the upper (or lower) steady state will shift to the lower (or upper) steady state suddenly at the bifurcation points. In this investigation the maximum energy-based hydrogen yield is about 15.58 moles of hydrogen produced per mole heptane fed at the lower steady state when steam to carbon feed ratio is very close to the bifurcation point of 1.444 mol/mol. It is about 70.82 % of the maximum theoretical hydrogen yield of 22 when the final steam reforming products are carbon dioxide and hydrogen.

Since the sweep gas in hydrogen permselective membranes is steam, if the exit mixture gas is cooled to 300K at down stream, most of steam is condensed and the hydrogen concentration can be high. For example, when the above maximum energy-based hydrogen yield of 15.58 is obtained in the autothermal process, the hydrogen concentration in hydrogen permselective membranes is about 56.95mol%. After condense the most steam, the hydrogen concentration is about 98.14mol%. If any remaining steam is further removed using some de-humidification agents, the hydrogen purity can be almost 100%, which is ideal for fuel cells.

CONCLUSION

This investigation highlights the great advantages associated with the novel circulating fluidized bed membrane reformer and catalyst regenerator system for the efficient production of pure hydrogen with minimum energy consumption. Due to the high mass flow ratio of solid catalyst to gas stream and continuous catalyst regeneration, the steady-state catalyst activity is high despite the large carbon deposition rate. When the heat generated from the burning of carbon and other exit flammable gases from the riser reformer is equal to the heat consumed for preheating the cold water (to generate steam), heptane and hot recycle catalyst to the feed temperature (in which it supplies the heat necessary for the endothermic steam reforming reactions in the reformer), the riser reformer and catalyst regenerator system can be operated under autothermal conditions. Under autothermal conditions, the reformer-regenerator system shows a bifurcation behavior in the range of steam to carbon feed ratio of 1.444-2.251 mol/mol. It is interesting to notice that the energy-based hydrogen yield decreases when steam to carbon feed ratio increases. In the multiplicity region, the energy-based hydrogen yield at the upper steady state (or high autothermal feed temperature to the reformer) is the lowest one and at the lower steady state is the highest one. This is due to the high carbon formation rate associated with the high feed temperature. The maximum energy-based hydrogen yield is about 15.58 moles of hydrogen per mole of heptane fed at the lower steady state when steam to carbon feed ratio is close to the bifurcation point of 1.444 mol/mol.

ACKNOWLEDGEMENT

This work is financially supported by Auburn University, Grant Number 2-12085.

REFERENCES

1. Armor J.N., Review: The Multiple Roles for Catalysis in the Production of H_2, *Appl. Catal. A,* 176, p159-176, (1999).
2. Rostrup-Nielsen, J., "Hydrogen via Steam Reforming of Naphtha", *Chemical Engineering Progress*, 9, 87 (1977).
3. Tottrup P.B., Evaluation of Intrinsic Steam Reforming Kinetic Parameters from Rate Measurments on Full Particle Size, *Appl.Catal.*, 4, p377-389, (1982).
4. Xu J and G.F. Froment, Methane Steam Reforming, Methanation and Water-Gas Shift: I. Intrinsic Kinetics, J. AIChE, 35 (1), 1989, p88-96.
5. Elnashaie, S.S.E.H.; Elshishini, S.S., *Modelling, simulation and optimization of industrial fixed bed catalytic reactors*, Gordon and Breach Science Publishers: London, UK, (1993).
6. Chen, Z., Y. Yan and S.S.E.H. Elnashaie, Modeling and Optimization of a Novel Membrane Reformer for Higher Hydrocarbons, 49 (5), *AIChE J.*, p1250-1265, (2003).
7. Borroni-Bird C.E., Fuel cell commercialization issures for light-duty vehicle applications. *J Power Sources* **61** (1996), p33–48.
8. Brown, L.F., "A Comparitive Study of Fuels for Onboard Hydrogen production for Fuel Cell Powered Automobiles", *Int. J of Hydrogen Energy*, 26(4), 381-397, 2001.
9. Twigg M. V., Catalyst Handbook, 2nd Edition, Wolfe Publishing Ltd, England, pp225-282, (1989).
10. Siminski V.J., F.J. Wright, R.B. Edelman, C. Economos and O.F. Fortune, Research on Methods of Improving the Combustion Characteristics of Liquid Hydrocarbon Fuels, *AFAPL* TR 72-74, I & II, Air Force Aeropropulsion Laboratory, Wright Patterson Air Force Base, OH, February, (1972).
11. Jin W., X.Gu, S.Li, P.Huang, N.Xu, J.Shi, Experimental and Simulation Study on a catalyst Packed Tubular Dense Membrane Reactor for Partial Oxidation of Methane to Syngas, *Chem.Eng.Sci.*, 55, p2617-2625, (2000).
12. Rostrup-Nielsen,J.R. Coking on Nickel Catalysts for Steam Reforming of Hydrocarbons, J. Catal., 33, pp184-201, (1974).
13. Snoeck J.W., G.F. Froment and M. Fowles, Kinetic study of the Carbon Filament Formation by Methane Cracking on a Nickel Catalyst, J. Catal., 169, pp250-262, (1997).
14. Tottrup P.B., Kinetics of Decomposition of Carbon Monoxide on a Supported Nickel Catalyst, J. Catal., 42, pp29-36, (1976).
15. Chen C.X., Masayuki Horio and Toshinori Kojima, Numerical Simulation of Entrained Flow Coal Gasifiers. Part I: Modeling of Coal Gasification in an Entrained Flow Gasifier, Chem. Eng. Sci., 55, pp3861-3874, (2000).
16. Chen, Z., Y. Yan and S.S.E.H. Elnashaie, Using Coking and Decoking Model in a Circulating Fast Fluidized Bed Membrane Reformer for Efficient Production of Pure Hydrogen by Steam Reforming of Higher Hydrocarbons, Proceedings of the Regional Symposium on Chemical Engineering and the 16th Symposium of Malaysian Chemical Engineers, p1239-1247, Kuala Lumpur, Malaysia, (2002).
17. Shu, B.P.A. Grandjean and S. Kaliaguine, Methane Steam Reforming in asymmetric Pd- and Pd-Ag/porous SS Membrane Reactor, Appl. Catal, A 119, 1994, p305-325.
18. Tsai, C.Y., A. G. Dixon, W.R. Moser and Y.H. Ma, "Dense Perovskite Membrane Reactors for Partial Oxidation of Membrane to Syngas", *Journal of AIChE,* 43(11A), 2741 (1997).
19. Elnashaie, S.S.E.H. and Elshishini,S.S. *Dynamic Modeling, Bifurcation and Chaotic Behavior of Gas-Solid Catalytic Reactors*, Gordon and Breach Science Publishers, London, UK, (1996).

DEFINITIONS

A_f	free cross-section area of the reactor (m^2)
C_k	concentration of deposited carbon (g/gcatalyst)
Cp_{cat}	specific heat of solid catalyst (J/g catalyst/K)
Cp_i	specific heat of component i (J/mol/K)
d_{H_2}	outside diameter of hydrogen membrane tubes (m)
d_{O_2}	outside diameter of oxygen membrane tubes (m)
F_i	molar flow rate of component i (mol/s)
G_{cat}	mass flow rate of catalyst (g/s)
ΔH_C	heat of combustion (J/mol)
ΔH_j	heat of reaction for the jth reaction (J/mol)
ΔH_{vap}	heat of vaporization (J/mol)
J_{H_2}	permeation fluxes of hydrogen ($mol/m^2/s$)
J_{O_2}	permeation fluxes of oxygen ($mol/m^2/s$)
l	length of reactor (m)
N_{H_2}	number of hydrogen permselective membrane tubes
N_{O_2}	number of oxygen permselective membrane tubes
$\dot{Q}$	rate of heating along the reactor length (J/s/m)
r_j	reaction rates (mol/gcatalyst/s)
r_{j0}	reaction rates with fresh catalyst (mol/gcatalyst/s)
T	temperature (K)
T_{exit}	riser reformer exit temperature (K)
T_0	autothermal feed temperature to reformer (K)
T_{room}	room temperature (K)

Greek letters

α_C	deactivation constant (gcatalyst/gcoke)
ε	void fraction
π	3.1415926
ρ_c	density of catalyst (kg/m^3)
$\sigma_{i,j}$	stoichiometric coefficient of component i for the jth reaction
ϕ	catalyst activity function

Storage of Hydrogen on Nanostructure Carbons: Experiments and Analyses

Li Zhou, Yan Sun
High Pressure Adsorption Laboratory (State Key Laboratory of Chemical Engineering), School of Chemical Engineering
Tianjin University, Tianjin 300072, China

Yaping Zhou
Group of Physical Chemistry, School of Science
Tianjin University, Tianjin 300072, China Tianjin University, Tianjin 300072, China

ABSTRACT

Superactivated carbon and carbon nanotubes are common in their nanostructures. Adsorption isotherms of H_2 on carbon AX-21 and multiwalled carbon nanotubes were collected by a volumetric method for the temperature range 77, 233-298 K and pressures up to 7 or 10 MPa. Limiting heat of adsorption was evaluated based on experimental data, and 7.6 and 1.8 kJ/mol were obtained with activated carbon and carbon nanotubes respectively for the range 233-298 K. Absolute adsorption was determined by a new strategy and the adsorption behavior of H_2 on carbon nanotubes was reasonably explained. A comparison was given for the storage capacity of a container that vacant or filled with powder or pellets of the two samples. It was concluded that the adsorption of H_2 on carbon nanotubes is too weak to increase the storage more than compression, but activated carbon enhances the storage capacity considerably. The weight percent of hydrogen stored in a container filled with carbon powder is as high as 10.8 wt % at 77 K and 6 MPa basing on the carbon mass filled, and 4.1 kg H_2 could be stored in a container of volume 100 liter filled with carbon pellets under the condition mentioned. This method of hydrogen storage is technically feasible and cost competitive. It is also advantageous with respect to safety and availability of fueling facilities.

KEY WORDS

Hydrogen energy, Storage, Adsorption, Activated carbon, Carbon nanotubes

INTRODUCTION

Fueling vehicles with hydrogen is a common target that the whole world tried to reach. It is not only for protecting the atmosphere from emission pollution, but also for the development of a renewable energy source. Fueling cars with hydrogen is feasible for power systems of either inner combustion engine or electric motor coupled with fuel cells. The bottleneck is on-board storage of hydrogen to satisfy the demand of both energy density and cost competition. Several ways of hydrogen storage, including liquid hydrogen, compressed (to 20 MPa) hydrogen, decomposed in situ from methanol or metal hydride, have not been proven as a practical technology of storage to compete with gasoline cars. Therefore, searching for a new way to store hydrogen on board is an urgent task.

BACKGROUND

Intensifying the storage of hydrogen by adsorption on activated carbon was proposed previously [1,2], yet it did not receive much attention of industries especially when carbon nanotubes were claimed to possess abnormal performance as a hydrogen carrier [3]. Systematic measurements on the adsorption capacity of activated carbon, powder and pellets, as well as on multiwalled carbon nanotubes, powder and pellets, were carried out in our laboratory. Presented is the result of experiments and their analyses based especially on the fundamental theory of supercritical adsorption. The storage capacity of a container, vacant or filled with different carbon materials, was evaluated and mutually compared.

EXPERIMENTAL

THE ADSORBENT – The critical temperature of hydrogen is 33.2 K (-240°C). Adsorptive storage of hydrogen, at ambient temperature or cooled by liquid nitrogen (77 K), is absolutely taking advantage of

[1] Noh, J. S., Agarwal, R. K. and Schwarz, J. A., " Hydrogen Storage Systems Using Activated Carbon," *Int. J. Hydrogen Energy*, **12**, 693 (1987).
[2] Zhou, L., "Impacts of the Adsorptive Storage of Hydrogen on Superactivated Carbons to Large-Scale Utilization of Hydrogen Energy," *Science & Technology Review* (in Chinese), **No.12**, 11 (1999).
[3] Dillon, A. C., Jones, K. M., Bekkedahl, T. A., Kiang, C. H., Bethune, D. S., Heben, M. J. "Storage of hydrogen in single-walled carbon nanotubes," *Nature*, **386**, 377 (1997).

supercritical adsorption. Multilayer or volume-filling mechanism of adsorption [4] can hardly happen at above-critical temperatures to the knowledge of physics or chemistry available, and the experimental isotherms were well explained by monolayer adsorption mechanism [5]. Therefore, to increase the capacity of adsorption one has to increase the specific surface area of adsorbents to as much as possible. Superactivated carbon of high surface and micropore structure was thus developed. Activated carbon (AC) Ax-21 is an example of such carbon although it is not commercially available presently. It was used in present study because the property of the carbon is well known. It has a BET surface area of about 3000 m^2/g and pore volume of about 1.3 cm^3/g. The bulk density of the carbon powder is only 0.3 g/cm^3. Carbon pellets were made using binder to increase the bulk density to 0.72 g/cm^3.

Although single walled carbon nanotubes (SWNT) were firstly noticed of extra ordinal capacity of hydrogen up taking [3], it was difficult, if not impossible, to have enough quantity of SWNT sample for a reliable volumetric measurement of adsorption. A product can hardly be applied industrially if it could not be produced economically in large scale. On the other hand, multiwalled carbon nanotubes (MWNT) was also claimed to have a high capacity of hydrogen storage [6], and could be produced in large scale, therefore, is worth of further study of its engineering potential. The sample of MWNT used was manufactured in a fluidized-bed reactor. The catalyst was pre-reduced $LaCoO_3$. The product was first washed in toluene and methanol to clean up the fullerenes and trace of organic chemicals. It was next washed in HCl and HNO_3 to clean up the residual catalyst and other inorganic impurities. Finally, amorphous carbons were oxidized in CO_2 stream. Both the wall thickness and the inner diameter of the multiwall tube were about 10 nm, which were estimated from the scale of a 10^5 times enlarged TEM photo. The BET surface area of the sample was determined based on the adsorption isotherm of N_2 at 77 K, and 137 m^2/g was obtained. Pore size of superactivated carbon is largely less than 2 nm, but the inner and outer diameter of MWNT is much larger than 2 nm, therefore, the specific surface area of MWNT cannot be as large as superactivated carbon. The bulk density of the MWNT powder is 0.35g/cm^3. Pellets were made without using any binder, and the bulk density of pellets was as high as 0.96 g/cm^3.

Samples experienced vacuum degassing at 393 K for 24 hours, and once more in-situ degassing for four hours at 393 K before adsorption measurement. The vacuum readings were kept constant indicating no more gases released after the treatment. Hydrogen and helium used in experiments is of purity 99.9995%.

THE METHOD USED FOR ADSORPTION MEASUREMENT – There are different methods that can be applied for high-pressure adsorption measurement [7]; however, gravimetric and volumetric methods are mostly applied in practice. The former uses a microbalance and the amount adsorbed was determined by weight increment during adsorption. The latter uses two containers (cells). One is empty and functions as a meter of gas quantity and was named as reference cell; the other is the sample container where adsorption happens and was thus named as adsorption cell. The reference cell was maintained in a thermostat whose temperature is as close to the room temperature as possible, while the adsorption cell was put in another thermostat or cryostat whose temperature was maintained at the temperature specified. The pressure of the reference cell was monitored and recorded. The amount adsorbed was determined from the difference of pressure readings before and after connecting the two cells based on an appropriate equation of state of real gases assuming the volume of the two cells was known. The volume of the reference cell was determined beforehand, and the volume of the void space in the adsorption cell with sample loaded was determined afterwards. Both methods need to know the skeleton volume of adsorbent. This volume has to be subtracted from the total volume of sample container to obtain the volume of void space, which is used for the calculation of the amount adsorbed in the volumetric method. The skeleton volume of sample was directly used in the calculation of buoyancy correction for gravimetric method. The skeleton volume was usually determined by helium expansion assuming the amount of helium adsorbed was negligible. If, however, helium adsorption could not be neglected, error would yield in the skeleton volume and, finally, in the calculated amount adsorbed. It is clear that the effect of helium adsorption would be much less in the volumetric method if the skeleton volume is considerably

---

[4] Gregg, S. J. and Sing, K. S. W. "Adsorption, Surface Area and Porosity," 2nd ed., Academic Press Inc., London (1982).

[5] Zhou, L., Zhou, Y. P., Li, M., Chen, P., Wang, Y. "Experimental and Modeling Study of the Adsorption of Supercritical Methane on a High Surface Carbon," *Langmuir*, **16**, 5955, (2000).

[6] Fan, Y. Y., Liao, B., Liu, M., Wei, Y. L., Lu, M. Q., Cheng, H. M. "Hydrogen uptake in vapor-grown carbon nanofibers," *Carbon*, **37**, 1649 (1999).

[7] Zhou, L. " *Adsorption Isotherms for the Supercritical Region*," Chapter 4 of " Adsorption: Theory, Modeling & Analysis " ed. J. Toth, Marcel Dekker, Inc. New York, May 2002.

less than the volume of void space. However, the volume of void space does not have any relevance in the gravimetric method. In fact, helium adsorption is not a major source of measurement error especially when the skeleton volume was determined at room temperature and pressures less than 15 MPa. The precision and constancy of pressure, temperature and the reliability of compressibility factor usually cause much larger relative error than does by helium adsorption [8]. To avoid the effect of helium adsorption, the skeleton volume (or density) was taken for a model parameter in some gravimetric measurements. However, the true value of skeleton volume (or density) can hardly be more reliable based on a fitted parameter than on a measured value. The consequence of helium adsorption in adsorption measurement should be estimated appropriately.

It is clear that the accuracy of the temperature, pressure and compressibility factor in the reference cell is of primary importance for volumetric method since they affect directly the resulted amount adsorbed. The accuracy of temperature of the adsorption cell is of secondary importance for adsorption measurement since it affects the result only through the dependence of adsorption on temperature. Compared to the measurement completed previously on AC AX-21 [9], many improvements were made recently for the measurements on carbon nanotubes. The reference cell was kept in double thermostats to keep its temperature constant within ±0.05 K. A pressure transducer model PAA-23/8465.1-200 manufactured by Keller Druckmesstechnik, Switzerland, detected the pressure of the reference cell. The deviation from linearity of the transducer output is less than 0.05 % for the whole range of 20 MPa. Compressibility factor of hydrogen was evaluated from the data of molar volume [10]. Besides, larger quantity of adsorbent sample was used because more H_2 adsorbed, larger difference in pressure readings would yield and smaller relative error of measurement would involve. Therefore, about 20 g of activated carbon AX-21 and 8-10 g of MWNT sample were used in our experiment.

RESULTS AND DISCUSSION

THE EXPERIMENTAL ISOTHERMS ON POWDER SAMPLES – Five isotherms are shown respectively in Figs.1 and 2 for MWNT and AC AX-21 samples. Although two sets of isotherms have similar trend of variation with pressure, the ordinate scale is more than ten times different. The adsorption isotherms at 77 K are highly above those for 233-298 K, which is normal for physical

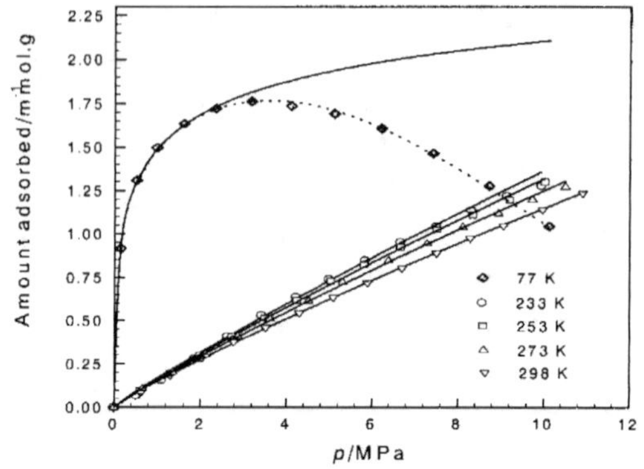

Figure 1. The excess and the absolute adsorption isotherms of H_2 on MWNT
Points: the excess; Solid curves: the absolute; Doted curve: model predicted excess isotherm at 77 K

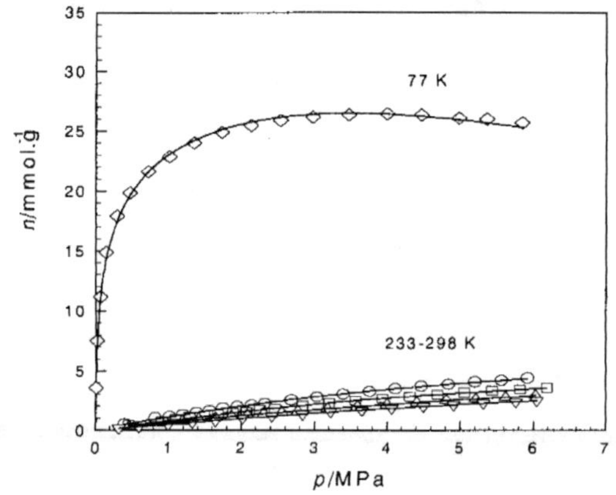

Figure 2. The excess adsorption isotherms of H_2 on AC AX-21. Points: experimental; Curves: model

adsorption. Maximum is shown on the isotherms of 77 K, and negative increment with pressure is more clearly shown on the isotherm of MWNT sample, which is the typical feature of supercritical adsorption. However, the absolute isotherm, determined subsequently, is an increasing function of pressure, and the absolute isotherm of 77 K is always above the others. The dotted curve of Fig.1 and whole Fig.2 show the fitness of isotherm model that subsequently established to the

---

[8] Bai, S. P. "Studies on Adsorption Behavior of CO_2 on Porous Solids near the Critical Temperature," Ph. D. Dissertation, School of Chemical Engineering, Tianjin University, Tianjin, China, 2003.
[9] Zhou, Y. P. and Zhou, L. "Experimental study on high-pressure adsorption of hydrogen on activated carbon," *Science China* (Series B), **39**, 598 (1996).
[10] Vargaftik, N. B. "Handbook of Physical Properties of Liquids and Gases, Pure Substances and Mixtures," 2nd ed., Hemisphere Publishing Corporation, New York, 1975.

excess isotherms. Comparing with previous modeling [9], remarkable improvement was made due to accounting for the difference between the excess and the absolute adsorption.

EVALUATION OF THE LIMITING HEAT OF ADSORPTION – Heat of adsorption reflects the change of molecular energy state before and after adsorption; therefore, is a measure of the strength of interaction between gas molecules and the adsorbent surface. There are isosteric and limiting heat of adsorption [11]and the latter is the one when the amount adsorbed approaches zero. The limiting heat of adsorption can be evaluated from the Henry's law constants of several temperatures. The value of Henry's law constant can be determined from the intercept of a plot of ln(p/n) versus n basing on the virial isotherm equation [11]. Such plot was shown in Figs.3 and 4 respectively for MWNT and AC AX-21, where p is pressure (MPa) and n is the excess amount adsorbed (mmol/g). Assuming the intercept is ξ, one can determine the Henry's law constant, K, as $K = \text{Exp}(-\xi)$. The variation of K with temperature follows the van't Hoff law, and the limiting heat of adsorption can be evaluated from the slope of a plot of ln$K$ versus $1/T$ [11]. Such plots are shown in Figs.5 and 6, and the limiting heat of adsorption evaluated from them is 1.8 and 7.6 kJ/mol respectively for MWNT and AC AX-21. The difference is about four times and so is the amount adsorbed under the condition of 233-298 K and pressures up to 6 MPa.

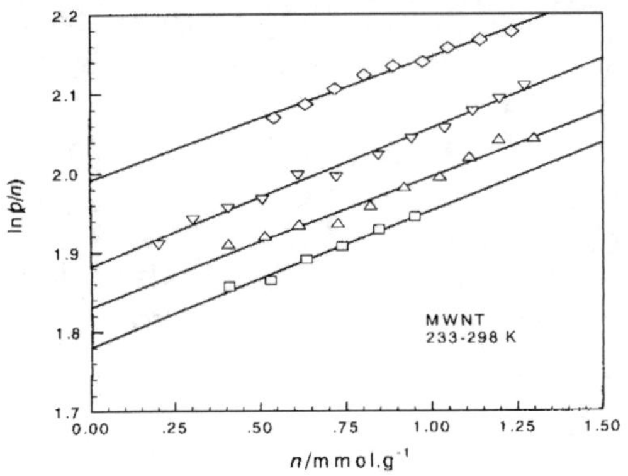

Figure 3. The virial plot for the adsorption of H_2 on MWNT

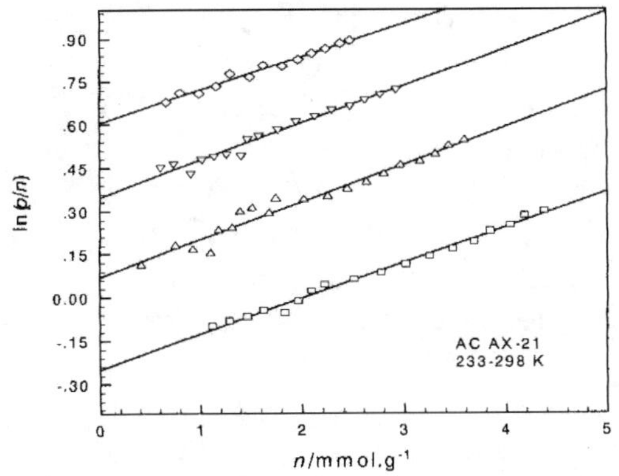

Figure 4. The virial plot for the adsorption of H_2 on AC AX-21

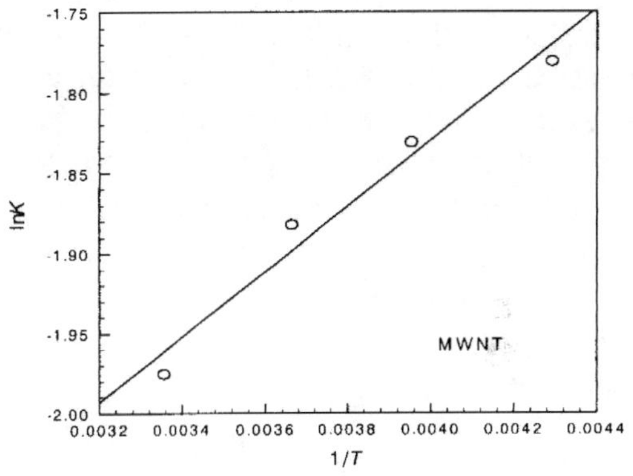

Figure 5. The van't Hoff plot for MWNT

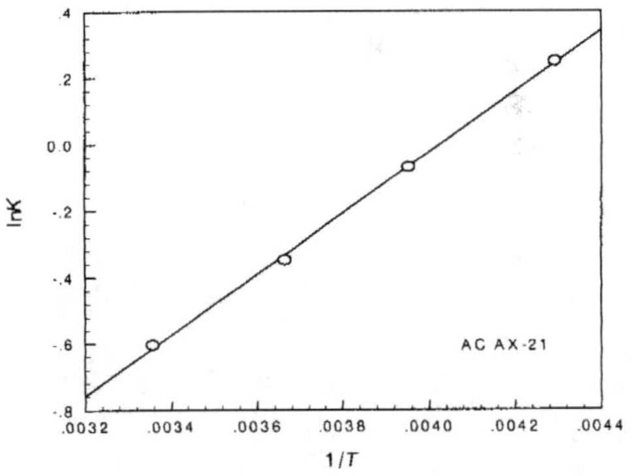

Figure 6. The van't Hoff plot for AC AX-21

GENERATION OF THE ABSOLUTE ADSORPTION ISOTHERMS – Adsorption itself is, according to the Gibbs definition, an *excess* quantity. The average number of molecules in an element volume near the surface is larger than in an element volume of equal size in the bulk gas due to the stronger interaction between adsorptive and solid surface than among adsorptive molecules. A density profile is thus established from the solid surface to the bulk gas. Both molecular simulation and

[11] Ruthven, D. M. "*Principles of Adsorption and Adsorption Processes*," John Wiley & Sons, New York (1984).

experimental proof reveal that this density profile vanishes over quite a short range. Therefore, adsorbate molecules are concentrated in a layer of definite thickness above solid surface. Such layer is referred to as "adsorption space" or "adsorbed phase" [12]. The amount adsorbed, n, can be determined if the density profile $\rho(z)$ is known:

$$n = \int_{\substack{all \\ adsorption \\ space}} (\rho(z) - \rho_g) dV \qquad (1)$$

where ρ is the density of adsorbate, ρ_g is the density of gas phase, V is volume and z is the distance to the solid surface. Equation (1) clearly indicates that n is a density excess amount of the adsorbed substance, which is usually referred to as "surface excess adsorption" or simply "excess adsorption". Although the density profile $\rho(z)$ can be determined by molecular simulation, it could not be experimentally measured; therefore, another way to express n is often cited in literature. Suppose the volume of the adsorption space (or the adsorbed phase) is V_a, the excess adsorption must be:

$$n = n^s - \rho_g V_a = V_a(\rho_a - \rho_g) \qquad (2)$$

where n^s is the total mass contained in the adsorbed phase, which is usually referred to as the absolute adsorption. That the density of the adsorbed phase ρ_a is uniform is an implicit assumption of Eq.(2). It is clear, the so-called amount adsorbed is the difference of the total mass contained in the adsorption space minus the quantity of adsorptive gas existed in the same space if adsorption were not happen. The subtrahend term is usually trivial at subcritical temperatures because the pressure cannot be higher than saturation pressure; however, it could become considerably large at supercritical temperatures since condensation could no longer happen. Maximum would appear on the isotherm when the ambient gas density becomes comparable to the density of the adsorbed phase, and even negative value of adsorption has been recorded experimentally at conditions $\rho_g > \rho_a$. Therefore, to explain the behavior of supercritical isotherm with maximum and negative increment the subtrahend term of Eq.(2) must be properly accounted for. On the other hand, it is the absolute adsorption, $n^s = V_a\rho_a$, that should be depended on for thermodynamic relations because it is the total mass of the adsorbed phase; therefore, it also determines the amount of gas released form the adsorbed phase upon releasing pressure. The problem is we can only experimentally measure the excess quantity of adsorption up to now. How to estimate the total adsorption from the excess adsorption data was regarded as a major challenge in the area of high-pressure adsorption [13]

Such a method was recently proposed [14,15] based on a simple yet very sound principle: the excess and the absolute adsorption are equal provided the surface concentration of adsorbate is dilute. The adsorbate volume in the adsorbed phase, V_a, must be trivial if its surface concentration is dilute. It is known from Eq.(2) that the quantity of absolute adsorption is equal to the quantity of excess adsorption if V_a is trivial and the subtrahend term could be neglected. Therefore, the isotherm of absolute adsorption could be formulated using the excess adsorption data under the condition of dilute surface concentration. To screen the data that comply with the constraint out of the whole set of experimental data, the excess isotherms of different temperatures were first reduced to a generalized isotherm. It was proven that all the data corresponding to low surface concentration locate on a single smooth isotherm no matter what kind of adsorbents was used [16]. Therefore, omitting the data that deviated from the smooth isotherm would sift out those data corresponding to higher surface concentration. Modeling the generalized isotherm with a Langmuir type equation and expanding the equation to Taylor series, one obtained a linear expression of isotherm by omitting the terms of higher order. Apparently, the data locating on the linear isotherm must correspond to the condition of even diluter surface concentration; therefore, formulated the isotherm of absolute adsorption. The above-mentioned procedure was named as IST (Isotherm Space Transformation) strategy [17] and yielded a plot of Ln$\ln n$ versus $1/\ln p$. Such plots are shown in Figs.7 and 8 for MWNT and AC AX-21 respectively. Because of linear, the formulation of absolute adsorption isotherm is a simple two-parameter equation:

---

[12] Everett, D. H. "*Manual of symbols and terminology for physicochemical quantities and units*," Appendix II, Part 1, Butterworths, London (1971).

[13] Humayun, R., Tomasko, D. L. High-resolution adsorption isotherms of supercritical carbon dioxide on activated carbon, *AIChE Journal*, **46**, 2065 (2000).

[14] Zhou, L. and Zhou, Y. P., "Linearization of Adsorption Isotherms for High Pressure Applications," *Chem. Eng. Sci.*, **53**, 2531 (1998).

[15] Zhou, L. and Zhou, Y. P., " A Mathematical Method for the Determination of Absolute Adsorption from Experimental Isotherms of Supercritical Gases " *Chn J. of Chem. Eng.*, **9**, 110 (2001).

[16] Zhou, Y. P., Zhou, L., Bai, S. P., Yang, B. "Experimental Studies of the Generalized Adsorption Isotherm for the Supercritical Region," *Adsorption Sci. & Technol.* **19**, 681 (2001).

[17] Zhou, L., Chen, P., Li, M., Sun, Y., Zhou, Y. P. "Prying the Nature of Supercritical Adsorption via Isotherm Space Transformations," *Proceedings, The 2nd Pacific Basin Conference on Adsorption Science and Technology*, Brisbane Australia, 717 (2000).

$$n^s = \frac{1}{\delta} Exp\left[Exp\left(\alpha + \frac{\beta}{Lnp} \right) \right] \quad (3)$$

Parameter δ is a multiplier of n to avoid evaluating the logarithm of negative numbers. The value of δ was set to 100 in Fig.7, but to 10 in Fig.8 depending on the magnitude of the amount adsorbed. The unit of p was set in kPa for the same reason. Knowing the values of α and β we can draw the absolute isotherm as those shown in Fig.1 as an example.

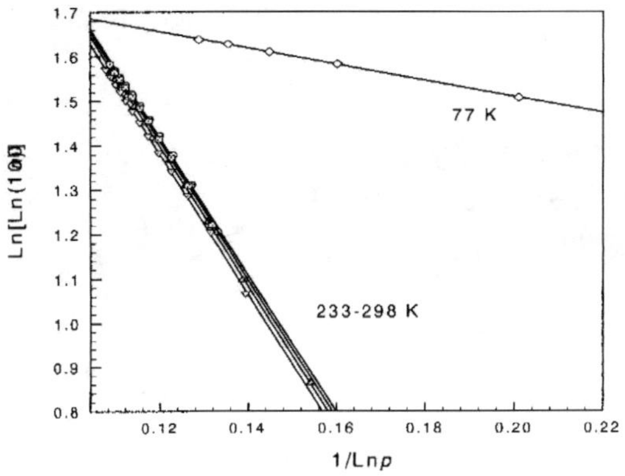

Figure 7. Linearization of isotherms for MWNT

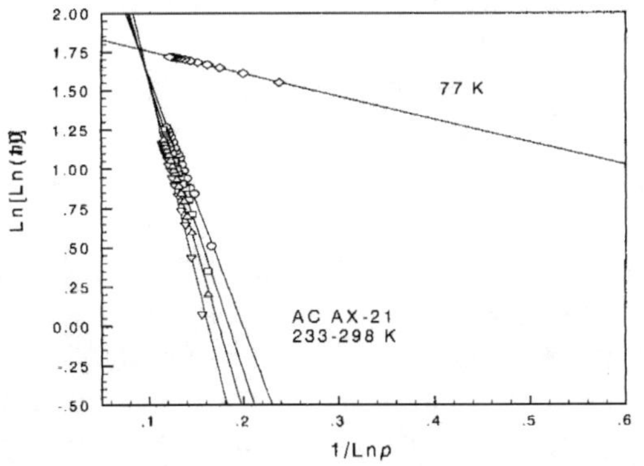

Figure 8. Linearization of isotherms for AC AX-21

MODELING THE EXCESS ADSORPTION ISOTHERMS

– Equation 2 is the definition of the surface excess adsorption; therefore, should be used to model the excess adsorption isotherms. Replacing n^s term in Eq.(2) with an isotherm equation, for example, Eq.4, which was proposed by the authors especially for supercritical adsorption [18]

[18] Zhou, L., Zhang, J. S., Zhou, Y. P. "A Simple Isotherm Equation for Modeling the Adsorption Equilibria on Porous Solids over Wide Range Temperatures," *Langmuir*, **17**, 5503 (2001).

$$n^s = n_t^0 \left[1 - \exp\left(-bp^q \right) \right] \quad (4)$$

Then we have the model for the excess adsorption isotherms:

$$n = n_t^0 \left[1 - \exp\left(-bp^q \right) \right] - V_a \rho_g \quad (5)$$

The volume of the adsorbed phase, V_a, is a function of pressure and temperature as shown in Fig.9, which was evaluated by

$$V_a = \frac{n^s - n}{\rho_g} \quad (6)$$

where n^s was yielded by Eq.(3) and n is the experimental data. Applying Eq.(5) to the excess adsorption isotherms with given values of V_a and ρ_g, the values of model parameters were evaluated through nonlinear regression. Fig.2 and the dotted curve of Fig.1 show that the model fits the experimental excess isotherms very well. Perfect fit of the model to experimental data was also observed at other adsorption systems [19,20,21,22,23].

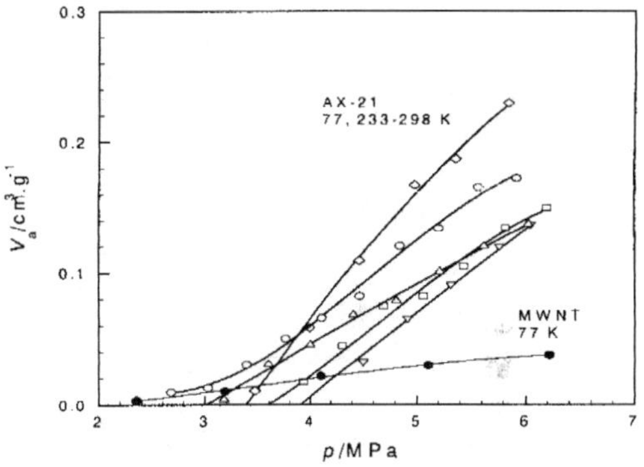

Figure 9. Variation of V_a with pressure and temperature

THE STORAGE CAPACITY OF H_2 ON ACTIVATED CARBON AT 77 K – Shown in Fig.10 are the adsorption isotherms of H_2 at 77 K on AC AX-21 powder and pellets

[19] Zhou, L., Zhou, Y. P., Bai, S. P., Lü, C. Z., Yang, B. "Determination of the Adsorbed Phase Volume and Its Application in Isotherm Modeling for the Adsorption of Supercritical Nitrogen on Activated Carbon," *J. Colloid & Interface Sci.* **239**, 33 (2001).

[20] Zhou, Y. P., Bai, S. P., Zhou, L., Yang, B. "Studies on the Physical Adsorption Equilibria of Gases on Porous Solids for a Wide Temperature Range Spanning the Critical Region--- Adsorption on Microporous Activated Carbon," *Chn J. Chem.*, **19**, 943 (2001).

[21] Zhou, L., Bai, S. P., Zhou, Y. P., Yang, B. "Adsorption of Nitrogen on Silica Gel over a Large Range of Temperatures," *Adsorption*, **8**, 79 (2002).

[22] Zhou, L., Zhou, Y. P., Bai, S. P., Yang B., "Studies on the Transition Behavior of Physical Adsorption from the Sub- to the Supercritical Region: Experiments on Silica Gel," *J. Colloid & Interface Sci.* **253**, 9 (2002).

[23] Zhou, L., Bai, S. P., Su, W., Yang, J., Zhou, Y. P. "Comparative Study of the Excess versus Absolute Adsorption of CO_2 on Superactivated Carbon for the Near-Critical Region," *Langmuir*, **19**, 2683 (2003).

respectively. Although the excess adsorption reached saturation at quite low pressure, the amount of hydrogen actually fixed on carbon surface, i.e., the amount of absolute adsorption, always increases with increasing pressure for the range tested. The bulk density of superactivated carbon is very low due to the abundance of micropores, e.g., the bulk density of AX-21 is only 0.3 g/cm^3. To store more gas in a limited space, more carbon is preferred to pack in; therefore, carbon powder was compressed to pellets. For example, the density of pellets shown in Fig.10 is 0.72 g/cm^3. The amount adsorbed per gram of carbon decreased on pellets due to losing some surface area during making pellets. However, the quantity of hydrogen stored in unit volume filled with pellets was increased as shown in Fig.11.

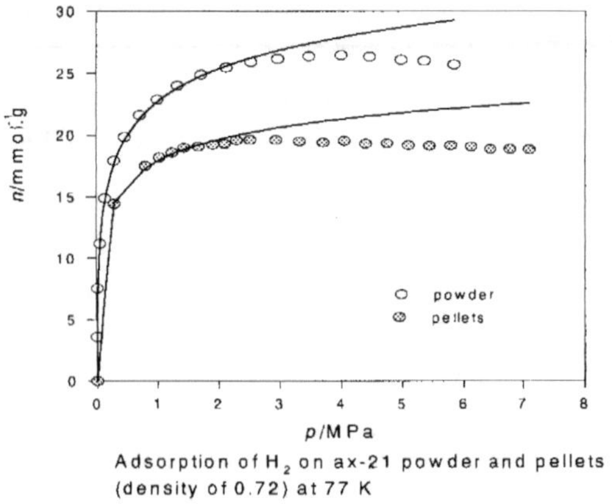

Figure 10. Adsorption isotherm on AC AX-21 at 77 K. Points: the excess; Curves: the absolute

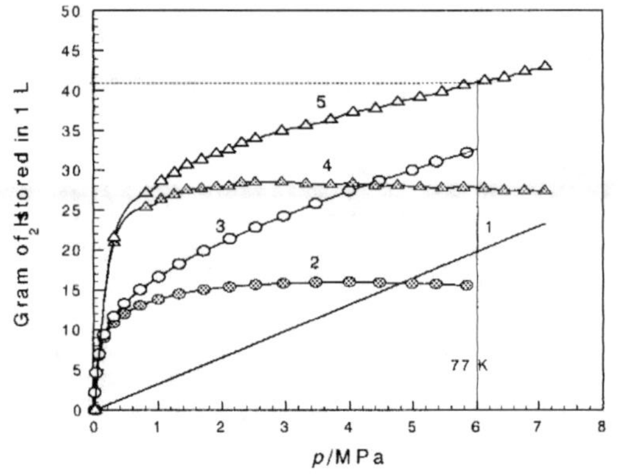

Figure 11. Storage capacity on AC AX-21

1. Compressed; 2. the excess amount adsorbed on powder; 3. Total quantity stored in unit volume of powder; 4. the excess amount adsorbed on pellets; 5. Total quantity stored in unit volume of pellets.

The storage capacity in a specified volume is the sum of the capacity of carbon surface due to adsorption and that in the void space due to compression. The volume of the void space before adsorption, V_{tv}, is the sum of pore volume inside adsorbent particles and the volume outside particles, and has been detected by helium expansion. However, the volume of the void space changed to V_g after adsorption since a part of V_a was occupied by the adsorbed phase and

$$V_{tv} = V_a + V_g \qquad (7)$$

Times both sides of equation 7 by the gas phase density, ρ_g

$$\rho_g V_{tv} = \rho_g V_a + \rho_g V_g \qquad (8)$$

Adding the excess adsorption, n, to both sides of equation 8, we have

$$\begin{aligned} n + \rho_g V_{tv} &= (n + \rho_g V_a) + \rho_g V_g \\ &= n^s + \rho_g V_g \qquad (9) \\ &= C_{tot} \end{aligned}$$

where C_{tot} is the total capacity of storage. Therefore, neither V_a nor V_g is needed for the evaluation of C_{tot}. Suppose the bulk density and the skeleton density of adsorbent is respectively ρ_b and ρ_s, then the volume occupied by "solid" or "skeleton" part of adsorbent is $V_s = \rho_b/\rho_s$, and the void space occupied $V_{tv} = 1 - \rho_b/\rho_s$. Determination of ρ_b is a common practice of laboratory, but that of ρ_s needs special technique. The density of graphite was usually taken for the skeleton density of activated carbon, but this may not be true because the existence of closed pores or small pores that inaccessible to helium molecules make the skeleton density of activated carbon lighter than graphite. The volume of the skeleton part of activated carbon was determined as the difference of the volume of adsorption cell without and with loading adsorbent, and the density of the solid carbon was calculated based on such volume.

The storage capacity of hydrogen on activated carbon AX-21 is shown in Fig.11 on the basis of 1 liter container. Curve 1 is the quantity of hydrogen stored just by compression; curves 2 and 4 are the excess adsorption data on powder and pellets respectively basing on the mass filled in 1 liter; curves 3 and 5 are the total storage capacity of 1-liter container filled with powder and pellets respectively. For example, 19.6 g hydrogen could be stored at 6 MPa in a vacant container of volume 1 liter, 32.5 g hydrogen stored in the same container but filled with carbon powder, and 41 g hydrogen could be stored if the container filled with pellets. The Department of Energy, USA sets up a criterion of 6.5 wt % for any qualified hydrogen carrier. The value of such index is 10.8 wt% for carbon powder at 6 MPa, and even higher at higher pressures; however, only 5.7 wt % was reached with carbon pellets as shown in Fig.12. Therefore, the storage capacity expressed by weight percentage is not an all-round index since much more hydrogen was stored in unit volume of container filled with carbon pellets as shown in Fig.11.

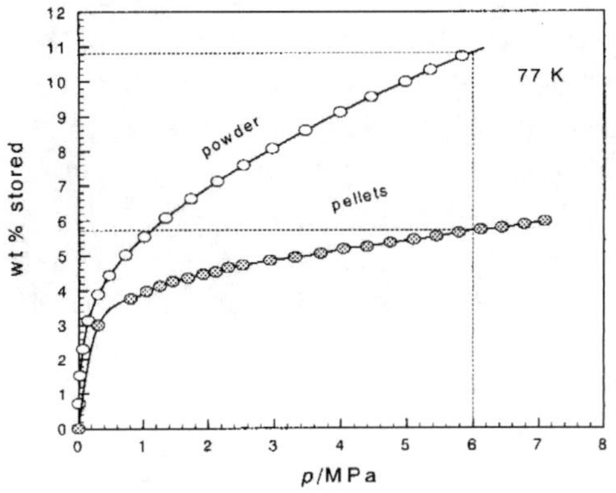

Figure 12. Weight percentage of H_2 stored on AC AX-21 powder and pellets

***Remark** on the supercritical adsorption mechanism*: As mentioned above, the volume of the adsorbed phase cannot be directly determined, but it is an important information since the absolute quantity of adsorption would be known if V_a is available. The value of V_a can also be evaluated via the calculation of storage capacity. As indicated by equation 9:

$$C_{tot} = n + \rho_g V_{tv}$$

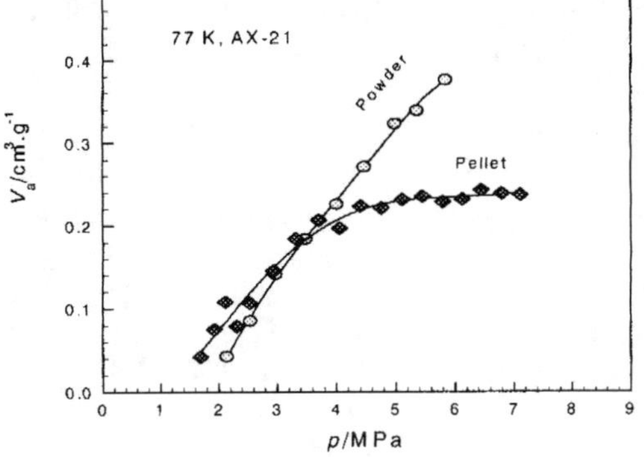

Figure 13. The volumes of the adsorbed phase evaluated based on the storage capacity of H_2

If, however, instead of n, n^s is used in the above equation, then the difference in the calculated storage capacity would be equal to $V_a\rho_g$, and the value of V_a could thus be determined. The results are shown in Fig.13 respectively for power and pellets. The volume is based on 1 gram, which becomes smaller in pellets as a consequence of compression and binder. Compared to pore volume, 1.3 cm^3/g, the volume of the adsorbed hydrogen is considerably smaller; therefore, the mechanism of volume filling can hardly happen.

THE STORAGE CAPACITY OF H_2 ON MWNT AT 77 K – The surface excess isotherms of H_2 at 77 K are shown in Fig.14 for MWNT powder and pellets respectively. It is interesting that the amount adsorbed based on unit mass of adsorbent increased remarkably due to making pellets. New micropores might be formed in between nanotubes due to pressing. The bulk and skeleton density of MWNT powder is respectively 0.35 and 2.23 g/cm^3; those of

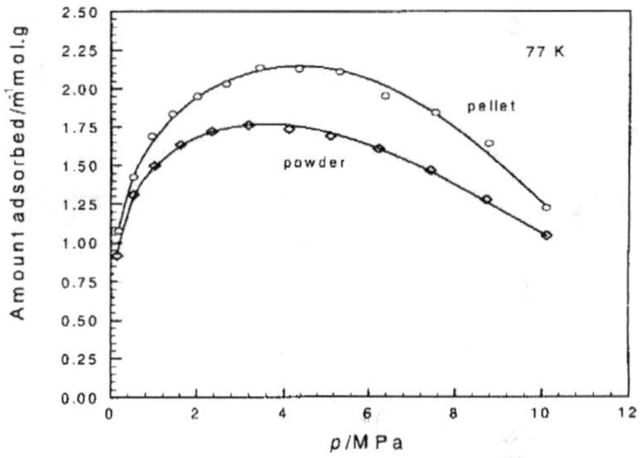

Figure 14. The surface excess isotherms on MWNT powder and pellets at 77 K
Points: experimental data; Curves: model predicted

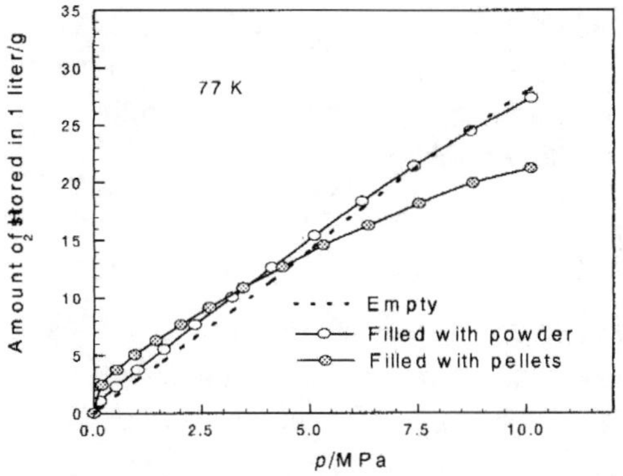

FIGURE 15. Storage capacity of 1 liter container filled with MWNT powder and pellets

MWNT pellets are 0.96 and 2.38 g/cm^3 respectively. The skeleton densities were determined basing on the difference of void space in the adsorption cell before and after loading sample. The storage capacity of a 1 liter container filled with MWNT powder or MWNT pellets was determined based on the above information and shown respectively in Fig.15. Shown by the dotted line is the quantity of hydrogen stored in an empty container at the same pressure as that filled with MWNT materials. The quantity of hydrogen stored in a container filled with MWNT powder is almost the same as an empty one; however, a container filled with MWNT pellets stores less hydrogen than being empty.

Remark *on the potential technology of hydrogen storage*: Carbon nanotubes are hopeless to be a hydrogen carrier since the adsorption on them is too weak to increase the storage capacity than by compression. Storing hydrogen in a container filled with pellets of activated carbon and cooled by liquid nitrogen enhanced the storage capacity considerably. This method bears important merits from the commercial point of view. First, the storage capacity is acceptable for driving practice. A container of volume 100 liter can store 4.1 kg H_2 at pressure 6 MPa. Such data can be repeated in any laboratory equipped with precise apparatus of adsorption measurement and operated by a qualified person. This is worth mentioning because the high storage capacity of carbon nanotubes claimed by one laboratory could not be repeated in another laboratory as usual. Second, superactivated carbon can be produced in large scale with reasonable cost. Third, hydrogen is stored in gas state, therefore, is much cheaper than liquid hydrogen, and cleaner and safer than methanol. Fourth, the storage facility is much simpler than that for on-board producing hydrogen. Fifth, instead of hydrogen, nitrogen is lost due to insulation loss, therefore, is safer and cheaper. Sixth, liquid nitrogen has widespread distribution even presently and available with low cost. It does not need to save a nitrogen cylinder on board, but to add some periodically due to the large heat capacity of the fuel tank.

SUMMARY

1. Both superactivated carbon and carbon nanotubes possess nanostructure and can adsorb hydrogen physically. Adsorption isotherms were experimentally collected for temperature 77, 233-298 K and pressures up to 6 and 10 MPa respectively on powder and pellets of the two adsorbents.
2. Adsorption data were analyzed by a new method to yield the absolute quantity of hydrogen fixed on adsorbent. This quantity was used to model the experimental isotherms, and perfect fit was obtained for the adsorption on both activated carbon and carbon nanotubes indicating the agreement of adsorption mechanism.
3. The limiting heat of adsorption was evaluated and the results of 1.8 and 7.6 kJ/mol were obtained respectively with MWNT and AC AX-21 for the temperature range 233-298 K.
4. The total amount of hydrogen stored in a container filled with adsorbent is the sum of that absolutely adsorbed on adsorbent surface and that compressed in the void space. The weight percentage of hydrogen stored in a container filled with AC AX-21 powder is 10.8 wt % at 6 MPa basing on the weight of carbon filled in container, which is much higher than the criterion of 6.5 wt% set up by DOE, USA.
5. The volume of the adsorbed phase was determined from the calculation of storage capacity, which was considerably less than the micropore volume of adsorbent; therefore, the adsorbed hydrogen cannot fully fill the space of micropores at above-critical temperatures.
6. Carbon nanotubes are hopeless to be a hydrogen carrier since they cannot have the total quantity of stored hydrogen more than by compression.
7. Storing hydrogen in a container filled with pellets of activated carbon and cooled by liquid nitrogen is an effective method, and 4.1 kg hydrogen gas could be stored in a container of volume 100 liter. Such method is technically feasible and cost competitive. It is also advantageous with respect to safety and availability of fueling facilities.

ACKNOWLEDGMENTS

This work is subsidized by the Special Funds of Major State Basic Research Projects (G2000026404) and supported by the National Natural Science Foundation of China (#29936100).

CONTACT

Li Zhou (surname), Professor and Director of the High Pressure Adsorption Laboratory, School of Chemical Engineering, Tianjin University; Studied in the Department of Chemical Engineering, Tsinghua University, Beijing, China for 1959-1965; visiting scholar in the Department of Chemical Engineering, University of Wisconsin-Madison, USA for 1981-1983; cooperation research in the Hydrogen Research Institute, University of Quebec, Canada for 1992-1995. E-mail address: zhouli@public.tpt.tj.cn

Hydrogen adsorption property of lithium doped mesoporous silica

Naotaka Chino, Masaru Ogura, Jun Izumi and Tatsuya Okubo

[1)] Department of Chemical System Engineering,
The University of Tokyo, Tokyo 113-8656, Japan
[2)] Mitsubishi Heavy Ind. Co. Ltd., Nagasaki R&D Ctr.,
Nagasaki 851-0392, Japan

okubo@chemsys.t.u-tokyo.ac.jp

Keywords Hydrogen Adsorption, Mesoporous Silica and Lithium

Prepared for presentation at the 2003 Annual Meeting, San Francisco, CA, Nov16-21

Abstract

Lithium, which is light in weight and reacts with hydrogen to form hydride, is one of the practical candidates for hydrogen storage. However, lithium hydride is deficient in decomposition and stability under ambient conditions. The hydrogen adsorption property of lithium-doped mesoporous silica has been investigated to create reversible adsorption sites. After impregnating mesoporous silica with LiCl/ethanol solutions followed by heat treatment in helium, hydrogen adsorption/desorption experiments are carried out at room temperature in a fixed-bed flow apparatus. The adsorption experiments reveal that heat treatment above 200°C induces the reversible hydrogen adsorption. The profiles on mass spectroscopy of the gases produced in the heat treatment show that CH_3CH_2Cl is released above 200°C, suggesting that LiCl reacts with ethanol. On the other hand, lithium is hardly released during the whole temperature range, indicating that the adsorption sites including Li are produced after the heat treatment. The hydrogen adsorption site is characterized by ^{7}Li MAS NMR and ESR.

1. Introduction

With fossil hydrocarbon resources becoming scarce, the concept of using hydrogen as a future energy source has been under discussion for at least the last thirty years. For this to become economically feasible, means for a large-scale, hydrogen storage has to be developed. The storage in porous materials with adequate capacity is an attractive proposal. In this study, the hydrogen adsorption property of Li-doped mesoporous silica (SBA-3 type) is investigated under ambient conditions to create stable and reversible adsorption sites.

2. Experimental section

2-1. Samples preparation

First, the pellets of SBA-3 was prepared by pressing the powder at 200kg/cm^2 for 1min. Li-doped SBA-3 was prepared by impregnating pellets of SBA-3 with lithium chloride (LiCl) / ethanol solutions, followed by filtration and drying at 90°C in air. Finally, heat treatment was carried out up to at the prescribed temperatures at the heating rate of 4 degree /min in He flow (50cm^3/min). The released gaseous species were analyzed by a quadruple mass spectrometer (QPMS). Until the release of variable gases was completed, the sample was kept at the predetermined temperature, and then cooled down to room temperature. X-ray diffraction and N_2 adsorption confirmed that the porous structures of SBA-3 were maintained after doping LiCl. Pore size distributions of all LiCl doped SBA-3 were narrow at ca.2.0nm. Surface areas of all LiCl supported SBA-3 were around 1000m^{2}g^{-1}. FT-IR spectra indicated that the number of silanol group was hardly changed by doping LiCl. The amounts of doped lithium and chloride were determined by ICP emission spectroscopy (ICP) and fluorescent X-ray spectroscopy (XRF).

2-2. Hydrogen adsorption/desorption measurement

The adsorption/desorption measurement of hydrogen was carried out in a flow-type apparatus with a fixed bed absorber as shown in Fig. 1. Adsorption and desorption runs were periodically controlled by solenoid valves (S1-4). The adsorbent of 0.5-1.0 g was placed in a stainless steel column (C1, 10mm inner diameter and 22-cm length), and was heated at the prescribed temperature for 3h under a helium stream (50cm^3/min) as a pretreatment just before the adsorption run. In the adsorption run, hydrogen gas was fed at rate of 100cm^3/min into column. After the adsorption run, pure He was introduced at rate of 100 cm^3/min to desorb hydrogen from the adsorbent. The temperature of the adsorbent was kept at room temperature, and the adsorption and desorption time was fixed at 10min. The effluent concentration and the composition were continuously monitored by an on-line QPMS, as shown in Fig. 1. The amount of adsorbed hydrogen was calculated with the difference in elution curves between samples and α-Al_2O_3.

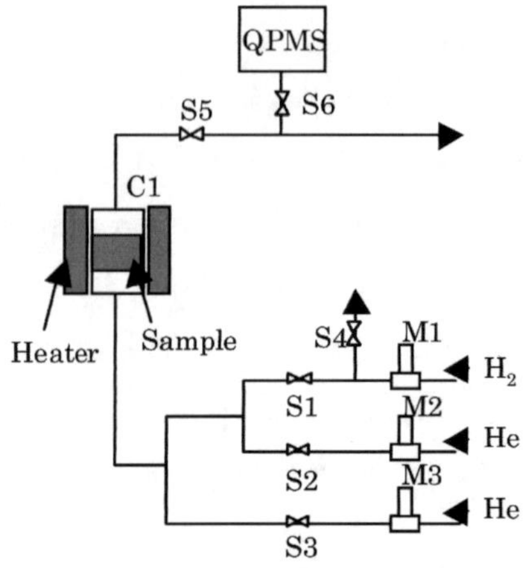

Fig. 1 Schematic diagram of the apparatus: C, adsorption column; M, mass flow controller; S, solenoid valve; QPMS, quadrupole mass spectrometer.

2-3. ESR and ^{7}Li MAS NMR measurement

ESR spectra were recorded on a JEOL JES-TE200 (X-band). The measurement temperature was 298K. Microwave frequency was around 9.20GHz, the modulation width was 0.32mT, time constant was 1sec, and the microwave power was 1.0mW. ESR measurements were performed with a quartz tube sealed in argon.

^{7}Li MAS NMR spectra were recorded on a JEOL JNX-ECX 300, operated at Larmor frequencies of 116.8MHz with LiCl as a reference. All the MAS NMR spectra were recorded at the spinning frequency of 3kHz, using the relaxation delay of 3sec, the receiver delay of 20.0 μsec, a receiver gain of 20.0 μsec.

3. Results and discussion

3-1. Hydrogen adsorption property

Fig. 2 shows the hydrogen uptakes of LiCl- doped SBA-3 along with that of untreated SBA-3. Hydrogen uptakes of untreated SBA-3 and LiCl-doped one after heating at 200°C were too little to be detected. However, hydrogen uptake of LiCl- doped ones after the heat treatment at 420 or 520°C was certainly detected. The heat treatment above 200°C induced reversible hydrogen adsorption. Hydrogen uptake was dependent on the concentration of hydrogen. The adsorbed hydrogen was completely desorbed in 10min, indicating that this adsorption has physisorption property.

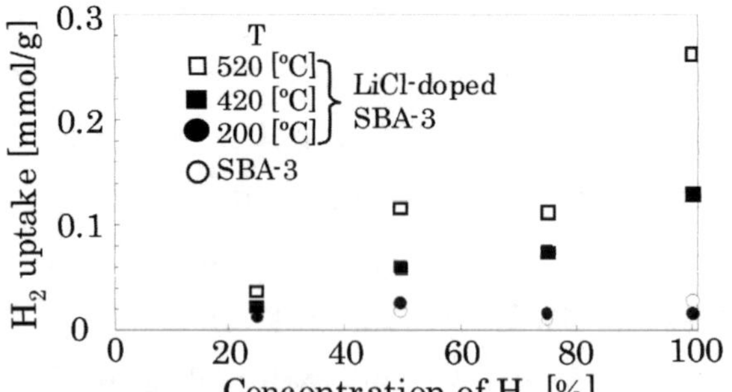

Fig. 2 Hydrogen uptakes at room temperature on LiCl-doped SBA-3 (□,■,●) and original SBA-3 (○) : T, heat treatment temperature [°C]

3-2. Heat treatment process

Above experiment indicates that hydrogen adsorption sites are produced during the heat treatment. Temperature programmed desorption (TPD) was performed in order to investigate reactions at elevated temperatures on LiCl- doped SBA-3. Fig. 3 shows the profiles on mass spectroscopy of the gaseous species produced in the heat treatment. Adsorbed water and ethanol were desorbed below 200°C. Ethyl chloride, hydrochloric acid and water were released above 200°C. These results suggest that LiCl reacts with ethanol as well as water. On the other hand, species including lithium could not be detected at all the temperature.

The quantitative analysis of lithium and chloride by ICP and XRF measurements was performed in order to determine the amount of released chloride and lithium. Fig.4 shows the amount of lithium and chloride on SBA-3 after the heat treatment. It was revealed that lithium was hardly released from the porous structure. However, chloride was certainly released at elevated temperature, and the amount of released chloride was dependent on the heating temperature. These observations support the results of TPD. It is suggested that the adsorption site including lithium species was produced during the heat treatment process.

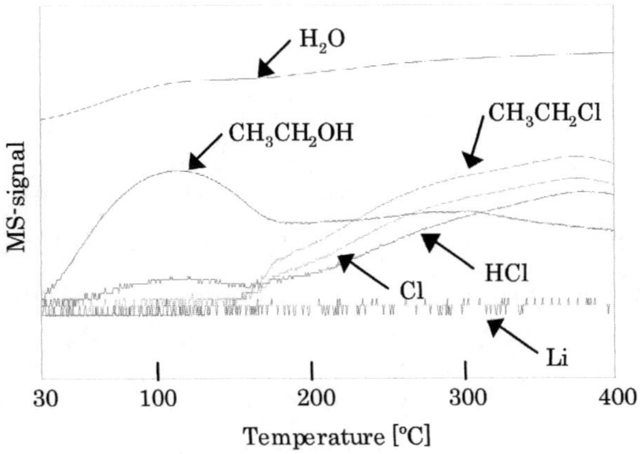

Fig. 3 MS profiles of the gaseous species produced during the heat treatment

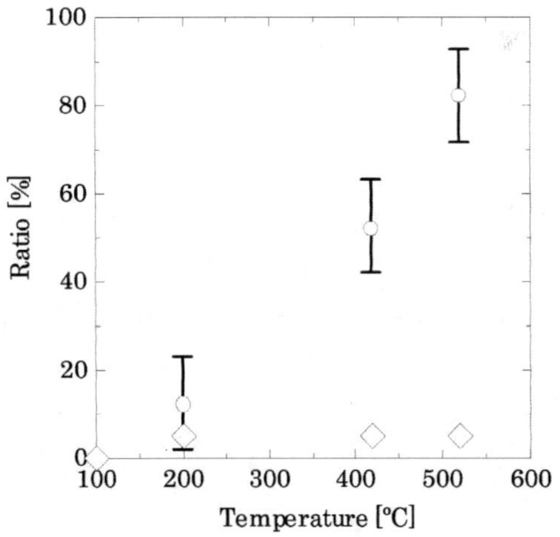

Fig. 4 The ratio of lithium(◇) and chloride(○) released from SBA-3 during the heat treatment

3-3. ^{7}Li MAS NMR and ESR measurement

The surface of LiCl-doped SBA-3 was investigated by ^{7}Li MAS NMR and ESR. A slight chemical shift (c.a. 1.0 ppm) to the reference was observed in ^{7}Li MAS NMR spectra. The chemical shift and the shape of signal were hardly changed by the heat treatment. It is suggested that most of lithium are not metallic. Fig.5 shows the ESR spectra of LiCl-doped SBA-3 at various heating temperatures. A sharp signal was observed on LiCl-doped samples after the heat treatment above 420°C, indicating that an ESR-active species having an unpaired electron is produced by the heat treatment. The ESR-active species was generated at the same temperature as that of the release of chloride. By the release of chloride, the slight collapse of the charge balance between lithium, chloride and oxygen might occur on the surface. A sharp signal was not detected on SBA-3 or LiCl even after heating at 400°C. By combining mesoporous silica with LiCl, the ESR-active species were generated on the surface. From the results of ^{7}Li MAS NMR and ESR, it is revealed that the ESR-active species with ionic lithium exists on the surface, suggesting that the specific surface produced by the heat treatment causes hydrogen adsorption capability of the mesoporous silica.

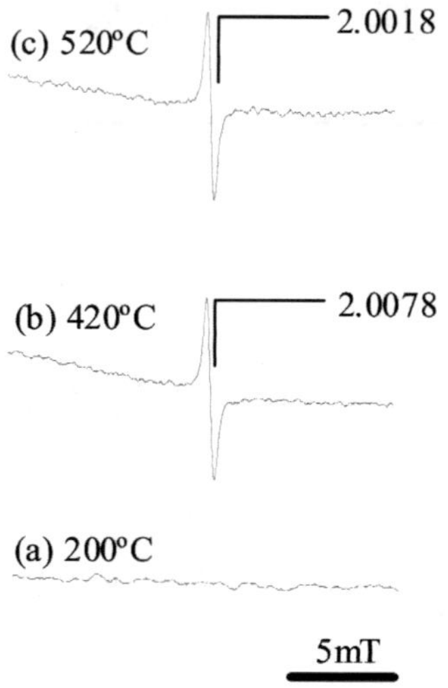

Fig. 5 ESR spectra of LiCl-doped SBA-3 at various heating temperatures;(a) 200°C, (b) 420°C and (c) 520°C.

4. Conclusions

By the reaction between lithium chloride and ethanol at elevated temperature, reversible adsorption sites for hydrogen under room temperature can be formed. It is revealed that the ESR-active surface with lithium could contribute to hydrogen adsorption.

High pressure adsorption equilibrium of natural gas in carbon adsorbents

Célio L. Cavalcante Jr., A. Eurico B. Torres, Diana C. S. Azevedo and J. Cássia S. Araújo
GPSA, Department of Chemical Engineering, Universidade Federal do Ceará
Campus do Pici, Bl. 709, 60455-760 Fortaleza, CE, BRAZIL

Emerson F. Jaguaribe
DTM, Universidade Federal da Paraíba
Campus I da UFPB, 58059 900 João Pessoa, PB, BRAZIL

ABSTRACT

The increase in natural gas (NG) usage in the Brazilian energy matrix is one of the main guidelines set by energy policy makers in the country, due to high availability of NG associated with oil offshore production. One of the main difficulties for large scale NG usage is the large distances between oil fields and consumption sites, which sometimes do not justify constructing pipelines and pumping accessories. In this study we have measured adsorption equilibrium of methane with commercial samples of activated carbon, as well as with laboratory samples obtained from Brazilian natural sources. Experiments were performed using a gravimetric setup with pressures between 1-70 atm, at temperatures between 20-60ºC. The studied samples have also been characterized using BET, Hg porosimetry and SEM. Results indicate possible usage of biomass natural sources for obtaining microporous activated carbon to be used in natural gas storage by adsorption.

KEY WORDS: Methane, storage, adsorption

INTRODUCTION

The adsorption of combustible gases on microporous solids has been intensively studied as an alternative for their storage and transportation (1-15). Storing gases in adsorbed form is thought to be safer and cheaper than in compressed form. In order to assess the performance of adsorbed natural gas as compared to compressed natural gas, it is necessary to fully determine the equilibrium adsorption properties of the adsorbent. Since the target pressure on storage vessels is 3.5 MPa, adsorption measurements must be carried out at relatively high pressures. Effects of temperature and buoyancy force may interfere in such measurements.

This research note shows partial results of a broader research project, with the objective of searching alternative means to store and transport natural gas to remote areas where no pipelines are available. Storing gases by adsorption on porous solids allows high packing densities at relatively low pressures and controlled delivery. Equilibrium adsorption properties of microporous carbons were measured using a gravimetric experimental setup. Samples of a commercial carbon adsorbent and of active carbons prepared from coconut shells were tested. The effect of the buoyancy force, which is significant at higher pressures, was taken into account using methodology from the literature (16).

The adsorption isotherms obtained for all samples tested were of IUPAC type I, showing perfect reproducibility and no hysteresis. Carbons prepared in laboratory reached superficial areas around 2000 m^2/g and their isotherm shape and capacity were very similar to those found for the commercial adsorbent.

EXPERIMENTAL

Microporous active carbons have been found to be a suitable material for natural gas storage. It can be prepared in a carbonization process (pyrolysis) from a large number of different carbon-containing source materials. Coconut shells have been used in this work as active carbon precursor. The starting material was previously ground and sieved in order to obtain a particle size between 1 and 2.36 mm. The particles were impregnated with either a strong base (KOH) or a metal chloride ($ZnCl_2$) and then subjected to pyrolysis at around 800ºC and 450ºC, respectively, under nitrogen flow. The samples with higher BET areas (1000 m^2/g and above) were tested for their methane adsorption properties.

Adsorption equilibrium measurements were carried out in a magnetic suspension balance by Rubotherm (Bochum, Germany). The experimental setup is shown in Figure 1. It consists of the balance chamber, a data acquisition unit, computer, thermostat bath, vacuum pump, and gas line.

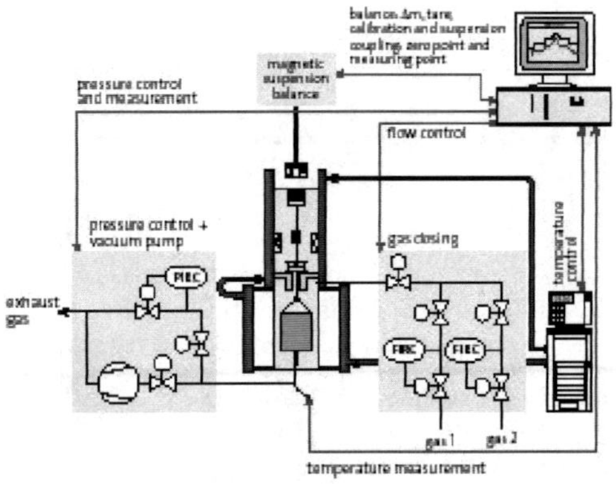

Figure 1. Magnetic suspension balance setup.

Adsorption isotherms were measured for methane in the pressure range of 1-70 bar (ca. 6 MPa) at temperatures between 20° and 60°C. In order to bench market the synthesized carbon samples (CSN active carbons), a commercial adsorbent by Sutcliffe Carbons (UK) was tested. This adsorbent is an activated carbon in granular form, labeled SRD, which was recommended by the supplier for methane storage.

RESULTS AND DISCUSSION

The BET surface areas and micropore areas of selected samples are shown in Table 1. We have included synthesized samples from Coconut Shells (CNS) with BET areas below and above the BET area of the commercial sample (SRD).

Methane adsorption isotherms of these active carbons were measured and compared. Figures 2-4 show the isotherms obtained for the 3 carbon samples at different temperatures. Note that the shape of all isotherms follows the pattern of IUPAC type I isotherms, which are typical in microporous solids. All experimental data could be well fitted using a Langmuir regression model.

Table 1. Properties of CNS active carbons and commercial sample

Sample	Activation agent	BET area (m^2/g)
CNS1	KOH	1043
CNS2	$ZnCl_2$	2099
SRD (commercial)	---	1972

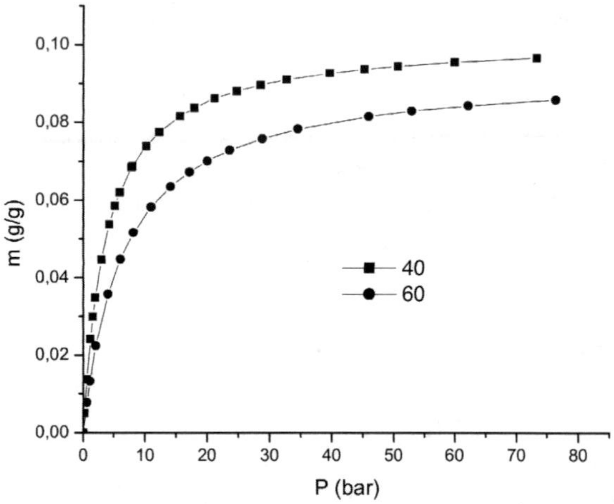

Figure 2. Adsorption isotherms of CNS1 at 40 and 60°C.

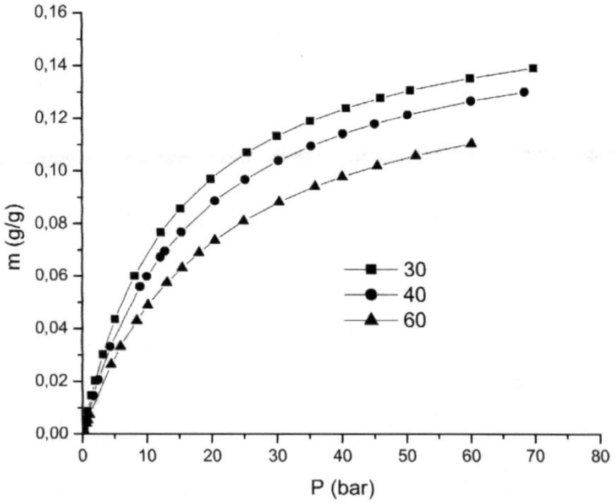

Figure 3. Adsorption isotherms of CNS2 at 30-60°C.

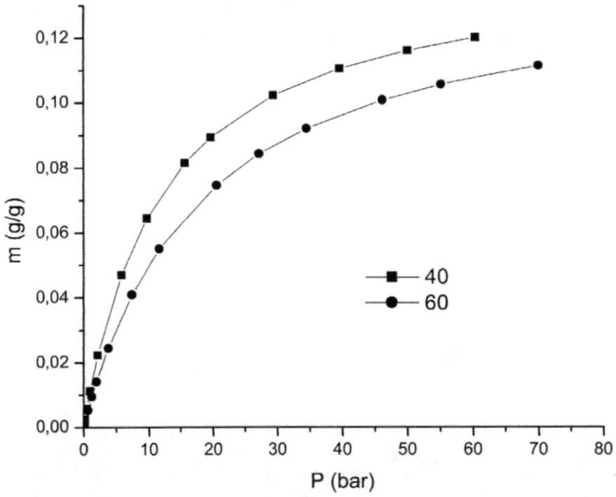

Figure 4. Adsorption isotherms of commercial sample (SRD) at 40 and 60°C.

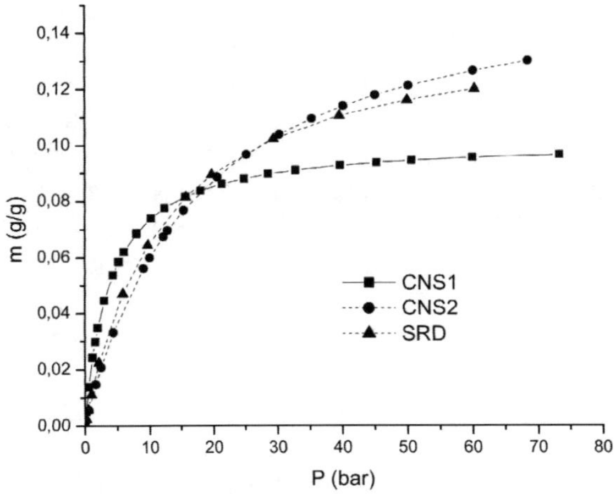

Figure 5. Comparison of adsorption isotherms for samples CNS1, CNS2 and SRD at 40 °C.

Figure 5 shows a comparison of the 3 samples at the same temperature (40°C). It may be observed that the sample CNS2 has an adsorption behavior that closely matches the SRD commercial sample at that temperature. The sample CNS1, however, has a much lower adsorption capacity (about 25% lower) than the commercial sample. Furthermore, at low gas pressures, the sample CNS1 adsorbs more methane than the other two samples, which is an undesired behavior for delivery purposes. The steep initial slope of the CNS1 isotherm may indicate higher energies of adsorption as compared to the other carbons. These results indicate that treatment with $ZnCl_2$ might be more appropriate for the chemical activation of the coconut shell for obtaining high microporous active carbon that could be used for methane storage and transportation.

ACKNOWLEDGEMENTS

The authors wish to acknowledge the Brazilian Petroleum Research Fund (FINEP/CTPETRO) and PETROBRAS for financial support for this research.

REFERENCES

1. D.Lozano-Castello, J.Alcaniz-Monge, M.A.de la Casa-Lillo, D.Cazorla-Amorós, A.Linares-Solano "Advances in study of methane storage in porous carbonaceous materials", *Fuel*, Vol. 81 (2002) p 1777-1803.
2. N.D. Parkyns, D.F. Quinn In: J.W. Patrick, Editor, *Porosity in Carbons*, Edward Arnold, London (1995), pp. 293–325.
3. D. Lozano-Castello, D. Cazorla-Amoros, A. Linares Solano, D.F. Quinn *Carbon* **40** 6 (2002), p. 989.
4. D. Lozano-Castello, D. Cazorla-Amoros, A. Linares Solano, D.F. Quinn. *Carbon* **40** 15 (2002), p. 2817.
5. J. Alcañiz-Monge, M.A. De la Casa-Lillo, D. Cazorla-Amorós, A. Linares-Solano *Carbon* **35** 2 (1997), p. 291.
6. L.L. Vasil'ev, L.E. Kanonchik, D.A. Mishkinis *J Engng Phys Thermophys* **72** 5 (1999), p. 884.
7. S. Biloe, V. Goetz, S. Mauran *Carbon* **39** (2001), p. 1653.
8. J. Alcañiz-Monge, D. Cazorla-Amorós, A. Linares-Solano, A. Oya, A. Sakamoto, K. Hoshi *Carbon* **37** (1997), p. 1079.
9. D. Lozano-Castelló, M.A. Lillo-Ródenas, D. Cazorla-Amorós, A. Linares-Solano *Carbon* **39** (2001), p. 741.
10. J.Wegrzyn, M.Gurevich, *Applied Energy*, Vol. 2, p 71-83.
11. J.P.B. Mota, *AIChE Journal*, Vol. 45 (1999) p.986-996.
12. J.P.B. Mota, A.E. Rodrigues, E. Saatdjian, D. Tondeur, *Adsorption*, Vol. 3 (1997) p.117-125.
13. J.P.B. Mota, E. Saatdjian, D. Tondeur, A.E. Rodrigues, *Adsorption*, Vol. 1 (1995) p.17-27.
14. L.L Vasiliev, L. E. Kanonchik, D.A. Mishkinis, M.I. Rabetsky, *Int. J. Therm. Sci.*, Vol. 39 (2000) p.1047-1055.
15. S. Biloe, V. Goetz, S. Mauran, *AIChE Journal*, Vol. 47 (2001) p.2819-2830.
16. F. Dreisbach and H.W. Losch, *Adsorption*, Vol. 8 (2002) p. 95-109.

Catalytic Generation of H_2 by Aqueous-phase Reforming of Biomass-Derived Oxygenated Hydrocarbons

J. W. Shabaker, G. W. Huber, R. R. Davda, R. D. Cortright, and J. A. Dumesic
University of Wisconsin
Madison, WI 53706

ABSTRACT

Aqueous phase reforming of oxygenated hydrocarbons (e.g., methanol, ethylene glycol, glycerol, sorbitol) is a promising new technology for the catalytic production of high-purity hydrogen for fuel cells. Previous studies have shown that silica-supported Pt catalysts exhibit high activity and selectivity for the production of H_2 at temperatures near 500 K. This paper will address the catalytic performance of Pt supported on a variety of materials and will present the results of detailed reaction kinetics studies over Pt/Al_2O_3 catalysts. We show that the production of H_2 by aqueous-phase reforming of ethylene glycol proceeds with high activity over Pt-black and over Pt supported on TiO_2, Carbon, and Al_2O_3 (i.e., turnover frequencies near 8 – 15 min^{-1} at 498 K); moderate activity is observed over Pt supported on SiO_2-Al_2O_3 and ZrO_2 (turnover frequencies near 5 min^{-1}); and lower activity is measured over Pt supported on CeO_2, ZnO, and SiO_2 (H_2 turnover frequencies lower than about 2 min^{-1}). Platinum supported on Al_2O_3, and to a lesser extent ZrO_2, exhibits high selectivity for production of H_2 and CO_2. An important bifunctional route for the formation of liquid-phase alkane-precursor compounds over less selective catalysts involves dehydration reactions on the catalyst support (or in the aqueous reforming solution), followed by hydrogenation reactions on Pt. Reaction kinetics studies of the aqueous reforming of methanol and ethylene glycol over Pt/Al_2O_3 catalysts show equilibration of the water gas shift reaction (WGS), fractional order in the feedstock, slight inhibition by the hydrogen product, and activation barriers between 100-140 kJ/mol. A reaction mechanism is proposed that explains these and other observations.

KEY WORDS

hydrogen production, ethylene glycol, reforming, supported platinum catalysts, support effects

INTRODUCTION

The catalytic production of hydrogen for fuel cells and other industrial applications from renewable resources is a major challenge. To this end, we have recently developed a catalytic process that generates H_2 from the aqueous-phase reforming of biomass-derived oxygenated compounds such as methanol, ethylene glycol, glycerol, sugars and sugar alcohols[1]. Unlike current technologies for producing hydrogen from non-renewable petroleum feedstocks methods (i.e., steam reforming[2]), the production of hydrogen by aqueous-phase reforming (APR) of renewable biomass resources generates no net atmospheric carbon dioxide because this greenhouse gas is consumed during further production of biomass. In addition, APR reactions take place at substantially lower temperatures (e.g., 500 K) than conventional alkane steam reforming processes (e.g., 900 K). The low temperatures at which aqueous-phase reforming reactions occur minimize undesirable decomposition reactions typically encountered when carbohydrates are heated to elevated temperatures[3]. Also the water-gas shift reaction is favorable at the same temperatures as ACR reactions, thus making it possible to generate H_2 and CO_2 in a single reactor with low amounts of CO according to the following overall reaction for ethylene glycol.

$$C_2H_6O_{2\,(l)} + 2H_2O_{(l)} \rightleftharpoons 2CO_{2\,(g)} + 5H_{2\,(g)} \qquad (1)$$

In contrast, typical steam reforming processes require multi-stage or multiple reactors to achieve low levels of CO in the product gas[4]. Another advantage of the APR process is that it eliminates the need to volatilize water, representing a major energy savings compared to conventional, vapor-phase steam-reforming processes.

BACKGROUND

Supported catalysts are prepared by incipient wetness impregnation of the dry support with aqueous solutions of tetraamine platinum nitrate ($Pt(NH_4)_4(NO_3)_2$, Strem Chemicals) (1.6 cm^3 solution per g of support), followed by treatment in an oven at 373 K for 12 hr. Most catalysts were subsequently heated to 533 K (1.3 K/min) in a mixture of 10% O_2/He (300 cm^3(STP)/min) in a Pyrex cell and held at 533 K for 2 hr. The calcined catalysts were reduced in flowing H_2 (100 cm^3(STP)/min) at 523K for 2 hr after heating at 0.5 K/min. Catalyst loadings were verified using atomic absorption spectroscopy (AAS/ICP) and the number of surface sites was estimated as the irreversible CO chemisorption uptake at 300 K. Properties of the catalysts are listed in Table 1.

Table 1. Catalyst characteristics and pretreatment conditions. Adapted from Ref. 6.

Catalyst	Support Supplier	Pt Wt %	Calc. K	Red. K	CO uptake umol/g	CO/Pt ratio
Pt/SiO_2	Cabot, EH-5	0.75	533	523	16.1	0.42
Pt-Black	Aldrich	99.90	none	348	27.9	0.01
Pt/ZrO_2	Alfa	0.69	533	523	30.2	0.85
Pt/C	Norit, SX 1 G	0.89	503 (He)	503	10.5	0.23
Pt/TiO_2	Degussa, P-25	0.85	533	523	16.8	0.39
Pt/SiO_2-Al_2O_3	Grace, MS-25	0.79	533	523	13.7	0.34
Pt/CeO_4[a]	Aldrich	0.59	533	523	20.1	0.66
Pt/γ-Al_2O_3[b]	Grace, Catapal B	3.63	533	523	58.1	0.31
Pt/γ-Al_2O_3	Grace, Catapal B	0.59	533	523	24.5	0.81
Pt/ZnO	Alfa	0.88	533	523	4.49	0.10

a) CeO_2 produced by oxidation of cerium (III) nitrate at 533 K for 2 h in flowing pure oxygen.
b) Sintering done by heating to 723 K over 30 minutes and holding for 12 h under 100 cm^3(STP)/min hydrogen saturated with water at 300 K.

The reactor is a 6.3 mm OD (¼") and 5 mm ID stainless steel tube filled with a packed bed of powder catalyst. Prior to collecting data, each reduced catalyst was allowed to cool below 323 K. The H_2 flow is then stopped and N_2 (500 sccm) is used to flush the reactor. A solution of the feed molecule in deionized water is then introduced in an up-flow configuration with an HPLC pump, and the reactor is heated to reaction temperature over 1 hr. The total pressure of the system is regulated just above the vapor pressure of water with N_2 carrier gas, which bubbles through the effluent to sweep gaseous products to gas chromatographs (GCs) equipped with thermal conductivity detectors (TCD) for analysis. Hydrogen is separated via a Porapak Q column, carbon monoxide by a molecular sieve (5A) column, and CO_2 and alkanes by a Hayesep D column. Liquid products are collected in the gas-liquid separator and analyzed after each run with a GC equipped with a flame-ionization detector (FID) and an Innowax column and an HPLC with an Aminex HPX 87H column and both refractive index (RI) and ultraviolet (UV) detectors. Reaction kinetics data are collected for up to 24 h for each set of reaction conditions, to assure that steady state is reached. Replicate runs with methanol and ethylene glycol give a standard deviation in the hydrogen turnover frequency of 10% at 483 K and 3% at 498 K. Carbon balances closed to within 3% for all runs, and a Madon-Boudart test was performed to verify operation in the kinetic regime[5].

SUPPORT EFFECTS AND REACTION NETWORK

Aqueous-phase reforming of 10 wt% ethylene glycol feed solutions produced primarily hydrogen and carbon dioxide at conversions less than 20%, although small amounts of liquid organic byproducts, gaseous alkanes, and carbon monoxide were also detected. Since it has been shown elsewhere[1] that aqueous-phase reforming of methanol over Pt leads primarily to H_2 and CO_2, we have included in Fig. 1 the potential hydrogen that could be generated by complete reforming of the methanol byproduct. It is clear, however, that this additional amount of hydrogen that could be produced from the residual methanol is small compared to the total amount of hydrogen produced by the catalyst.

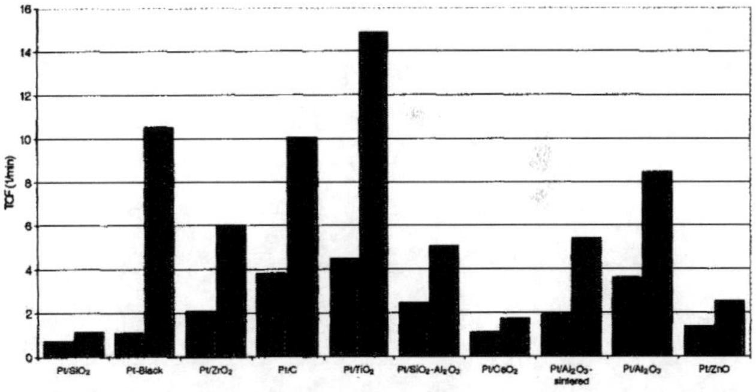

Figure 1. Hydrogen turnover frequencies for reforming of 10 wt% aqueous ethylene glycol at 483 (left columns) and 498 K (right columns) over supported Pt catalysts. The top portion of each bar is the amount of hydrogen that could be generated by total reforming of the methanol byproduct. Adapted from Ref. 6.

It can be seen that Pt/TiO_2 is the most active catalyst for the production of hydrogen from aqueous-phase reforming of ethylene glycol. In addition, unsupported Pt-black, as well as catalysts comprised of Pt supported on Al_2O_3 and activated carbon, also show high activity. Moderate catalytic activity for production of hydrogen is demonstrated by Pt supported on SiO_2-Al_2O_3 and ZrO_2, whereas lower catalytic activity is exhibited by Pt supported on CeO_2, ZnO, and SiO_2. In general, the supported Pt catalysts (and Pt-black) can be ranked in the following order with respect to rate of hydrogen production:

TiO_2 > Al_2O_3, Carbon, Black > SiO_2-Al_2O_3, ZrO_2 > CeO_2, ZnO, SiO_2

In addition to H_2 and CO_2, all catalysts produced smaller amounts of gaseous alkanes (methane and ethane) and liquid-phase products including alcohols (methanol and ethanol), organic acids (acetic and glycolic acids), and aldehydes (acetaldehyde and glycolaldehyde)[6]. Liquid products having a carbon atom that is not bonded to an oxygen atom, such as ethanol, acetic acid, and acetaldehyde, can be considered to be precursors for the subsequent formation of gaseous alkanes, because these liquid species cannot be reformed to H_2 and CO_2 with high selectivity (greater than 50%)[7]. In this respect, Fig. 2 shows the turnover frequencies for formation of carbon in gaseous alkanes and in liquid-phase products that will subsequently lead to alkanes at higher conversions (i.e., alkane precursor compounds in the liquid-phase). It can be seen in this figure that the rates of production of gaseous alkanes and liquid alkane-precursor compounds are dependent on the support. For example, platinum supported on activated carbon shows the highest rates of production of gaseous alkanes and liquid alkane-precursor compounds. These alkanes are not formed by gasification of the support because these products were not observed when the Pt/Carbon catalyst was treated in flowing water at 498 K.

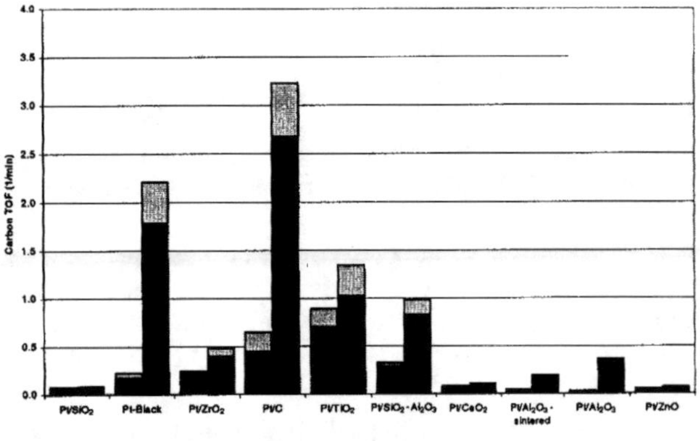

Figure 2. Production of alkanes for reforming of 10 wt% aqueous ethylene glycol at 483 (left columns) and 498 K (right columns) over supported Pt catalysts. The bottom portion of each column represents the carbon TOF for alkane precursors (acetaldehyde, ethanol, and acetic acid) while the top portion of each column represents the carbon TOF for alkanes (methane and ethane). Adapted from Ref. 6.

Comparison of Figs 1 and 2 shows that Pt/Al_2O_3, and to a lesser extent Pt/ZrO_2 and Pt/TiO_2, are the most active and selective catalysts we tested for the production of H_2 from liquid-phase reforming of ethylene glycol. Interestingly, Pt/Al_2O_3, Pt/ZrO_2 and Pt/TiO_2 produce lower amounts of alkanes compared to unsupported Pt-Black. The higher rate of alkane production over Pt-black does not appear to be caused by an effect of Pt particle size, since the rates of alkane production are similar on two Pt/Al_2O_3 catalysts with different Pt dispersions. In particular, the catalytic activity for production of hydrogen decreases by less than a factor of two and the selectivity remains nearly unchanged on alumina-supported catalysts when the CO/Pt ratio is decreased from 81% to 31%.

Effective catalysts for production of hydrogen by liquid-phase reforming of ethylene glycol, such as Pt/Al_2O_3, must facilitate specific pathways toward reforming and water gas shift. Initial dehydrogenation of ethylene glycol to form an adsorbed reaction intermediate species must be followed by rapid cleavage of the C-C bond, to avoid parallel reaction pathways leading to alkanes, such as pathways involving dehydration and hydrogenation reactions. This parallel selectivity route leads to C-O bond cleavage and the formation of alkanes through aldehyde and alcohol intermediates. After cleavage of the C-C bond, adsorbed C_1-species, such as formyl or methoxy species, may undergo further dehydrogenation to form strongly adsorbed carbon monoxide. Finally, the water gas shift reaction may take place, leading to the formation of weakly adsorbed CO_2 species that can readily desorb from the Pt surface. In general, subsequent reaction pathways exist for the production of alkanes from methanation and Fischer-Tropsch reactions (leading to series selectivity challenges for the production of H_2 by liquid-phase reforming). Although alkanes are the thermodynamically favorable reaction products under the conditions of this study, Pt catalysts are not generally effective for methanation and Fisher-Tropsch reactions, and the series selectivity challenge for production of H_2 is not an issue in this study. A simple reaction network illustrating these selectivity trends is presented in Fig. 3.

One of the most likely routes for the formation of liquid-phase alkane-precursor compounds, such as ethanol, involves dehydration of ethylene glycol followed by hydrogenation of the acetaldehyde intermediate. While both of these reactions may take place on Pt, it is also possible that the formation of ethanol may occur by a bifunctional pathway wherein dehydration is catalyzed by a Brønsted acid and hydrogenation takes place on Pt (i.e. the metal/support pathway in Fig. 3). The acid-catalyzed dehydration may take place on the surface of the support or in the aqueous solution.

After compounds such as ethanol and acetic acid have been formed from ethylene glycol, it is likely that these compounds will subsequently undergo cleavage of

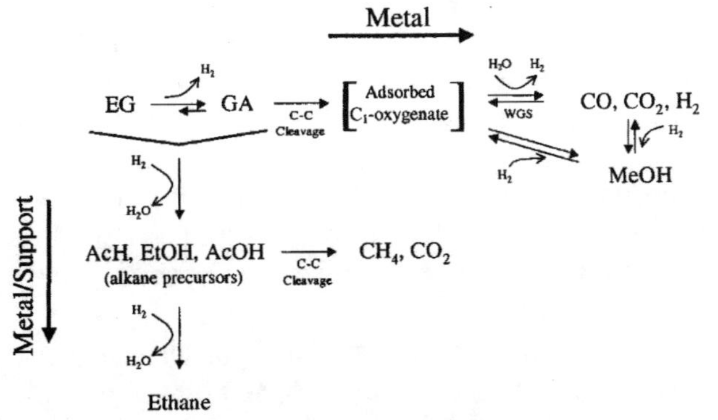

Figure 3. Schematic reaction network for aqueous-phase reforming of ethylene glycol near 500 K. Metal-catalyzed reactions move toward the right, while bifunctional metal-support reactions move downward. Adapted from Ref. 6.

the C-C bond, such that one of the carbon atoms leads to methane and the other carbon atom leads eventually to CO_2. For example, in the gas phase, ethanol and acetic acid are known to be reactive over Pt catalysts[7], and hydrogenolysis of ethanol and acetic acid over Pt lead to methane and carbon monoxide (or CO_2). Conversion between ethanol and acetic acid may also take place in the presence of hydrogen and water on Pt surfaces.

The initial dehydrogenation of ethylene glycol to form glycolaldehyde can lead to production of a range of byproducts, in addition to the desirable C-C bond cleavage of this intermediate to eventually form H_2 and CO_2. One likely pathway, which has very favorable reaction thermodynamics, is the rearrangement of glycolaldehyde to form acetic acid. In addition, glycolaldehyde can be converted to form glycolic acid, analogous to the conversion of acetataldehyde to acetic acid. Cleavage of the C-C bond in glycolic acid may lead to methanol and CO_2 over some catalysts, and this route allows for the direct formation of CO_2 without first passing through an adsorbed CO species followed by subsequent water-gas shift. This reaction pathway may possibly occur over Pt/ZnO since this catalyst appears to make significant amounts of glycolaldehyde and glycolic acid. The production of acids over ZnO-supported catalysts has precedence in the literature. Takezawa and Iwasa detected formic acid intermediates in methanol decomposition over Group VIII metals, while Cu/ZnO showed no acid byproducts[8].

An important aspect of the present study is that very low levels of CO (<500 ppm) are detected in the gaseous effluents from aqueous-phase reforming of ethylene glycol over the supported Pt catalysts of this study. This observation suggests that the water-gas shift reaction may be an important aspect of the liquid-phase reforming process, as seen in steam reforming of methanol over supported Pt[8-9] and Ni[10] catalysts. In particular, it appears that reforming and water-gas shift reactions take place in the same reactor at the temperatures of the present study (e.g., at 500 K). The equilibrium constant, K_{WGS}, for the water gas shift reaction at 500 K is equal to 3100 bar, when the reaction is written in terms of liquid-phase water (i.e., $K_{WGS} = \frac{P_{CO_2}P_{H_2}}{P_{CO}a_{H_2O}}$). For the dilute solutions of the present study, the activity of liquid water is essentially equal to unity. Thus, as the water-gas shift reaction becomes quasi-equilibrated, the corresponding ratio of pressures Q_P ($=\frac{P_{CO_2}P_{H_2}}{P_{CO}}$) approaches the value of K_{WGS}. Reaction kinetics experiments conducted at higher conversions (e.g., 10-15%) where detectable amounts of CO can be observed indicate that the value of Q/K_{WGS} is near 0.5, indicating that the water gas shift reaction is nearly equilibrated regardless of support. The presence of CeO_2, a common promoter of the water-gas shift reaction[11], appears to enhance the water-gas shift reaction, since the value of Q_p/K_{WGS} is approximately equal to 0.9 over Pt/CeO_2.

Calculation of the partial pressure of H_2, CO_2, and other species in the reactor is done by assuming that gas bubbles are formed within the liquid-phase flow reactor that are composed of reaction products. These bubbles are not diluted by nitrogen carrier gas from the separator, because nitrogen is not soluble in the aqueous reactor effluent that stands between the catalyst bed and the separator. Accordingly, the total pressure inside the bubbles can be approximated as the total pressure of the system. Assuming vapor-liquid equilibrium, the bubbles should contain water and feed partial pressures commensurate with their vapor pressures in the feed solution. The remaining pressure must be the sum of the partial pressures of the product gases.

$$P_{system} \approx P_{bubble} = \sum_{feed} P_i + \sum_{products} P_j \qquad (2)$$

Since the relative concentrations of the product gases (i.e., the reaction selectivity) is known from analysis of the nitrogen-diluted stream ($P_{j,\,diluted}$), the partial pressure of each product gas (P_j) in the bubbles can be found.

$$P_j = \frac{P_{j,diluted}}{\sum_{products} P_{j,diluted}} \left(P_{bubble} - \sum_{feed} P_i \right) \qquad (3)$$

KINETICS OVER PT/ALUMINA CATALYSTS

Because it shows high activity, selectivity, and stability for the production of H_2, 0.59 wt% Pt/Al_2O_3 was selected for further studies of the intrinsic kinetics of methanol and ethylene reforming. Fig. 4 shows the rates of hydrogen production at 498 K from ethylene glycol and methanol reforming versus the liquid-phase concentrations of these oxygenated hydrocarbons. The reaction order with respect to ethylene glycol concentration decreases from about 0.5 between 1 and 4 wt% ethylene glycol in the feed to about 0.3 between 4 and 10 wt% ethylene glycol in the feed. In contrast, the reaction order with respect to the liquid-phase methanol concentration is about 0.8 over the entire concentration range from 1 to 10 wt%. Since the selectivity to hydrogen is essentially 100%, the rate of CO_2 production can be determined by the reforming reaction stoichiometry.

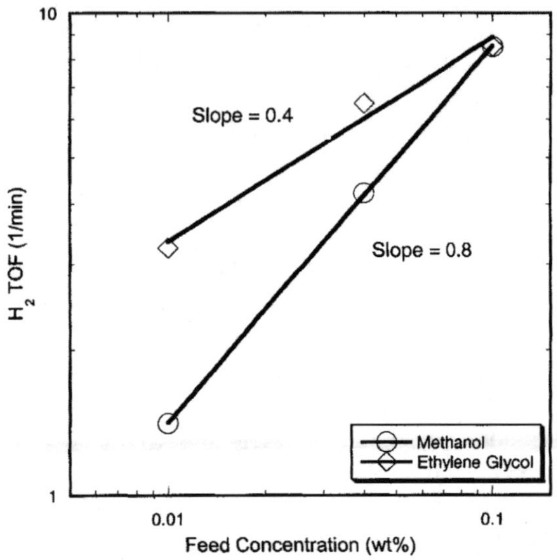

Figure 4. The effect of methanol and ethylene glycol inlet concentration on the rate of hydrogen production over the range of 1-10 wt% in an aqueous feed solution. Experiments were conducted at 498 K and a total pressure of 29.3 bar. Adapted from Ref. 5.

Fig. 4 shows that ethylene glycol reforming approaches low-order kinetics (i.e., 0.3) as the liquid-phase concentration approaches 10 wt%. These results suggest that the fractional surface coverage by adsorbed species derived from ethylene glycol is significant under the experimental conditions of the present study. In contrast to ethylene glycol reforming, the methanol reforming reaction is nearly first order (i.e., 0.8) with respect to methanol over the liquid-phase concentration range from 1 to 10 wt%. The higher reaction order for methanol reforming may be due to the weaker adsorption of methanol on Pt surfaces compared to bidentate adsorbed ethylene glycol[7]. Thus, at high feed concentrations, the rates of reforming of methanol and ethylene glycol are similar, but the rate of methanol reforming is slower compared to ethylene glycol at lower feed concentrations, due to weaker adsorption and thus lower surface coverages of adsorbed species derived from methanol.

Fig. 4 also shows that the rate of aqueous-phase reforming of methanol is similar to the rate of ethylene glycol reforming over Pt/alumina at temperatures near 500 K. This observation suggests that cleavage of the C-C bond in ethylene glycol is not a rate-limiting step in the reforming process, otherwise the rate of ethylene glycol reforming would have been slower than the rate of methanol reforming. Thus, it appears that aqueous-phase reforming reactions of methanol and ethylene glycol may be limited by the same surface processes, such as C-H and O-H bond cleavage steps leading to the formation of dehydrogenated surface intermediates, and steps involving the removal of adsorbed CO from the Pt surface by water-gas shift.

Fig. 5 shows the effect of the system total pressure on the rates of methanol and ethylene glycol reforming at 483 K. It can be seen that the total pressure has a strong inhibiting effect on the rate of hydrogen production. As the total pressure of the system is lowered closer to the bubble-point of the feed solution (i.e., approximately equal to the vapor pressure of water plus the vapor pressure of the oxygenated hydrocarbon times its mole fraction), the rate becomes approximately negative third order in the total pressure. The effect of total pressure effect is less significant at pressures well above the bubble-point.

The effect of total pressure may also be caused in part by partial vaporization of the oxygenated hydrocarbon reactant, especially for the case of methanol reforming. Alternately, if the chemistry is dominated by liquid-phase reforming, changes in pressure may change the wetting efficiency of the catalyst. Catalyst wetting is of importance in gas-liquid-solid reactors[12], and system pressure affects the density and bubble size distribution of gases bubbling through the liquid[13].

According to the procedure outlined above, we calculate the hydrogen partial pressure in the catalyst bed. Fig. 6 shows a plot of the rate of hydrogen production from methanol and ethylene glycol reforming at 483 K versus the calculated hydrogen partial pressure. It can be seen that the rate of hydrogen production is inhibited by the hydrogen partial pressure, and the reaction rate is approximately negative one-half order with respect to the hydrogen pressure for both methanol and ethylene glycol

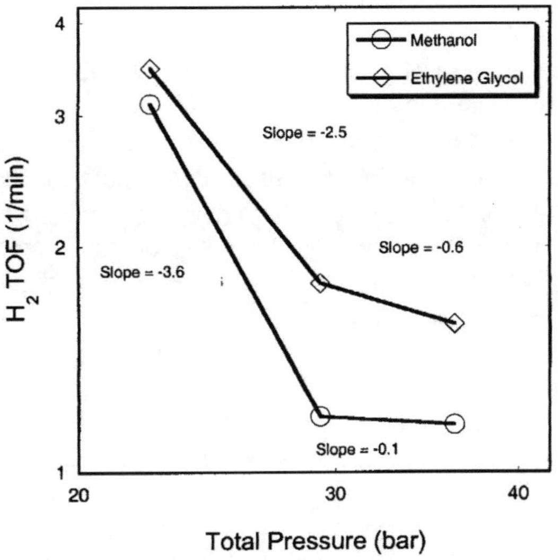

Figure 5. The effect of total system pressure on the rate of hydrogen production. Experiments were conducted with a 10 wt% aqueous methanol or ethylene glycol feed solution at 483 K over a total pressure range of 22.4-36.2 bar. Adapted from Ref. 5.

feed-stocks. This inhibiting effect of hydrogen on the rate of reforming reactions could be caused by the blocking of surface sites by adsorbed hydrogen atoms. In addition, hydrogen could inhibit the rate by decreasing the surface concentrations of reactive intermediates formed from dehydrogenation of the oxygenated hydrocarbon reactants. The effect of temperature on the rate of reforming for methanol and ethylene glycol was studied while maintaining the total pressure of the system at a value 3 bar above the vapor pressure of water at each temperature. In this way, the hydrogen partial pressure was maintained at a nearly constant value for each temperature. Accordingly, measurements of the rate of hydrogen production were made at 483 K and 22.4 bar total pressure, and at 498 K and 29.3 bar total pressure (corresponding to hydrogen partial pressures of 1.5 and 3.0 bar for experiments with methanol and ethylene glycol, respectively). These measurements give apparent activation energy barriers of approximately 140 and 100 kJ/mol for methanol and ethylene glycol reforming, respectively.

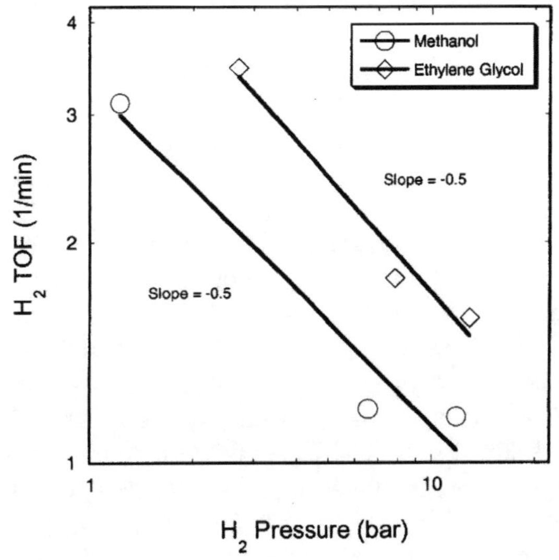

Figure 6. The effect of hydrogen partial pressure on the rate of hydrogen production. Experiments were conducted with a 10 wt% aqueous methanol or ethylene glycol feed solution at 483 K over a total pressure range of 22.4-36.2 bar. Adapted from Ref. 5.

REACTION MECHANISM

We now speculate about the form of the rate expression for aqueous-phase reforming of an oxygenated hydrocarbon such as ethylene glycol. We suggest that the following reaction scheme may capture the essential surface chemistry:

$$C_2O_2H_6 + * \underset{}{\overset{K_{EG}}{\rightleftarrows}} C_2O_2H_6^* \qquad \text{quasi-equilibrated}$$

$$H_2O + 2^* \overset{K_{H_2O}}{\rightleftarrows} OH^* + H^* \qquad \text{quasi-equilibrated}$$

$$H_2 + 2^* \overset{K_{H_2}}{\rightleftarrows} 2H^* \qquad \text{quasi-equilibrated}$$

$$CO_2 + * \overset{K_{CO_2}}{\rightleftarrows} CO_2^* \qquad \text{quasi-equilibrated}$$

$$C_2O_2H_6^* + * \xrightarrow{k_{rxn}} C_2O_2H_5^* + H^*$$

Irreversible steps leading to $C_2O_2H_x^*$

$$C_2O_2H_x^* + * \xrightarrow{k_{C-C}} 2\,COH_y^*$$

Irreversible steps leading to CO^*

$$CO^* + OH^* \xrightarrow{k_{WGS}} CO_2^* + H^*$$

where * represents a surface Pt site. We assume in this scheme that adsorption steps involving ethylene glycol, water, H_2 and CO_2 are quasi-equilibrated. We also assume that reforming takes place through a series of irreversible steps, since the equilibrium constants for these steps are probably rather large in view of the very favorable thermodynamics for the overall reforming reaction leading to H_2 and CO_2. These reforming reactions involve dehydrogenation of adsorbed ethylene glycol to form adsorbed $C_2O_2H_x^*$ species (rate constant k_{rxn}), cleavage of the C-C bond to form adsorbed COH_y^* species (rate constant $k_{C\text{-}C}$), further dehydrogenation leading to adsorbed CO*, and removal of adsorbed CO* by the water-gas shift reaction (rate constant k_{WGS}). We have assumed that CO_2 is formed from CO by the water-gas shift reaction; however, we note it is also possible that reforming forms CO_2 by reaction of water or hydroxyl groups with more hydrogenated surface intermediates (i.e., the reaction scheme does not pass through adsorbed CO*). This development is similar to the vapor phase decomposition of methanol over Pt/SiO_2 and other catalysts[14]. If we now assume that the most abundant surface species are adsorbed $C_2O_2H_6^*$, H*, OH* and CO*, then the following rate expression for aqueous-phase reforming of ethylene glycol is obtained:

$$r = \frac{k_{rxn} K_{EG} P_{EG}}{\left[1 + \sqrt{K_{H_2} P_{H_2}} \left(1 + \frac{k_{rxn} K_{EG} P_{EG}}{k_{WGS} K_{H_2O} P_{H_2O}}\right) + K_{EG} P_{EG} + \frac{K_{H_2O} P_{H_2O}}{\sqrt{K_{H_2} P_{H_2}}}\right]^2} \qquad (4)$$

The terms in the denominator of this rate expression correspond (from left to right) to surface coverage by vacant sites, H*, CO*, $C_2O_6H_6^*$ and OH*. This rate expression is consistent with the observed negative reaction order with respect to hydrogen, and this negative order is explained by the blocking of surface sites by adsorbed H* or CO*. In addition, the rate expression is consistent with the observed fractional reaction order with respect to ethylene glycol, and this fractional order is explained by the blocking of surface sites by adsorbed $C_2O_2H_6^*$ or CO*.

Aqueous-phase reforming of methanol may take place by a reaction scheme similar to that described above for ethylene glycol, without the step involving cleavage of the C-C bond. The negative reaction order with respect to hydrogen can explained by the blocking of surface sites by adsorbed H* or CO*. The nearly first order reaction kinetics with respect to methanol can be explained by noting results from DFT calculations which show that molecular adsorption of methanol on Pt is weaker compared to molecular adsorption of ethylene glycol[7]. We also note that the observed reaction kinetics for aqueous-phase reforming of methanol may be complicated by the possible contribution of vapor-phase methanol reforming, since methanol is more volatile compared to ethylene glycol.

SUMMARY

Platinum catalysts show high activity and selectivity for production of hydrogen by aqueous reforming of ethylene glycol feed stocks. Turnover frequencies for production of hydrogen are highest over Pt-black and Pt supported on TiO_2, Carbon, and Al_2O_3 (i.e., 8 – 15 min^{-1} at 498 K for 10 wt% ethylene glycol). Moderate catalytic activity for production of hydrogen is demonstrated by Pt supported on SiO_2-Al_2O_3 and ZrO_2 (near 5 min^{-1}), and lower turnover frequencies are exhibited by Pt supported on CeO_2, ZnO, and SiO_2 (lower than about 2 min^{-1}). Catalysts consisting of Pt supported on Carbon, TiO_2, SiO_2-Al_2O_3 and Pt-black also lead to the production (about 1 – 3 min^{-1}) of gaseous alkanes and liquid-phase compounds that would lead to alkanes at higher conversions (e.g., ethanol, acetic acid, acetaldehyde). Thus, we conclude that Pt/Al_2O_3, and to a lesser extent Pt/ZrO_2 and Pt/TiO_2, are active as well as selective catalysts for the production of H_2 from liquid-phase reforming of ethylene glycol.

Selectivity for the production of hydrogen versus alkanes appears to be controlled by competition between C-C and C-O bond scission reactions, i.e., with cleavage of the C-C bond leading to subsequent production of H_2 and CO_2, and cleavage of the C-O bond leading eventually to formation of alkanes. Cleavage of the C-C bond appears to be an irreversible process, whereas steps following cleavage of the C-C bond appear to be reversible, i.e., water gas shift and the conversion between methanol and CO_2 plus H_2. An important bifunctional route for the formation of liquid-phase alkane-precursor compounds over less selective catalysts involves dehydration reactions on the catalyst support (or in the aqueous reforming solution) followed by hydrogenation reactions on Pt.

Reaction kinetics measurements on Pt/Al_2O_3 catalysts with different metal loadings were carried out (employing the Madon-Boudart test) to determine kinetically-controlled rates for aqueous-phase reforming of methanol and ethylene glycol at 483 and 498 K. These results show that methanol and ethylene glycol have similar reactivity for aqueous-phase reforming over Pt/Al_2O_3 at consistent C/H_2O feed ratios, indicating that C-C bond cleavage is not rate limiting for ethylene glycol reforming. The rate of hydrogen production is higher order in methanol (0.8) than ethylene glycol (0.3-0.5), while the reaction is weakly inhibited by hydrogen for both feed-stocks (-0.5 order). Thus, it appears that the surface coverage by species derived from ethylene glycol is higher than from methanol under aqueous-phase reforming reaction conditions. The inhibiting effect of hydrogen on the rate of reforming could be caused by the blocking of surface sites by adsorbed hydrogen atoms. In addition, hydrogen could inhibit the rate by decreasing the surface concentrations of reactive intermediates formed from dehydrogenation of the oxygenated hydrocarbon reactants. In addition, the effect of total pressure on the reaction rate may be caused by physical phenomena such as gas-liquid holdup and catalyst wetting. Finally, the rate of aqueous-phase reforming is more temperature dependent for methanol than ethylene glycol, with apparent activation barriers of 140 and 100 kJ/mol, respectively, from 483 K and 22.4 bar total pressure to 498 K and 29.3 bar total pressure.

The liquid-phase environment of the present study favors the water gas shift reaction, thereby mitigating the potential blocking of surface sites by adsorbed CO species. Accordingly, low levels of CO (<300 ppm) are detected in the gaseous effluents from aqueous-phase reforming of methanol and ethylene glycol over alumina-supported Pt catalysts, and aqueous-phase reforming of both oxygenated hydrocarbons over Pt/Al_2O_3 leads to nearly 100% selectivity for the formation of H_2 (compared to the formation of alkanes). Since the selectivity for hydrogen production is essentially independent of conversion, it appears that the series hydrogenation of CO/CO_2 to alkanes is not significant over Pt/Al_2O_3 under the conditions of the present study.

The observed reaction kinetics of this study for ethylene glycol reforming can be explained by a reaction scheme involving quasi-equilibrated adsorption of ethylene glycol, water, H_2 and CO_2, combined with irreversible steps involving dehydrogenation of adsorbed ethylene glycol to form adsorbed $C_2O_2H_x{}^*$ species, cleavage of the C-C bond to form adsorbed $COH_y{}^*$ species, further dehydrogenation leading to adsorbed CO*, and removal of adsorbed CO* by water-gas shift. Aqueous-phase reforming of methanol may take place by a reaction scheme, without the step involving cleavage of the C-C bond. The nearly first order reaction kinetics with respect to methanol can be explained by weaker adsorption of methanol compared to molecular adsorption of ethylene glycol. The observed reaction kinetics for aqueous-phase reforming of methanol may also be complicated by the possible contribution of vapor-phase methanol reforming.

ACKNOWLEDGMENTS

This work was supported by U.S. Department of Energy (DOE), Office of Basic Energy Sciences, Chemical Sciences Division. We also acknowledge funding from the Energy Center of Wisconsin and the University of Wisconsin Office of University-Industrial Relations.

REFERENCES

1. R.D. Cortright, R.R. Davda and J.A. Dumesic, Nature 418 (2002) 64.
2. J.R. Rostrup-Nielsen, *Steam Reforming Catalysts* (Danish Technical Press, Inc., Copenhagen, 1975).
3. G. Eggleston and J.R. Vercellotti, Journal of Carbohydrate Chemistry 19 (2000) 1305.
4. F.D. Lomax, Jr., J.P. Reardon and J.P. Barbour, in *PCT Int. Appl.* (2002).
5. J. W. Shabaker, G. W. Huber, R. R. Davda, R. D. Cortright, and J. A. Dumesic, J. Catal. 215(2) (2003) 344.
6. J. W. Shabaker, G. W. Huber, R. R. Davda, R. D. Cortright, and J. A. Dumesic, Catal. Lett. 88(1-2) (2003) 1.
7. R.R. Davda, R. Alcala, J.W. Shabaker, G. Huber, R.D. Cortright, M. Mavrikakis and J.A. Dumesic, Dft and Experimental Studies of C-C and C-O Bond Cleavage in Ethanol and Ethylene Glycol on Pt Catalysts, Fourth Tokyo Conference on Advanced Catalytic Science and Technology, Tokyo, Japan (2002).
8. N. Takezawa and N. Iwasa, Catal. Today 36 (1997) 45.
9. K.T. Takahashi, N.; Kobayashi, H., Chem. Lett. (1985) 759.
10. K. Mizuno, N. Yoshioka, Y. Wakejima, Y. Takeuchi and A. Watanabe, Chem. Lett. (1986) 1969.
11. J. Kaspar, P. Fornasiero and M. Graziani, Catal. Today 50 (1999) 285.
12. M.P. Dudukovic, F. Larachi and P.L. Mills, Chem. Eng. Sci. 54 (1999) 1975.
13. I. Iliuta and F.C. Thyrion, Chem. Eng. Sci. 52 (1997) 4045.
14. S.H. Imamura, T.; Saito, Y.; Aritani, H.; Kanai, H.; Matsumura, Y.; Tsuda, N., Catalysis Today 50 (1999) 369.

CONTACT

Corresponding author: James A. Dumesic (dumesic@engr.wisc.edu)

Comparison of Steam Reforming, Autothermal Reforming, and Membrane Reactor Fuel Processors using Methanol and Liquid Hydrocarbon Fuels in a PEM Fuel Cell System

James R. Lattner and Michael P. Harold
Department of Chemical Engineering
University of Houston
Houston, TX 77204-4004

ABSTRACT

Complete fuel processor / fuel cell systems were modeled for the purpose of comparing fuel processor reactor configurations and selection of methanol versus liquid hydrocarbon fuels. Kinetic and heat transfer constraints were included to compare fuel processor volume requirements. The reactor models were incorporated into an overall process model that includes the PEM fuel cell in order to compare overall system efficiencies. Six combinations of fuel choice (methanol versus C_{14} alkane), method of reforming (steam versus autothermal), and method of hydrogen purification (preferential oxidation versus dense membranes) were evaluated. Each system was optimized with respect to reformer temperature, water:carbon ratio, and oxygen:carbon ratio. For a given fuel selection, each system, when optimized, achieves roughly the same overall system efficiency. Methanol achieves a higher efficiency than tetradecane in the fuel cell system, but this is offset by the higher energy requirement for producing methanol compared to diesel or gasoline fuels. The adiabatic autothermal membrane reactor with countercurrent steam sweep using hydrocarbon fuel has the potential for the smallest overall fuel processor volume, as removal of hydrogen by the membrane in the reformer drives the equilibrium for the steam reforming and water gas shift reactions, eliminating the water gas shift and preferential oxidation reaction steps. The use of a membrane reactor for the reforming of methanol reduces the fuel processor volume by eliminating the preferential oxidation step. However, the reforming of methanol is not equilibrium limited, so the membrane reactor does not reduce the reformer volume required, as is the case with the hydrocarbon membrane reactor.

INTRODUCTION

Development of fuel processors for a PEM fuel cells used for transportaion presents many challenges. Variables include choice of fuel, type of reformer, reactor design, reforming reactor conditions, method of hydrogen purification, and heat integration and water management schemes. Many investigations have been made that explore one or more of these variables. For example, a complete process model using thermodynamic equilibrium reactors has been used to explore the effects of water:carbon, oxygen:carbon, and reformer temperature and pressure in a gasoline autothermal reformer (ATR) system [1]. The effect of temperature and pressure have been studied in the steam reforming of methanol [2]. A palladium membrane was integrated into a fuel processor in the steam reforming of diesel, and the effects of pressure and temperature were studied in an overall process model [3]. Fuel processors utilizing partial oxidation and autothermal reforming of liquid hydrocarbon fuels have been compared [4]. The non-isothermal effects of an adiabatic hydrocarbon-fueled autothermal reformer have been tested [5]. The factors affecting water management in a fuel cell system have been studied [6]. The reformer, water gas shift, and preferential oxidation reactors have been sized for a 50 kW PEM fuel cell system utilizing liquid hydrocarbon fuels [7]. Reaction kinetic modeling of an isooctane ATR was performed at isothermal conditions [8].

In an effort to provide research guidance that will ultimately lead to an optimum fuel processor design, we have evaluated reactor and process designs for several fuel/reactor combinations, utilizing a method that enables direct comparisons of reactor volumes and system efficiencies. Reaction kinetics were applied to the steam reforming, oxidation, and water gas shift reactions under non-isothermal adiabatic conditions to allow comparison of reactor volumes. The kinetic simulations were integrated into an overall process model including the PEM fuel cell. Heat integration was applied on a consistent basis using a constant log-mean temperature difference (LMTD) in each exchanger. Exhaust temperature and pressures were adjusted to keep each system in water balance. The reformer temperature, water:carbon ratio, and oxygen:carbon ratios were optimized for each case to achieve the maximum overall system efficiencies.

Six different fuel processor systems were compared with respect to overall system efficiency and fuel processor reactor volumes for 50 kW of net power production:

1. Methanol Steam Reformer (SR) with Preferential Oxidation (PrOx) of carbon monoxide
2. Methanol adiabatic Autothermal Reformer (ATR) with PrOx
3. Methanol adiabatic ATR Pd Membrane Reactor with countercurrent steam sweep
4. Liquid hydrocarbon adiabatic ATR with Water Gas Shift (WGS) and PrOx
5. Liquid hydrocarbon adiabatic ATR Pd Membrane Reactor with countercurrent steam sweep
6. Liquid hydrocarbon adiabatic ATR Dense Oxide Membrane Reactor with countercurrent steam sweep

Below we highlight the modeling approach. Details will be presented elsewhere [9].

MODELING BASES

Kinetic models derived from literature were utilized in the study. The bases for these models are listed in Table 1. A model component, n-tetradecane (nC_{14}), was used to represent a diesel fuel. The reforming of nC_{14} occurs at high temperatures (700-1000°C) where the reactions are very fast and mass transfer limited. Apparent kinetics were applied in these systems. The reforming of methanol occurs at much lower temperatures (200-300°C), where the intrinsic kinetics are known. In these systems, the catalyst effectiveness factors were calculated and incorporated into the model. The oxidation reactions are much faster than the reforming reactions in both systems, and the reactor size is not very sensitive to the form of the oxidation rate expression.

In the reforming of nC_{14}, the high temperature drives the water gas shift reaction to approach equilibrium. At the reformer exit temperature of ~800°C, the equilibrium CO content is about 10 mol%. The CO content is reduced to below 1 mol% in a water gas shift (WGS) reactor using a tubular reactor cooled with boiling water. Two catalysts are required; a high temperature Re catalyst in the first half of the bed and low temperature Cu catalyst in the last half of the bed.

Table 1. Reactions simulated in the fuel processor kinetic models.		
Reaction	**Catalyst**	**Reference**
(1) Oxidation of n-tetradecane $nC_{14}H_{30} + 21.5O_2 \rightarrow 14CO_2 + 15H_2O$	Supported Ni	[10, 11]
(2) Steam reforming on n-tetradecane $nC_{14}H_{30} + 14H_2O \leftrightarrow 14CO + 29H_2$	Supported Ni	[11, 12]
(3) Water gas shift in n-C_{14} reformer $CO + H_2O \leftrightarrow CO_2 + H_2$	Supported Ni	[12]
(4) Water gas shift reactor $CO + H_2O \leftrightarrow CO_2 + H_2$	Supported Re, supported Cu	[13]
(5) Oxidation of methanol $CH_3OH + 1.5O_2 \rightarrow CO_2 + 2H_2O$	$CuO/ZnO/Al_2O_3$	[14]
(6) Steam reforming of methanol $CH_3OH + H_2O \leftrightarrow CO_2 + 3H_2$	$CuO/ZnO/Al_2O_3$	[15]
(7) Methanol decomposition $CH_3OH \leftrightarrow CO + 2H_2$	$CuO/ZnO/Al_2O_3$	[15]
(8) Water gas shift in methanol reformer $CO + H_2O \leftrightarrow CO_2 + H_2$	$CuO/ZnO/Al_2O_3$	[15]

In the methanol reforming reactor, the temperature must be kept low to minimize the extent of both the methanol decomposition reaction (7) and the reverse water gas shift reaction (8), as either of these reactions produce CO. By proper design of this reactor, an intermediate WGS reactor for CO reduction is avoided. Moreover, the Cu/ZnO catalyst must be kept below ~300°C to avoid irreversible deactivation.

The final reformate product must be essentially free of CO, as it will poison the Pt anode catalyst in the PEM fuel cell. The CO content can be reduced from ~1 mol% to less than 20 ppm using a preferential oxidation (PrOx) reaction step. A selectivity for CO oxidation of 50% is assumed. No suitable kinetic model of these reactions is available in the literature; the size of the PrOx reactor was set at 12.3 liters based on work done at Argonne National Laboratory on a pelletized precious metal catalyst [16].

In the membrane reactor cases, the hyrogen flux is assumed to be limited by diffusion of hydrogen atoms through the dense membrane, and thus follows a square root of the partial pressure driving force. The permeability of palladium as a function of temperature is taken from Holleck [17], and the permeability of an experimental high temperature dense oxide (dual phase proton conducting) membrane is based on work at Argonne National Laboratory and the National Energy Technology Laboratory [18]. The assumed thickness is 10 microns for the Pd membrane and 20 microns for the oxide membrane.

System efficiencies are maximized at minimum reforming pressures. The minimum reforming pressure is therefore set by practical considerations, primarily pressure drop. The reforming pressure for the steam and autothermal reforming cases was set at 3 bar. Higher reforming pressures are generally utilized for membrane reactors to improve the driving force for hydrogen flux. Pressures exceeding 10 bar may be needed if the permeate stream is pure hydrogen at 1-2 bar [3]. The use of a sweep gas allows the reformer pressure to be lowered considerably; high hydrogen recoveries are attained at a reformer pressure of 5 bar by use of countercurrent steam sweep on the permeate side of the membrane. The presence of steam in the hydrogen feed to the fuel cell anode is beneficial, as it helps provide humidification for the polymer membrane.

FUEL PROCESSOR SIMULATIONS

All fuel processing reactors were simulated using one-dimensional species and energy balance equations [11]. No isothermal assumptions were made; the internal reformer temperature profiles were calculated. An ordinary differential equation solver in MATLAB® was used to obtain the solutions.

1. Steam Reforming of Methanol – The steam reforming of methanol is conducted in a shell and tube reactor. Methanol and steam are mixed and fed to the shell side of the reactor where the endothermic reforming reaction occurs over a $CuO/ZnO/Al_2O_3$ catalyst. Fuel is catalytically combusted on the tube side of the reactor to provide the reaction heat. For simplicity, the tube side is taken to be a constant temperature equal to the exit temperature of the exhaust gases. A one-dimensional heat transfer coefficient of 200 $J.m^{-2}.s^{-1}.K^{-1}$ was used. The temperature and partial pressure profiles for this reactor are shown in Figures 1 and 2 using an H_2O:C ratio of 1.1.

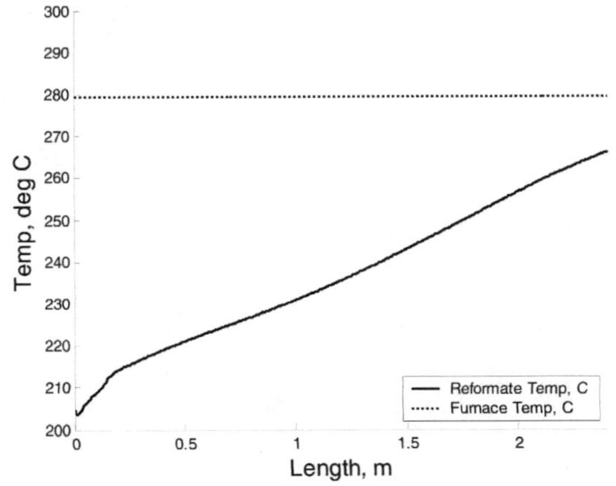

Figure 1. Temperature profiles for the steam reforming of methanol kinetic model.

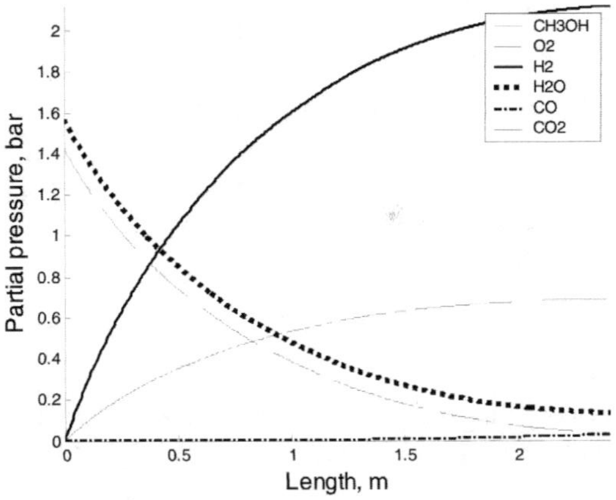

Figure 2. Partial pressure profiles for the steam reforming of methanol kinetic simulation

2. Autothermal Reforming of Methanol – Air and steam are mixed with methanol in the adiabatic autothermal reforming reactor over a $CuO/ZnO/Al_2O_3$ catalyst. The oxidation (combustion) of methanol is very rapid and exothermic, and generates a "hot spot" within the reactor. No hydrogen is generated over the oxidized copper catalyst. After the oxygen is depleted, the catalyst becomes reduced and the endothermic reforming reactions occur [14].

When all of the air is injected at the front of the bed, a hot spot of about 370°C is generated. This temperature is not only too high for stability of the copper catalyst, but it also accelerates the methanol decomposition and water gas shift reactions so that unacceptably high levels of CO are generated. We have evaluated staged injection of air to the catalyst bed assuming a uniform feed profile. This design is successful in lowering the peak temperatures and keeping the CO content below 1 mol%. Figures 3 and 4 show the temperature and partial pressure profiles for the adiabatic autothermal reforming of methanol with staged air injection, with an overall O_2:C ratio of 0.135 and an H_2O:C ratio of 0.8.

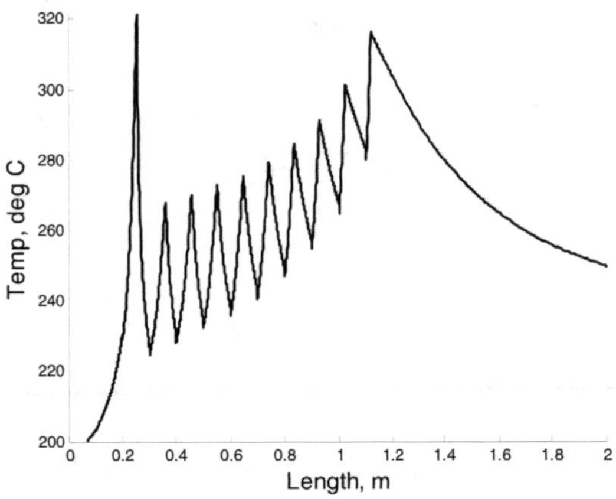

Figure 3. Temperature profile for the adiabatic autothermal reforming of methanol with staged air injection kinetic simulation.

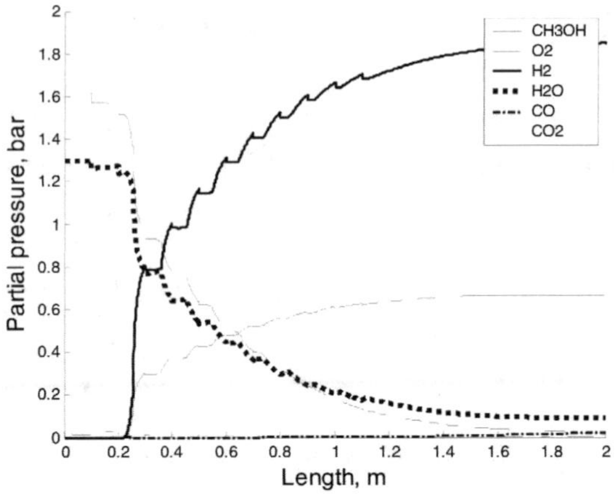

Figure 4. Partial pressure profile for the adiabatic autothermal reforming of methanol with staged air injection kinetic simulation.

3. Autothermal Membrane Reactor for Methanol Reforming – One of the advantages of a membrane reactor is for selective removal of hydrogen from the reforming mixture to improve the thermodynamic driving force for equilibrium-limited reactions. Moreover, a defect-free membrane will deliver a high purity hydrogen stream to the fuel cell without the need for reformate purification steps. This requires the incorporation of membrane tubes with sufficient area. The addition of countercurrent steam sweep on the permeate side adds an additional complication. As discussed in the previous case, distributed air addition to the catalyst is needed to avoid over-heating of the catalyst. Figure 5 shows a conceptual reactor design that meets all of the above requirements. The oxidation and membrane separation zones do not overlap due to practical concerns, although in theory there is no need to keep them separate.

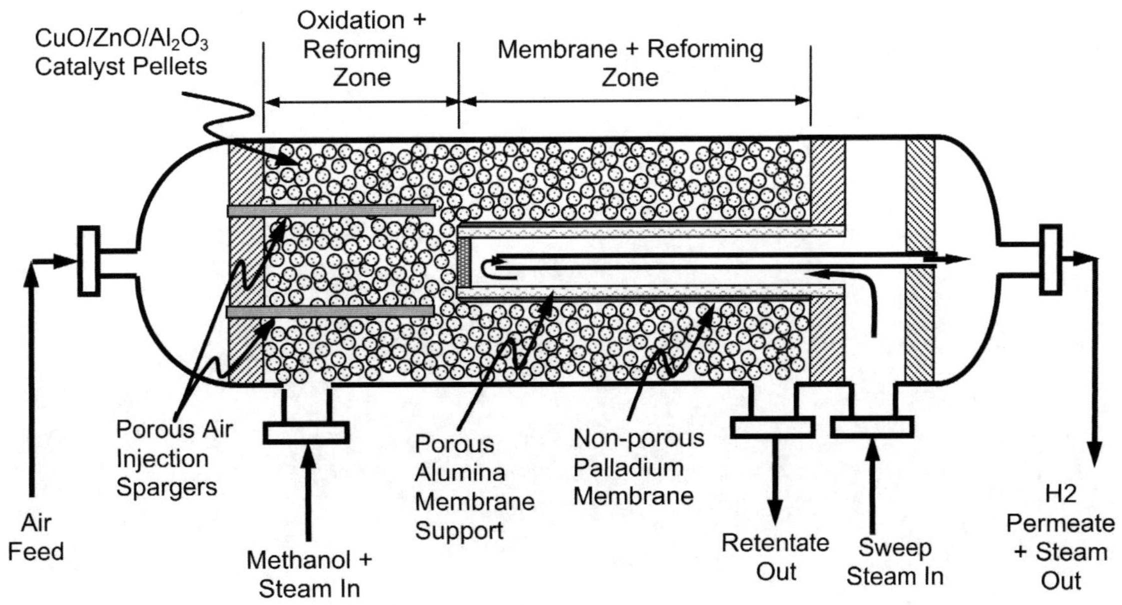

Figure 5. Schematic of an adiabatic ATR membrane reactor with air distributor tubes, membrane tubes, and provisions for countercurrent sweep gas.

Figures 6, 7, and 8 show the temperature and partial pressure profiles for the retentate and permeate sides of the membrane reactor. The oxidation zone covers the first 1.2 m of the reactor length and the membrane zone covers the final 0.9 m of reactor length. On the reforming side, the O_2:C ratio is 0.135 and the H_2O:C ratio is 0.60. The total retentate pressure is 5 bar and the permeate pressure is 1.1 bar. Convective heat transfer across the membrane is included in the simulation; the relatively cool steam sweep gas lowers the retentate exit temperature. The slower reforming kinetics at the cool exit limit the methanol conversion to about 90%. A high hydrogen recovery of 98% is achieved with 1.7 m^2 of membrane surface requiring 204 grams of palladium.

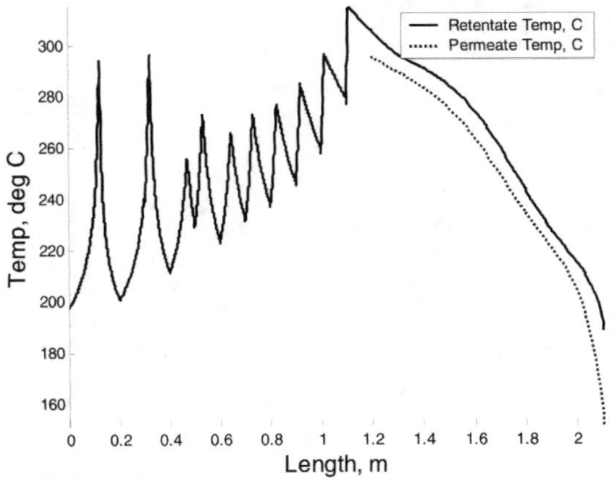

Figure 6. Temperature profiles for the methanol ATR membrane reactor kinetic simulation.

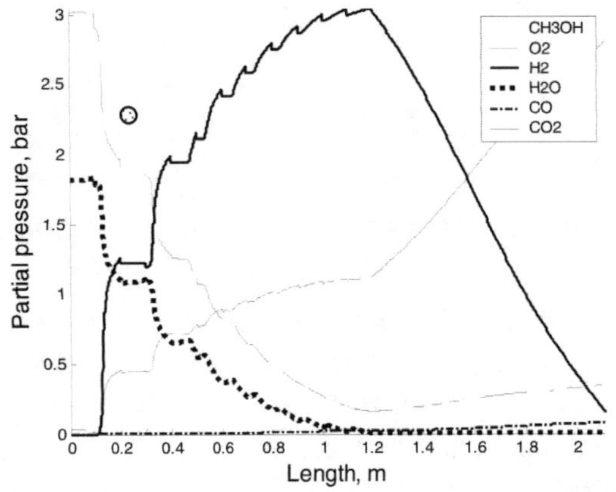

Figure 7. Partial pressure profiles on the retentate side of the methanol ATR membrane reactor kinetic simulation.

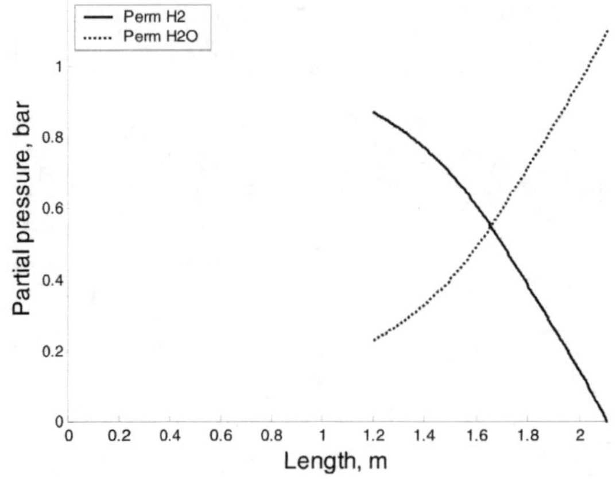

Figure 8. Partial pressure profiles on the permeate side of the methanol ATR membrane reactor kinetic simulation

4. Adiabatic ATR for Liquid Hydrocarbons with Water Gas Shift Reactor – Air and steam are mixed with the vaporized hydrocarbon and fed to the adiabatic ATR reactor. As with methanol, a "hot spot" is developed near the bed entrance due the very fast oxidation reaction. After oxygen is depleted, the endothermic steam reforming reaction takes over. The configuration of the combined reformer and water gas shift reactors is shown in Figure 9. The effluent from the reformer is cooled by exchange of heat with the reformer feed and then enters the tubular water gas shift reactor. This reactor is cooled by boiling water on the shell side.

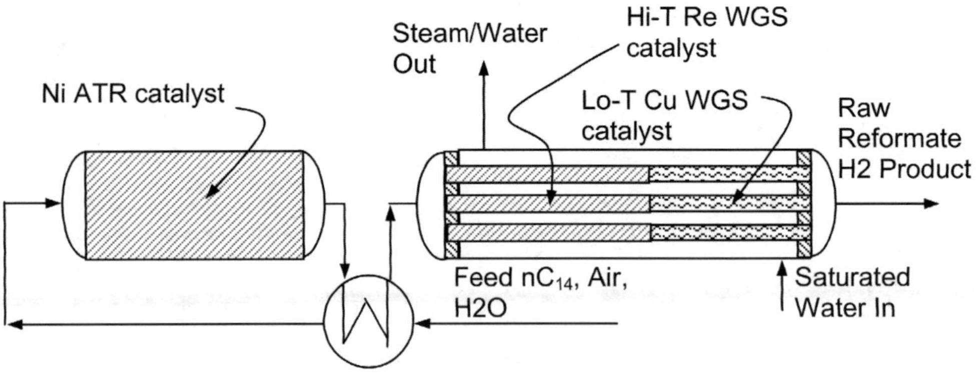

Figure 9. Adiabatic ATR of nC_{14} followed by cooled WGS reactor.

The temperature and partial pressure profiles for this reactor configuration are shown in Figures 10 and 11. The H_2O:C ratio is 2.0 and the O_2:C ratio is 0.36. The peak temperature attained in the ATR reactor approaches 1200°C. In practice, this peak temperature should be mitigated somewhat by axial conduction and radiation of heat. The reformer outlet temperature of 814°C yields a CO content of about 10 mol%, which is reduced to about 1 mol% in the WGS reactor. The overall conversion of feed is 96.8%.

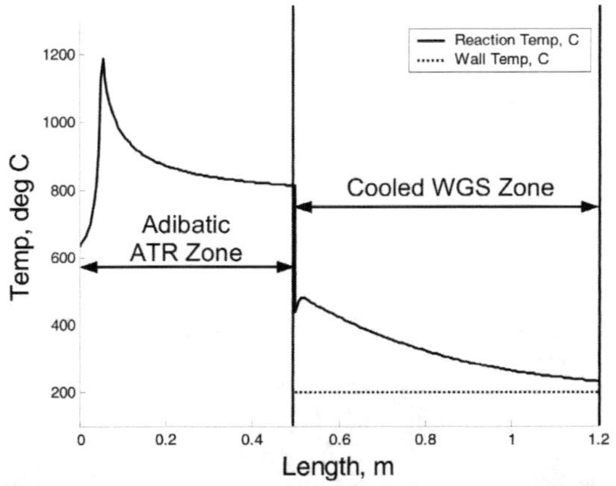

Figure 10. Temperature profiles for the adiabatic ATR of nC14 followed by cooled WGS reactor kinetic simulation.

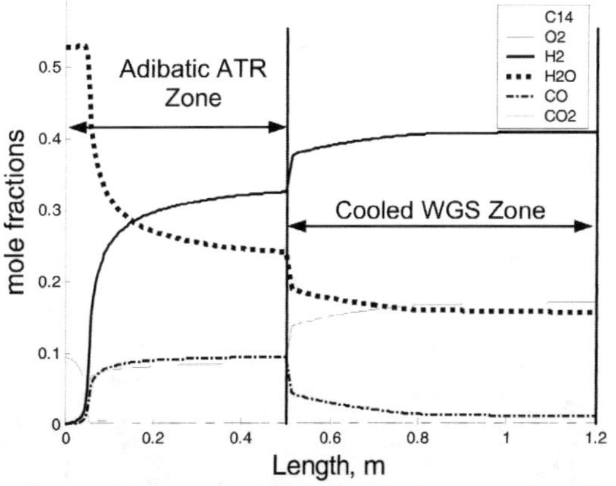

Figure 11. Composition profile for the adiabatic ATR of nC14 followed by cooled WGS reactor kinetic simulation.

5. Autothermal Pd Membrane Reactor for Liquid Hydrocarbon Fuel Reforming – The ATR membrane reactor for hydrocarbon feed contains an adiabatic oxidation zone, followed by the Pd membrane reactor zone. The Pd membrane must not extend into the oxidation zone for two reasons: (i) The Pd membrane is unlikely to be stable at temperatures in excess of about 900°C, and (ii) the membrane should not start at the beginning of the reformer, to avoid back permeation of hydrogen into the reformer. Figures 12, 13 and 14 show the temperature and partial pressure profiles for the retentate and permeate sides of the membrane reactor. The adiabatic ATR zone covers the first 0.32 m of the reactor length and the membrane zone covers the final 0.5 m of reactor length. On the reforming side, the O_2:C ratio is 0.36 and the H_2O:C ratio is 1.5. The total retentate pressure is 5 bar and the permeate pressure is 1.1 bar. Convective heat transfer across the membrane is included in the simulation. The cooling effect of the steam sweep does not reduce the reforming temperature significantly, as the exothermic water gas shift reaction is occurring in the membrane reactor zone. A high hydrogen recovery of 97% is achieved with this design using 0.8 m^2 of membrane surface requiring 94 grams of palladium.

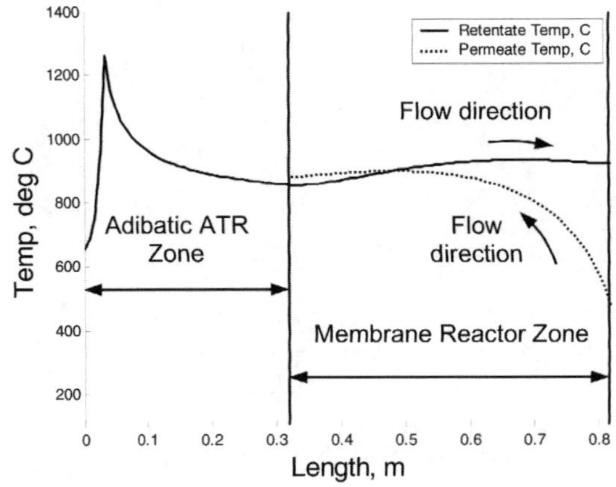

Figure 12. Temperature profiles for the ATR membrane reactor with nC_{14} feed and countercurrent steam sweep kinetic simulation.

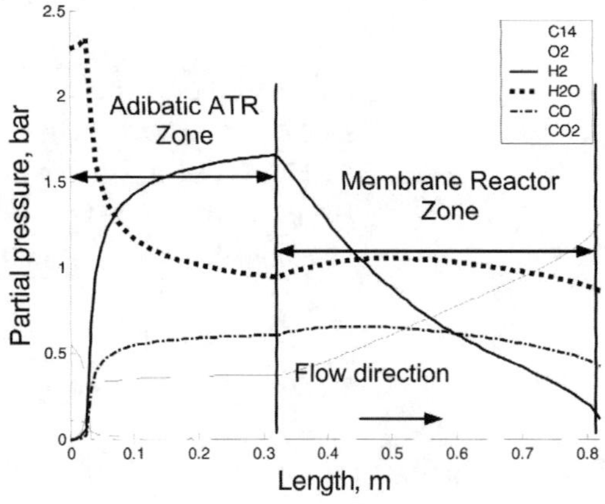

Figure 13. Partial pressure profiles for retentate in the ATR membrane reactor with nC_{14} feed and countercurrent steam sweep kinetic simulation.

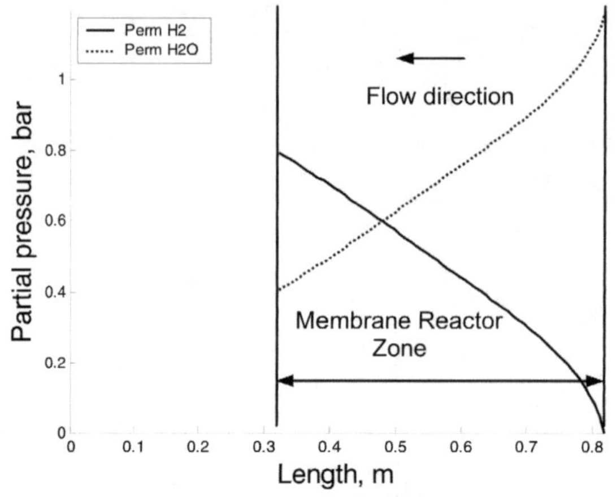

Figure 14. Partial pressure profiles on the retentate side for the ATR membrane reactor with nC_{14} feed and countercurrent steam sweep kinetic simulation.

OVERALL PROCESS SIMULATIONS

The kinetic simulations of the fuel processors were linked within the framework of an overall HYSYS® process model to carry out the complete heat and material balance of the combined fuel processor / fuel cell system. The assumptions for the fuel cell and auxiliary equipment in this model are listed in Table 2. In order to keep the comparisons on a consistent basis, the heat integration exchangers each maintained a consistent Log-Mean Temperature Difference (LMTD) of 100°C. The fuel cell operating pressure was selected to achieve 100% water balance by recovery of condensate from the exhaust from ambient cooling. More details of this analysis will be presented elsewhere [9].

Table 2. Assumptions in the Fuel Cell and Auxiliary Equipment

Fuel cell operating temperature	80°C
Cell voltage	0.75 V
LHV efficiency of fuel cell	60%
Air stoichiometry in cathode	2.0
Hydrogen utilization from reformate in anode	85%
Hydrogen utilization from permeate in anode	95%
Minimum exhaust temperature from ambient cooling	48°C
Fuel Cell operating pressure	1.1-1.2 bar
LMTD of heat integration exchangers	100°C
Air compressor efficiency	75%

The process flow diagrams for the complete fuel processor / fuel cell systems are shown in Figures 15 through 19.

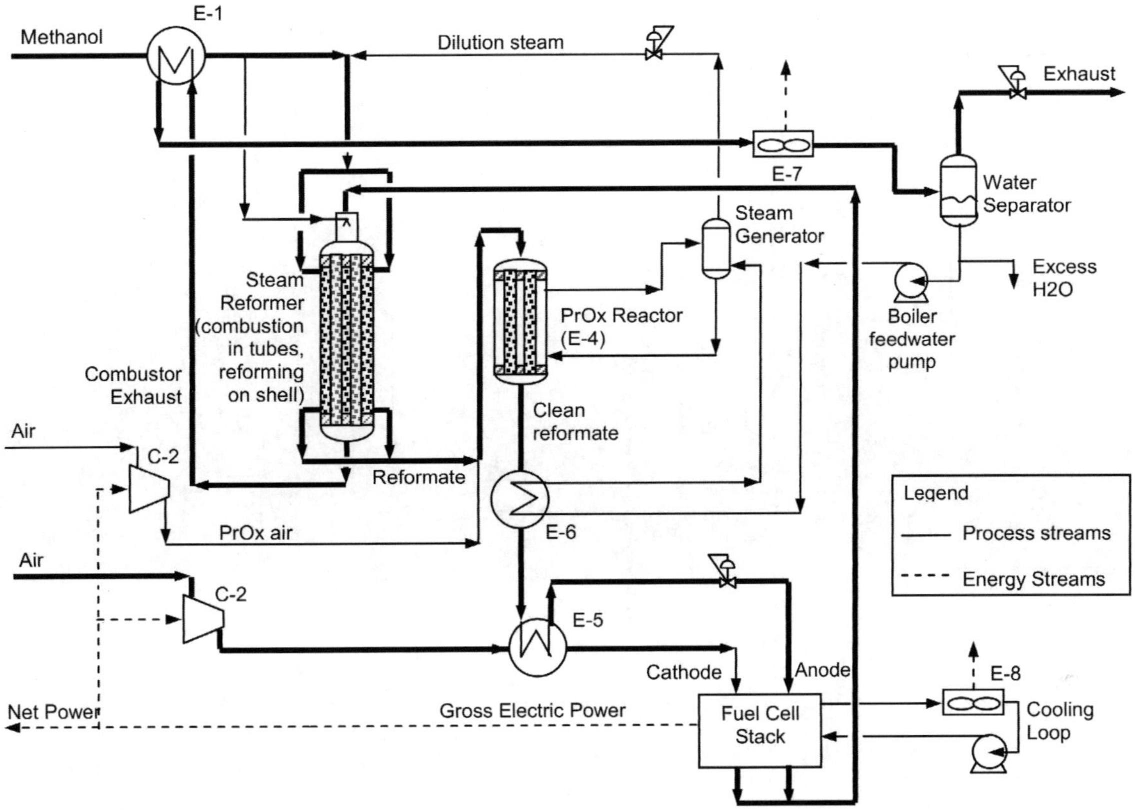

Figure 15. Overall process flow diagram for the steam reforming of methanol in a catalytically-fired tubular reformer with CO removal by PrOx.

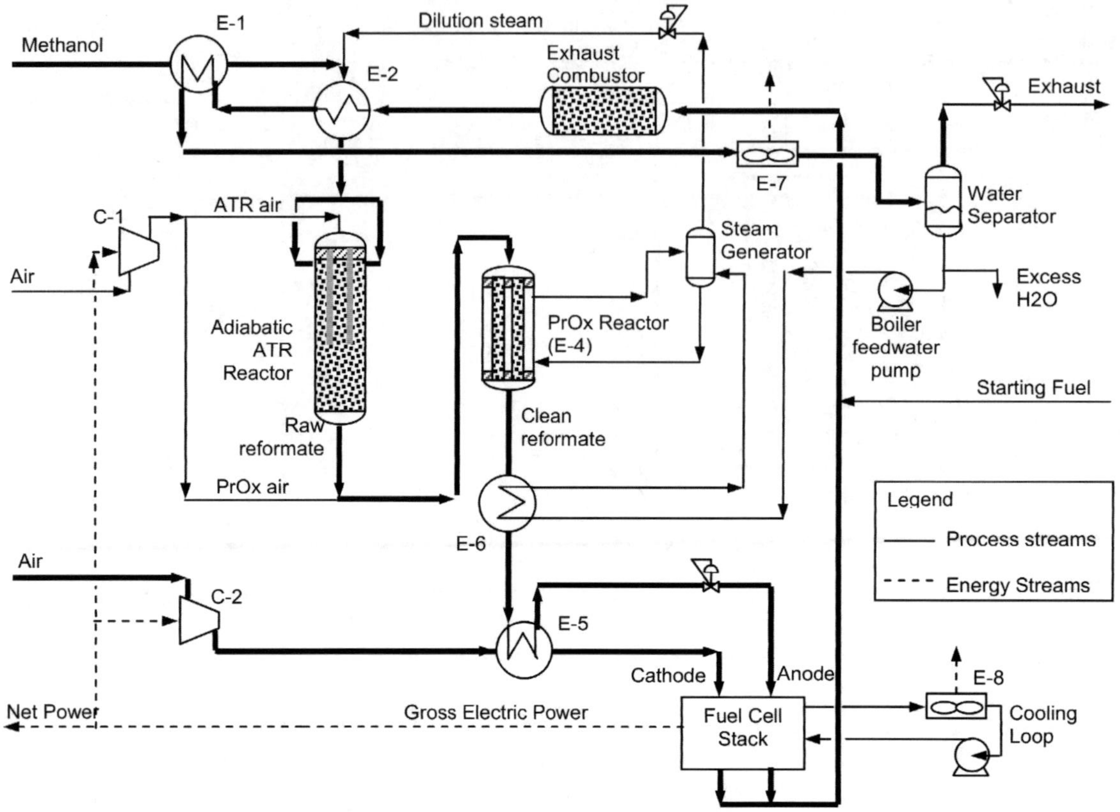

Figure 16. Overall process flow diagram for the adiabatic autothermal reforming of methanol with CO removal by PrOx.

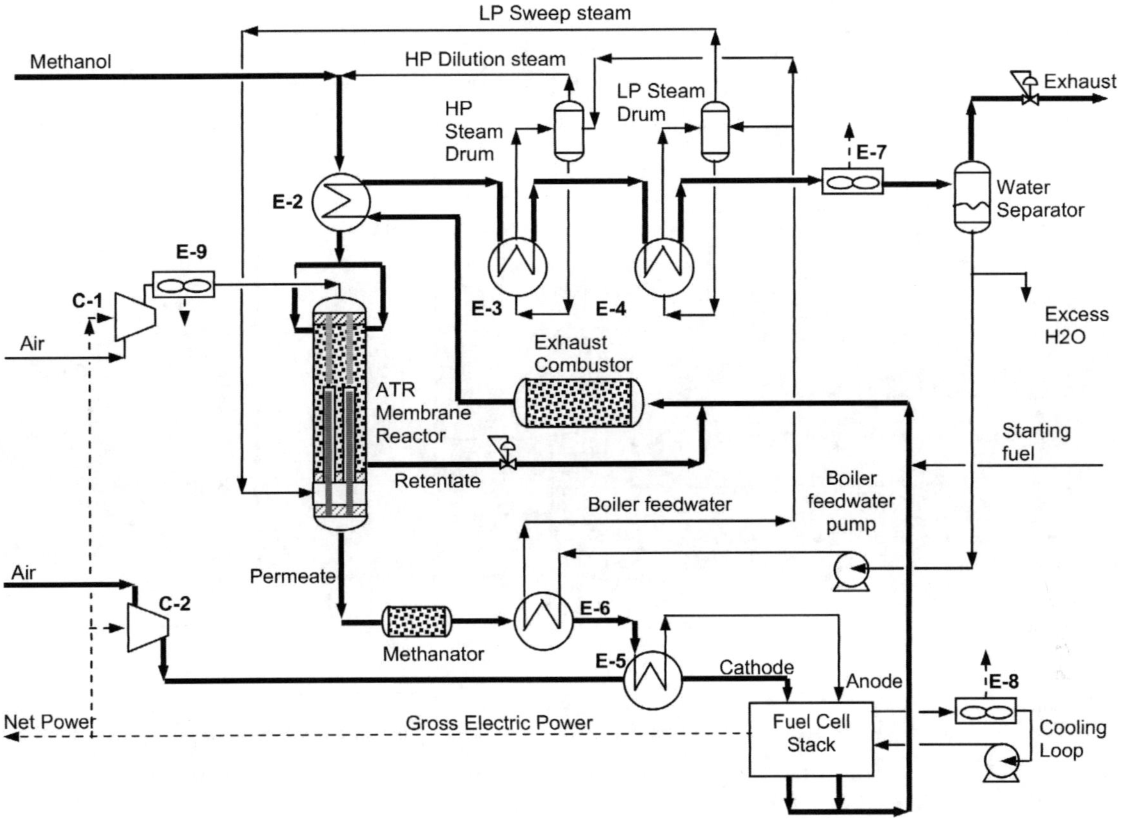

Figure 17. Overall process flow diagram for the Adiabatic ATR membrane reactor for methanol feed with countercurrent steam sweep.

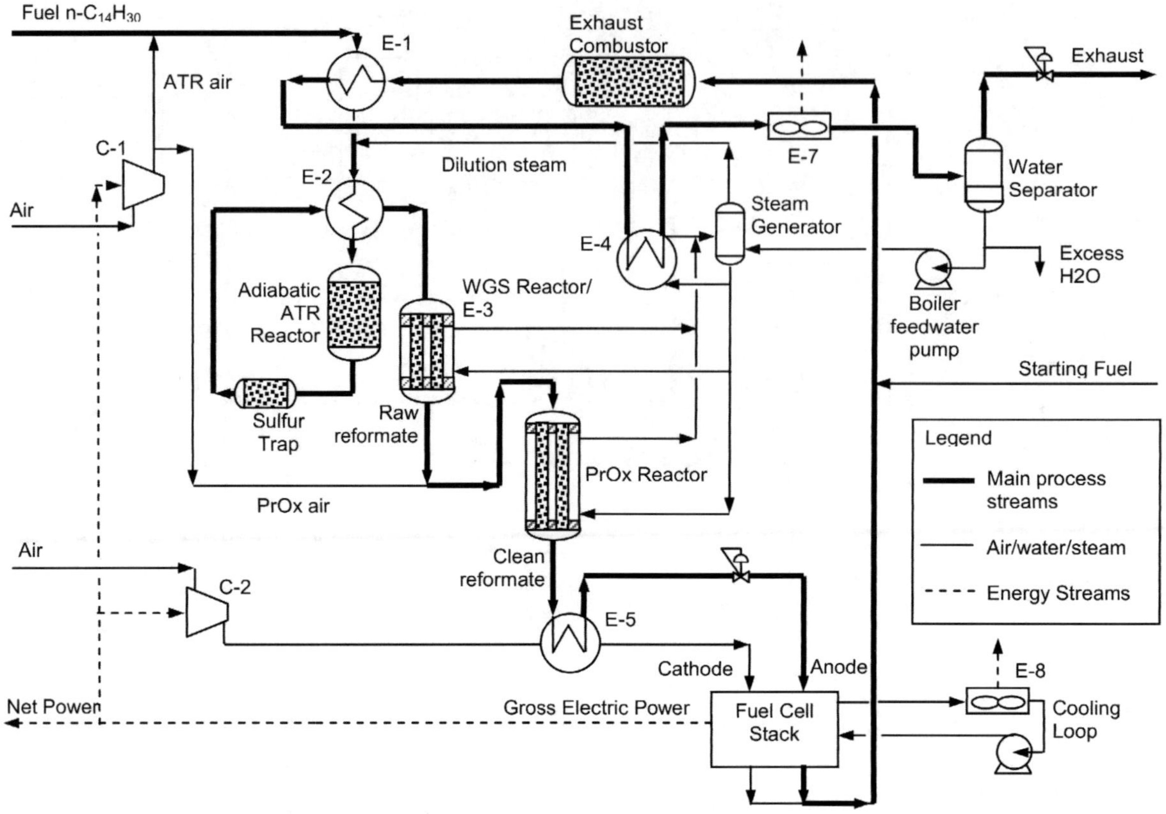

Figure 18. Overall process flow diagram for the adiabatic ATR of nC_{14} with WGS and CO removal by PrOx.

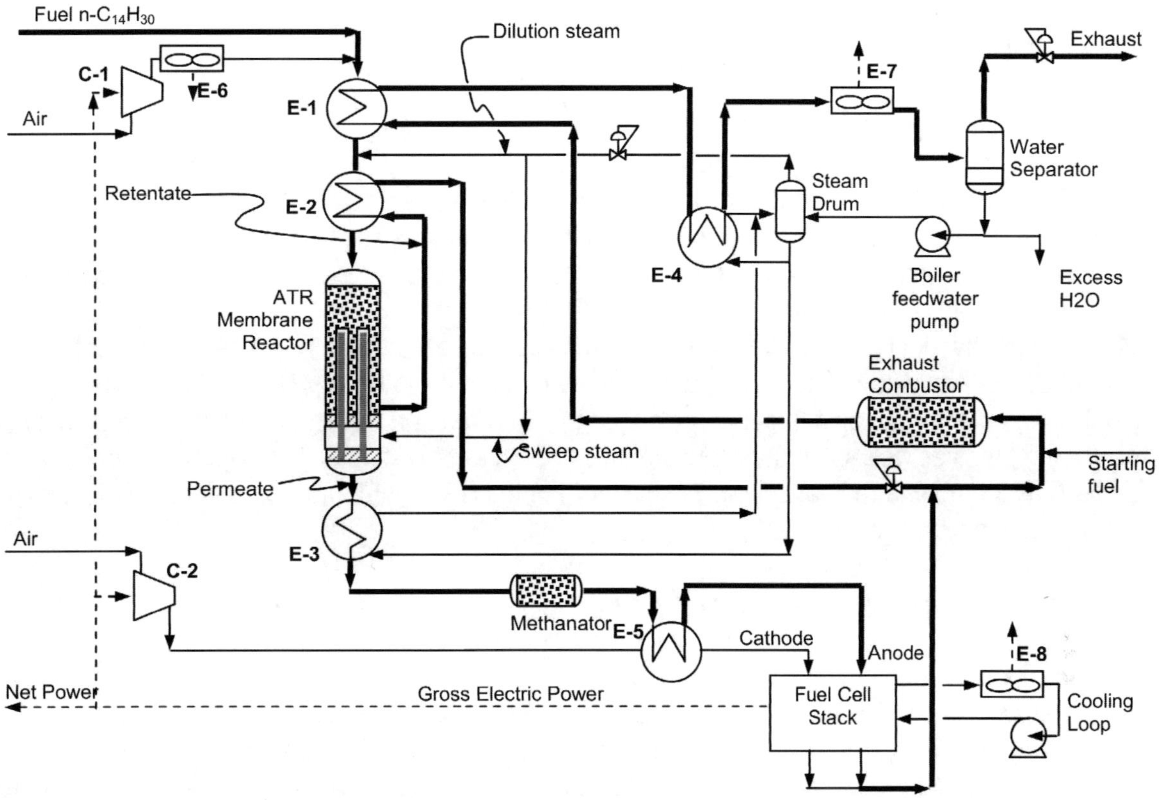

Figure 19. Overall process flow diagram for the adiabatic ATR membrane reactor with nC14 feed and countercurrent steam sweep.

RESULTS

The results of the combined fuel processor / fuel cell simulations are summarized in Table 3. Shown are estimates for the overall system efficiencies (LHV basis) and total fuel procesor (reactor) volumes.

Table 3. Overall system efficiencies and volume requirements for various fuel processor designs of combined fuel processor / fuel cell systems for 50 kW net power production.

Case	LHV Efficiency	Fuel Processor Sizes, liters Reformer	WGS	PrOx	Total
(1) Methanol Steam Reformer w/ PrOx	50.2%	11.2	0	12.3	**23.5**
(2) Methanol ATR w/ PrOx	50.3%	9.1	0	12.3	**21.4**
(3) Methanol ATR Membrane Reactor	48.8%	13.1	0	0	**13.1**
(4) Liquid HC ATR w/ WGS & PrOx	39.7%	4.7	6.5	12.3	**23.5**
(5) Liquid HC ATR Pd Membrane Reactor	40.7%	4.3	0	0	**4.3**
(6) Liquid HC ATR Oxide Membrane Reactor	37.9%	17.3	0	0	**17.3**

CONCLUSIONS

Methanol-fueled systems – These systems each achieved overall efficiencies of about 50%. Since the assumed fuel cell efficiency was 60%, the fuel processors alone achieve an efficiency of about 83%. The steam reformer with external combustion achieved essentially the same efficiency as the adiabatic autothermal reformer. The ATR membrane reactor was slightly less efficient at about 49%, which is probably due to the cooling effect of the steam sweep and the inefficiencies associated with generation of both reforming and sweep steam. The catalyst volume required for the steam reformer is slightly larger than that required for the ATR; which is an indication that in-situ generation of heat is more effective than external heat generation from a reactor sizing basis. The methanol ATR membrane reactor requires a larger volume than the ATR reactor, due to the volume occupied by the membrane tubes as well as from the cooling effect of the sweep steam. The overall fuel processor size for the ATR membrane reactor system is smaller than the SR or ATR fuel processors due to elimination of the PrOx reactor.

Hydrocarbon-fueled systems – These systems each achieved efficiencies of about 40%, for a fuel processor efficiency of about 67%. The ATR Pd membrane reactor system was slightly more efficient than the ATR system, primarily due to the elimination of hydrogen losses associated with the PrOx step. The fuel processor volume reductions for the Pd membrane reactor are very dramatic. The selective removal of hydrogen by the membrane increases the thermodynamic driving force for the steam reforming and water gas shift reactions. The extent of CO conversion by the WGS reaction is sufficient for elimination of the separate cooled WGS reactor. The purification of hydrogen through the dense Pd membrane eliminates the need for a PrOx reactor (although a small methanation reactor may be required to handle CO leakage through the membrane from defects). The Pd membrane reactor fuel processor requires just 4.3 liters of volume, or just 18% of the volume required of the ATR/WGS/PrOx fuel processor.

Comparison of Fuels – The efficiency of producing various consumer fuels from their fossil fuel sources has been studied at Argonne National Laboratory [19]. These efficiencies can be added to the fuel processor and fuel cell efficiencies to give overall "well-to-wheel" efficiencies, as shown in Table 4. Interestingly, the differences in fuel production efficiencies essentially even out the differences in fuel processor efficiencies. This calculation shows the diesel hydrocarbon system to be slightly more efficient than the methanol-based system.

Table 4. "Well-to-Wheels" efficiency comparison of methanol and diesel hydrocarbon based systems.

Fuel	Fuel production efficiency [19]	Fuel processor efficiency	Fuel cell efficiency	"Well-to-Wheels" efficiency
Methanol	68%	83%	60%	**34%**
Diesel	87%	67%	60%	**35%**

FINAL REMARKS

This study provides a comparison of several combinations of fuel processor designs and fuel types. Our approach underscores the importance of synthesizing and evaluating process flowsheets in concert with the reactor system. The results reveal that comparable efficiencies for different fuel processor configurations can be achieved through proper design and optimization of the parameters. Kinetic simulations show that a substantial reduction in processor volume may be achieved using a membrane fuel processor. This may have attractive implications with dynamic performance (e.g. startup time). On the other hand, the cost and stability of Pd-based membranes must obviously be considered.

REFERENCES

1. Doss, E.D., R. Kumar, R.K. Ahluwalia, and M. Krumpelt, *Fuel processors for automotive fuel cell systems: a parametric analysis.* Journal of Power Sources, 2001. **102**: p. 1-15.
2. Amphlett, J.C., K.A.M. Creber, J.M. Davis, R.F. Mann, B.A. Peppley, and D.M. Stokes, *Hydrogen production by steam reforming of methanol for polymer electrolyte fuel cells.* Int. J. Hydrogen Energy, 1994. **19**(2): p. 131-137.
3. Amphlett, J.C., R.F. Mann, B.A. Peppley, P.R. Roberge, A. Rodrigues, and J.P. Salvador, *Simulation of a 250 kW diesel fuel processor/PEM fuel cell system.* Journal of Power Sources, 1998. **71**: p. 179-184.
4. Docter, A. and A. Lamm, *Gasoline fuel cell systems.* Journal of Power Sources, 1999. **84**: p. 184-200.
5. Flytzani-Stephanopoulos, M. and G.E. Voecks, *Autothermal reforming of aliphatic and aromatic hydrocarbon liquids.* Int. J. Hydrogen Energy, 1983. **8**(7): p. 539-548.
6. Ahmed, S., J. Kopasz, R. Kumar, and M. Krumpelt, *Water balance in a polymer electrolyte fuel cell system.* J. Power Sources, 2002. **112**: p. 519-530.
7. Zalc, J.M. and D.G. Loffler, *Fuel processing for PEM fuel cells: transport and kinetic issues of system design.* J. Power Sources, 2002. **111**: p. 58-64.
8. Pacheco, M., J. Sira, and J. Kopasz, *Reaction kinetics and reactor modeling for fuel processing of liquid hydrocarbons to produce hydrogen: isooctane reforming.* Applied Catalysis A: General, 2003. **250**: p. 161-175.
9. Lattner, J.R. and M.P. Harold, *Comparison of steam reforming, autothermal reforming, and membrane reactor fuel processors using methanol and liquid hydrocarbon fuels in a PEM fuel cell.* AIChE Journal, 2003. **To be submitted**.
10. de Smet, C.R.H., M.H.J.M.d. Croon, R.J. Berger, G.B. Marin, and J.C. Schouten, *Design of adiabatic fixed-bed reactors for the partial oxidation of methane to synthesis gas. Application to production of methanol and hydrogen for fuel cells.* Chem. Eng. Sci., 2001. **56**: p. 4849-4861.
11. Lattner, J.R. and M.P. Harold, *Comparison of conventional and membrane reactor fuel processors for hydrocarbon-based PEM fuel cell systems.* Int. J. Hydrogen Energy, 2003: p. submitted for publication.
12. Numaguchi, T. and K. Kikuchi, *Intrinsic kinetics and design simulation in a complex reaction network; steam-methane reforming.* Chem. Eng. Sci., 1988. **43**(8): p. 2295-2301.
13. Grenoble, D.C., M.M. Estadt, and D.F. Ollis, *The chemistry and catalysis of the water gas shift reaction.* Journal of Catalysis, 1981. **67**: p. 90-102.

14. Reitz, T.L., S. Ahmed, M. Krumpelt, R. Kumar, and H.H. Kung, *Characterization of CuO/ZnO under oxidizing conditions for the oxidative methanol reforming reaction.* Journal of Molecular Catalysis A: Chemical, 2000. **162**: p. 275-285.
15. Peppley, B.A., J.C. Amphlett, L.M. Kearns, and R.F. Mann, *Methanol-steam reforming on Cu/ZnO/Al2O3 catalysts Part 2: A comprehensive kinetic model.* Applied Catalysis A: General, 1999. **179**: p. 31-49.
16. Ahluwalia, R.K., E.D. Doss, and R. Kumar, *Performance of high-temperature polymer electrolyte fuel cell systems.* J. Power Sources, 2003. **117**: p. 45-60.
17. Holleck, G.L., *Diffusion and solubility of hydrogen in palladium and palladium-silver alloys.* The Journal of Physical Chemistry, 1970. **74**(5): p. 503-511.
18. Balachandran, U., T.H. Lee, S. Wang, G. Zhang, J.J. Picciolo, J.T. Dusek, S.E.Dorris, and K.S. Rothenberger. *Recent Advances in the development of dense ceramic membranes for hydrogen separation.* in *ACS Meeting.* 2002. Boston, MA.
19. Wang, M., *Fuel choices for fuel-cell vehicles: well-to-wheels energy and emission impacts.* J. Power Sources, 2002. **112**: p. 307-321.

Development of a Combined Catalyst and Sorbent for Hydrocarbon Reforming

J. A. Satrio, B. H. Shanks, and T. D. Wheelock
Department of Chemical Engineering
Iowa State University, Ames, IA 50011

ABSTRACT

A combined catalyst and calcium-based sorbent is being developed to improve the efficiency of steam reforming of hydrocarbons, such as methane and propane. The material is made in the form of small spherical pellets having a layered structure such that each pellet consists of a highly reactive lime core enclosed within a porous but strong protective shell made of alumina on which a nickel catalyst is loaded. The material serves two functions in catalyzing the reaction of hydrocarbons with steam to produce hydrogen while simultaneously absorbing carbon dioxide formed by the reaction. The *in-situ* removal of CO_2 shifts the steam reforming reaction equilibrium towards increased hydrogen concentration and production.

The preparation and characterization of catalyst/ sorbent, core-in-shell pellets is reviewed. Also presented are preliminary results achieved with the catalyst /sorbent pellets when used for reforming propane in a thermogravimetric analysis (TGA) system.

KEY WORDS

Fuel reforming, carbon dioxide sorbents, hydrogen production.

INTRODUCTION

The advancement of fuel cell technology for the future production of clean and efficient electric power requires the concomitant advancement of inexpensive, efficient and reliable hydrogen production technology. The conventional method of producing hydrogen from hydrocarbons, such as methane, consists of several steps, i.e. steam reforming, CO-shift (usually 2 steps), and CO_2 removal. A technology that can reduce the number of steps, thus making the production of hydrogen simpler, more efficient and less costly, will be very attractive.

To produce a hydrogen-rich gas with a very low CO_2 and CO content, a new material that combines a catalyst and a sorbent for promoting the steam reforming of hydrocarbons, such as methane and propane, is being developed. The material will simultaneously improve the reaction rate through catalytic action and separate the products (H_2 and CO_2) by selective absorption of the carbon dioxide. The carbon dioxide absorption will also drive the reaction to produce more hydrogen since the reaction is otherwise limited by thermodynamic equilibrium

BACKGROUND

In the future hydrogen will be supplied to fuel cells on a large scale to generate electric power according to the concept of the Vision 21 program of the US Department of Energy. This initiative aims to greatly reduce the emission of CO_2 by increasing the efficiency of producing power from fossil fuels. Although H_2 can be generated by coal gasification, the H_2 content of the raw gas is generally too low to be used directly in fuel cells. To increase the H_2 content, methane reforming and the water gas shift reaction will be required.

Both methane reforming and the water gas shift reaction can be limited by thermodynamic equilibrium as well as the rate of reaction. To overcome these limitations, a new material is proposed that combines a catalyst to increase the rate of reaction and a chemical sorbent for CO_2. Removing CO_2 by reaction with CaO to form $CaCO_3$ results in shifting the reforming and/or water gas shift reactions toward increased hydrogen production.

This principle was demonstrated previously by conducting the water gas shift reaction in a fixed-bed reactor packed with dolomite particles (1) and by reforming methane in a similar reactor packed with a mixture of catalyst and CO_2-acceptor particles (2-4). While these demonstrations were valuable, they did not culminate in the development of a material that combines both a catalyst and a CO_2 acceptor in one durable pellet.

The integration of catalyst and sorbent into a single material potentially can make the process for converting fuels into hydrogen much simpler, more efficient, and less costly.

MATERIAL PREPARATION AND APPLICATION

Catalyst/Sorbent Preparation and Characterization - Core-in-shell spherical pellets were prepared in two stages using a small bench-scale revolving drum pelletizer which has a diameter of 25 cm and can be operated at various speeds. In the first stage powdered limestone or other calcium-bearing material was pelletized to make the pellet cores of desired sizes. In the second stage the cores were coated with a mixture of powdered alumina and limestone to make the shells of desired thickness. Typically, the core-in-shell pellets had an average outer diameter of 4.5 mm and a shell thickness of 0.5 mm. The core-in-shell pellets were subsequently calcined at 1100°C to partially sinter the shell material and convert $CaCO_3$ to CaO.

Incorporation of limestone or pure calcium carbonate in the shell material formulation proved very important to produce a shell material with the necessary compressive strength. The compressive strength of the shell material increased as limestone was incorporated in the formulation. However, when the limestone content exceeded an optimum level, the shell compressive strength decreased rapidly. The results suggest that with the optimum level of limestone, calcium aluminate is formed upon sintering, which serves as a particle bonding agent. However, when excess limestone is incorporated in the formulation, free CaO is present that weakens the calcined material. Table 1 shows the compressive strength of core-in-shell pellets with shells made of alumina mixtures with limestone in various concentrations. It was also noted that the particle size of the limestone powder significantly affects the compressive strength of the shell material.

The prepared pellets were impregnated with a nickel catalyst by immersing them in either an organic or aqueous solution of nickel nitrate under atmospheric pressure or under a vacuum. While it was presumed desirable to load only the shells with the catalyst, the capacity of the CaO cores to absorb CO_2 was found not to be affected by the presence of nickel in the cores. BET analysis indicated that the shells made of an alumina mixture with 20% limestone had a surface area of approximately 5 m^2/g after calcination at 1100°C. Prior to impregnation the pellets were treated with CO_2 at 650°C to convert the CaO cores to $CaCO_3$. The carbonation step was necessary since CaO forms $Ca(OH)_2$ in contact with moisture, which caused pellet cracking during impregnation. The impregnated pellets were placed in a drying oven to vaporize the liquid before being calcined to decompose the nickel nitrate. The resulting nickel oxide then was reduced with hydrogen to elemental nickel. The impregnation step was repeated to increase the nickel content. Table 2 shows representative values of the nickel content of prepared core-in-shell pellets after impregnation with nickel nitrate solutions under different conditions.

Table 1. Compressive strength of core-in-shell pellets after calcination at 1100°C for 2 hours

Shell Material	Compressive Strength[a] (N/mm)
Alumina mix with 0% limestone	1.6 ± 0.3
Alumina mix with 5% limestone	16.0 ± 3.5
Alumina mix with 10% limestone	10.0 ± 2.4
Alumina mix with 20% limestone	5.0 ± 0.4

[a] Breaking force/pellet diameter.

Table 2. Nickel content of core-in-shell pellets after treatment with $Ni(NO_3)_2$ solutions under different conditions.

$Ni(NO_3)_2$ Solution	Impregnation Condition	Impregnation Cycles	Ni Content (wt%)
$Ni(NO_3)_2$ in THF	Atmospheric	1 x	3.7
		2 x	5.1
$Ni(NO_3)_2$ + urea in water	Atmospheric	1 x	4.4
		2 x	6.3
	Vacuum	1 x	5.8
		2 x	7.4

CO_2 absorption characteristics - The absorption characteristics of the pellets were determined by employing a thermogravimetric analysis (TGA) system to measure changes in pellet weight over time as the material was treated with an appropriate gas mixture at a selected temperature. The reaction of CO_2 with CaO is reversible and highly dependent on the temperature and

CO_2 concentration of the gas. Table 3 shows the minimum concentration required for CO_2 to react with CaO to form $CaCO_3$ under a total pressure of 1 atm.

Table 3. Minimum CO_2 concentration required for CO_2 to react with CaO to form $CaCO_3$

Temperature (°C)	Minimum CO_2 Concentration (vol%)
650	2
750	25
800	45

Depending on the reaction temperature, the concentration of CO_2 produced in steam reforming of fuels is between 10 and 15 vol%. In order for CO_2 to be continually absorbed the steam reforming reaction will need to be conducted at temperature as low as 650°C or at higher pressures.

Figure 1 is a plot of the carbonation and calcination behavior of a core-in-shell pellet which had been loaded with nickel. The core material was dolomite and the shell material was alumina mixed with 20% limestone. The catalyst/sorbent pellet was loaded with nickel by twice impregnating the pellet with a nickel nitrate-urea solution, which resulted in a pellet with a nickel content of 7.4%.

The carbonation step, using 5% CO_2, took place at 650°C. It can be seen that the rate of CO_2 absorption was fast in the first 10 minutes after CO_2 was introduced before the rate decreased abruptly, which suggests that the CO_2 absorption takes place by two different mechanisms. When the impregnated pellet was subsequently calcined in pure N_2 at 650°C, the rate was too slow to be practical. Therefore the calcination step had to be conducted at a higher temperature. At 750°C the CaO core was completely regenerated in 10 minutes.

The CO_2 absorption behavior of the pellet was compared with that of a pellet impregnated with nickel only one time and of a pellet not loaded with nickel. Since the absorption behavior of those pellets was very similar, it appeared that the nickel loading did not affect the pellet capacity to absorb CO_2.

Pellet Catalytic Activity – Thermogravimetric analysis was also utilized to evaluate the catalytic activity of the catalyst/sorbent pellets. In a typical experiment the progress of the steam reforming reaction was observed by measuring the increase in weight of a single pellet as it absorbed the CO_2 produced. The results of such an experiment conducted with the same catalyst-sorbent pellet used to produce Figure 1 will be described. The catalyst/sorbent pellet was first activated by calcination for 1 hour at 750°C followed by hydrogen reduction of the nickel oxide reduction for 12 hours. Propane was used for the initial reforming studies. The composition of the gas (excluding steam) was 9% propane and 91% nitrogen. The mole ratio of steam to carbon was 3.0. The reactant gas mixture was passed around the single catalyst/sorbent pellet at 650°C.

Figure 2 is a plot of the pellet weight changes during two cycles of reforming and pellet regeneration. It can be seen that the pellet gained weight by absorbing CO_2 produced by the reforming reaction. The weight gain due to CO_2 absorption was similar to that shown in Figure 1. While the reactant gases were kept flowing, increasing the reaction temperature to 750°C caused the pellet to lose weight due to CO_2 desorption until the core was converted back to CaO. CO_2 absorption resumed when the reaction temperature was reduced back to 650°C.

To verify the catalytic activity of the catalyst/ sorbent pellet, an identical experiment was conducted with a core-in-shell pellet from the same batch that had not been loaded with nickel. Using the same propane reforming conditions as before, no gain in pellet weight was observed which indicates that no CO_2 was formed nor absorbed by the pellet.

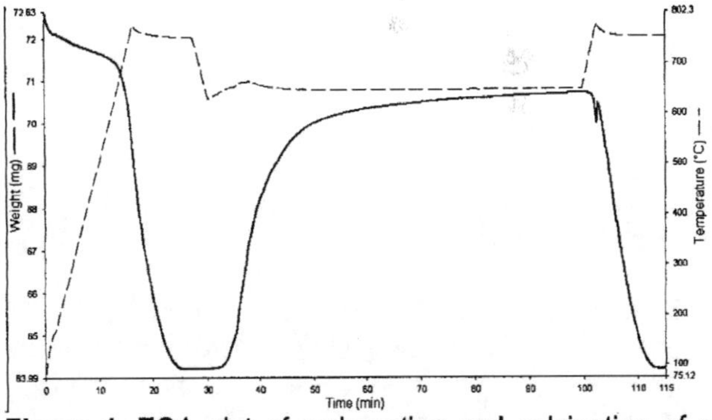

Figure 1. TGA plot of carbonation and calcination of a catalyst/sorbent pellet.

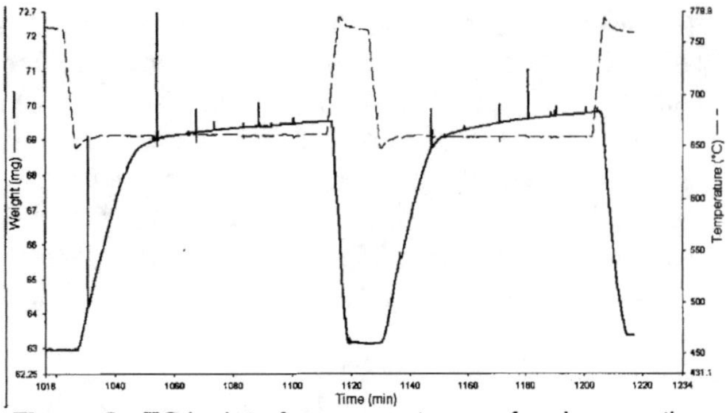

Figure 2. TGA plot of propane-steam reforming reaction using a catalyst/sorbent pellet.

SUMMARY

The preparation and characterization of a material that combines a catalyst and a calcium-based sorbent for steam reforming hydrocarbons, such as methane and propane, have been discussed. The material was made in the form of small spherical pellets with a layered structure such that each pellet consists of a highly reactive lime core enclosed within a porous but strong protective shell made of alumina on which a nickel catalyst was deposited. Evaluation of the catalytic activity and the sorbent characteristics of the pellets was conducted by using thermogravimetric analysis. The results showed that the material was able to catalyze the steam reforming reaction as well as to absorb the CO_2 produced during the reaction. Further testing of the catalyst/ sorbent pellets will be conducted in the future utilizing a fixed bed reactor.

ACKNOWLEDGMENTS

This conference presentation was prepared with the support of the U.S. Department of Energy, under Award No. DE-FG26-02NT41549. However, any opinions, findings, conclusion, or recommendations expressed herein are those of the authors and do not necessarily reflect the views of DOE. The project has been administered by the Center for Sustainable Environmental Technologies in the Institute for Physical Research and Technology.

REFERENCES

1. Han, C. and Harrison, D.P. (1994). "Simultaneous Shift Reaction and Carbon Dioxide Separation for the Direct Production of Hydrogen", *Chem. Eng. Sci.*, **49** (24B), 5875-5883.
2. Balasubramanian, B., Ortiz, A. L., Kaytakoglu, S., and Harrison, D. P., (1999). "Hydrogen from Methane in a Single-step Process", *Chem. Eng. Sci.*, **54**, 3543-3552.
3. Hufton, J. R., Mayorga, S., and Sircar, S. (1999). "Sorption-Enhanced Reaction Process for Hydrogen Production", *AIChE. J.*, **45**(2), 248-256.
4. Ding, Y. and Alpay, E. (2000). "Adsorption-enhanced Steam-Methane Reforming", *Chem. Eng. Sci.*, **55**, 3929-3940.

Catalytic Hydrogen Generation Based On Autothermal Cyclic Reforming For Fuel Cells And Vehicle Refueling

Parag P. Kulkarni, Ravi V. Kumar, Court Moorefield, Shawn Barge, and Vladimir M. Zamansky
General Electric Company – GE Global Research – Fuel Conversion Laboratory
18 Mason, Irvine, CA 92618

ABSTRACT

General Electric (GE) is developing a novel hydrogen generating technology that has the potential to meet the Department of Energy (DOE) performance and cost targets for refueling fuel cell vehicles. The GE-patented hydrogen generation system is based on a novel, highly efficient technology called Autothermal Cyclic Reforming (ACR). The ACR process is a unique technology that can be applied to the production of hydrogen from many fuels, including natural gas, diesel fuel, and renewable feedstock, such as biogas.

The ACR process represents a significant technological advancement in comparison with conventional reforming technologies such as steam methane reforming (SMR) and autothermal reforming (ATR). A 50 kWe natural gas fuel processor system based on the ACR technology has been fabricated and is currently undergoing testing at GE Global Research, Fuel Conversion Laboratory (FCL).

KEY WORDS

Catalytic Hydrogen Generation, Autothermal Cyclic Reforming, Fuel Processor

INTRODUCTION

In his 2003 State of the Union address, President Bush announced a $1.2 billion FreedomCAR and Fuel initiative to reverse America's growing dependence on foreign oil by developing the technology for commercially viable hydrogen generation and hydrogen-powered fuel cells. The U.S. Department of Energy (DOE) is supporting this effort with programs focusing on cost-effective, fuel cell powered vehicles and developing a cost-effective infrastructure needed to make hydrogen an available fuel source. The currently favored vehicle engine is a Polymer Electrolyte Membrane (PEM) fuel cell using hydrogen fuel. GE is developing a hydrogen generating technology, called Autothermal Cyclic Reforming (ACR), which has the potential to meet the DOE performance and cost targets for refueling fuel cell vehicles.

BACKGROUND

The ACR technology is a novel reforming process to convert fossil fuels to hydrogen. The target application is electric power generation in buildings and H_2 for automobiles using fuel cells. The ACR process has been demonstrated in a packed bed reactor using diesel fuel and natural gas as the feed-stock. The average concentration of hydrogen in the product stream was around 70%, with the balance being primarily methane, carbon monoxide and carbon dioxide.

In ACR process, the catalyst is not deactivated by coke or sulfur. The coke is burnt off during a regeneration step. A significant portion of the sulfur is rejected as sulfur dioxide. The catalyst has been demonstrated to be active for 2000 h without significant loss in activity and selectivity for the ACR process.

Under a DOE funded project a 50 kWe fuel processor system based on the ACR technology has been fabricated and is currently undergoing testing and optimization.

DESIGN OF THE ACR BASED FUEL PROCESSOR

The ACR-based fuel processor is designed to be a self-contained system to supply hydrogen for a range of applications including powering stationary fuel cells and refueling hydrogen vehicles. The fuel processor includes four major subsystems: reformer and low temperature shift (LTS) reactor, pressure swing adsorber (PSA), compression and storage system, and hydrogen dispenser (Figure 1). The reformer and LTS reactor convert the hydrocarbon feed gas into a hydrogen-rich reformate stream. The PSA (hydrogen purifier) separates the hydrogen from the reformate gas. The compression and storage system is designed for vehicle refueling up to 345 bar (5000 psig). Finally the dispenser is a user friendly, safe system to quickly refuel a hydrogen car. Purified hydrogen can also be delivered to a stationary PEM fuel cell for generation of combined heat and power

(CHP). The electricity from the fuel cell can be used to power various components of the fuel processor.

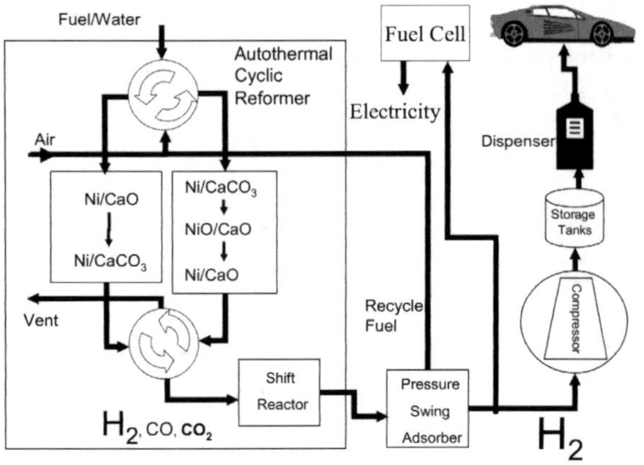

Figure 1. ACR-based fuel processor.

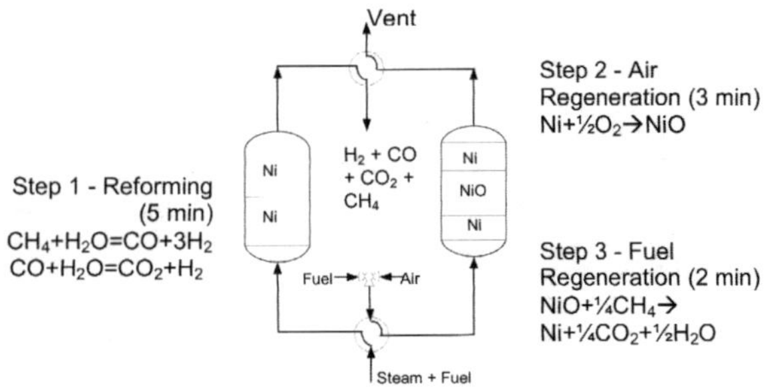

Figure 2. Steps and reactions in the ACR process.

DESCRIPTION OF THE ACR PROCESS - ACR is an autothermal cyclic steam reforming technology for converting hydrocarbons to a hydrogen-rich stream. The ACR process operates in a three-step cycle that includes steam reforming of the hydrocarbon fuel on a Ni catalyst (Steam Reforming Step), heating the reactor through oxidizing the Ni catalyst (Air Regeneration Step), and finally reducing the catalyst to its original state (Fuel Reduction Step) as shown in Figure 2. The heat required for the endothermic steam reforming reaction is provided during the exothermic oxidation of Ni to NiO.

The function of the LTS reactor is to convert CO and steam to H_2 and CO_2 via the water gas shift reaction over a Cu/ZnO based catalyst:

$$CO + H_2O = CO_2 + H_2$$

The LTS reactor reduces the CO in the reformate steam to less than 1% (on a dry basis). New catalysts for the LTS reactor including non-pyrophoric catalysts are under evaluation for application with ACR.

The hydrogen is being purified in a pressure swing absorber (PSA). The PSA operates using an adsorbent bed that traps and releases gases through a pressure cycle. The adsorbent materials are typically alumina, carbon, and zeolite. The major design issues for a PSA are bed size, number of adsorbing vessels, the number and placement of surge tanks and purge tanks, adsorption pressure, switching valving options, cyclic process operation, and maintaining 99.99% hydrogen purity. Integration of cyclic reformer (ACR) and cyclic PSA may be synergetic and can improve the overall performance of the fuel processor.

The hydrogen compression and storage system design needs to address many technical issues. Hydrogen has a very low viscosity and easily leaks through standard compressor seals. Hydrogen embrittlement of carbon steel can result in shorter material life due to loss of ductility. Various options of compression systems are being investigated.

The dispenser system should address both technical issues and consumer preferences. From a technical standpoint the dispenser must be safe and fast fill a vehicle in less than 5 minutes without over-pressurizing the vehicle storage system. The dispenser must be easy to operate and dispel any concerns the consumer has about hydrogen.

GE has developed an extensive cost model to analyze the hydrogen delivery cost from the ACR-based fuel processor. The results indicate that in future a commercial ACR-based fuel processor can meet the DOE cost of hydrogen targets if mass-produced, mainly because of the high efficiency and lower metal temperatures as compared to the conventional reforming technologies.

RESULTS

A 50 kWe pre-commercial fuel processor has been fabricated and is undergoing testing. The fuel processor includes the ACR, LTS reactor, fully automated control

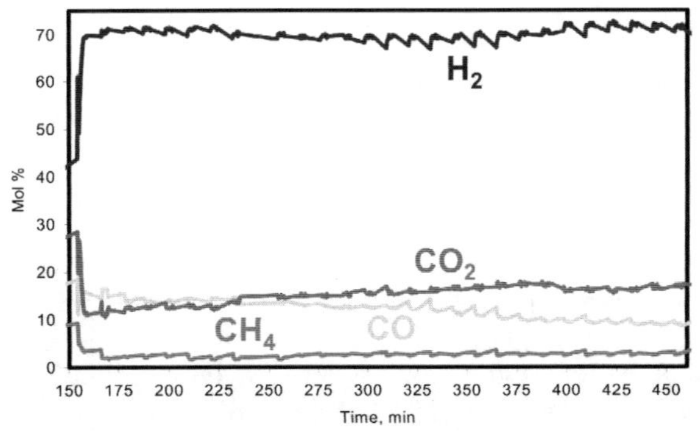

Figure 3. Product composition (dry %) at the outlet of the ACR reactor.

system, and all required subsystems. The fuel processor will be integrated with a PSA.

Figure 3 shows typical product composition out of the ACR based reformer reactor. Typically 70% H_2 is obtained on the dry basis. The methane slip is typically < 5% out of the reactor. The downstream purifier system (LTS + PSA) will increase the H_2 concentration to 99.99% and decreases the CO concentration below 10 ppm.

SUMMARY

The ACR process is a unique technology that can be applied to the production of hydrogen from many fuels, including natural gas, diesel fuel, and renewable feedstock, such as biogas. The ACR process represents a technological advancement in comparison with Autothermal Reforming, as the ACR-produced syngas is not diluted with nitrogen. When compared to conventional Steam Methane Reforming, the ACR process has significantly lower capital costs, higher conversion efficiency, and lower emissions. In addition, ACR process is fuel flexible and has been successfully demonstrated using high-sulfur fuels.

ACKNOWLEDGMENTS

The authors gratefully acknowledge the U.S. DOE and General Electric Company for the financial support of this project.

Autothermal Reforming of Iso-Octane and Simulated Gasoline Using Nickel-Based Catalysts

Andrew R. Tadd, Ben Gould, Johannes Schwank
Department of Chemical Engineering, University of Michigan
2300 Hayward, Ann Arbor MI 48109

ABSTRACT

Autothermal reforming (ATR) of gasoline is a promising process to deliver hydrogen streams to portable fuel cells via a compact fuel processor. Autothermal reforming combines exothermic partial oxidation with endothermic steam reforming, and is considered the most attractive option for on-board reforming of gasoline. Catalysts are the key factor determining the size, weight, and cost of a fuel processor, as well as controlling the dynamic behavior and durability of a fuel processor under varying operating conditions. Catalysts comprised of nickel supported on the solid electrolytes yttria-stabilized-zirconia and Na β-Al_2O_3, as well as on Ce-ZrO_2 mixed oxides, silica, and alumina have been prepared and evaluated for autothermal reforming of iso-octane and simulated gasoline using a high-throughput reactor system.

KEY WORDS

Autothermal reforming, ceria, nickel.

INTRODUCTION

Most types of fuel cells require a feed stream of pure hydrogen. For portable applications of proton exchange membrane (PEM) fuel cells, including automotive, hydrogen might be produced from gasoline in fuel processors. In the fuel processor, hydrocarbons undergo a series of reactions, including desulphurization and reforming. The reforming reaction produces carbon monoxide, carbon dioxide, and hydrogen. Since carbon monoxide poisons the electrodes of PEM fuel cells, water-gas shift (WGS) and preferential oxidation of CO to CO_2 are carried out after the reforming step to reduce the CO content to less than 100 ppm. An autothermal reformer (ATR), which combines partial oxidation and steam reforming processes in a single reactor, can serve as a first stage in catalytic fuel processors for hydrogen production. The heat released from partial oxidation supports steam reforming of the remaining fuel. The reaction temperatures of ATR tend to be lower than in partial oxidation. Compared to steam reforming, ATR requires less catalyst, starts faster, and has better dynamic response to control actions or changes in feed conditions. Typical catalysts used for ATR of hydrocarbons include Ni, Pt, and Pt-Ni on Al_2O_3 and ZrO_2-Al_2O_3 supports. Reforming of gasoline requires catalysts that are flexible enough to process fuel of varying composition, as gasoline formulations vary geographically and seasonally. Improved autothermal reforming catalyst formulations are required to address the need of required fuel flexibility, high activity, transient response, and resistance to coking and deactivation.

Ionic conducting solids have shown increased activity over traditional supports such as silica and alumina. The most revolutionary potential benefit of using solid ionic conductors in catalysis lies in electrochemical promotion of activity, which offers the potential of 'tunable promotion' by ionic species in the support. Ceria and mixed oxides of cerium and other metals have been used as supports for nickel catalysts in methane oxidation and steam reforming catalysts. The ability of these catalysts to interact with steam and oxygen via a redox mechanism makes them interesting for autothermal reforming.

EXPERIMENTAL

Na β-Al_2O_3 was used as received from Ionotech, Inc. (England). $Ce_{0.75}Zr_{0.25}O_2$ mixed oxide was prepared by coprecipitation. The appropriate quantities of $Ce(NO_3)_3$-$6H_2O$ and $ZrOCl_2$-$8H_2O$ were dissolved in de-ionized water and precipitated using 1 M NH_4OH. The resulting solids were filtered, washed with de-ionized water, and dried at 100 °C. Calcination at 800 °C for one hour yielded the final mixed oxide. X-ray diffraction indicated that the desired solid solution was obtained. The surface areas of both supports were measured using the BET single point method, and were found to be 6 and 30 m^2/g respectively. Supports were impregnated with a quantity of $Ni(NO_3)_3$ dissolved in ethanol to achieve the desired weight loading (5-15 wt% Ni). After impregnation, the catalysts were dried at 100 °C and calcined at 600 °C. Pulse titration of reduced catalysts with H_2 showed that nickel dispersions were low, less than 2%.

A single flow reactor system with two interchangeable reactors is used for testing catalyst activity. One reactor consists of a six quartz tubes, each of which holds a separate catalyst sample and receives an equal portion of an autothermal feed mixture. The feed mixture is composed of air, steam, and iso-octane. Air is delivered by a mass flow controller. Steam and iso-octane are metered by Eldex HPLC pumps, and pumped through a heated section of tubing to accomplish vaporization. The six-sample reactor may be replaced by a single quartz tube, which has thermocouples located above and below the catalyst sample. Catalyst screening is done using the six-sample reactor, and interesting formulations found there are studied in more detail using the single sample reactor. Product analysis is performed using a Varian gas chromatograph, equipped with three columns and three TCD dectectors operating in parallel. For screening studies, the columns are identical, and products are measured by alternating between samples 1-3 and 4-6. For the single sample reactor, two detectors with different columns are used to analyze a wider range of products. Catalysts are reduced under flowing 5% H_2 in N_2 at 600 °C for one hour prior to reaction.

RESULTS

Catalysts of 5, 10, and 15 wt% Ni supported on $Ce_{0.75}Zr_{0.25}O_2$ and Na β-Al_2O_3 were prepared. These were evaluated for autothermal reforming of iso-octane using the six-sample reactor. Reaction conditions were $H_2O/C = 2$, $O/C = 1$, and a GHSV of 100,000 h^{-1}. Figure 1 shows the mole % of H_2 in the product for the Ni/$Ce_{0.75}Zr_{0.25}O_2$ catalysts. The support itself shows minimal activity, but above 600 °C there is some homogeneous reaction. Experiments run with no catalyst indicate that the homogeneous reaction produces primarily CO, with low H_2/CO ratios, and some residual hydrocarbons (CH_4, C_3H_8). The Ni/β-Al_2O_3 catalysts showed H_2 product concentrations similar to the Ni/$Ce_{0.75}Zr_{0.25}O_2$, but also showed more evidence of coking. Product composition was influenced by temperature and corresponded well with equilibrium concentrations predicted using ASPEN®, particularly for the Ni/$Ce_{0.75}Zr_{0.25}O_2$ catalysts. Increasing temperature led to increases in H_2 and CO concentration, with a decrease in CH_4 concentration. Over the course of 24 hours on stream, no loss of activity was observed for the 10 wt% Ni/$Ce_{0.75}Zr_{0.25}O_2$ catalyst (see Figure 2).

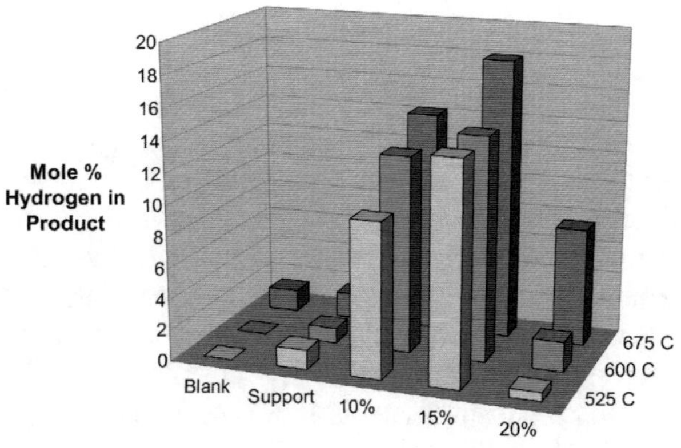

Figure 1. H_2 product concentration for Ni/$Ce_{0.75}Zr_{0.25}O_2$ catalysts at $H_2O/C = 2/1$, $O/C = 1/1$, SV = 100,000 h^{-1}.

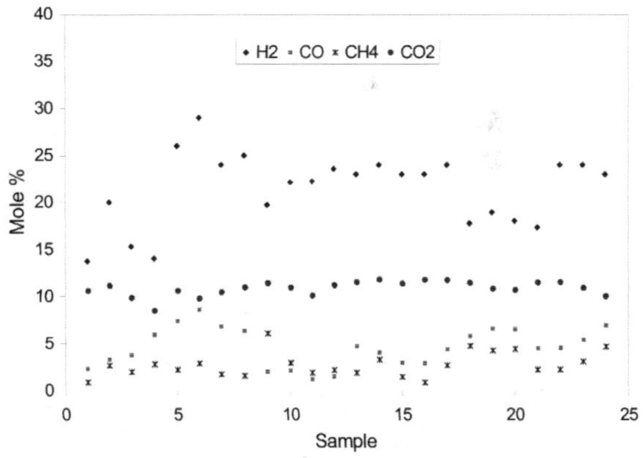

Figure 2. Product composition using 10 wt% Ni/$Ce_{0.75}Zr_{0.25}O_2$ over 24 hours of reaction time at $H_2O/C = 2/1$, $O/C = 1/1$, SV = 100,000 h^{-1}.

As a result of experiments with the parallel reactor, the 10 wt% Ni/$Ce_{0.75}Zr_{0.25}O_2$ was studied further in the single tube reactor. Reaction conditions were $H_2O/C = 2$, $O/C = 1$, and a GHSV of 250,000 h^{-1}. It was found that with feed temperatures as low as 375 °C, autothermal reforming began, resulting in a large temperature increase in the catalyst bed and the product gas. The typical product gas temperature at the bed exit was 650 to 700 °C. The product composition closely matched that predicted using ASPEN® with an equilibrium temperature of 800 °C. Varying feed temperature from 375 to 500 °C resulted in no change in product composition, possibly because the catalyst temperature was dominated by the heat of reaction. Additional experiments were undertaken to investigate catalyst performance under conditions which might be encountered in a portable fuel processor. The catalyst was tested for autothermal reforming without pre-reduction in H_2. The catalyst started reacting rapidly and products were indistinguishable from experiments which included pre-reduction. A total of three cycles of startup and shutdown with no reduction treatments were conducted, with no obvious decrease in performance based on product composition. A separate experiment was conducted to investigate startup time for the catalyst. The catalyst was heated to 375 °C in flowing air, after which point steam and iso-octane were introduced. The exit gas temperature reached 650 °C within 90 seconds of starting fuel and steam, and remained near that temperature for the duration of the experiment.

SUMMARY

$Ni/Ce_{0.75}Zr_{0.25}O_2$ and $Ni/\beta\text{-}Al_2O_3$ have been identified as active catalysts for autothermal reforming of iso-octane at moderate temperatures and high space velocities using our parallel flow reactor. 10 wt% $Ni/Ce_{0.75}Zr_{0.25}O_2$ was stable as well as active, showing no deactivation over 24 hours of reaction. Experiments with single samples indicated non-isothermal behavior, with the partial oxidation reaction supporting the steam reforming reaction to give high hydrogen yields at low temperatures. In addition, it was found that the 10 wt% $Ni/Ce_{0.75}Zr_{0.25}O_2$ catalyst was active for autothermal reforming of iso-octane without the need for pre-reduction using H_2, and is capable of rapid start up.

ACKNOWLEDGMENTS

Financial Support by DoE (DE-FC04-02AL67630)

CONTACT

Johannes Schwank
Professor of Chemical Engineering
(734) 764-3374
Fax: (734) 763-0459
schwank@umich.edu

Preparation and Properties of Bimetallic Carbide Catalysts and Their Performance for Dry Reforming of Methane

Huifeng Shao, Edwin L. Kugler, Mahesh V. Iyer[1] and Dady B. Dadyburjor
West Virginia University
Department of Chemical Engineering, Morgantown WV 26506-6102

ABSTRACT

Work in our laboratory has concentrated on the use of the η-carbide Co_6W_6C as a bimetallic carbide catalyst for dry reforming of methane to syngas. Such catalysts, when properly pretreated, are remarkably active and have been shown to be stable for 150 h in the laboratory. Initially, the η-carbide was obtained from a commercial source. However, we have been able to prepare the material in-house. We have also been able to prepare more-active forms of the catalyst directly, without recourse to the η phase. One form of the material results in H_2/CO ratios much greater than the stoichiometric value (1.0) at least at unsteady state.

KEY WORDS

Dry reforming, Methane, Cobalt, Tungsten, Carbide, Nickel

INTRODUCTION

Dry reforming of methane is a viable option for natural gas reserves that are in inhospitable and/or arid locations. Here methane can be reacted with carbon dioxide (often found in appreciable amounts in natural gas) to form synthesis gas with a theoretical H_2/CO ratio of unity. Metallic catalysts such as Ni are preferred over noble metals for economic reasons, but Ni is susceptible to coking. Carbide catalysts are moderately resistant to carbon deposition.

BACKGROUND

We have shown[2] that a very active and stable catalyst can be produced by pretreating commercially obtained η-Co_6W_6C (Nanodyne, Inc.) with CH_4 and CO_2 at 850°C for a 24-hour period. Pretreatment consists of heating at 850°C in the presence of methane. Under these conditions, there is a net laydown of carbon on the catalyst, and it is converted to a mixture of Co and WC. After pretreatment, the catalyst does not deactivate for at least 100 hours. At the end of the run, the material removed from the reactor is a mixture of Co, WC and C. Micrographs[3] show that the carbon deposited is filamentous.

A reaction model was developed[4] involving Langmuir-Hinshelwood mechanics with carbon deposition, the (reverse) water-gas-shift reaction and the (reverse) Boudouard reaction as well as dry reforming. This was fit to kinetic data over a range of temperatures, pressures and relative amounts of CH_4 to CO_2 at the inlet to obtain rate constants and activation energies for this catalytic process.

Because commercial bimetallic carbides are difficult to obtain, and also because we wish to have more control over the phases present in our catalyst, we have started to develop expertise in preparing our own catalysts. Our approach has been to start with the patent literature[5,6] then modify as necessary to obtain different phases. Three different catalysts have been characterized before and after reaction. The reaction results have been correlated with the structure of the catalytic materials.

EXPERIMENTAL

All the materials produced start with 14.5 g tungstic acid, H_2WO_4, added to a mixture containing 400 ml of an aqueous (1:1) NH_4OH solution and 300 ml of ethylenediamine, followed by heating to 50°C and stirring to get a clear solution. Then, 13.8 g $CoCl_2{\cdot}6H_2O$ are dissolved in 50 ml distilled water and added into the above solution. A precipitate, $Co(en)_4WO_4$, forms when the volume of the mixture is reduced by 50 percent by evaporating and slow overhead stirring. The product is then filtered, washed and dried. The $Co(en)_4WO_4$ is then heated in an Ar-H_2 mixture to 650°C in 3h, then heated in Ar to 850°C in 0.5h, reduced in a CO_2-CO mixture at 850°C for 24h and finally cooled back to room temperature in flowing Ar.. The heating/cooling process is carried out at ambient pressure in a downflow reactor.

As mentioned above, three different materials were prepared. These corresponded to CO_2/CO ratio (≡ x) values of 0.75, 0.2 and 0.1. This process yields 3-5 g of material of surface areas 2 – 30 m^2/g and size 2-3μ. Some of this material was characterized by x-ray powder diffraction (XRD). Some material was placed in the reactor system to catalyze the CH_4 and CO_2.

The reactor system is computer-controlled with automated data logging. Details can be found in, for example, Iyer *et al.*[4] In order to verify the stability of the catalysts formed, and to compare these with the commercial material, the reactor was run for at least 100 h at the highest temperature used previously, 850°C. The material showed stable catalytic behavior well before this time on stream. After the reaction, the catalyst was again analyzed by XRD. The difference between the pre-reaction XRD and the post-reaction XRD indicates the stability of the original formulation as a catalyst. The initial reactivity behavior can be correlated with the initial XRD measurement, and the final reactivity with the final XRD.

RESULTS

The XRD pattern for x = 0.75 corresponds closely with that of pure η-Co_6W_6C (and that of the Nanodyne material). With lower values of x, i.e., mixtures which are more reducing, the patterns change. For x = 0.2, the material appears to consist of tungsten carbide (WC) and Co_3W_3C, an intercalated carbide. This is consistent with the greater reduction of the bimetallic carbide by the CO_2/CO stream. Finally, for x = 0.1, the pattern indicates the presence of Co_3W_3C as well as elemental Co, WC and elemental C (graphite). This material is still further reduced. It contains the components found in the Nanodyne η-Co_6W_6C after it had been pretreated and shown to be an active and stable catalyst, but it also contains some of the intercalated (and incompletely reduced) bimetallic carbide.

Figures 1-3 show the results of reactivity studies with these three materials. The "carbon balance" is a measure of the carbon being deposited on the catalyst and closure of the material balance (with a fraction carbon balance of 1.0 implying no carbon being deposited and a complete material balance of carbon). All three materials go through an unsteady state, as can be seen by the changes in conversions. While the material with x = 0.75 (Figure 1) requires about 40 h to reach steady state, the other materials require approximately 20 hours. Again, this is because the latter are more reduced during preparation.

The reactivity behavior of the material with x = 0.2 (Figure 2) is similar to that of x = 0.75. In both cases, the ratio H_2/CO increases from the stoichiometric value of 1 to approximately 1.6 with time on stream, before decreasing back to around 1.0. The increase in the relative amount of H_2 present is consistent with the dip in the carbon balance, implying more carbon is being deposited at that time. This further implies less reverse-water-gas-shift and less reverse-Bouduard reactions. However, at steady state, the conversions of CH_4 and CO_2 are equal, consistent with the stoichiometric value of the ratio H_2/CO at that point.

The behavior of the material prepared with x = 0.1 (Figure 3) is different. The maximum value of the ratio H_2/CO is greater than 2.0. Initially, the CO_2 conversion is less than that of CH_4, as is the case for the other two materials; however, at steady state, the CO_2 conversion is somewhat greater than that for CH_4. At least in the unsteady-state portion of the data, the catalyst is even less selective to the reverse-water-gas-shift and to the reverse-Bouduard reactions.

Given that the initial bulk structures of the materials using x = 0.2 and x = 0.1 are similar to each other, and different from those with x = 0.75, and given that the reactivity behavior of the material with x = 0.2 is more akin to that with x = 0.75 to that with x = 0.1, it would appear that the active surface structure of the material with x = 0.2 must be closer to that for the material with x = 0.75 than to that for the material with x = 0.1. This hypothesis remains to be proved.

The XRD patterns of the two extreme materials, corresponding to x = 0.75 and x = 0.1, were also obtained after 100 h on stream. For the former material, the phases present are WC, Co and C, different from what is present before the reaction and consistent with those present after the Nanodyne η-Co_6W_6C is on stream. For the material prepared with x = 0.1, the pattern appears different. However, the same peaks are present, but in different relative amounts: smaller amounts of Co_3W_3C, and larger amounts of WC, Co and C. The reduction of the catalytic material has continued during the reaction, but is not as complete as that for the material prepared with x = 0.75. The differences in the structure of these materials mirror the changes in their catalytic effectiveness.

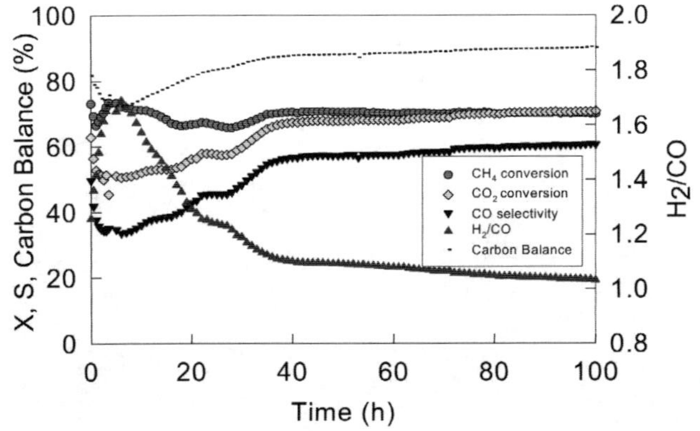

Figure 1. Reactivity of bimetallic-carbide-based catalysts with x = 0.75. T = 850°C, P = 3.4 atm with $P_{CH4} = P_{CO2}$, GHSV = 9000 h^{-1}. X, Conversion; S, Selectivity.

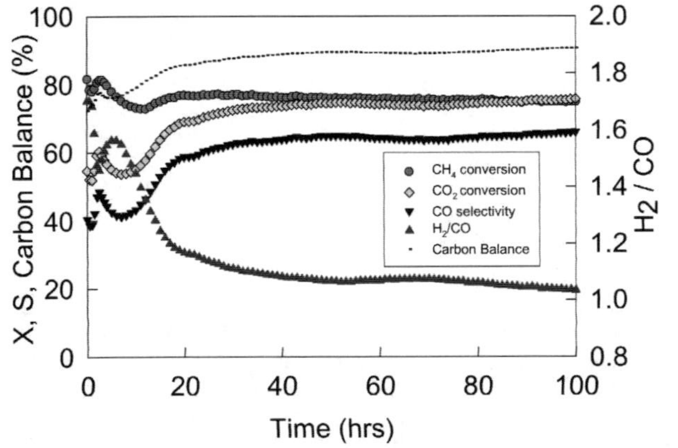

Figure 2. Reactivity of bimetallic-carbide-based catalysts with x = 0.2. T = 850°C, P = 3.4 atm with $P_{CH4} = P_{CO2}$, GHSV = 9000 h^{-1}. X, Conversion; S, Selectivity.

SUMMARY

The material prepared in-house with x = 0.75 is as active, selective and stable as the Nanodyne η-phase material, and passes through a similar unsteady-state period. This material has the structure of the commercially obtained material both before and after reaction, Co_6W_6C and Co/WC/C respectively. Materials prepared using more-reducing gas streams ($x \leq 0.2$) show the intercalated carbide Co_3W_3C as part of the bulk structure, as well as Co, WC and C. After reaction, all four materials are still seen in XRD patterns, but there appears to be less of the intercalated carbide and more of Co, WC and C. The reactivity behavior of material prepared using x = 0.2 is different from that using x = 0.1. The behavior of the former material is closer to that using x = 0.75, but with perhaps a shorter unsteady-state period. Using the material with x = 0.1 results in an unsteady-state period with much larger H_2/CO ratios and a steady-state period with CO_2 conversion greater than CH_4 conversion. At least during the unsteady-state period, the material with x = 0.1 is less selective to the reverse-water-gas-shift and to the reverse-Bouduard reactions. The difference in the reactivities of the materials with x = 0.2 and x = 0.l are hypothesized to be due to different species present on the surface of the two materials.

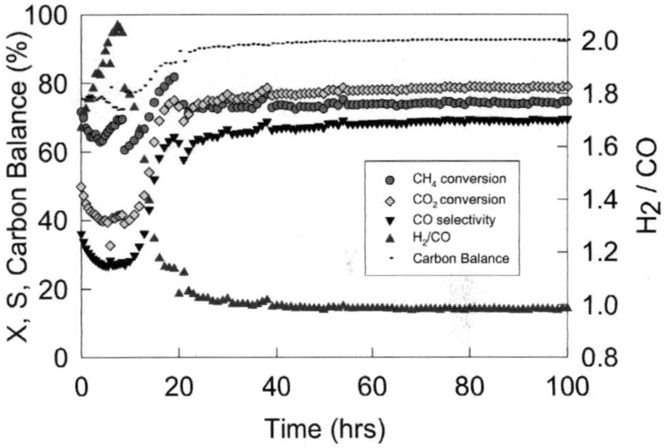

Figure 3. Reactivity of bimetallic-carbide-based catalysts with x = 0.1. T = 850°C, P = 3.4 atm with $P_{CH4} = P_{CO2}$, GHSV = 9000 h^{-1}. X, Conversion; S, Selectivity.

ACKNOWLEDGMENTS

XRD patterns were obtained in the laboratory of Professor M.S. Seehra, Department of Physics, WVU. Financial support from the US Department of Energy under Cooperative Agreement DE-AC22-99FT40540 with the Consortium for Fossil Fuel Science is gratefully acknowledged.

REFERENCES

1. Current address: Dept. Chemical Engg., Ohio State University, Columbus OH 43210.
2. Iyer, M.V., L.P. Norcio, A. Punnoose, E.L. Kugler, M.S. Seehra and D.B. Dadyburjor, Catal. Lett., accepted.
3. Iyer, M. V. *New Catalysts for Syngas Production from Carbon Dioxide and Methane.* M.S.Ch.E. Thesis: West Virginia University, Morgantown WV, 2001.
4. Iyer, M.V., L.P. Norcio, E.L. Kugler and D.B. Dadyburjor, Ind & Eng Chem. Res. 42(12) 2712 (2003).
5. E. L. Kugler, L. E. McCandlish, A. J. Jacobson, R. R. Chinaselli, *Eta Phase Materials, Methods of Producing the Same, and Use Thereof as Catalyst for Alcohol Synthesis, Hydrocarbon Synthesis, Hydrocarbon Hydrogenation and Hydrocarbon Conversion.* US Patent 5,138,111, August 11, 1992.
6. R.S. Polizzotti, L.E. McCandlish and E.L. Kugler, *Multiphase Composite Particle Containing a Distribution of Nonmetallic Compound Particles.* US Patent 5,338,330 August 16, 1994..

American Institute of Chemical Engineers
3 Park Avenue
New York, New York 10016-5991

ISBN: 978-1-60423-483-1